Modern Birkhäuser Classics

Many of the original research and survey monographs in pure and applied mathematics published by Birkhäuser in recent decades have been groundbreaking and have come to be regarded as foundational to the subject. Through the MBC Series, a select number of these modern classics, entirely uncorrected, are being re-released in paperback (and as eBooks) to ensure that these treasures remain accessible to new generations of students, scholars, and researchers.

T0205762

Linear Algebraic Groups

Second Edition

T.A. Springer

Reprint of the 1998 Second Edition

Birkhäuser
Boston • Basel • Berlin

T.A. Springer
Rijksuniversiteit Utrecht
Mathematisch Instituut
Budapestlann 6
3584 CD Utrecht
The Netherlands
springer@math.uu.nl

Originally published as Volume 9 in the series *Progress in Mathematics*

ISBN: 978-0-8176-4839-8   e-ISBN: 978-0-8176-4840-4
DOI: 10.1007/978-0-8176-4840-4

Library of Congress Control Number: 2008938475

Mathematics Subject Classification (2000): 14XX, 20Gxx

Printed on acid-free paper

birkhauser.com

T.A. Springer

Linear Algebraic Groups
Second Edition

Birkhäuser
Boston • Basel • Berlin

T.A. Springer
Mathematisch Instituut
3584 CD Utrecht
The Netherlands

Library of Congress Cataloging-in-Publication Data

Springer, T.A. (Tonny Albert), 1926-
    Linear algebraic groups / T.A. Springer. -- 2nd ed.
       p.   cm. -- (Progress in mathematics ; v. 9)
    Includes bibliographical references and index.
    ISBN 0-8176-4021-5 (alk. paper). -- ISBN 3-7643-4021-5 (alk.
paper)
    1. Linear algebraic groups. I. Title. II. Series: Progress in
mathematics (Boston, Mass.) ; vol. 9.
QA179.S67   1998   98-9333
512'.55--DC21   CIP

1991 AMS Subject Classification: 14XX, 20Gxx

Printed on acid-free paper
© 1998 Birkhäuser Boston, 2nd edition   *Birkhäuser*
© 1981 Birkhäuser Boston, 1st edition

ISBN 0-8176-4021-5
ISBN 3-7643-4021-5

Typeset in LATEX 2ε by TEXniques, Inc., Boston, MA
Printed in the United States of America

9 8 7 6 5 4 3 2 1

Contents

Preface to the Second Edition

This volume is a completely new version of the book under the same title, which appeared in 1981 as Volume 9 in the series "Progress in Mathematics," and which has been out of print for some time. That book had its origin in notes (taken by Hassan Azad) from a course on the theory of linear algebraic groups, given at the University of Notre Dame in the fall of 1978. The aim of the book was to present the theory of linear algebraic groups over an algebraically closed field, including the basic results on reductive groups. A distinguishing feature was a self-contained treatment of the prerequisites from algebraic geometry and commutative algebra.

The present book has a wider scope. Its aim is to treat the theory of linear algebraic groups over arbitrary fields, which are not necessarily algebraically closed. Again, I have tried to keep the treatment of prerequisites self-contained.

While the material of the first ten chapters covers the contents of the old book, the arrangement is somewhat different and there are additions, such as the basic facts about algebraic varieties and algebraic groups over a ground field, as well as an elementary treatment of Tannaka's theorem in Chapter 2. Errors – mathematical and typographical – have been corrected, without (hopefully) the introduction of new errors. These chapters can serve as a text for an introductory course on linear algebraic groups.

The last seven chapters are new. They deal with algebraic groups over arbitrary fields. Some of the material has not been dealt with before in other texts, such as Rosenlicht's results about solvable groups in Chapter 14, the theorem of Borel of Tits on the conjugacy over the ground field of maximal split torus in an arbitrary linear algebraic group in Chapter 15 and the Tits classification of simple groups over a ground field in Chapter 17.

The prerequisites from algebraic geometry are dealt with in Chapter 11.

I am grateful to many people for comments and assistance: P. Hewitt and Zhe-Xian Wang sent me several years ago lists of corrections of the second printing of the old book, which were useful in preparing the new version. A. Broer, Konstanze Rietsch and W. Soergel communicated lists of comments on the first part of the present book and K. Bongartz, J. C. Jantzen. F. Knop and W. van der Kallen commented on points of detail. The latter also provided me with pictures, and W. Casselman provided Dynkin and Tits diagrams. A de Meijer gave frequent help in coping with the mysteries of computers.

Lastly. I thank Birkhäuser – personified by Ann Kostant– for the help (and patience) with the preparation of this second edition.

T. A. Springer

Linear Algebraic Groups

Chapter 1

Some Algebraic Geometry

This preparatory chapter discusses basic results from algebraic geometry, needed to deal with the elementary theory of algebraic groups. More algebraic geometry will appear as we go along. More delicate results involving ground fields are deferred to Chapter 11.

1.1. The Zariski topology

1.1.1. Let k be an algebraically closed field and put $V = k^n$. The elements of the polynomial algebra $S = k[T_1, \ldots, T_n]$ (abbreviated to $k[T]$) can be viewed as k-valued functions on V. We say that $v \in V$ is a *zero* of $f \in k[T]$ if $f(v) = 0$ and that v is a zero of an ideal I of S if $f(v) = 0$ for all $f \in I$. We denote by $\mathcal{V}(I)$ the set of zeros of the ideal I. If X is any subset of V, let $\mathcal{I}(X) \subset S$ be the ideal of the $f \in S$ with $f(v) = 0$ for all $v \in X$.

Recall that the *radical* or nilradical $\sqrt{I}$ of the ideal I (see [Jac5, p. 392]) is the ideal of the $f \in S$ with $f^n \in I$ for some integer $n \geq 1$. A *radical ideal* is one that coincides with its radical. It is obvious that all $\mathcal{I}(X)$ are radical ideals.

We shall need Hilbert's Nullstellensatz in two equivalent formulations.

1.1.2. Proposition. *(i) If I is a proper ideal in S then $\mathcal{V}(I) \neq \emptyset$;*
(ii) For any ideal I of S we have $\mathcal{I}(\mathcal{V}(I)) = \sqrt{I}$.

For a proof see for example [La2, Ch. X, §2] . The proposition can also be deduced from the results of 1.9 (see Exercise 1.9.6 (2)).

1.1.3. Zariski topology on V. The function $I \mapsto \mathcal{V}(I)$ on ideals has the following properties:
(a) $\mathcal{V}(\{0\}) = V$, $\mathcal{V}(S) = \emptyset$;
(b) If $I \subset J$ then $\mathcal{V}(J) \subset \mathcal{V}(I)$;
(c) $\mathcal{V}(I \cap J) = \mathcal{V}(I) \cup \mathcal{V}(J)$;
(d) If $(I_\alpha)_{\alpha \in A}$ *is a family of ideals and* $I = \sum_{\alpha \in A} I_\alpha$ *is their sum, then* $\mathcal{V}(I) = \bigcap_{\alpha \in A} \mathcal{V}(I_\alpha)$.

The proof of these properties is left to the reader (*Hint*: for (c) use that $I.J \subset I \cap J$). It follows from (a), (c) and (d) that there is a topology on V whose closed subsets are the $\mathcal{V}(I)$, I running through the ideals of S. This is the *Zariski topology*. The induced topology on a subset X of V is the Zariski topology of V. A closed set in V is called an *algebraic set*.

1.1.4. Exercises. (1) Let $V = k$. The proper algebraic sets are the finite ones.

(2) The Zariski closure of $X \subset V$ is $\mathcal{V}(\mathcal{I}(X))$.

(3) The map $\mathcal{I}$ defines an order reversing bijection of the family of Zariski closed subsets of V onto the family of radical ideals of S. Its inverse is $\mathcal{V}$.

(4) The Euclidean topology on $\mathbf{C}^n$ is finer than the Zariski topology.

1.1.5. Proposition. *Let $X \subset V$ be an algebraic set.*

(i) The Zariski topology of X is T_1, i.e., points are closed;

(ii) Any family of closed subsets of X contains a minimal one;

(iii) If $X_1 \supset X_2 \supset \ldots$ is a descending sequence of closed subsets of X, there is an h such that $X_i = X_h$ for $i \geq h$;

(iv) Any open covering of X has a finite subcovering.

If $x = (x_1, \ldots, x_n) \in X$ then x is the unique zero of the ideal of S generated by $T_1 - x_1, \ldots, T_n - x_n$. This implies (i). (ii) and (iii) follow from the fact that S is a Noetherian ring [La2, Ch. VI, §1], using 1.1.4 (3).

To establish assertion (iv) we formulate it in terms of closed sets. We then have to show: if $(I_\alpha)_{\alpha \in A}$ is a family of ideals such that $\bigcap_{\alpha \in A} \mathcal{V}(I_\alpha) = \emptyset$, there is a finite subset B of A such that $\bigcap_{\alpha \in B} \mathcal{V}(I_\alpha) = \emptyset$. Using properties (a), (d) of 1.1.3 and 1.1.4 (3) we see that $\sum_{\alpha \in A} I_\alpha = S$. There are finitely many of the I_α, say $I_1, \ldots, I_h$, such that 1 lies in their sum. It follows that $I_1 + \ldots + I_h = S$, which implies that $\bigcap_{i=1}^h \mathcal{V}(I_i) = \emptyset$. $\square$

A topological space X with the property (ii) is called *noetherian*. Notice that (ii) and (iii) are equivalent properties (compare the corresponding properties in noetherian rings, cf. [La2, p. 142]. X is *quasi-compact* if it has the property of (iv).

1.1.6. Exercise. A closed subset of a noetherian space is noetherian for the induced topology.

1.2. Irreducibility of topological spaces

1.2.1. A topological space X (assumed to be non-empty) is *reducible* if it is the union of two proper closed subsets. Otherwise X is *irreducible*. A subset $A \subset X$ is irreducible if it is irreducible for the induced topology. Notice that X is irreducible if and only if any two non-empty open subsets of X have a non-empty intersection.

1.2.2. Exercise. An irreducible Hausdorff space is reduced to a point.

1.2.3. Lemma. *Let X be a topological space.*

(i) $A \subset X$ is irreducible if and only if its closure $\bar{A}$ is irreducible;

(ii) Let $f : X \to Y$ be a continuous map to a topological space Y. If X is irreducible then so is the image fX.

Let A be irreducible. If $\bar{A}$ is the union of two closed subsets A_1 and A_2 then A is the union of the closed subsets $A \cap A_1$ and $A \cap A_2$. Because of the irreducibility of A, we have (say) $A \cap A_1 = A$, and $A \subset A_1$, $\bar{A} \subset A_1$. So $\bar{A}$ is irreducible.

Conversely, assume this to be the case. If A is the union of two closed subsets $A \cap B_1$, $A \cap B_2$, where B_1, B_2 are closed in X, then $\bar{A} \subset B_1 \cup B_2$. It follows that $\bar{A} \cap B_1 = \bar{A}$, whence $A \cap B_1 = A$. The irreducibility of A follows.

The proof of (ii) is easy and can be omitted. □

1.2.4. Proposition. *Let X be a noetherian topological space. Then X has finitely many maximal irreducible subsets. These are closed and cover X.*

It is clear from 1.2.3 (i) that maximal irreducible subsets of X are closed.

Next we claim that X is a union of finitely many irreducible closed subsets. Assume this to be false. Then the noetherian property 1.1.5 (ii) and 1.1.6 imply that there is a minimal non-empty closed subset A of X which is not a finite union of irreducible closed subsets. But A must be reducible, so it is a union of two proper closed subsets. Because of the minimality of A these have the property in question, and a contradiction emerges. This establishes the claim.

Let $X = X_1 \cup ... \cup X_s$, where the X_i are irreducible and closed. We may assume that there are no inclusions among them. If Y is an irreducible subset of X then $Y = (Y \cap X_1) \cup ... \cup (Y \cap X_s)$ and by the definition of irreducibility we must have $Y \subset X_i$ for some i, i.e., any irreducible subset of X is contained in one of the X_i. This implies that the X_i are the maximal irreducible subsets of X. The proposition follows. □

The maximal irreducible subsets of X are called the (irreducible) *components* of X.

We now return to the Zariski topology on $V = k^n$.

1.2.5. Proposition. *A closed subset X of V is irreducible if and only if $\mathcal{I}(X)$ is a prime ideal.*

Let X be irreducible and let $f, g \in S$ be such that $fg \in \mathcal{I}(X)$. Then

$$X = (X \cap \mathcal{V}(fS)) \cup (X \cap \mathcal{V}(gS))$$

and the irreducibility of X implies that (say) $X \subset \mathcal{V}(fS)$, which means that $f \in \mathcal{I}(X)$. It follows that $\mathcal{I}(X)$ is a prime ideal.

Conversely, assume this to be the case and let $X = \mathcal{V}(I_1) \cup \mathcal{V}(I_2) = \mathcal{V}(I_1 \cap I_2)$. If $X \neq \mathcal{V}(I_1)$, then there is $f \in I_1$ with $f \notin \mathcal{I}(X)$. Since $fg \in \mathcal{I}(X)$ for all $g \in I_2$ it follows from the primeness of $\mathcal{I}(X)$ that $I_2 \subset \mathcal{I}(X)$, whence $X = \mathcal{V}(I_2)$. So X is irreducible. □

1.2.6. Exercises. (1) Let X be a noetherian space. The components of X are its maximal irreducible closed subsets.

(2) Any radical ideal I of S is an intersection $I = P_1 \cap \dots \cap P_s$ of prime ideals. If there are no inclusions among them, they are uniquely determined, up to order.

1.2.7. Recall that a topological space is *connected* if it is not the union of two disjoint proper closed subsets. An irreducible space is connected. The following exercises give some results on connectedness and the relation with the notion of irreducibility.

1.2.8. Exercises. (1) (a) A noetherian space X is a disjoint union of finitely many connected closed subsets, its connected components. They are uniquely determined.

(b) A connected component of X is a union of irreducible components.

(2) A closed subset X of $V = k^n$ is not connected if and only if there are two ideals I_1, I_2 in S with $I_1 + I_2 = S$, $I_1 \cap I_2 = \mathcal{I}(X)$.

(3) Let $X = \{(x, y) \in k^2 \mid xy = 0\}$. Then X is a closed subset of k^2 which is connected but not irreducible.

1.3. Affine algebras

1.3.1. We now turn to more intrinsic descriptions of algebraic sets. Let $X \subset V$ be one. The restriction to X of the polynomial functions of S form a k-algebra isomorphic to $S/\mathcal{I}(X)$, which we denote by $k[X]$. This algebra has the following properties:

(a) $k[X]$ *is a k-algebra of finite type*, i.e., there is a finite subset $\{f_1, \dots, f_r\}$ of $k[X]$ such that $k[X] = k[f_1, \dots, f_r]$;

(b) $k[X]$ *is reduced*, i.e., 0 is the only nilpotent element of $k[X]$.

A k-algebra with these two properties is called an *affine k-algebra*. If A is an affine k-algebra, then there is an algebraic subset X of some k^r such that $A \simeq k[X]$. For $A \simeq k[T_1, \dots, T_r]/I$, where I is the kernel of the homomorphism sending the T_i to the generator f_i of A (as in (a)), then A is reduced if and only if I is a radical ideal. We call $k[X]$ the *affine algebra* of X.

1.3.2. We next show that the algebraic set X and its Zariski topology are determined by the algebra $k[X]$.

If I is an ideal in $k[X]$ let $\mathcal{V}_X(I)$ be the set of the $x \in X$ with $f(x) = 0$ for all $f \in I$. If Y is a subset of X let $\mathcal{I}_X(Y)$ be the ideal in $k[X]$ of the f such that $f(y) = 0$ for all $y \in Y$. If A is any affine algebra, let $\mathrm{Max}(A)$ be the set of its maximal ideals. If X is as before and $x \in X$, then $M_x = \mathcal{I}_X(\{x\})$ is a maximal ideal (because $k[X]/M_x$ is isomorphic to the field k).

1.3.3. Proposition. *(i) The map $x \mapsto M_x$ is a bijection of X onto $\mathrm{Max}(k[X])$, moreover $x \in \mathcal{V}_X(I)$ if and only if $I \subset M_x$;*

(ii) The closed sets of X are the $\mathcal{V}_X(I)$, I running through the ideals of $k[X]$.

Since $k[X] \simeq S/\mathcal{I}(X)$ the maximal ideals $k[X]$ correspond to the maximal ideals of S containing $\mathcal{I}(X)$. Let M be a maximal ideal of S. Then 1.1.4 (3) and 1.1.5 (ii) imply that M is the set of all $f \in S$ vanishing at some point of k^n. From this the first point of (i) follows, and the second point is obvious. (ii) is a direct consequence of the definition of the Zariski topology of X. $\qquad\square$

From 1.3.3 we see that the algebra $k[X]$ completely determines X and its Zariski topology.

1.3.4. Exercises. (1) For any ideal I of $k[X]$ we have $\mathcal{I}_X(\mathcal{V}_X(I)) = \sqrt{I}$; for any subset Y of X we have $\mathcal{V}_X(\mathcal{I}_X(Y)) = \bar{Y}$.
(2) The map $\mathcal{I}_X$ defines a bijection of the family of Zariski-closed subsets of X onto the family of radical ideals of $k[X]$, with inverse $\mathcal{V}_X$.
(3) Let A be an affine k-algebra. Define a bijection of $\mathrm{Max}(A)$ onto the set of homomorphisms of k-algebras $A \to k$.
(4) Let X be an algebraic set.

(a) X is irreducible if and only if $k[X]$ is an integral domain (i.e., does not contain zero divisors $\neq 0$).

(b) X is connected if and only if the following holds: if $f \in k[X]$ and $f^2 = f$, $f \neq 0$ then $f = 1$.

(c) Let $X_1, \ldots, X_s$ be the irreducible components of X. If $X_i \cap X_j = \emptyset$ for $1 \leq i, j \leq s$, $i \neq j$, then there is an isomorphism $k[X] \to \bigoplus_{1 \leq i \leq s} k[X_i]$, defined by the restriction maps $k[X] \to k[X_i]$.

1.3.5. We shall have to consider locally defined functions on X. For this we need special open subsets of X, which we now introduce.

If $f \in k[X]$ put

$$D_X(f) = D(f) = \{x \in X \mid f(x) \neq 0\}.$$

This is an open set, namely the complement of $\mathcal{V}(fk[X])$. We have

$$D(fg) = D(f) \cap D(g), \quad D(f^n) = D(f) \ (n \geq 1).$$

The $D(f)$ are called *principal open subsets* of X.

1.3.6. Lemma. *(i) If $f, g \in k[X]$ and $D(f) \subset D(g)$ then $f^n \in gk[X]$ for some $n \geq 1$;*
(ii) The principal open sets form a basis of the topology of X.

Using 1.1.4 (3) we see that $D(f) \subset D(g)$ if and only if $\sqrt{(fk[X])} \subset \sqrt{(gk[X])}$, which implies (i). (ii) is equivalent to the statement that every closed set in X is an intersection of sets of the form $\mathcal{V}_X(fk[X])$. This is obvious from the definitions. $\qquad\square$

1.3.7. F-structures. Let F be a subfield of k. We say that F is a *field of definition* of the closed subset X of $V = k^n$ if the ideal $\mathcal{I}(X)$ is generated by polynomials with coefficients in F. In this situation we put $F[X] = F[T]/\mathcal{I}(X) \cap F[T]$. Then the inclusion $F[T] \rightarrow S$ induces an isomorphism of $F[X]$ onto an F-subalgebra of S and an isomorphism of k-algebras $k \otimes_F F[X] \rightarrow k[X]$. But this notion of field of definition is not intrinsic, as it depends on a particular choice of generators of $k[X]$.

The intrinsic way to proceed is as follows. Let $A = k[X]$ be an affine algebra. An F-*structure* on X is an F-subalgebra A_0 of A which is of finite type over F and which is such that the homomorphism induced by multiplication

$$k \otimes_F A_0 \rightarrow k[X]$$

is an isomorphism. We then write $A_0 = F[X]$. The set $X(F)$ of F-*rational points* for our given F-structure is the set of F-homomorphisms $F[X] \rightarrow F$. More generally, if W is any vector space over k (not necessarily finite dimensional), an F-*structure* on W is an F-vector subspace W_0 of W such that the canonical homomorphism

$$k \otimes_F W_0 \rightarrow W$$

is an isomorphism. A subspace W' of W is *defined over* F if it is spanned by $W' \cap W_0$. Then $W' \cap W_0$ is an F-structure on W'.

1.3.8. A closed subset Y of X is F-*closed* (relative to our F-structure on X) if the ideal $\mathcal{I}_X(Y)$ is defined over F. A subset is F-*open* if its complement is F-closed. The F-open sets define a topology, the F-*topology*. An example of an F-open set is a principal open set $D(f)$ with $f \in F[X]$. These form a basis of the F-topology.

1.3.9. Exercise. Let $k = \mathbf{C}$, $F = \mathbf{R}$, let $k[X] = \mathbf{C}[T, U]/(T^2 + U^2 - 1)$ and let a, b be the images in $k[X]$ of T and U. Show that $\mathbf{R}[a, b]$ and $\mathbf{R}[ia, ib]$ define two different $\mathbf{R}$-structures on X. (*Hint*: consider the sets of rational points.)

1.4. Regular functions, ringed spaces

1.4.1. Notations are as in 1.3. Let $x \in X$. A k-valued function f defined in a neighborhood U of x is called *regular in* x if there are $g, h \in k[X]$ and an open neighborhood $V \subset U \cap D(h)$ of x such that $f(y) = g(y)h(y)^{-1}$ for $y \in V$.

A function f defined in a non-empty open subset U of X is *regular* if it is regular in all points of U. So for each $x \in U$ there exist g_x, h_x with the properties stated above. Denote by $\mathcal{O}_X(U)$ or $\mathcal{O}(U)$ the k-algebra of regular functions in U. The following properties are obvious:

(A) *If U and V are non-empty open subsets and $U \subset V$, restriction defines a k-algebra homomorphism $\mathcal{O}(U) \rightarrow \mathcal{O}(V)$;*

(B) *Let $U = \bigcup_{\alpha \in A} U_\alpha$ be an open covering of the open set U. Suppose that for each*

$\alpha \in A$ we are given $f_\alpha \in \mathcal{O}(U_\alpha)$ such that if $U_\alpha \cap U_\beta$ is non-empty, f_α and f_β restrict to the same element of $\mathcal{O}(U_\alpha \cap U_\beta)$. Then there is $f \in \mathcal{O}(U)$ whose restriction to U_α is f_α, for all $\alpha \in A$.

1.4.2. Sheaves of functions. Now let X be an arbitrary topological space. Suppose that for each non-empty open subset U of X, a k-algebra of k-valued functions $\mathcal{O}(U)$ is given such that (A) and (B) hold. The function $\mathcal{O}$ is called a *sheaf of k-valued functions on X*. (We shall not need the general notion of a sheaf on a topological space.) A pair $(X, \mathcal{O})$ consisting of a topological space and a sheaf of functions is called a *ringed space*.

Let $(X, \mathcal{O})$ be a ringed space. If Y is a subset of X, we define an induced ringed space $(Y, \mathcal{O}|_Y)$ as follows. Y is provided with the induced topology. If U is an open subset of Y then $\mathcal{O}|_Y(U)$ consist of the functions f on U with the following property: there exists an open covering $U \subset \bigcup U_\alpha$ of U by open sets in X, and for each α, an element $f_\alpha \in \mathcal{O}(U_\alpha)$ such that the restriction of f_α to $U \cap U_\alpha$ coincides with the restriction of f.

We leave it to the reader to show that $\mathcal{O}|_Y$ is a sheaf of functions. Notice that if Y is open we have $\mathcal{O}|_Y(U) = \mathcal{O}(U)$ for all open sets U of Y.

1.4.3. Affine algebraic varieties. The ringed spaces $(X, \mathcal{O}_X)$ of 1.4.1 are the *affine algebraic varieties* over k, which we also call *affine k-varieties*. In the sequel we shall usually drop the $\mathcal{O}_X$ and speak of an algebraic variety X.

We denote by $\mathcal{O}_{X,x}$ or $\mathcal{O}_x$ the k-algebra of functions regular in $x \in X$. By definition these are functions defined and regular in some open neighborhood of x, two such functions being identified if they coincide on some neighborhood of x. A formal definition is

$$\mathcal{O}_x = \text{limind } \mathcal{O}(U),$$

where U runs through the open neighborhoods of x, ordered by inclusion and limind denotes inductive limit. We write $\mathbf{A}^n$ for the affine variety defined by k^n. This is *affine n-space*.

1.4.4. Exercises. (1) $\mathcal{O}_{X,x}$ is a local ring, i.e., has only one maximal ideal (namely the ideal of functions vanishing in x).
(2) Let $M_x \subset k[X]$ be the ideal of functions vanishing in $x \in X$. Show that $\mathcal{O}_{X,x}$ is isomorphic to the localization $k[X]_{M_x}$. (If A is a commutative ring and S a subset that is closed under multiplication, the ring of fractions $S^{-1}A$ is the quotient of $S \times A$ by the following equivalence relation: $(s, a) \sim (s', a')$ if there is $s_1 \in S$ with $s_1(s'a - sa') = 0$. The equivalence class of (s, a) is written as a fraction $s^{-1}a$ and these are added and multiplied in the usual way. If P is a prime ideal in A and S is the complement $A - P$, then $S^{-1}A$ is written A_P and is called the *localization* of A at P. See [La2, Ch. II, §3].)

Let $(X, \mathcal{O}_X)$ be an algebraic variety. It follows from the definitions that there is a homomorphism $\phi : k[X] \to \mathcal{O}(X)$.

1.4.5. Theorem. ϕ *is an isomorphism.*

It is obvious that ϕ is injective. We have to prove surjectivity. Let $f \in \mathcal{O}(X)$. For each $x \in X$ there exists an open neighborhood U_x of x and $g_x, h_x \in k[X]$ such that h_x does not vanish at any point of U_x and that for $y \in U_x$

$$f(y) = g_x(y)h_x(y)^{-1}.$$

By 1.3.6 (ii) we may assume that there is $a_x \in k[X]$ with $U_x = D(a_x)$. Then $D(a_x) \subset D(h_x)$ and by 1.3.6 (i) there exist $h'_x \in k[X]$ and an integer $n_x \geq 1$ with

$$a_x^{n_x} = h_x h'_x.$$

The restriction of f to U_x equals $g_x h'_x (a_x^{n_x})^{-1}$. Observing that $D(a_x) = D(a_x^{n_x})$ we see that we may assume that $h_x = a_x$.

Since X is quasi-compact (1.1.5 (iv)) there are finitely many of the $h_x, h_1, \ldots, h_s$, such that the open sets $D(h_i)$ $(1 \leq i \leq s)$ cover X. Let $g_i \in k[X]$ be such that the restriction of f to $D(h_i)$ equals $g_i h_i^{-1} (1 \leq i \leq s)$. Since $g_i h_i^{-1}$ and $g_j h_j^{-1}$ coincide on $D(h_i) \cap D(h_j)$ whereas $h_i h_j$ vanishes outside this set, we have $h_i h_j (g_i h_j - g_j h_i) = 0$. Since the $D(h_i)$ cover X, the ideal generated by $h_1^2, \ldots, h_s^2$ is $k[X]$. So there exist $b_i \in k[X]$ with

$$\sum_{i=1}^{s} b_i h_i^2 = 1.$$

Let $x \in D(h_j)$. Then

$$h_j^2(x) \sum_{i=1}^{s} b_i(x) g_i(x) h_i(x) = \sum_{i=1}^{s} b_i(x) h_i^2(x) h_j(x) g_j(x) = h_j^2(x) f(x).$$

It follows that $f = \phi(\sum_{i=1}^{s} b_i g_i h_i)$, which proves surjectivity. $\square$

1.4.6. Exercise. Let $D(f)$ be a principal open subset of X. Show that there is an isomorphism onto $\mathcal{O}_X(D(f))$ of the algebra $k[X]_f = k[X][T]/(1 - fT)$. ($k[X]_f$ is isomorphic to the ring of fractions $S^{-1}k[X]$, where $S = (f^n)_{n \geq 0}$; see 1.4.4 (2).)

1.4.7. Morphisms. Let $(X, \mathcal{O}_X)$ and $(Y, \mathcal{O}_Y)$ be two ringed spaces. Let $\phi : X \to Y$ be a continuous map. If f is a function on an open set $V \subset Y$, denote by $\phi_V^* f$ the function on the open subset $\phi^{-1}V$ of X which is the composite of f and the restriction of ϕ to that set. We say that ϕ is a *morphism of ringed spaces* if, for each open

$V \in Y$, we have that ϕ_V^* maps $\mathcal{O}_Y(V)$ into $\mathcal{O}_X(\phi^{-1}V)$. If, moreover, $(X, \mathcal{O}_X)$ and $(Y, \mathcal{O}_Y)$ are affine algebraic varieties, a morphism of ringed spaces $X \to Y$ is called a *morphism of affine algebraic varieties*. It is clear how to define an isomorphism of affine varieties.

If X is a subset of Y, ϕ the injection $X \to Y$ and $\mathcal{O}_X = \mathcal{O}_Y|_X$ (see 1.4.2) then ϕ defines a morphism of ringed spaces in the previous sense. In this case we say that $(X, \mathcal{O}_X)$ is a *ringed subspace* of $(Y, \mathcal{O}_Y)$.

A morphism $\phi : X \to Y$ of affine varieties defines an algebra homomorphism

$$\mathcal{O}_Y(Y) \to \mathcal{O}_X(X),$$

whence by 1.4.5 a homomorphism $\phi^* : k[Y] \to k[X]$. Conversely, an algebra homomorphism $\psi : k[Y] \to k[X]$ defines a continuous map $\psi\ : X \to Y$ with $(\psi\ )^* = \psi$ (view the points of X and Y as homomorphisms, cf. 1.3.4 (3)). Then $\psi\ $ is a morphism of affine varieties. If ϕ is as before, we have $(\phi^*)\ = \phi$. The upshot of this section is that affine k-varieties and their morphisms can be described in algebraic terms.

1.4.8. Exercises. (1) Complete the proofs of the statements of the last paragraph.
(2) Make affine k-algebras and affine algebraic varieties over k into categories. Show that these categories are anti-equivalent. (For categories see [La2, Ch. I, §7] or [Jac5, Ch. 1].)
(3) A morphism of affine varieties $\phi : X \to Y$ is an isomorphism if and only if the algebra homomorphism ϕ^* is an isomorphism.

1.4.9. Affine F-varieties. Let F be a subfield of k and let $(X, \mathcal{O}_X)$ be an affine k-variety. An F-*structure* on this affine variety is given by the following data:
(a) an F-structure on X, in the sense of 1.3.7;
(b) for each F-open subset U of X we are given an F-subalgebra $\mathcal{O}_X(U)(F)$ of $\mathcal{O}_X(U)$ such that the homomorphism induced by multiplication

$$k \otimes_F \mathcal{O}_X(U)(F) \to \mathcal{O}_X(U)$$

is an isomorphism and that properties like (A) and (B) of 1.4.1 hold. An affine variety over k with an F-structure will be called an *affine F-variety*. It is clear how to define morphisms of these (called *F-morphisms*).

The proof of 1.4.5 carries over to the case of F-varieties and gives that $F[X] \simeq \mathcal{O}_X(X)(F)$. We conclude that affine F-varieties and their morphisms can be described in algebraic terms. An instance of an affine F-variety is affine n-space $\mathbf{A}^n$ ($n \geq 0$), whose algebra is $k[T_1, \dots, T_n]$. If X is an affine F-variety, its set $X(F)$ of F-rational points (1.3.7) can be viewed as the set of F-morphisms $\mathbf{A}^0 \to X$.

1.4.10. Exercise. Complete details in 1.4.9.

1.5. Products

1.5.1. Let X and Y be two affine algebraic varieties over k. In accordance with the general notion of product in a category [La2, Ch. I, §7] we say that a product of X and Y is an affine variety Z over k, together with morphisms $p : Z \to X, q : Z \to Y$ such that the following holds: for any triple (Z', p', q') of an affine variety Z' together with morphisms $p' : Z' \to X$, $q' : Z' \to Y$, there exists a unique morphism $r : Z' \to Z$ such that $p' = p \circ r$, $q' = q \circ r$. Put $A = k[X]$, $B = k[Y]$, $C = k[Z]$. Using 1.4.7 we see that C has the following property: there exist k-algebra homomorphisms $a : A \to C$, $b : B \to C$ such that for any triple (C', a', b') of an affine k-algebra and k-algebra homomorphisms $a' : A \to C', b' : B \to C'$, there is a unique k-algebra homomorphism $c : C \to C'$ with $a' = c \circ a$, $b' = c \circ b$.

Working in the category of all k-algebras, i.e., forgetting the condition that $k[Z]$ is an affine algebra, it follows from familiar properties (see e.g. [La2, Ch. XVI, §4]) that $C = A \otimes_k B$ and $a(x) = x \otimes 1$, $b(y) = 1 \otimes y$ satisfy our requirements.

1.5.2. Lemma. *Let A and B be k-algebras of finite type. If A and B are reduced (respectively, integral domains) then the same holds for $A \otimes_k B$.*

Assume that A and B are reduced. Let $\sum_{i=1}^{n} a_i \otimes b_i$ be a nilpotent element of $A \otimes B$. We may assume the b_i to be linearly independent over k. For any homomorphism $h : A \to k$ we have that $h \otimes \mathrm{id}$ is a homomorphism $A \otimes B \to B$. Then $\sum_{i=1}^{n} h(a_i)b_i$ is a nilpotent element of B, which must be zero since B is reduced. As the b_i are linearly independent, all $h(a_i)$ are zero, for any h. This means that the a_i lie in all maximal ideals of A. It follows that $a_i = 0$ for all i (apply 1.3.4 (1) with $k[X] = A$, $I = (a_i)$), which shows that $A \otimes B$ is reduced.

Next let A and B be integral domains. Let $f, g \in A \otimes B$, $fg = 0$. Write $f = \sum_i a_i \otimes b_i$, $g = \sum_j c_j \otimes d_j$, the sets $\{b_i\}$ and $\{d_j\}$ being linearly independent. An argument similar to the one just given then shows that $a_i c_j = 0$, from which it follows that f or g equals 0. $\square$

1.5.3. Exercises. (1) Show that $\mathbf{C} \otimes_{\mathbf{R}} \mathbf{C}$ is not an integral domain.
(2) Show that in 1.5.2 the assumption that A and B are of finite type can be omitted.

1.5.4. Theorem. *Let X and Y be two affine k-varieties.*
(i) A product variety $X \times Y$ exists. It is unique up to isomorphism;
(ii) If X and Y are irreducible then so is $X \times Y$.

From the discussion of 1.5.1 it is clear that it suffices to show that if A and B are affine k-algebras (respectively, affine k-algebras that are integral domains), the same is true for $A \otimes_k B$. This follows from 1.5.2. The uniqueness statement of (i) follows formally from the definition of products. $\square$

1.5.5. Exercises. X and Y are affine varieties.
(1) Show that the set underlying $X \times Y$ can be identified with the product of the sets underlying X and Y.
(2) With the identification of (1), the Zariski topology on the $X \times Y$ is finer than the product topology. Give an example where these topologies do not coincide.
(3) Let F be a subfield of k. A product of two affine F-varieties exists and is unique up to F-isomorphism.

1.6. Prevarieties and varieties

1.6.1. Prevarieties. A *prevariety* over k is a quasi-compact ringed space $(X, \mathcal{O}_X)$ (or simply X) such that any point of X has an open neighborhood U with the property that the ringed subspace $(U, \mathcal{O}|_U)$ (see 1.4.7) is isomorphic to an affine k-variety. Such a U is called an *affine open subset* of X. A morphism of prevarieties is a morphism of the ringed spaces. A subprevariety of a prevariety is a ringed subspace which is isomorphic to a prevariety.

1.6.2. Exercises. (1) A prevariety is a noetherian topological space.
(2) If X is an irreducible prevariety and U an affine open subset, then U is irreducible.
    The notion of a product of prevarieties is defined in the categorical manner; see 1.5.1.

1.6.3. Proposition. *A product of two prevarieties exists and is unique up to isomorphism.*

Let X and Y be prevarieties and let $X = \bigcup_{i=1}^{m} U_i$, $Y = \bigcup_{j=1}^{n} V_j$ be finite coverings by affine open sets. The underlying set of the product $X \times Y$ will be the set theoretic product $X \times Y$, which is covered by the sets $U_i \times V_j$. On these sets we have a structure of affine variety (by 1.5.4 and 1.5.5 (1)). We declare a set $U \subset X \times Y$ to be open if $U \cap (U_i \times V_j)$ is an open subset of the algebraic set $U_i \times V_j$, for all i, j. This defines a topology on $X \times Y$. A function f, defined in an open neighborhood U of $x \in U_i \times V_j$, is defined to be regular in x if its restriction to $U \cap (U_i \times V_j)$ is regular in x, for the structure of affine variety on $U_i \times V_j$. This defines a structure of ringed space on $X \times Y$. One verifies that it has the required properties. The uniqueness statement is proved in the standard manner. $\square$

1.6.4. Exercise. Fill in the details of the proof of 1.6.3.

1.6.5. Separation axiom. Let X be a prevariety, denote by Δ_X the diagonal subset of $X \times X$, i.e.,

$$\Delta_X = \{(x, x) \mid x \in X\},$$

and denote by $i : X \to \Delta_X$ the obvious map. We provide Δ_X with the induced topology.

1.6.6. Example. Let X be an affine k-variety. Then Δ_X is a closed subset of $X \times X$, namely the set $\mathcal{V}_{X \times X}(I)$ (see 1.3.2), where I is the kernel of the homomorphism

$$k[X \times X] = k[X] \otimes_k k[X] \to k[X]$$

defined by the product of the algebra $k[X]$. The ideal I is generated by the elements $f \otimes 1 - 1 \otimes f$ $(f \in k[X])$. Since $k[X \times X]/I \simeq k[X]$, we conclude that i now defines a homeomorphism of topological spaces $X \simeq \Delta_X$.

1.6.7. Exercise. Prove the assertions made in 1.6.6.

1.6.8. Lemma. $i : X \to \Delta_X$ *defines a homeomorphism of topological spaces for any prevariety* X.

Cover Δ_X by open sets of the form $U \times U$, with U affine open in X, and use that the result holds for affine varieties (1.6.6).   □

1.6.9. The prevariety X is defined to be a *variety* (or an algebraic variety over k or a k-variety) if the following holds:

Separation axiom. Δ_X *is closed in* $X \times X$.

By 1.6.6 this holds if X is an affine variety. See 1.6.13 (1) for an example of a prevariety which is not a variety.

It is clear how to define morphisms of varieties.

1.6.10. Exercises. (1) Show that a topological space X is Hausdorff if and only if the diagonal Δ_X is closed in $X \times X$ for the product topology.
(2) The product of two varieties is a variety.
(3) A subprevariety of a variety is a variety.
(4) Let X be a variety. Define an induced variety structure on open and closed subsets of X.

One needs the separation axiom to establish the following results.

1.6.11. Proposition. *Let X be a variety and Y a prevariety.*
(i) If $\phi : Y \to X$ is a morphism, then its graph $\Gamma_\phi = \{(y, \phi(y)) \mid y \in Y\}$ is closed in $Y \times X$;
(ii) If $\phi, \psi : Y \to X$ are two morphisms which coincide on a dense set, then $\phi = \psi$.

In the situation of (i), consider the continuous map $Y \times X \to X \times X$ sending (y, x) to $(\phi(y), x)$. Then Γ_ϕ is the inverse image of the closed set Δ_X, hence is closed. This proves (i). In the situation of (ii) it follows similarly that $\{y \in Y \mid \phi(y) = \psi(y)\}$ is closed, whence (ii).   □

The following result gives a useful criterion for a prevariety to be a variety.

1.6.12. Proposition. *(i) Let X be a variety and let U, V be affine open sets in X. Then $U \cap V$ is an affine open set and the images under restriction of $\mathcal{O}_X(U)$ and $\mathcal{O}_X(V)$ in $\mathcal{O}_X(U \cap V)$ generate the last algebra;*
(ii) Let X be a prevariety and let $X = \bigcup_{i=1}^{m} U_i$ be a covering by affine open sets. Then X is a variety if and only if the following holds: for each pair (i, j) the intersection $U_i \cap U_j$ is an affine open set and the images under restriction of $\mathcal{O}_X(U_i)$ and $\mathcal{O}_X(U_j)$ in $\mathcal{O}_X(U_i \cap U_j)$ generate the last algebra.

In the situation of (i), we have that $\Delta_X \cap (U \times V)$ is closed in $U \times V$. Now i induces an isomorphism of ringed spaces $U \cap V \simeq \Delta_X \cap (U \times V)$. It follows that $U \cap V$ is affine and that regular functions on $\Delta_X \cap (U \times V)$ are restrictions of regular functions on $U \times V$, i.e., of elements of $k[U] \otimes k[V]$. Then (i) follows from the fact that this algebra is generated by $k[U] \otimes 1$ and $1 \otimes k[V]$.

The necessity of the condition of (ii) follows from (i). If it is satisfied, then for each pair (i, j) the intersection $\Delta_X \cap (U_i \times V_j)$ is an affine algebraic variety whose algebra is a quotient of $\mathcal{O}_{X \times X}(U_i \times U_j)$. This implies that $\Delta_X \cap (U_i \cap U_j)$ is closed in $U_i \times U_j$. Hence Δ_X is closed in $X \times X$. $\square$

1.6.13. Exercises. (1) Define the 'line with doubled point 0' X as follows. As a set X is the union of $\mathbf{A}^1$ and a point $0'$, moreover $\mathbf{A}^1$ is an affine open set, with its usual variety structure. Define $\phi : \mathbf{A}^1 \to X$ by $\phi(x) = x \in \mathbf{A}^1$ if $x \neq 0$, $\phi(0) = 0'$. Then Im ϕ can be given a structure of affine variety isomorphic to $\mathbf{A}^1$. Define Im ϕ to be an affine open subset. Show that X is a prevariety which is not a variety.
(2) Define the projective line $\mathbf{P}^1$ in a similar way: $\mathbf{P}^1 = \mathbf{A}^1 \cup \{\infty\}$, with $\phi(x) = x^{-1}$ for $x \neq 0$, $\phi(0) = \infty$. Show that $\mathbf{P}^1$ is a variety. Show that $\mathcal{O}_{\mathbf{P}^1}(\mathbf{P}^1) = k$ and deduce that any morphism of $\mathbf{P}^1$ to an affine variety is constant.
(3) (a) Let $U \subset \mathbf{A}^n$ be open and non-empty. If $f \in \mathcal{O}_{\mathbf{A}^n}(U)$, there exist $g, h \in k[T_1, \dots, T_n]$ such that h does not vanish on U and that $f(x) = g(x)h(x)^{-1}$ for $x \in U$. (*Hint:* cover U by principal open sets and use 1.4.6).
   (b) Let $X = \mathbf{A}^n - \{0\}$, with the induced structure of variety. Show that X is not an affine variety if $n \geq 2$.

1.6.14. F-structures. Let F be a subfield of k. We say that the k-variety X has an F-*structure* or that X is an F-*variety* if we are given a family of open subsets of X, called F-*open subsets*, with the following properties:
(a) the F-open subsets form a topology;
(b) the F-open subsets, which are also affine open, cover X (these sets will be called affine F-open sets);
(c) an affine F-open set has a structure of affine F-variety;
(d) if U and V are two affine F-open sets with $V \subset U$, then the inclusion morphism of affine varieties $V \to U$ is defined over F.

If X and Y are two F-varieties, a morphism $\phi : X \to Y$ is *defined over F* if for any F-open set $V \subset Y$, the set $U = \phi^{-1}U$ is F-open in X and if, moreover, the induced morphism $U \to V$ is defined over F. We also say that ϕ is an *F-morphism*. The notion of an F-isomorphism is the obvious one. Likewise the notion of an F-subvariety.

If a k-variety X has an F-structure, we shall say that F is a *ground field* for X.

1.6.15. To define an F-structure on a k-variety X, the following data suffice: a covering (U_i) of X by affine open subsets, together with F-structures on the U_i and their intersections, such that all inclusion morphisms $U_i \cap U_j \to U_i$ are defined over F. An F-open set is defined to be an open set whose intersection with U_i is F-open in U_i, for all i. Then the properties of 1.6.14 hold.

1.6.16. If X is an F-variety, the set $X(F)$ of its *F-rational points* is the set of F-morphisms $\mathbf{A}^0 \to X$ (cf. 1.4.9).

If X and Y are two F-varieties, there is a unique structure of F-variety on the product $X \times Y$ of 1.5 such that the projection morphisms to X and Y are defined over F.

1.6.17. Exercises. (1) Check the statements made in 1.6.15 and 1.6.16.
(2) A morphism of F-varieties $\phi : X \to Y$ is defined over F if and only if its graph (1.6.11 (i)) is a closed F-subvariety of the F-variety $X \times Y$ of 1.6.16.
(3) Let X be a k-variety. There exists a subfield F of k which is an extension of finite type of the prime field in k such that X has an F-structure.

1.7. Projective varieties

The most important example of non-affine varieties, and practically the only ones that we shall encounter, are the projective spaces and their closed subvarieties, to be discussed in the present section.

1.7.1. $\mathbf{P}^n$. The underlying set of projective n-space $\mathbf{P}^n$ is the set of all one dimensional subspaces of the vector space k^{n+1} or, equivalently, $k^{n+1} - \{0\}$ modulo the equivalence relation: $x \sim y$ if there is $a \in k^* = k - \{0\}$ such that $y = ax$. Write x^* for the equivalence class of x. If $x = (x_0, x_1, \dots, x_n)$, we call the x_i *homogeneous coordinates* of x^*.

For $0 \le i \le n$, put

$$U_i = \{(x_0, \dots, x_n)^* \in \mathbf{P}^n \mid x_i \ne 0\}.$$

Define a bijection $\phi_i : U_i \to \mathbf{A}^n$ by

$$\phi_i(x_0, \dots, x_n) = (x_i^{-1}x_0, \dots, x_i^{-1}x_{i-1}, x_i^{-1}x_{i+1}, \dots, x_i^{-1}x_n),$$

and transport the structure of the affine variety on $\mathbf{A}^n$ to U_i via ϕ_i. Then $\phi_i(U_i \cap U_j)$ is a principal open set $D(f)$ in $\mathbf{A}^n$. In fact, identifying $k[\mathbf{A}^n]$ with $k[T_1, \dots, T_n]$, we may take $f = T_{j+1}$ if $j < i$, $f = 1$ if $j = i$ and $f = T_j$ if $j > i$.

We make $\mathbf{P}^n$ into a prevariety in the following manner (cf. 1.6.15). A subset U is defined to be open if $U \cap U_i$ is open in the affine variety U_i, for $0 \le i \le n$. Let $x \in \mathbf{P}^n$ and assume $x \in U_i$. A function f, defined in a neighborhood of x, is declared to be regular in x if the restriction of f to U_i is regular in x for the structure of affine variety of U_i which was introduced above. As in 1.4.1 we obtain a sheaf $\mathcal{O}_{\mathbf{P}^n}$ and a ringed space $(\mathbf{P}^n, \mathcal{O}_{\mathbf{P}^n})$, which is a prevariety (check the details).

This is, in fact, a variety. From the definitions we see that $\mathcal{O}(U_i \cap U_j)$ $(0 \le i, j \le n)$ is the k-algebra of functions whose value in $(x_0, \dots, x_n)^*$ is a polynomial function of $x_i^{-1}x_0, \dots, x_i^{-1}x_n, x_j^{-1}x_0, \dots, x_j^{-1}x_n$. It is then clear that the condition of 1.6.12 (ii) is satisfied.

The variety thus obtained is a *projective n-space*. For $n = 1$ we recover the variety of 1.6.13 (2) (check this). A *projective variety* is a closed subvariety of some $\mathbf{P}^n$, i.e., a closed subset with the induced structure of a variety. A *quasi-projective variety* is an open subvariety of a projective variety.

1.7.2. Exercises. (1) An invertible linear map of k^{n+1} induces an isomorphism of $\mathbf{P}^n$.

(2) Let V be a finite dimensional vector space over k. Define a variety $\mathbf{P}(V)$ whose underlying point set is the set of one-dimensional subspaces of V and which is isomorphic to $\mathbf{P}^{n-1}$, where $n = \dim V$.

(3) Let F be a subfield of k. Define an F-structure on $\mathbf{P}^n$, inducing on each U_i the F-structure obtained by transporting the F-structure of $\mathbf{A}^n$.

1.7.3. Closed sets in $\mathbf{P}^n$. We shall now give a concrete description of closed sets in $\mathbf{P}^n$. Let $S = k[T_0, \dots, T_n]$ be the polynomial algebra in $n + 1$ indeterminates. An ideal I in S is *homogeneous* if it is generated by homogeneous polynomials or, equivalently, if $f_0 + \dots + f_h \in I$, where f_i is homogeneous of degree i, then all f_i lie in I.

If I is a proper homogeneous ideal in S, then if $x \in k^{n+1}$ is a zero of I, the same is true for all $ax, a \in k^*$. Hence we can define a set $\mathcal{V}^*(I) \in \mathbf{P}^n$ by

$$\mathcal{V}^*(I) = \{x^* \in \mathbf{P}^n \mid x \in \mathcal{V}_{k^{n+1}}(I)\}.$$

1.7.4. Proposition. *The closed sets in $\mathbf{P}^n$ coincide with the sets $\mathcal{V}^*(I)$, I running through the homogeneous ideals of S.*

Let I be a homogeneous ideal. It is easy to see that $\mathcal{V}^*(I) \cap U_i$ is closed for all i, from which it follows that $\mathcal{V}^*(I)$ is closed.

Let U be open in $\mathbf{P}^n$. To prove 1.7.4 it suffices to show that U is the complement of some $\mathcal{V}^*(I)$, and by an analogue of 1.1.3 (d), it suffices to do this if

$U = \phi_i^{-1}(D(f)) \subset U_i$ (for some i), where $f \in k[T_1, \ldots, T_n]$. There is a homogeneous polynomial $f^* \in k[T_0, \ldots, T_n]$ which is divisible by T_i, such that

$$f^*(x_1, \ldots, x_{i-1}, 1, x_i, \ldots, x_n) = f(x_1, \ldots, x_n).$$

Then U is the complement of $V^*(f^*S)$.   □

1.7.5. Exercises. (1) Let $f \in S$ be homogeneous and $\neq 0$. Let S_f^* be the algebra of rational functions gf^{-h}, where $g \in S$ is homogeneous of degree $\deg g = h \deg f$. Show that $D^*(f) = \mathbf{P}^n - V^*(fS)$ is an affine open subset of $\mathbf{P}^n$ and that $\mathcal{O}_{\mathbf{P}^n}(D^*(f)) \simeq S_f^*$.
(2) Let I be a homogeneous ideal in S.
   (a) Show that $V^*(I) = \emptyset$ if and only if there exists $N > 0$ such that $T_i^N \in I$ for $0 \leq i \leq n$.
   (b) Show that $V^*(I)$ is irreducible if and only $\sqrt{I}$ is a prime ideal.
(3) Let F be a subfield of k. The F-closed subsets of $\mathbf{P}^n$, for the F-structure of 1.7.2 (3), are the $V^*(I)$, where the homogeneous ideal I is generated by polynomials with coefficients in F.
(4) (a) Define a map of sets $\phi : \mathbf{P}^m \times \mathbf{P}^n \to \mathbf{P}^{mn+m+n}$ by

$$\phi((x_0, \ldots, x_m)^*, (y_0, \ldots, y_n)^*) = (x_i y_j)^*.$$

Show that the image of ϕ is a closed subset $V^{m,n}$ of $\mathbf{P}^{mn+m+n}$ and that ϕ defines an isomorphism of varieties of $\mathbf{P}^m \times \mathbf{P}^n$ onto the projective variety defined by $V^{m,n}$.

   (b) The product of two projective F-varieties is isomorphic to a projective F-variety.

1.8. Dimension

1.8.1. Let X be an irreducible variety. First assume that X is affine. Then $k[X]$ is an integral domain (1.2.5). Let $k(X)$ be its quotient field [La2, p. 69]. If $U = D(f)$ is a principal open subset of X, then $k[U] = k[X]_f$ (1.4.6) from which it follows that the quotient field $k(U)$ is isomorphic to $k(X)$. Using 1.3.6 (ii) we conclude that the same holds for any affine open subset U.

   Now let X be arbitrary. Using 1.6.12 (ii) the preceding remarks imply that for any two affine open sets U, V of X, the quotient fields $k(U), k(V)$ can be canonically identified. It follows that we can speak of the *quotient field* $k(X)$ of X. If X is an irreducible F-variety, we define similarly the F-quotient field $F(X)$. The *dimension* $\dim X$ of X is the transcendence degree of $k(X)$ over k (see [La2, Ch. X, §1]). If X is reducible and if $(X_i)_{1 \leq i \leq m}$ is the set of its components, we define $\dim X = \max_i (\dim X_i)$.

   If X is affine and $k[X] = k[x_1, \ldots, x_r]$ then $\dim X$ is the maximal number of elements among the x_i that are algebraically independent over k.

1.8.2. Proposition. *If X is irreducible and if Y is a proper irreducible closed subvariety of X, then $\dim Y < \dim X$.*

We may assume X to be affine. Let $k[X] = A = k[x_1, \ldots, x_r]$, $k[Y] = A/P$, where P is a non-zero prime ideal. Let y_i be the image in $k[Y]$ of x_i and put $d = \dim X$, $e = \dim Y$. We may assume that $y_1, \ldots, y_e$ are algebraically independent. Clearly, $x_1, \ldots, x_e$ are also algebraically independent, whence $e \le d$. Assume that $d = e$. Let f be a non-zero element of P. There is a relation $H(f, x_1, \ldots, x_e) = 0$, with $H \in k[T_0, \ldots, T_e]$. We may assume that H is not divisible by T_0. But then we would have a non-trivial relation $H(0, y_1, \ldots, y_e) = 0$. It follows that we must have $d < e$.   □

1.8.3. Proposition. *Let X and Y be irreducible varieties. Then $\dim X \times Y = \dim X + \dim Y$.*

We may assume X and Y to be affine. Let $x_1, \ldots, x_d$ and $y_1, \ldots, y_e$ be maximal sets of algebraically independent elements in $k[X]$, respectively $k[Y]$. Then

$$\{x_1 \otimes 1, \ldots, x_d \otimes 1, 1 \otimes y_1, \ldots, 1 \otimes y_e\}$$

is such a set in $k[X] \otimes k[Y] = k[X \times Y]$.   □

1.8.4. Exercises. (1) $\dim \mathbf{A}^n = \dim \mathbf{P}^n = n$.
(2) A zero dimensional variety is finite.
(3) Let $f \in k[T_1, \ldots, T_n]$ be an irreducible polynomial. The set of its zeros is an $(n-1)$-dimensional irreducible subvariety of $\mathbf{A}^n$.
(4) Let X be an irreducible F-variety. Then $\dim X$ equals the transcendence degree of $F(X)$ over F.

1.9. Some results on morphisms

1.9.1. Lemma. *Let $\phi : X \to Y$ be a morphism of affine varieties and let $\phi^* : k[Y] \to k[X]$ be the associated algebra homomorphism.*
(i) If ϕ^* is surjective, then ϕ maps X onto a closed subset of Y;
(ii) ϕ^* is injective if and only if ϕX is dense in Y;
(iii) If X is irreducible, then so is the closure $\overline{\phi X}$ and $\dim \overline{\phi X} \le \dim X$;
(iv) Let F be a subfield of k. If X and Y are F-varieties and ϕ is defined over F, then $\overline{\phi X}$ is an F-subvariety of Y.

Put $I = \mathrm{Ker}\, \phi^*$. If ϕ^* is surjective, then $\phi X = \mathcal{V}_Y(I)$, whence (i). Also, ϕ^* is injective if and only if $\mathcal{I}_Y(\phi X) = \{0\}$, which implies (ii). The first point of (iii) follows from 1.2.3 and the last point from (ii), applied to the restriction morphism $X \to \overline{\phi X}$, using 1.8.2. Notice that $\overline{\phi X} = \mathcal{V}_Y(I)$. In the situation of (iv), the ideal I is spanned by the kernel of the homomorphism $F[Y] \to F[X]$ induced by ϕ. This implies (iv). $\square$

The image ϕX need not be closed. As an example, let $X = \{(x, y) \in \mathbf{A}^2 \mid xy = 1\}$ and define $\phi : X \to \mathbf{A}^1$ by $\phi(x, y) = x$. Then ϕX is the open set $\mathbf{A}^1 - \{0\}$.

1.9.2. For the proof of the main result 1.9.5 of this section, we need some algebraic results.

Let B be a reduced ring and A a subring such that B is of finite type over A. Suppose that we are given a homomorphism of A to an algebraically closed field. We want to extend it to a homomorphism of B. We first assume that we have the special case that $B = A[b]$ is generated over A by one element. Then $B \simeq A[T]/I$, where I is the ideal of the $f \in A[T]$ with $f(b) = 0$. It does not contain non-zero constant polynomials. Denote by $\mathcal{J}(I)$ the union of the set of leading coefficients of the non-zero polynomials in I and $\{0\}$. This is an ideal in A.

1.9.3. Lemma. *Let K be an algebraically closed field and let $\phi : A \to K$ be a homomorphism such that $\phi \mathcal{J}(I) \neq \{0\}$. Then ϕ can be extended to a homomorphism $B \to K$.*

Let $f = f_0 + f_1 T + ... + f_m T^m \in I$ be such that $\phi f_m \neq 0$. We may assume that m is minimal. We shall proceed by induction on m. First extend ϕ to the obvious homomorphism $A[T] \to K[T]$, also denoted by ϕ. Assume that ϕI does not contain a non-zero constant. Then it generates a proper ideal of $K[T]$. Let $z \in K$ be a zero of that ideal. It is immediate that $\phi b = z$ then defines an extension of ϕ to B.

We claim that the assumption always holds. If not then I contains a polynomial $g = g_0 + ... + g_n T^n$ with $\phi(g_0) \neq 0, \phi(g_i) = 0$ $(i > 0)$. The division algorithm shows that there exist $q, r \in A[T]$ and an integer $d \geq 0$ such that $f_m^d g = qf + r$ and that $\deg r < m$. Then $\phi(f_m^d)\phi(g_0) = \phi q.\phi f + \phi r$. Since ϕf has degree $m > 0$, we have that ϕr is also a non-zero constant. This means that we may assume $n < m$. Then g cannot exist if $m = 1$, proving the claim in that case. Assume that $m > 1$ and that the assertion of the lemma is true for smaller values of m.

If $h = h_0 + ... + h_s T^s \in A[T]$ and $h_s \neq 0$, put $\tilde{h} = T^s h(T^{-1}) = h_s + ... + h_0 T^s$. Let $\tilde{I}$ be the ideal in $A[T]$ generated by the $\tilde{h}$ with $h \in I$. If $a \in \tilde{I} \cap A$, there is an integer $s \geq 0$ such that $aT^s \in I$, whence $(aT)^s \in I$. Since B is reduced, we have $aT \in I$. We conclude that $\tilde{I} \cap A$ is the ideal $J = \{a \in A \mid aT \in I\}$. If $\phi J \neq \{0\}$ we have $m = 1$, contradicting the assumption $m > 1$. So $\phi J = \{0\}$. Put $\tilde{A} = A/J$, $\tilde{B} = A[T]/\tilde{I} = \tilde{A}[\tilde{b}]$. Then $\tilde{B}$ is reduced. In fact, if $f \in A[T]$ and $f^t \in \tilde{I}$, then $T^u \tilde{f}^t \in I$ for some $u \geq 0$. Since B is reduced we can conclude that $T\tilde{f} \in I$, whence $f \in \tilde{I}$, proving that $\tilde{B}$ is reduced.

Now ϕ defines a homomorphism $\tilde{\phi} : \tilde{A} \to K$. Notice that $\tilde{I}$ contains $\tilde{g} = g_n + ... + g_0 T^n$, with $\phi(g_0) \neq 0$. Since $n < m$ the induction assumption shows that $\tilde{\phi}$ extends to a homomorphism $\tilde{B} \to K$. As $\phi(g_i) = 0$ for $i > 0$, we have $\tilde{\phi}(\tilde{b}) = 0$. But $\tilde{I}$ also contains $\tilde{f}$ and $\tilde{\phi}(\tilde{f}) \neq 0$. This contradiction establishes our claim. $\square$

1.9.4. Proposition. *Let B be an integral domain and let A a subring such that B is of finite type over A. Given $b \neq 0$ in B there exists $a \neq 0$ in A such that any homomorphism ϕ of A to the algebraically closed field K with $\phi(a) \neq 0$ can be extended to a homomorphism $\phi : B \to K$ with $\phi(b) \neq 0$.*

We have $B = A[b_1, \ldots, b_n]$. By an easy induction we may assume that $n = 1$, i.e., that $B = A[b_1] \simeq A[T]/I$, as before. First assume that $I \neq \{0\}$. Let $f \in I$ be non-zero and of minimal degree. Denote by a_1 its leading coefficient. The division algorithm shows that $g \in I$ if and only if for some $d \geq 0$ we have that $a_1^d g$ is divisible by f. Let $h \in A[T]$ represent the given element b, then $h \notin I$. Since f is irreducible over the quotient field of A, it follows that f and h are coprime over that field. Hence there exist $u, v \in A[T]$ and $a_2 \in A - \{0\}$ such that $uf + vh = a_2$. Then $a = a_1 a_2$ is as required. For if ϕ is as in the proposition, it follows from the preceding lemma that ϕ can be extended to B. Then $\phi(v(b_1))\phi(b) = \phi a_2 \neq 0$, whence $\phi(b) \neq 0$. This settles the case $I \neq \{0\}$. The easy case $I = \{0\}$ is left to the reader. $\square$

1.9.5. Theorem. *Let $\phi : X \to Y$ be a morphism of varieties. Then ϕX contains a non-empty open subset of its closure $\overline{\phi X}$.*

Using a covering of Y by affine open sets, we reduce the proof to the case that Y is affine. Using 1.3.6 (ii) and 1.4.6 we see that X may also be taken to be affine. If $X_1, \ldots, X_s$ are the irreducible components of X, we have $\overline{\phi X} = \bigcup_i \overline{\phi X_i}$, from which we see that we may also assume X to be irreducible. Replace Y by $\overline{\phi X}$. Then the assertion of the theorem is a consequence of 1.9.4, with $A = k[Y]$, $B = k[X]$, $b = 1$. $\square$

1.9.6. Exercises. (1) Let X be a variety. A subset of X is *locally closed* if it is the intersection of an open and a closed subset. A union of finitely many locally closed sets is a *constructible* set.

(a) The complement of a constructible subset of X is constructible.

(b) Let $\phi : X \to Y$ be a morphism. Deduce from 1.9.5 that the image ϕX is a constructible subset of Y. (*Hint:* proceed by induction on dim X, using 1.8.2.)

(c) If C is a constructible subset of X then ϕC is constructible.

(2) (a) Let E be a field and let F be a field extension of F which is an F-algebra of finite type. Then E is a finite algebraic extension of F. (*Hint:* use 1.9.4.)

(b) Let F be a field. If M is a maximal ideal in the polynomial ring $S = F[T_1, \ldots, T_n]$, then S/I is a finite algebraic extension of F.

(c) Let I be an ideal in S. Then the radical $\sqrt{I}$ is the intersection af the maximal ideals containing I. (*Hint:* Let f be a non-zero element in that intersection and let $\bar{f}$ be its image in S/I. Show that $(S/I)_{\bar{f}} = \{0\}$.)

(d) Prove the Nullstellensatz 1.1.2 (note that the proof of 1.9.4 does not use results from the previous sections).

Notes

This first chapter contains standard material from algebraic geometry and needs few comments. We have also included the definition of algebraic varieties over a ground field that is not algebraically closed, as well as some simple results on such varieties. More delicate results will be taken up in Chapter 11. In another terminology our algebraic varieties are the schemes of finite type over a field which are absolutely reduced. For more about algebraic geometry we refer to Hartshorne's book [Har] or Mumford's notes [Mu2].

In 1.4 we have introduced only sheaves of functions. We did not discuss more general sheaves, as they will not be needed in the sequel. Generalities about sheaf theory can be found in [God].

In the literature 1.9.4 is often proved by using valuations. We used an elementary approach that goes back to Chevalley and Weil [We3, p. 30-31]. The auxiliary result 1.9.3 will also be used in 5.2.

Chapter 2

Linear Algebraic Groups, First Properties

In this chapter algebraic groups are introduced. We establish a number of basic results, which can be handled with the limited amount of algebraic geometry dealt with in the first chapter. k is an algebraically closed field and F a subfield. All algebraic varieties are over k.

2.1. Algebraic groups

2.1.1. An *algebraic group* is an algebraic variety G which is also a group such that the maps defining the group structure $\mu : G \times G \to G$ with $\mu(x, y) = xy$ and $i : x \mapsto x^{-1}$ are morphisms of varieties. (We may view the set of points of the variety $G \times G$ as a product set, see 1.5.5 (i) and 1.6.3). If the underlying variety is affine, G is a *linear algebraic group*. These are the ones we shall be concerned with. It is usual to use the adjective 'linear' instead of 'affine' (this is explained by 2.3.7 (i)).

Let G and G' be algebraic groups. A *homomorphism of algebraic groups* $\phi : G \to G'$ is a group homomorphism is also a morphism of varieties. The notions of isomorphism and automorphism are clear.

The product variety $G \times G'$, provided with the direct product group structure is an algebraic group, the *direct product* of the algebraic groups G and G' (check this).

A *closed subgroup* H of the algebraic group G is a subgroup that is closed in the Zariski topology. Then there is a structure of algebraic group on H such that the inclusion map $H \to G$ is a homomorphism of algebraic groups.

We say that the algebraic group G is an *F-group* if G is an F-variety, if the morphisms μ and i are defined over F, and if the identity element e is an F-rational point (see 1.6.14). The notions of an F-homomorphism of F-groups and of an F-subgroup are clear.

If G is an F-group, the set $G(F)$ of F-rational points (1.6.16) has a canonical group structure.

2.1.2. Let G be a linear algebraic group and put $A = k[G]$. By 1.4.7 the morphisms μ and i are defined by an algebra homomorphism $\Delta : A \to A \otimes_k A$ (called *comultiplication*) and an algebra isomorphism $\iota : A \to A$ (called *antipode*). Moreover, the identity element is a homomorphism $e : A \to k$. Denote by $m : A \otimes A \to A$ the multiplication map (so $m(f \otimes g) = fg$) and let ϵ be the composite of e and the inclusion map $k \to A$.

The group axioms are expressed by the following properties:
(associativity) the homomorphisms $\Delta \otimes \mathrm{id}$ and $\mathrm{id} \otimes \Delta$ of A to $A \otimes A \otimes A$ coincide;
(existence of inverse) $m \circ (\iota \otimes \mathrm{id}) \circ \Delta = m \circ (\mathrm{id} \otimes \iota) \circ \Delta = \epsilon$;
(existence of identity element) $(e \otimes \mathrm{id}) \circ \Delta = (\mathrm{id} \otimes e) \circ \Delta = \mathrm{id}$ (we identify $k \otimes A$ and $A \otimes k$ with A).

The properties can also be expressed by commutativity properties of the following diagrams:

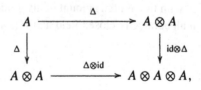

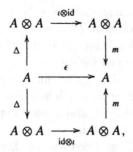

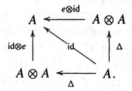

Let G be an F-group and put $A_0 = F[G]$. Then Δ, ι and e come from F-algebra homomorphisms $A_0 \to A_0 \otimes_F A_0$, $A_0 \to A_0$, $A_0 \to F$, respectively, which we denote by the same symbols. They have properties similar to the ones listed above.

2.1.3. Exercises. (1) Check the translation of the group axioms stated in 2.1.2.
(2) Define the notion of a prealgebraic group, based on the notion of a prevariety (see 1.6.1). Show that a prealgebraic group is an algebraic group.

2.1.4. Examples. The notations are as in 2.1.1.
(1) $G = \mathbf{A}^1 = k$ with addition as group operation. This defines a linear algebraic group, with $A = k[T]$. The homomorphism $\Delta : k[T] \to k[T] \otimes k[T] \simeq k[T, U]$ is given by $\Delta T = T + U$ and $\iota : k[T] \to k[T]$ by $\iota T = -T$. Moreover, the algebra homomorphism $e : k[T] \to k$ sends T to 0. We denote this algebraic group by $\mathbf{G}_a$; it is the *additive group*. It is obvious that $F[T]$ defines an F-structure on $\mathbf{G}_a$, for any subfield F of k.

(2) $G = \mathbf{A}^1 - \{0\} = k^*$, with multiplication as group operation. Now $A = k[T, T^{-1}]$ and $\Delta : k[T, T^{-1}] \to k[T, T^{-1}] \otimes k[T, T^{-1}] \simeq k[T, T^{-1}, U, U^{-1}]$ is given by $\Delta T = TU$. Moreover, $\iota T = T^{-1}$ and e send T to 1. This algebraic group is denoted by $\mathbf{G}_m$ or by $\mathbf{GL}_1$; it is the *multiplicative group*. It has the F-structure defined by $F[T, T^{-1}]$.

If n is any non-zero integer then $\phi x = x^n$ defines a homomorphism of algebraic groups $\mathbf{G}_m \to \mathbf{G}_m$. If k has characteristic $p > 0$ and n is a power of p then ϕ is an isomorphism of abstract groups but not of algebraic groups (since $\phi^* : A \to A$ is not surjective, see 1.4.8 (3)).

(3) View the set $\mathbf{M}_n$ of all $n \times n$-matrices as k^{n^2} in the obvious manner. For $X \in \mathbf{M}_n$ let $D(X)$ be its determinant. Then D is a regular function on $\mathbf{M}_n$. The *general linear group* $\mathbf{GL}_n$ is the principal open set $\{X \in \mathbf{M}_n \mid D(X) \neq 0\}$, with matrix multiplication as group operation.

We have $A = k[T_{ij}, D^{-1}]_{1 \leq i, j \leq n}$, with $D = \det(T_{ij})$. Now Δ is given by

$$\Delta T_{ij} = \sum_{h=1}^{n} T_{ih} \otimes T_{hj}.$$

and ιT_{ij} is the (i, j)-entry of the matrix $(T_{ab})^{-1}$. The identity e sends T_{ij} to δ_{ij} (Kronecker symbol). For $n = 1$ we recover the previous example. Since $\mathbf{M}_n$ is an irreducible variety so is $\mathbf{GL}_n$ (by 1.2.3 (i)). Its dimension is n^2. Finally notice that $F[T_{ij}, D^{-1}]$ defines an F-structure.

(4) Any subgroup of $\mathbf{GL}_n$ which is closed in the Zariski topology of $\mathbf{GL}_n$ defines a linear algebraic group. Here are a number of examples:

(a) a finite subgroup;

(b) the group $\mathbf{D}_n$ of non-singular diagonal matrices;

(c) the group $\mathbf{T}_n$ of upper triangular matrices $X = (x_{ij}) \in \mathbf{GL}_n$ with $x_{ij} = 0$ for $i > j$;

(d) the group $\mathbf{U}_n$ of unipotent upper triangular matrices, i.e., the subgroup of the previous group whose elements have diagonal entries 1;

(e) the special linear group $\mathbf{SL}_n = \{X \in \mathbf{GL}_n \mid \det(X) = 1\}$;

(f) the orthogonal group $\mathbf{O}_n = \{X \in \mathbf{GL}_n \mid {}^t X.X = 1\}$, where ${}^t X$ denotes the transpose of X;

(g) the special orthogonal group $\mathbf{SO}_n = \mathbf{O}_n \cap \mathbf{SL}_n$;

(h) the symplectic group $\mathbf{Sp}_{2n} = \{X \in \mathbf{GL}_{2n} \mid {}^t X J X = J\}$, where J is the matrix

$$\begin{pmatrix} 0 & 1_n \\ -1_n & 0 \end{pmatrix}$$

(5) As examples of non-linear algebraic groups (not needed in the sequel) we mention the *elliptic curves*. These are closed subsets of the projective plane $\mathbf{P}^2$. Assuming for convenience that the characteristic is not 2 or 3, such a group G can be defined to be the set of $(x_0, x_1, x_2)^* \in \mathbf{P}^2$ (notations of 1.7.1) such that

$$x_0 x_2^2 = x_1^3 + a x_1 x_0^2 + b x_0^3,$$

where $a, b \in k$ are such that the polynomial $T^3 + aT + b$ has no multiple roots.

The neutral element e is $(0, 0, 1)^*$. The group operation of G is commutative and is written as addition. It is such that if three points of G are collinear in $\mathbf{P}^2$ (i.e., if their homogeneous coordinates satisfy a non-trivial linear relation $c_0 x_0 + c_1 x_1 + c_2 x_2 = 0$), then their sum is e. This defines addition. It is easy to check that, if $x = (x_0, x_1, x_2)^* \in G$, we must have $-x = (x_0, x_1, -x_2)^*$.

The addition can be described by explicit formulas which, however, are not very enlightening. A proof of the associativity of the group operation based on these formulas would be clumsy. There are better, more geometric, ways to deal with the group structure on such a curve. We refer to [Har, p. 321].

2.1.5. Exercises. (1) Let V be a finite dimensional vector space over k.
(a) Define a linear algebraic group $GL(V)$ whose underlying abstract group is the group of all invertible linear maps of V and which is isomorphic to $\mathbf{GL}_{\dim V}$.
(b) An F-structure V_0 on V (1.3.7) defines a structure of F-group on $GL(V)$. The corresponding group of rational points $GL(V)(F)$ is the group $GL(V_0)$ of invertible F-linear maps of V_0.
(2) Check that the subgroups of $\mathbf{GL}_n$ listed in 2.1.4 (4) are indeed closed.
(3) We have $A = k[\mathbf{SL}_2] = k[T_1, T_2, T_3, T_4]/(T_1 T_4 - T_2 T_3 - 1) = k[t_1, t_2, t_3, t_4]$ (t_i denoting the image of T_i). Let B be the subalgebra of A generated by the products $t_i t_j (1 \leq i, j \leq 4)$.

(a) Let Δ and ι define the group structure of $\mathbf{SL}_2$. Show that $\Delta B \subset B \otimes B$, $\iota B = B$ and deduce that there is an algebraic group $\mathbf{PSL}_2$ whose algebra is B. Show that the inclusion map $B \to A$ defines a homomorphism of algebraic groups $\mathbf{SL}_2 \to \mathbf{PSL}_2$ with kernel of order at most two.

(b) If char$(k) \neq 2$, then B is the algebra of functions $f \in A$ such that $f(-X) = f(X)$ for all $X \in \mathbf{SL}_2$.

(c) If char$(k) = 2$, the homomorphism of (a) defines an isomorphism of underlying abstract groups but is not an isomorphism of algebraic groups.
(4) Show that the group $\mathbf{T}_n$ of 2.1.4 (c) is solvable.
(5) Show that the automorphisms of $\mathbf{G}_a (= k)$ are the multiplications by non-zero elements of k.

2.1.6. Generalizations of algebraic groups. (a) The description of the notion of an algebraic group in terms of algebra homomorphisms, given in 2.1.2, leads to the following generalization. Let R be a commutative ring and A a commutative R-algebra. Assume given homomorphisms of R-algebras $\Delta : A \to A \otimes_R A$, $\iota : A \to A$, $e : A \to R$, such that we have the properties of 2.1.2, with k replaced by R. We say that the set of data $G = (A, \Delta, \iota, e)$ defines a *group scheme* over R (more precisely, an affine group scheme.) We shall occasionally encounter this notion. But we shall not go into the theory of group schemes. It is dealt with, for example, in [DG, SGA3].
(b) Let G be a group scheme over R. It follows from the axioms of 2.1.2, that for

each R-algebra S, the set $G(S)$ of R-algebra homomorphisms $A \to S$ has a canonical group structure. In fact, $S \mapsto G(S)$ defines a functor from the category of R-algebras to the category of groups. Such a functor is an *R-group functor*. Group functors generalize group schemes. For more about group functors we refer to [loc.cit.].

(c) A more recent generalization that has become quite important is obtained by admitting in (a) non-commutative R-algebras A. In this case Δ and e are as before, but ι is required to be an anti-automorphism. We impose the same axioms as before. Moreover, we require that the opposite algebra A^{opp} (i.e., A with reversed multiplication) has the same properties, relative to Δ, ι^{-1}, e. Now the set of data $G = (A, \Delta, \iota, e)$ defines a *quantum group* over R. We refer the reader to [Jan2, Kas] for more about the theory of quantum groups and for examples.

2.2. Some basic results

Let G be an algebraic group. If $g \in G$, the maps $x \mapsto gx$ and $x \mapsto xg$ define isomorphisms of the variety G. We shall frequently use this observation.

2.2.1. Proposition. *(i) There is a unique irreducible component G^0 of G that contains the identity element e. It is a closed normal subgroup of finite index;*
(ii) G^0 is the unique connected component of G containing e;
(iii) Any closed subgroup of G of finite index contains G^0.

Let X and Y be irreducible components of G containing e. If μ and i are as in 2.1.1, it follows from 1.2.3 that $XY = \mu(X \times Y)$ and its closure $\overline{XY}$ are irreducible. Since X and Y are contained in $\overline{XY}$, it follows from 1.2.6 (1) that $X = Y = \overline{XY}$. It also follows that X is closed under multiplication. Since i is a homeomorphism, we see that X^{-1} is an irreducible component of G containing e, so must coincide with X. We conclude that X is a closed subgroup. Using that inner automorphisms define homeomorphisms, one sees that for $x \in G$, we have $xXx^{-1} = X$, so that X is a normal subgroup. The cosets xX must be the components of G, and by 1.2.4 the number of cosets is finite. We have proved that $G^0 = X$ has the properties of (i).

We also see that the irreducible components of G are mutually disjoint. It then follows that the irreducible components must coincide with the connected components (use 1.2.8 (i)). This implies (ii).

If H is a closed subgroup of G of finite index, then H^0 is a closed subgroup of finite index of G^0. Then H^0 is both open and closed in G^0. Since G^0 is connected, we have $H^0 = G^0$, which proves (iii).   □

The proposition shows that, for algebraic groups, the notions of irreducibility and connectedness coincide. In the sequel we shall, as usual, speak of connected algebraic groups, and not of irreducible ones. We always denote by G^0 the component of the algebraic group G containing the identity (briefly: the *identity component*).

Dimensions being defined as in 1.8.1, we see from 2.2.1 that all components of G have dimension $\dim G$.

2.2.2. Exercises. (1) The groups $\mathbf{G}_a$, $\mathbf{G}_m$, $\mathbf{GL}_n$, $\mathbf{D}_n$, $\mathbf{T}_n$, $\mathbf{U}_n$, $\mathbf{SL}_n$ of 2.1.4 are connected.

(2) Assume that k has characteristic $\neq 2$.

    (a) The group $\mathbf{O}_n$ is not connected.

    (b) Let V be the set of skew symmetric $n \times n$-matrices. Then $x \mapsto (1+x)^{-1}(1-x)$ defines an isomorphism of a non-empty open subset of $\mathbf{SO}_n$ onto an open subset of V. Show that $\mathbf{SO}_n$ is the identity component of $\mathbf{O}_n$.

(3) The variety of 1.2.8 (3) cannot be the underlying variety of an algebraic group.

(4) Let G be a connected algebraic group and let N be a finite normal subgroup. Then N lies in the center of G. (*Hint*: for $n \in N$ consider the map $x \mapsto xnx^{-1}$ of G to N.)

2.2.3. Lemma. *Let U and V be dense open subsets of G. Then $UV = G$.*

Notice that a subset of G is open and dense if and only if it intersects any component of G in a non-empty open subset. The intersection of two such subsets is one with the same properties.

Let $x \in G$. Then xV^{-1} and U are both dense open subsets. They have a non-empty intersection (1.2.1), which means that $x \in UV$. $\square$

Notice that if G is connected we need only require U and V to be open and non-empty.

2.2.4. Lemma. *Let H be a subgroup of G.*

(i) The closure $\overline{H}$ is a subgroup of G;

(ii) If H contains a non-empty open subset of $\overline{H}$ then H is closed.

Let $x \in H$. Then $H = xH \subset x\overline{H}$. Since $x\overline{H}$ is closed we have $\overline{H} \subset x\overline{H}$ and $x^{-1}\overline{H} \subset \overline{H}$, whence $H\overline{H} \subset \overline{H}$. Now let $x \in \overline{H}$. Then $Hx \subset \overline{H}$, and a similar argument shows that $\overline{H}$ is closed under multiplication. Since $(\overline{H})^{-1} = \overline{H^{-1}} = \overline{H}$, we conclude that $\overline{H}$ is a group. If $U \subset H$ is open in $\overline{H}$ and non-empty, then H, being a union of translates of U, is open in $\overline{H}$. By 2.2.3 we conclude that $\overline{H} = H.H = H$. $\square$

2.2.5. Proposition. *Let $\phi : G \to G'$ be a homomorphism of algebraic groups.*

(i) $\mathrm{Ker}\,\phi$ is a closed normal subgroup of G;

(ii) $\phi(G)$ is a closed subgroup of G';

(iii) If G and G' are F-groups and ϕ is defined over F then $\phi(G)$ is an F-subgroup of G';

(iv) $\phi(G^0) = (\phi G)^0$.

$\mathrm{Ker}\,\phi = \phi^{-1}e$ is closed in G, whence (i). By 1.9.5, $\phi(G)$ contains a non-empty open subset of its closure. Then (ii) follows from 2.2.4 (ii) and (iii) is a consequence of 1.9.1 (iv). $\phi(G^0)$ is a closed subgroup of G' by (ii), which is connected (by 1.2.3 (ii)) and which is of finite index in ϕG. Using 2.2.1 (iii) we obtain (iv). $\square$

2.2.6. Proposition. *Let $(X_i, \phi_i)_{i \in I}$ be a family of irreducible varieties together with morphisms $\phi_i : X_i \to G$. Denote by H the smallest closed subgroup of G containing the images $Y_i = \phi_i X_i$ ($i \in I$). Assume that all Y_i contain the identity element e.*

(i) H is connected;
(ii) There exist an integer $n \geq 0$, $a = (a(1), \dots, a(n)) \in I^n$ and $\epsilon(h) = \pm 1$ $(1 \leq h \leq n)$ such that $H = Y_{a(1)}^{\epsilon(1)} \dots Y_{a(n)}^{\epsilon(n)}$;
(iii) Assume, moreover, that G is an F-group, that all X_i are F-varieties and that the morphisms ϕ_i are defined over F. Then H is an F-subgroup.

We may assume that the sets Y_i^{-1} occur among the Y_j. For each element $a = (a(1), \dots, a(n))$ of some I^n, we write $Y_a = Y_{a(1)} \dots Y_{a(n)}$. This is an irreducible sub-set of G, and so is the closure $\overline{Y_a}$ (see 1.2.3 (i)). With obvious notations we have $Y_b.Y_c = Y_{(b,c)}$. An argument as in the proof of 2.2.4 then shows that $\overline{Y_b}.\overline{Y_c} \subset \overline{Y_{(b,c)}}$. Now take a such that $\dim \overline{Y_a}$ is maximal. For any b, we then have $\overline{Y_a} \subset \overline{Y_a.Y_b} \subset \overline{Y_{(a,b)}}$. From 1.8.2 we conclude that $\overline{Y_{(a,b)}} = \overline{Y_a}$ and $\overline{Y_b} \subset \overline{Y_a}$, for all b. It also follows that $\overline{Y_a}$ is closed under multiplication, which implies that $\overline{Y_a}$ is a group. By 1.9.5 and 2.2.3 we have $\overline{Y_a} = Y_a.Y_a$. It follows that $H = Y_a$ has the properties stated in (i) and (ii). Now (iii) is a consequence of 1.9.1 (iv). $\quad\square$

2.2.7. Corollary. *(i) Assume that $(G_i)_{i \in I}$ is a family of closed, connected, subgroups of G. Then the subgroup H generated by them is closed and connected. There is an integer $n \geq 0$ and $a = (a(1), \dots, a(n)) \in I^n$ such that $H = G_{a(1)} \dots G_{a(n)}$;*
(ii) If, moreover, G is an F-group and all G_i are F-subgroups then H is an F-subgroup.

If H and K are subgroups of G, we denote by (H, K) the subgroup generated by the commutators $xyx^{-1}y^{-1}$ with $x \in H$, $y \in K$.

2.2.8. Corollary. *(i) If H and K are closed subgroups of G one of which is con-nected, then (H, K) is connected;*
(ii) If, moreover, G is an F-group and H, K are F-subgroups then (H, K) is a con-nected F-subgroup.

Assume that H is connected. Then (i) follows by applying 2.2.6 with $I = K$, all X_i being H, with $\phi_i(x) = xix^{-1}i^{-1}$ $(i \in K)$. The statement of (ii) follows from 1.9.1 (iv), using that by 2.2.6 (H, K) is the image under an F-morphism $(H \times K)^n \to G$. In particular, the commutator subgroup (G, G) of a connected F-group is a connected F-subgroup. $\quad\square$

2.2.9. Exercises. (1) (a) Give another proof of the connectedness of $\mathbf{SO}_n$ in character-istic $\neq 2$ (see 2.2.2 (2)) using 2.2.7 and the fact that $\mathbf{O}_n$ is generated by 'symmetries' (see [Jac4, p. 353].

(b) Prove by a similar argument that $\mathbf{Sp}_{2n}$ is connected for arbitrary k, using that $\mathbf{Sp}_{2n}$ is generated by 'symplectic transvections', see [loc. cit., p. 373]).
(2) (char(k) $\neq 2$) The complement of $\mathbf{SO}_n$ in $\mathbf{O}_n$ is irreducible and generates $\mathbf{O}_n$. Deduce that in 2.2.6 the condition $e \in Y_i$ cannot be omitted.
(3) Let G be a connected F-group and let n be an integer ≥ 2. The subgroup $G^{(n)}$ of

G generated by the n-th powers of its elements is a connected normal F-subgroup.
(4) Show by a counterexample that 2.2.8 (i) is not true if neither H nor K is connected. (*Hint*: take H and K finite.)

2.3. G-spaces

2.3.1. A *G-variety*, or a *G − space*, is a variety X on which G acts as a permutation group, the action being given by a morphism of varieties. More precisely, there is a morphism of varieties $a : G \times X \to X$, written $a(g, x) = g.x$, such that $g.(h.x) = (g.h).x$, $e.x = x$. If, moreover, G is an F-group and X is an F-variety, then X is a *G-space over F* if a is defined over F. A *homogeneous space* for G is a G-space on which G acts transitively.

Let X and Y be G-spaces. A morphism $\phi : X \to Y$ is a *G-morphism*, or is said to be *equivariant*, if $\phi(g.x) = g.\phi(x)(g \in G, x \in X)$.

Let X be a G-space and let $x \in X$. The *orbit* of x is the set $G.x = \{g.x \mid g \in G\}$. The *isotropy group* of x is the closed subgroup $G_x = \{g \in G \mid g.x = x\}$ (check that G_x is closed).

2.3.2. Examples. (1) $X = G$ and G acts by inner automorphisms: $a(g, x) = gxg^{-1}$. The orbits are the conjugacy classes of G and the isotropy groups are the centralizers of elements of G.
(2) $X = G$ and G acts by left (right) translations: $a(g, x) = gx$ (resp. xg^{-1}). This is an example of a homogeneous space, even of a *principal homogeneous space* or *torsor*, where the isotropy groups are trivial (which means that the action of G is simply transitive).
(3) Let V be a finite dimensional vector space over k. A *rational representation* of G in V is a homomorphism of algebraic groups $r : G \to GL(V)$ (see 2.1.5 (1)). We also say that V is a *G-module* (it being understood that r is also given). In this case we can view V as an affine algebraic variety isomorphic to $\mathbf{A}^{\dim V}$, with a G-action defined by $g.v = r(g)v$ ($g \in G, v \in V$). We also have a structure of G-variety on the projective space $\mathbf{P}(V)$ of 1.7.2 (2).

If G is an F-group, a *rational representation over F* is a homomorphism of algebraic groups $G \to GL(V)$ which is defined over F, where now V is a finite dimensional vector space with an F-structure, the F-group $GL(V)$ being as in 2.1.5 (1).

Another version of the definition of a rational representation of G is: a homomorphism of algebraic groups $r : G \to \mathbf{GL}_n$, for some $n \geq 1$.

Assume the situation of 2.3.1.

2.3.3. Lemma. *(i) An orbit $G.x$ is open in its closure;*
(ii) There exist closed orbits.

Application of 1.9.5 to the morphism $g \mapsto g.x$ of G to X shows that $G.x$ contains a non-empty open subset U of its closure. Since $G.x$ is the union of the open sets $g.U \, (g \in G)$, the assertion (i) follows. It implies that for $x \in X$, the set $S_x = \overline{G.x} - G.x$ is closed. It is a union of orbits. By 1.1.5 (ii) there is a minimal set S_x. Because of (i) it must be empty. Hence the orbit $G.x$ is closed, proving (ii). $\qquad\qquad\qquad\qquad\qquad\qquad\qquad\qquad\qquad\qquad\qquad\qquad\qquad\qquad$ $\Box$

The lemma implies that an orbit $G.x$ is a locally closed subset of X (see 1.9.6 (1)), i.e., an open subset of a closed subset of X. It then has a structure of algebraic variety (use 1.6.10 (4)). It is immediate that this is a homogeneous space for G.

2.3.4. Exercises. (1) Let G be a closed subgroup of $\mathbf{GL}_n$. Then $\mathbf{A}^n$ has a structure of G-space. Determine the orbits for $G = \mathbf{GL}_n, \mathbf{D}_n, \mathbf{SL}_n$ (see 2.1.4 (4)).
(2) There is an action of $G = \mathbf{GL}_2$ on the projective line $\mathbf{P}^1$ (see 2.3.2 (3)), which makes $\mathbf{P}^1$ into a homogeneous G-space. Describe the isotropy group of a point. The diagonal action of G on $\mathbf{P}^1 \times \mathbf{P}^1$ is not homogeneous. In fact, there are two orbits.
(3) Generalize the results of the previous exercise to $\mathbf{GL}_n$, acting on $\mathbf{P}^{n-1}$.

2.3.5. From now on we assume that G is a linear algebraic group. Let X be an affine G-space, with action $a : G \times X \to X$. We have $k[G \times X] = k[G] \otimes_k k[X]$ and a is given by an algebra homomorphism $a^* : k[X] \to k[G] \otimes k[X]$ (see 1.4.7). For $g \in G, x \in X, f \in k[X]$ define

$$(s(g))f(x) = f(g^{-1}x).$$

Then $s(g)$ is an invertible linear map of the (in general infinite dimensional) vector space $k[X]$ and s is a representation of abstract groups $G \to GL(k[X])$. The next result will imply that s can be built up from rational representations (see 2.3.9 (1)).

2.3.6. Proposition. *Let V be a finite dimensional subspace of $k[X]$.*
(i) There is a finite dimensional subspace W of $k[X]$ which contains V and is stable under all $s(g)$ $(g \in G)$;
(ii) V is stable under all $s(g)$ if and only if $a^*V \subset k[G] \otimes V$. If this is so, s defines a map $s_V : G \times V \to V$ which is a rational representation of G ;
(iii) If, moreover, G is an F-group, X is an F-variety, V is defined over F (see 1.3.7) and a is an F-morphism then in (i) W can be taken to be defined over F.

It suffices to prove (i) in the case that $V = kf$ is one dimensional. Let

$$a^* f = \sum_{i=1}^{n} u_i \otimes f_i \, (u_i \in k[G], f_i \in k[X]).$$

Then

$$(s(g)f)(x) = f(g^{-1}x) = \sum_{i=1}^{n} u_i(g^{-1})f_i(x),$$

and we see that all $s(g)f$ lie in the subspace W' of $k[X]$ spanned by the f_i. The subspace W of W' spanned by the $s(g)f$ then has the properties of (i).

By a similar argument we see that if $a^*V \subset k[G] \otimes V$ the space V is $s(G)$-stable. Now assume that V is $s(G)$-stable. Let (f_i) be a basis of V and extend it to a basis $(f_i) \cup (g_j)$ of $k[X]$. Let $f \in V$ and write

$$a^*f = \sum u_i \otimes f_i + \sum v_j \otimes g_j,$$

where $u_i, v_j \in k[G]$. Then

$$s(g)f = \sum u_i(g^{-1})f_i + \sum v_j(g^{-1})g_j.$$

Our assumption implies that $v_j(g^{-1}) = 0$ for all g, hence all v_j vanish. This proves the first point of (ii). The second point is now immediate. The proof of (iii) is a copy of that of (i) and can be skipped.   □

We now consider the case that G acts by left or right translations on itself (see 2.3.2 (2)). For $g, x \in G$, $f \in k[G]$ define

$$(\lambda(g)f)(x) = f(g^{-1}x), \quad (\rho(g)f)(x) = f(xg).$$

Then λ and ρ are representations (of abstract groups) of G in $GL(k[G])$, even in the group of algebra automorphisms of $k[G]$. If ι is the automorphism of $k[G]$ defined by inversion (see 2.1.2), then $\rho = \iota \circ \lambda \circ \iota^{-1}$. The representations λ and ρ are faithful, i.e., have trivial kernel. If, for instance, $\lambda(g) = \mathrm{id}$, then $f(g^{-1}) = f(e)$ for all $f \in k[G]$, whence $g = e$.

2.3.7. Theorem. *(i) There is an isomorphism of G onto a closed subgroup of some GL_n;*
(ii) If G is an F-group the isomorphism of (i) may be taken to be defined over F.

By 2.3.6 (i) we may assume that $k[G] = k[f_1, \dots, f_n]$, where (f_i) is a basis of a $\rho(G)$-stable subspace V of $k[G]$.

By 2.3.6 (ii) there exist elements $(m_{ij})_{1 \le i, j \le n}$ in $k[G]$ with

$$\rho(g)f_i = \sum_{j=1}^{n} m_{ji}(g)f_j \ (g \in G).$$

Then $\phi(g) = (m_{ij}(g))_{1 \leq i,j \leq n}$ defines a group homomorphism $\phi : G \to \mathbf{GL}_n$. If $\phi(g) = e$ then $\rho(g) f_i = f_i$ for all i. Since $\rho(g)$ is an algebra homomorphism and since the f_i generate $k[G]$ it follows that $\rho(g)f = f$ for all $f \in k[X]$ and $g = e$. Hence ϕ is injective. Also, ϕ is a morphism of affine varieties. The corresponding algebra homomorphism

$$\phi^* : k[\mathbf{GL}_n] = k[T_{ij}, D^{-1}] \to k[G]$$

(notations of 2.1.4 (3)) is given by $\phi^* T_{ij} = m_{ij}$, $\phi^*(D^{-1}) = \det(m_{ij})^{-1}$. From

$$f_i(g) = \sum_j m_{ji}(g) f_j(e)$$

it follows that ϕ^* is surjective. By 2.2.5 (ii), $\phi(G)$ is a closed subgroup. Its algebra is isomorphic to $k[\mathbf{GL}_n]/\mathrm{Ker}\,\phi^*$, hence is isomorphic to $k[G]$. It follows that ϕ defines an isomorphism of algebraic groups $G \simeq \phi(G)$. This proves (i). The easy proof of (ii) is omitted. □

2.3.8. Lemma. *Let H be a closed subgroup of G. Then*

$$H = \{g \in G \mid \lambda(g)\mathcal{I}_G(H) = \mathcal{I}_G(H)\} = \{g \in G \mid \rho(g)\mathcal{I}_G(H) = \mathcal{I}_G(H)\}.$$

The notations are as in 1.3.2. It suffices to prove this for λ. If $g, h \in H, f \in \mathcal{I}_G(H)$ then $(\lambda(g)f)(h) = f(g^{-1}h) = 0$, whence $\lambda(g)f \in \mathcal{I}_G(H)$. Conversely, if this is so then $f(g^{-1}) = (\lambda(g)f)(e) = 0$ for all $f \in \mathcal{I}_G(H)$ and $g \in H$. □

2.3.9. Exercises. (1) In the situation of 2.3.5 there exists an increasing sequence of finite dimensional subspaces (V_i) of $k[X]$ such that (a) each V_i is stable under $s(G)$ and s defines a rational representation of G in V_i, and (b) $k[X] = \bigcup_i V_i$.
(2) Let X be an affine G-variety. There is an isomorphism ϕ of X onto a closed subvariety of some $\mathbf{A}^n$ and a rational representation $r : G \to \mathbf{GL}_n$ such that $\phi(g.x) = r(g)\phi(x)$ $(g \in G, x \in X)$ (*Hint*: adapt the proof of 2.3.7 (i)).

2.4. Jordan decomposition

2.4.1. We begin by recalling some results from linear algebra. Let V be a finite dimensional vector space over k. An endomorphism a of V is *semi-simple* if there is a basis of V consisting of eigenvectors of a. So with respect to this basis, a is represented by a diagonal matrix. We say that an endomorphism a is *nilpotent* if $a^s = 0$ for some integer $s \geq 1$ and that a is *unipotent* if $a - 1$ is nilpotent. Notice that if the characteristic p of k is non-zero, a is unipotent if and only if $a^{p^s} = 1$ for some integer $s \geq 1$.

We denote by $End(V)$ the algebra of endomorphisms of V. The group of its invertible elements is $GL(V)$. Choosing a basis of V we may identify $End(V)$ with an algebra $\mathbf{M}_n$ of $n \times n$-matrices with entries in k and $GL(V)$ with $\mathbf{GL}_n$.

2.4.2. Lemma. *Let $S \subset \mathbf{M}_n$ be a set of pairwise commuting matrices.*
(i) There exists $x \in \mathbf{GL}_n$ such that xSx^{-1} consists of upper triangular matrices;
(ii) If all matrices of S are semi-simple there is $x \in \mathbf{GL}_n$ such that xSx^{-1} consists of diagonal matrices.

The assertions are obvious if all elements of S are multiples of the identity. If not, there is $s \in S$ with an eigenspace that is a non-trivial subspace W of $V = k^n$. Because of our assumptions, W is S-stable. By induction on n, we may assume that an assertion like (i) holds for the endomorphisms induced by S in W and V/W. Then (i) follows. (ii) is proved similarly, writing V as a direct sum of eigenspaces of s. □

2.4.3. Lemma. *(i) The product of two commuting semi-simple (nilpotent, unipotent) endomorphisms of V is semi-simple (respectively: nilpotent, unipotent);*
(ii) If $a \in End(V)$, $b \in End(W)$ are semi-simple (nilpotent, unipotent) then the same is true for $a \oplus b \in End(V \oplus W)$, $a \otimes b \in End(V \otimes W)$;
(iii) If $a \in End(V)$, $b \in End(W)$ are semi-simple (nilpotent) then the same is true for $a \otimes 1 + 1 \otimes b \in End(V \otimes W)$.

The assertion about semi-simple endomorphisms of (i) follows from 2.4.2 (ii). The easy proofs of the other assertions are left to the reader. □

2.4.4. Proposition. *Let $a \in End(V)$.*
(i) There are unique elements $a_s, a_n \in End(V)$ such that a_s is semi-simple, a_n is nilpotent, $a_s a_n = a_n a_s$ and $a = a_s + a_n$ (additive Jordan decomposition of a);
(ii) There are polynomials $P, Q \in k[T]$ without constant term such that $a_s = P(a)$, $a_n = Q(a)$;
(iii) If $W \subset V$ is an a-stable subspace of V, then W is also stable under a_s and a_u and $a|_W = a_s|_W + a_u|_W$ is the additive Jordan decomposition of the restriction $a|_W$. A similar result holds for the endomorphism of V/W induced by a;
(iv) Let $\phi : V \to W$, $b \in End(W)$ be linear maps. If $\phi \circ a = b \circ \phi$ then $\phi \circ a_s = b_s \circ \phi$, $\phi \circ a_n = b_n \circ \phi$.

Let $\det(T.1 - a) = \Pi(T - a_i)^{n_i}$ be the characteristic polynomial of a, the a_i being the distinct eigenvalues of a. Put

$$V_i = \{x \in V \mid (a - a_i)^{n_i} x = 0\}.$$

The V_i are non-zero a-stable subspaces. By the Chinese remainder theorem there exists $P \in k[T]$ with

$$P(T) \equiv 0 \pmod{T}, \quad P(T) \equiv a_i \pmod{(T - a_i)^{n_i}} \text{ for all } i.$$

Put $a_s = P(a)$. Since $P(a_i) = a_i$ the eigenvalues of a_s are the same as those of a. Also, a_s stabilizes all V_i and the restriction of a_s to V_i is scalar multiplication by a_i. It follows that the V_i are the eigenspaces of a_s and that V is their direct sum. We conclude that a_s is semi-simple and that $a - a_s$ is nilpotent. The assertions of (i) and (ii) now readily follow, except for the uniqueness statement. To prove this, let $a = b_s + b_n$ be a second decomposition with the properties of (i). From (ii) we infer that b_s and b_n commute with a_s and a_n. From 2.4.3 (i) we conclude that $a_s - b_s = b_n - a_n$ is both semi-simple and nilpotent, hence must be zero, establishing the uniqueness.

If W is as in (iii) it is clear from (ii) that a_s and a_n stabilize W. Now observe that the characteristic polynomial of $a|_W$ divides the characteristic polynomial of a. It follows that the polynomial P introduced above can also serve for $a|_W$. The assertions of (iii) about V follow. For V/W the arguments are similar.

To prove (iv) we use that ϕ has a factorization

$$V \to V \oplus W \to W.$$

The first map is $v \mapsto (v, \phi(v))$; it is injective. The second map is projection, it is surjective. An easy argument shows that it suffices to prove (iv) if ϕ is either injective or surjective. These cases are taken care of by (iii). $\quad\square$

2.4.5. Corollary. *Let $a \in GL(V)$. There are unique elements $a_s, a_u \in GL(V)$ such that a_s is semi-simple, a_u is unipotent and $a = a_s a_u = a_u a_s$ (multiplicative Jordan decomposition of a). We have properties similar to those of 2.4.4 (iii) with a_u instead of a_n.*

Let $a = a_s + a_n$ be the additive Jordan decomposition of a. Since a is invertible it has no eigenvalue 0. From the proof of 2.4.4 we see that a_s is invertible. It follows from 2.4.4 that a_s and $a_u = 1 + a_s^{-1} a_n$ have the required properties. Conversely, if a_s and a_u are as in 2.4.5 then $a = a_s + a_s(a_u - 1)$ is the additive Jordan decomposition of a. The corollary follows from these observations. $\quad\square$

We call $a_s, a_n (a_u)$ the *semi-simple, nilpotent, (unipotent)* parts of $a \in End(V)$ (resp. $GL(V)$).

2.4.6. Corollary. *Let $a = a_s a_u$, $b = b_s b_u$ be the Jordan decompositions of $a \in GL(V)$ and $b \in GL(W)$. Then $a \oplus b = (a_s \oplus b_s) + (a_u \oplus b_u)$ is the Jordan decomposition of $a \oplus b \in GL(V \oplus W)$ and $a \otimes b = (a_s \otimes b_s)(a_u \otimes b_u)$ that of $a \otimes b \in GL(V \otimes W)$.*

This follows from 2.4.3 (ii). $\quad\square$

2.4.7. Let V be a not necessarily finite dimensional vector space over k. We denote again by $End(V)$ and $GL(V)$ the algebra of endomorphisms of V and the group of its invertible endomorphisms. We say that $a \in End(V)$ is *locally finite* if V is a union of finite dimensional a-stable subspaces. We say that a is *semi-simple* (resp. *locally nilpotent*) if its restriction to any finite dimensional a-stable subspace is semi-simple (resp. nilpotent). If a is semi-simple, it is also semi-simple according to the definition of 2.4.1 (check this). For a locally finite $a \in End(V)$ we again have an additive Jordan decomposition $a = a_s + a_n$ as in 2.4.4 (i), with locally finite a_s and a_n such that a_s is semi-simple and a_n locally nilpotent. To define $a_s x$ and $a_u x$ for $x \in V$, take a finite dimensional a-stable subspace W containing x and put

$$a_s x = (a|_W)_s x, \ a_u x = (a|_W)_u x.$$

It follows from 2.4.4 (iii) that this definition is independent of the choice of W and that a_s and a_n are as required. Similarly, we have a multiplicative Jordan decomposition $a = a_s a_u$ if $a \in GL(V)$ is locally finite. Here a_u is *locally unipotent*, i.e., $a_u - 1$ is locally nilpotent. We have obvious analogues of 2.4.4 (iii), (iv) and 2.4.6 (for locally finite b).

Now let G be a linear algebraic group and put $A = k[G]$. Let $g \in G$. By 2.3.9 (1) the right translation $\rho(g)$ is a locally finite element of $GL(A)$. So we have a Jordan decomposition $\rho(g) = \rho(g)_s \rho(g)_u$.

2.4.8. Theorem. *(i) (Jordan decomposition in G) There are unique elements $g_s, g_u \in G$ such that $\rho(g)_s = \rho(g_s)$, $\rho(g)_u = \rho(g_u)$ and $g = g_s g_u = g_u g_s$;*
(ii) If $\phi : G \to G'$ is a homomorphism of algebraic groups, then $\phi(g)_s = \phi(g_s)$, $\phi(g)_u = \phi(g_u)$;
(iii) If $G = GL_n$, then g_s and g_u are the semi-simple and unipotent parts of 2.4.5 (with $V = k^n$).

The elements g_s and g_u of (i) are called the *semi-simple part* and the *unipotent part* of $g \in G$.

As in 2.1.2 let $m : A \otimes A \to A$ be the homomorphism defined by multiplication. Since $\rho(g)$ is an algebra automorphism we have

$$m \circ (\rho(g) \otimes \rho(g)) = \rho(g) \circ m.$$

From 2.4.4 (iv) (applied to m) and 2.4.6 it follows that

$$m \circ (\rho(g)_s \otimes \rho(g)_s) = \rho(g)_s \circ m.$$

This means that $\rho(g)_s$ is also an automorphism of A, hence $f \mapsto (\rho(g)_s f)(e)$ defines a homomorphism $A \to k$, i.e., a point g_s of G. Since $\rho(g)$ commutes with all left translations $\lambda(x)$, $x \in G$ (which are locally finite) we have by 2.4.4 (iv) for $f \in A$

$$(\rho(g)_s f)(x) = (\lambda(x^{-1})\rho(g)_s f)(e) = (\rho(g)_s \lambda(x^{-1})f)(e) = (\lambda(x^{-1})f)(g_s) = f(xg_s),$$

and we see that $\rho(g)_s = \rho(g_s)$.

One obtains in the same way an element $g_u \in G$ with $\rho(g)_u = \rho(g_u)$. The remaining assertions of (i) now follow from the fact that ρ is a faithful representation of G.

A homomorphism of algebraic groups $\phi : G \to G'$ can be factored:

$$G \to \mathrm{Im}\,\phi \to G'.$$

Using 2.2.5 (ii) it follows that it suffices to prove (ii) in two special cases:
(a) G is a closed subgroup of G' and ϕ is the inclusion map. Let $k[G] = k[G']/I$. By 2.3.8

$$G = \{g \in G' \mid \rho(g)I = I\}.$$

The assertion (ii) now follows from 2.4.4 (iii).
(b) ϕ is surjective. In this case $k[G']$ can be viewed as a subspace of $k[G]$ (see 1.9.1 (ii)), which is stable under all $\rho(g)$ ($g \in G$). Again, the assertion of (ii) follows from 2.4.4 (iii).
Let $G = GL(V)$ with $V = k^n$. Let f be a nonzero element of the dual space $V^\vee$. For $v \in V$ define $\tilde{f}(v) \in k[G]$ by

$$\tilde{f}(v)(g) = f(gv).$$

Then $\tilde{f}$ is an injective linear map $V \to k[G]$ and it is immediate that for $g \in G$, $v \in V$

$$\tilde{f}(gv) = \rho(g)\tilde{f}(v).$$

From 2.4.4 (iv) we conclude that

$$\tilde{f}(g_s v) = \rho(g)_s \tilde{f}(v),$$

and similarly for g_u. This implies (iii). $\qquad\square$

2.4.9. Corollary. $x \in G$ is semi-simple (resp. unipotent) if and only if for any isomorphism ϕ of G onto a closed subgroup of some $\mathbf{GL}_n$ we have that $\phi(x)$ is semi-simple (resp. unipotent).

2.4.10. Exercises. Notations of 2.4.8.
(1) Show that $\lambda(x)_s = \lambda(x_s)$, $\lambda(x)_u = \lambda(x_u)$.
(2) The set G_u of unipotent elements of G is closed.
(3) Show by an example that the set G_s of semi-simple elements of G is not necessarily open or closed.

2.4.11. Let F be a subfield of k and G an F-group. If $x \in G(F)$ then x_s and x_u need not lie in $G(F)$. Here is an example.

Assume that char$(k) = 2$ and that $F \neq F^2$, so F is non-perfect. Let $G = \mathbf{GL}_2$. If $a \in F - F^2$ then

$$\begin{pmatrix} 0 & 1 \\ a & 0 \end{pmatrix}$$

has the semi-simple part

$$\begin{pmatrix} \sqrt{a} & 0 \\ 0 & \sqrt{a} \end{pmatrix},$$

which does not lie in $G(F)$. If, however, F is a perfect field, then the semi-simple and unipotent parts of an element of $G(F)$ also lie in $G(F)$. We postpone the proof (see 12.1.7 (c)).

The linear algebraic group G is *unipotent* if all its elements are unipotent. An example is the group $\mathbf{U}_n$ of 2.1.4 (4). The next result implies that any unipotent group is isomorphic to a closed subgroup of some $\mathbf{U}_n$.

2.4.12. Proposition. *Let G be a subgroup of $\mathbf{GL}_n$ consisting of unipotent matrices. There is $x \in \mathbf{GL}_n$ such that $xGx^{-1} \subset \mathbf{U}_n$.*

Put $V = k^n$. Then G is a group of unipotent linear maps of V. Use induction on n. If there is a non-trivial subspace W of V with $G.W = W$, then the statement follows by induction. We are left with the case that such a W does not exist, i.e., that G acts irreducibly in V. In this case we know by Burnside's theorem [La2, Ch. XVII, §3] that the elements of G span the vector space $End(V)$ of endomorphisms.

If $g \in G$ we have $Tr(g) = n$. It follows that $Tr((1 - g)h) = 0$ for $g, h \in G$. But then this equality holds for all $h \in End(V)$, which can only be if $g = 1$. Hence $G = \{1\}$ and the assertion is trivial.   □

Recall that a group H is *nilpotent* if there is an integer n such that all iterated commutators $(x_1, (...(x_{n-1}, x_n)...))$ equal e (as before, $(x, y) = xyx^{-1}y^{-1}$). Such a group is solvable (see [Jac4, p. 243, ex. 6].

2.4.13. Corollary. *A unipotent linear algebraic group is nilpotent, hence solvable.*

Using 2.3.7 (i) the proposition reduces the proof to verifying that a group $\mathbf{U}_n$ of unipotent upper triangular matrices is nilpotent. Verification is left to the reader.   □

A consequence of 2.4.12 is the fact that if G is a unipotent linear algebraic group and $G \to GL(V)$ a rational representation, there is a non-zero vector in V fixed by all of G. This fact is used to prove the following geometric result (theorem of Kostant-Rosenlicht).

2.4.14. Proposition. *Let G be a unipotent linear algebraic group and X an affine G-space. Then all orbits of G in X are closed.*

Let O be an orbit. Replacing X by the closure $\overline{O}$ we may assume that O is dense in X. By 2.3.3 (i), O is also open. Let Y be its complement. The group G acts locally finitely on the ideal $\mathcal{I}_X(Y)$ (by 2.3.6 (i)), and there is a non-zero function f in this ideal fixed by the elements of G. But then f is constant on O. Since O is dense, f is constant on all of X. Hence $\mathcal{I}_X(Y) = k[X]$ and $O = X$, as asserted. □

2.4.15. Exercise. Let G be a subgroup of $\mathbf{GL}_n$ which acts irreducibly in $V = k^n$. Show that the only normal unipotent subgroup of G is the trivial one.

2.5. Recovering a group from its representations

2.5.1. The results of this section, which will not be used in the sequel, illustrate the elementary theory of linear algebraic groups.

We keep the notations of the preceding sections. Let G be a linear algebraic group. Recall (see 2.3.2 (3)) that a rational representation of G is a finite dimensional vector space V (called a G-module) together with a homomorphism of algebraic groups $r_V : G \to GL(V)$. We denote by I the trivial G-module: $I = k$ and $r_I(g) = 1$ for all $g \in G$.

A homomorphism of G-modules $V \to W$ is a linear map ϕ which is equivariant, i.e., satisfies $\phi \circ r_V(g) = r_W(g) \circ \phi$ for all $g \in G$.

If V is a G-module the dual vector space $V^\vee$ is a G-module: if $\langle\,,\,\rangle$ (or $\langle\,,\,\rangle_V$) is the duality pairing, then for $g \in G$, $x \in V$, $u \in V^\vee$

$$\langle r_V(g)(x), r_{V^\vee}(g)(u) \rangle = \langle x, u \rangle.$$

If V and W are G-modules the tensor product $V \otimes W$ has a natural structure of G-module, with $r_{V \otimes W} = r_V \otimes r_W$.

As in 2.4.7 we define locally finite G-modules and their G-homomorphisms. An example is $A = k[G]$, the representation of G being ρ, the action by right translations.

2.5.2. Lemma. *Let V be a G-module. For $v \in V$, $u \in V^\vee$, $g \in G$ define $\phi_u(v)(g) = \langle r_V(g)v, u \rangle$. Then $\phi_u(v) \in A$ and ϕ_u is a homomorphism of G-modules $V \to A$.*

$\phi_u(v)$ can be viewed as a matrix element of a representation $G \to \mathbf{GL}_{\dim V}$, so lies in A. The equivariance of ϕ_u is obvious from the definitions. □

2.5.3. Theorem. [Tannaka's theorem] *Assume given for any finite dimensional G-module V an element $\alpha_V \in GL(V)$ such that the following holds:*
(a) If V and W are G-modules then $\alpha_{V \otimes W} = \alpha_V \otimes \alpha_W$,

(b) if $\phi : V \to W$ is a homomorphism of G-modules then $\phi \circ \alpha_V = \alpha_W \circ \phi$,
(c) $\alpha_I = 1$.
Then there is $x \in G$ with $\alpha_V = r_V(x)$ for all G-modules V.

We begin the proof by defining α_V for a locally finite G-module V. Let $v \in V$ and let W be a finite dimensional G-stable subspace of V containing v. Define $\alpha_V(v) = \alpha_W(v)$. Then α_V is well-defined. It is an invertible linear map of V. These maps have the properties (a) and (b) (check these facts).

Consider the locally finite G-modules A (for the representation ρ) and $A \otimes A$ (for $\rho \otimes \rho$). The multiplication map $m : A \otimes A \to A$ of 2.1.2 is G-equivariant. Put $\alpha = \alpha_A$. By (a) and (b) we have $m \circ (\alpha \otimes \alpha) = \alpha \circ m$. This means that α is an automorphism of the algebra A. It follows that there is an automorphism ϕ of the affine variety G such that $(\alpha f)(g) = f(\phi g)$ for $f \in A, g \in G$.

Next consider the homomorphism $\Delta : A \to A \otimes A$ of 2.1.2. Then $(\Delta f)(g, h) = f(gh)$ $(f \in A, g, h \in G)$. It follows that $\Delta \circ \rho(g) = (\mathrm{id} \otimes \rho(g)) \circ \Delta$ $(g \in G)$. Let B be the locally finite G-module $A \otimes A$, with $r_B = \mathrm{id} \otimes \rho$. Then $\Delta : A \to B$ is equivariant. By the properties (a), (b), (c) we have $\alpha_B = \mathrm{id} \otimes \alpha$. We conclude from (b) that $\Delta \circ \alpha = (\mathrm{id} \otimes \alpha) \circ \Delta$, which means that for $g, h \in G$ we have $\phi(gh) = g\phi(h)$. Put $x = \phi(e)$. Then $\phi(g) = gx$ and $\alpha = \rho(x)$.

Now let V be a finite dimensional G-module. For any u in the dual $V^\vee$ we have the G-homomorphism $\phi_u : V \to A$ of 2.5.2. By property (b) we know that

$$\phi_u \circ \alpha_V = \alpha \circ \phi_u = \rho(x) \circ \phi_u.$$

This means that for $g \in G, v \in V$ we have

$$\langle r_V(g)\alpha_V(v), u \rangle = \langle r_V(gx)v, u \rangle.$$

Since $u \in V^\vee$ is arbitrary we can conclude that $\alpha_V = r_V(x)$. $\square$

2.5.4. The theorem implies that one can recover the algebra A from the rational representations of G. We now discuss the explicit description of A in terms of representations.

Let V be a finite dimensional G-module. For $v \in V, u \in V^\vee$ define a linear map $\psi_V : V \otimes V^\vee \to A$ by $\psi_V(v \otimes u) = \phi_u(v)$, where ϕ_u is as in 2.5.2. We have the following properties.
(a) *If $\phi : V \to W$ is a homomorphism of G-modules then the maps $\psi_V \circ (\mathrm{id} \otimes \phi^\vee)$ and $\psi_W \circ (\phi \otimes \mathrm{id})$ of $V \otimes W^\vee$ to A coincide.*

Here $\phi^\vee : W^\vee \to V^\vee$ is the transpose of ϕ. This property is a reformulation of the equality $\langle v, \phi^\vee(u) \rangle = \langle \phi(v), u \rangle$, where $v \in V, u \in W^\vee$.

Next let V and W be two G-modules. There are canonical isomorphisms $(V \otimes W)^\vee \simeq V^\vee \otimes W^\vee$ and $c : (V \otimes W) \otimes (V \otimes W)^\vee \simeq (V \otimes V^\vee) \otimes (W \otimes W^\vee)$. As before, $m : A \otimes A \to A$ is the multiplication map.
(b) $\psi_{V \otimes W} = m \circ (\psi_V \otimes \psi_W) \circ c$.

Property (b) means that for $v \in V, w \in W, u \in V^\vee, t \in W^\vee$ we have

$$\langle v \otimes w, u \otimes t \rangle_{V \otimes W} = \langle v, u \rangle_V \langle w, t \rangle_W.$$

This relation describes the canonical isomorphism between $(V \otimes W)^\vee$ and $V^\vee \otimes W^\vee$.

2.5.5. Let $\mathcal{F}$ be the direct sum of the vector spaces $V \otimes V^\vee$, where V runs through all finite dimensional G-modules V. Denote by $i_V : V \otimes V^\vee \to \mathcal{F}$ the canonical injections and by $\mathcal{R}$ the subspace of $\mathcal{F}$ spanned by the elements $i_V(\mathrm{id} \otimes \phi^\vee)(z) - i_W(\phi \otimes \mathrm{id})(z)$. Here $\phi : V \to W$ is any homomorphism of G-modules and z runs through $V \otimes W^\vee$. Put $\mathcal{A} = \mathcal{F}/\mathcal{R}$. For $v \in V, u \in V^\vee$ denote by $a_V(v \otimes u)$ the image in $\mathcal{A}$ of $i_V(v \otimes u)$. If ϕ is as before, then for $v \in V, u \in W^\vee$

$$a_V(v \otimes \phi^\vee(u)) = a_W(\phi(v) \otimes u). \tag{1}$$

We define a multiplication of these elements. Let $v \in V, w \in W, u \in V^\vee, t \in W^\vee$. Then

$$a_V(v \otimes u)a_W(w \otimes t) = a_{V \otimes W}((v \otimes w) \otimes (u \otimes t)). \tag{2}$$

This product is well-defined, and defines a structure of commutative, associative algebra on $\mathcal{A}$. The identity element is provided by the trivial representation I. We leave it to the reader to check these statements. Notice that the properties of a_V reflect the properties (a) and (b) of 2.5.4 of the maps ψ_V.

2.5.6. The homomorphisms of algebraic groups $G \to \mathbf{G}_m$ form an abelian group X, the *character group* of G. If V is a one dimensional G-module, there is a unique $\chi \in X$ such that $g.v = \chi(g)v$ for $g \in G, v \in V$. Then $V \otimes V^\vee$ has a canonical basis element $e_{v \otimes v^\vee}$, namely $v \otimes u$, where $v \in V, u \in V^\vee$ and $\langle v, u \rangle = 1$. We write $a(\chi) = a_V(e_{v \otimes v^\vee})$. Then a is a homomorphism of X onto a subgroup of the set of invertible elements of $\mathcal{A}$.

If V is a finite dimensional G-module of dimension d we have the canonical homomorphism $\phi : \bigotimes^d V \to \bigwedge^d V$ with $\phi(v_1 \otimes \ldots \otimes v_d) = v_1 \wedge \ldots \wedge v_d$. Identify $(\bigwedge^d V)^\vee$ with $\bigwedge^d(V^\vee)$ via the pairing

$$\langle v_1 \wedge \ldots \wedge v_d, u_1 \wedge \ldots \wedge u_d \rangle = \det(\langle u_i, v_j \rangle).$$

Then

$$\phi^\vee(u_1 \wedge \ldots \wedge u_d) = \sum_{s \in S_d} (\mathrm{sgn}\ s)u_{s(1)} \otimes \ldots \otimes u_{s(d)},$$

where $\mathrm{sgn}\ s$ is the sign of the permutation s. Let $\det_V$ be the character defined by the one dimensional G-module $\bigwedge^d V$. It follows from (1) and (2) that for $v_i \in V, u_j \in V^\vee$ ($1 \leq, ij, \leq d$) we have

$$\det(a_V(v_i \otimes u_j)) = \det(\langle v_i, u_j \rangle)a(\det_V). \tag{3}$$

2.5.7. Theorem. *(i)* $\mathcal{A}$ *is a k-algebra of finite type;*
(ii) There is a surjective algebra homomorphism $\Phi : \mathcal{A} \to A$ *such* $\Phi \circ a_V = \psi_V$ *for all G-modules V;*
(iii) The kernel of Φ *is the ideal of nilpotent elements of* $\mathcal{A}$.

Let V be a G-module. Fix a basis $(u_i)_{1 \le i \le d}$ of $V^\vee$. It follows from 2.5.2 and the local finiteness of A that there is a finite dimensional G-submodule W of A such that, for all i, the images of the maps ϕ_{u_i} of 2.5.2 lie in W. These maps then define an injective G-homomorphism $\phi : V \to W^d$. Since $\phi^\vee$ is surjective, an element $a_V(v \otimes u)$ is a sum $\sum_{i=1}^n a_W(w_i \otimes t_i)$. It follows that $\mathcal{A}$ is generated by the images of the a_V, where V is a finite dimensional G-submodule of A. Fix such a V which generates A as an algebra and contains 1. The product in A defines for any integer $n \ge 1$ a G-equivariant surjective linear map ϕ_n of $V^{\otimes n}$ onto a finite dimensional G-submodule V_n of A. We have $V_n \subset V_{n+1}$ and $A = \bigcup_{n \ge 1} V_n$.

Now let W be any finite dimensional G-submodule of A. It is contained in a V_n, whence

$$\text{Im } a_W \subset \text{Im } a_{V_n} \subset \text{Im } a_{V^{\otimes n}}$$

(the second inclusion follows from the surjectivity of ϕ_n). The multiplication rules in $\mathcal{A}$ show that Im $a_{V^{\otimes n}}$ lies in the subalgebra generated by Im a_V. It follows that this subalgebra must be the full algebra $\mathcal{A}$. This proves (i).

Using property (b) of 2.5.4 we see that there exists an algebra homomorphism $\mathcal{A} \to A$ which maps $a_V(v, u)$ to $\psi_V(v, u)$, for all V. This implies (ii).

Fix an algebra homomorphism $\xi : \mathcal{A} \to k$. Let V be a G-module. Fix bases (v_i) and (u_j) of V resp. $V^\vee$. There is an endomorphism α_V of V such that $\langle \alpha_V v_i, u_j \rangle = \xi(a_V(v_i \otimes u_j))$. From (3) we see that $\det(a_V(v_i \otimes u_j))$ is an invertible element of $\mathcal{A}$. Hence α_V is an invertible linear map. It now follows that the α_V satisfy the assumptions of Tannaka's theorem. The conclusion of that theorem shows that there is a homomorphism $x : A \to k$ (i.e., a point x of G) such that $\xi = x \circ \Phi$. Let $\mathcal{N} = \sqrt{\{0\}}$ be the ideal of nilpotent elements of $\mathcal{A}$. There is an affine variety X with $k[X] = \mathcal{A}/\mathcal{N}$. Then Φ induces a morphism of G onto a closed subvariety of X. But we have seen that these varieties have the same points. This can only be if $\mathcal{A}/\mathcal{N}$ is isomorphic to A, which establishes (iii). $\square$

2.5.8. Remarks. (1) We can also recover the comultiplication $\Delta : A \to A \otimes A$ and the antipode $\iota : A \to A$ of 2.1.2 from similar homomorphisms for $\mathcal{A}$.

Let V be a G-module. Let (v_i) be a basis of V and denote by (u_i) the dual basis of $V^\vee$ (so $\langle v_i, u_j \rangle = \delta_{ij}$). Denote by $\tilde{\Delta}$ the linear map $\mathcal{A} \to \mathcal{A} \otimes \mathcal{A}$ sending an element $a_V(v \otimes u)$ to $\sum_i a_V(v \otimes u_i) \otimes a_V(v_i \otimes u)$. It can be checked that $\tilde{\Delta}$ is an algebra homomorphism and that $(\Phi \otimes \Phi) \circ \tilde{\Delta} = \Delta \circ \Phi$.

Define a homomorphism $\tilde{\iota} : \mathcal{A} \to \mathcal{A}$ by $\tilde{\iota}(a_V(v \otimes u)) = a_{V^\vee}(u \otimes v)$. Then $\tilde{\iota}$ induces the antipode ι of A.

(2) In fact, $\tilde{\Delta}$ and $\tilde{\iota}$ have the properties of 2.1.2. This means that $\mathcal{A}$ and these homomorphisms define a group scheme over k (2.1.6 (a)). If char $k = 0$ the algebra $\mathcal{A}$ is reduced by a theorem of Cartier [DG, p. 255]. In that case the homomorphism Φ of 2.5.7 is an isomorphism.

2.5.9. Exercises. (1) Let F be a subfield of k. If G is an F-group then the algebra $\mathcal{A}$ of 2.5.6 can be given an F-structure.
(2) Check the statements of 2.5.8.

Notes

The name 'algebraic group' (or rather 'groupe algébrique') seems first to occur in the work of E. Picard on the Galois theory of linear differential equations (around 1880). The Galois groups occurring in that work are, indeed, linear algebraic groups over **C**. Kolchin's work on algebraic groups of 1948 (in [Kol1, Kol2]) was also motivated by the Galois theory of linear differential equations. His results were taken up by Borel in his fundamental paper [Bo1]. Here the emphasis is on the analogy between Lie groups and linear algebraic groups.

The results of Sections 2 and 3 are contained in [Bo1, Ch. 1]. Several of the results (for example 2.2.1) go back to Kolchin (see [Kol1, Kol2]). The useful result 2.2.6 is due to Chevalley (see [Ch1, Ch. II, §7]).

The existence of closed orbits in a G-space (see (2.2.3 (ii)) was proved in [Bo1, 15.4]. This simple result is a cornerstone of the theory of linear algebraic groups. For example, it is needed in the proof of the crucial fixed point theorem 6.2.6. It should be noted that the algebraicity of the action is essential: a complex Lie group acting on a complex variety need not have closed orbits.

Theorem 2.4.8 about the Jordan decomposition in a linear algebraic group was proved in [Bo1, Ch. 2]. It is also a consequence of 3.1.1, which was first proved in [Kol2, §3]. The proof given here shows that the theorem is, essentially, a formal consequence of the functorial property 2.4.4 (iii) of the Jordan decomposition of a linear map.

Tannaka's theorem 2.5.3 goes back to [Tan], where a similar result is proved for compact Lie groups. The aim of [Tan] was to establish an analogue for such groups of Pontryagin duality in abelian topological groups. The results of 2.5 should nowadays be viewed in the context of the theory of tensor categories, which we did not go into. See [DelM, Del].

Chapter 3

Commutative Algebraic Groups

This chapter deals with results about commutative linear algebraic groups which are basic for the theory of the later chapters. The important tori are introduced in 3.2, and we prove the classification theorem 3.4.9 of connected one dimensional groups. The notations are as in the previous chapter.

3.1. Structure of commutative algebraic groups

3.1.1. Theorem. *Let G be a commutative linear algebraic group.*
(i) The sets G_s and G_u of semi-simple and unipotent elements are closed subgroups;
(ii) The product map $\pi : G_s \times G_u \to G$ is an isomorphism of algebraic groups.

We may assume that G is a closed subgroup of $\mathbf{GL}_n$, for some n. From 2.4.3 (i) we see that G_s and G_u are subgroups and by 2.4.10 (2) G_u is closed. It follows from 2.4.2 (ii) that there exists a direct sum decomposition of $V = k^n$, say $V = \bigoplus V_i$, together with homomorphisms $\phi_i : G_s \to k^*$ such that $g.v = \phi_i(g)v$ if $g \in G$, $v \in V_i$. The V_i are G-stable. Applying 2.4.2 (i) to the groups induced by G in the V_i we arrange things such that $G \subset \mathbf{T}_n$, $G_s = G \cap \mathbf{D}_n$ (notations of 2.1.4). This shows that G_s is closed, whence (i).

The uniqueness of the Jordan decomposition in G implies that π is an isomorphism of abstract groups. It is also a morphism of algebraic varieties. The map sending $x \in G$ to its semi-simple part x_s is also a morphism, since it maps x to a set of its matrix elements. Hence $\pi^{-1} : x \mapsto (x_s, x_s^{-1}x)$ is a morphism, which proves (ii). $\qquad\square$

3.1.2. Corollary. *If moreover G is connected then the same holds for G_s and G_u.*

G_s and G_u are images of the connected space G under continuous maps. $\qquad\square$

3.1.3. Proposition. *Let G be a connected linear algebraic group of dimension one.*
(i) G is commutative;
(ii) Either $G = G_s$ or $G = G_u$;
(iii) If G is unipotent and $p = \operatorname{char} k > 0$ then the elements of G have order dividing p.

Fix $g \in G$ and consider the morphism $\phi : x \mapsto xgx^{-1}$ of G to itself. By 1.2.3 (ii) the closure $\overline{\phi G}$ is an irreducible closed subset of G. Using 1.8.2 we conclude that this set is either $\{g\}$ or G. Assume we have the latter case. By 1.9.5 the complement $G - \phi G$ is finite. View G as a closed subgroup of some $\mathbf{GL}_n$. Then there are only finitely many possibilities for the characteristic polynomial $\det(T.1 - x)$ of $x \in G$. The connectedness of G implies that this characteristic polynomial is constant, and equal to $(T - 1)^n$. This means that G is unipotent, hence solvable (see 2.4.13). But the commutator subgroup G' is a connected, closed, subgroup (see 2.2.8 (i)), and can only be $\{e\}$. Since $g^{-1}\phi G \subset G'$ we get a contradiction. It follows that G is commutative.

(ii) follows from 3.1.1 and 1.8.2. Assume that G is unipotent and $p > 0$. Consider the subgroups $G^{(p^h)}$ generated by the p^h-th powers of elements of G. They are closed and connected (2.2.9 (3)), so must be either G or $\{e\}$. Viewing G as a subgroup of $\mathbf{GL}_n$ we see that $G^{(p^h)} = \{e\}$ if $p^h \geq n$. This can only be if $G^{(p)} = \{e\}$. $\qquad\square$

In the rest of this chapter we shall first discuss the abelian algebraic groups whose elements are semi-simple and then the abelian unipotent groups with the property of 3.1.3 (iii).

3.2. Diagonalizable groups and tori

3.2.1. Let G be a linear algebraic group. A homomorphism of algebraic groups $\chi : G \to \mathbf{G}_m$ is called a *rational character* (or simply a character). The set of rational characters is denoted by $X^*(G)$. It has a natural structure of abelian group, which we write additively. The characters are regular functions on G, so lie in $k[G]$. By Dedekind's theorem [La2, Ch. VIII, §4] the characters are linearly independent elements of $k[G]$.

A homomorphism of algebraic groups $\lambda : \mathbf{G}_m \to G$ is called a *cocharacter* (or multiplicative one-parameter subgroup) of G. We denote by $X_*(G)$ the set of cocharacters. If G is commutative $X_*(G)$ also has a structure of abelian group (written additively). If G is arbitrary we still have in $X_*(G)$ multiplication by integers, defined by $(n.\lambda)(a) = \lambda(a)^n$ (for $\lambda \in X_*(G)$, $n \in \mathbf{Z}$, $a \in k^*$). We write $-\lambda = (-1).\lambda$.

A linear algebraic group G is *diagonalizable* if it is isomorphic to a closed subgroup of some group $\mathbf{D}_n$ of diagonal matrices. G is an *algebraic torus* (or simply a torus) if it is isomorphic to some $\mathbf{D}_n$.

3.2.2. Example. $G = \mathbf{D}_n$. Write an element $x \in \mathbf{D}_n$ as $\mathrm{diag}(\chi_1(x), \ldots, \chi_n(x))$. Then χ_i is a character of $\mathbf{D}_n$ and we have $k[\mathbf{D}_n] = k[\chi_1, \ldots, \chi_n, \chi_1^{-1}, \ldots, \chi_n^{-1}]$. From Dedekind's theorem we see that the monomials $\chi_1^{a_1} \ldots \chi_n^{a_n}$ with $(a_1, \ldots, a_n) \in \mathbf{Z}^n$ form a basis of $k[\mathbf{D}_n]$ and that any character of $\mathbf{D}_n$ is such a monomial. It follows that $X^*(\mathbf{D}_n) \simeq \mathbf{Z}^n$. A homomorphism $\mathbf{G}_m \to \mathbf{D}_n$ is of the form $x \mapsto \mathrm{diag}(x^{a_1}, \ldots, x^{a_n})$ ($x \in k^*$), where the a_i are integers. It follows that $X_*(\mathbf{D}_n) \simeq \mathbf{Z}^n$.

3.2.3. Theorem. *The following properties of a linear algebraic group G are equivalent:*

(a) G is diagonalizable;

(b) $X^*(G)$ is an abelian group of finite type. Its elements form a k-basis of $k[G]$;
(c) Any rational representation of G is a direct sum of one dimensional such representations.

If G is a closed subgroup of $\mathbf{D}_n$ then $k[G]$ is a quotient of $k[\mathbf{D}_n]$. Since the restriction of a character of $\mathbf{D}_n$ to G is a character of G, we see that the restrictions to G of the characters of $\mathbf{D}_n$ span $k[G]$. By Dedekind's theorem they form a basis and any character of G must be a linear combination of these restrictions. So the restriction homomorphism of abelian groups $X^*(\mathbf{D}_n) \to X^*(G)$ is surjective. As $X^*(\mathbf{D}_n) \simeq \mathbf{Z}^n$, the group $X^*(G)$ is of finite type. We have proved that (a) implies (b).

Assume (b). Put $X = X^*(G)$. Denote by $\phi : G \to GL(V)$ a rational representation in a finite dimensional vector space V. We define linear maps A_χ of V ($\chi \in X$) by

$$\phi(x) = \sum_{\chi \in X} \chi(x) A_\chi.$$

Then $A_\chi = 0$ for all but finitely many χ. From $\phi(xy) = \phi(x)\phi(y)$ $(x, y \in G)$ we infer (using Dedekind's theorem for $G \times G$) that, for $\chi, \psi \in X$, we have $A_\chi A_\psi = \delta_{\chi,\psi} A_\chi$. We also have $\sum_\chi A_\chi = 1$. Put $V_\chi = \operatorname{Im} A_\chi$. The properties of the A_χ express that V is the direct sum of the non-zero V_χ and that $x \in G$ acts in V_χ as scalar multiplication by $\chi(x)$. This implies (c).
That (c) implies (a) is immediate if one views G as a closed subgroup of some $GL(V)$ (see 2.3.7 (i)). □

3.2.4. Corollary. *If G is diagonalizable then $X^*(G)$ is an abelian group of finite type, without p-torsion if $p = \operatorname{char} k > 0$. The algebra $k[G]$ is isomorphic to the group algebra of $X^*(G)$.*

The first point follows from (b), using that k does not contain p-th roots of unity $\neq 1$ if $p > 0$. The second point is implicit in the first part of the proof of 3.2.3. We shall make it more explicit. □

3.2.5. Let M be an abelian group of finite type. The group algebra $k[M]$ is the algebra with basis $(e(m))_{m \in M}$, the multiplication being defined by $e(m)e(n) = e(m + n)$. If M_1 and M_2 are two such groups we have

$$k[M_1 \oplus M_2] \simeq k[M_1] \otimes_k k[M_2]. \tag{4}$$

Define homomorphisms $\Delta : k[M] \to k[M] \otimes k[M], \iota : k[M] \to k[M], e : k[M] \to k$ by $\Delta e(m) = e(m) \otimes e(m), \iota e(m) = e(-m), e(e(m)) = 1$. Assume that M has no p-torsion if $p = \operatorname{char} k > 0$.

3.2.6. Proposition. *(i) $k[M]$ is an affine algebra, and there is a diagonalizable linear algebraic group $\mathcal{G}(M)$ with $k[\mathcal{G}(M)] = k[M]$, such that Δ, ι and e are comultiplication, antipode and identity element of $\mathcal{G}(M)$ (see 2.1.2);*

(ii) There is a canonical isomorphism $M \simeq X^*(\mathcal{G}(M))$;
(iii) If G is a diagonalizable group there is a canonical isomorphism $\mathcal{G}(X^*(G)) \simeq G$.

It is well-known that M is a direct sum of cyclic groups. By (4) and 1.5.2 it suffices to prove the first point of (i) in the cyclic case. If M is infinite cyclic then $k[M] \simeq k[T, T^{-1}]$, an integral domain. If M is finite of order d then p does not divide d (if $p > 0$) and $k[M] \simeq k[T]/(T^d - 1)$. Since the polynomial $T^d - 1$ does not have multiple roots, this is a reduced algebra. To complete the proof of (i) we have to verify the properties of 2.1.2. We leave the verification to the reader.

The map $x \mapsto e(m)(x)$ ($m \in M, x \in \mathcal{G}(M)$) defines a character of $\mathcal{G}(M)$, i.e., $e(m)$ is a character of $\mathcal{G}(M)$. By Dedekind's theorem these characters are distinct and, since they form a basis, any character of $\mathcal{G}(M)$ is an $e(m)$. It follows that the map $m \mapsto e(m)$ is an isomorphism, whence (ii). A similar map induces an algebra isomorphism $k[G] \to k[X^*(G)]$, whence (iii). $\square$

3.2.7. Corollary. *Let G be a diagonalizable group.*
(i) G is a direct product of a torus and a finite abelian group of order prime to p, where p is the characteristic exponent of k;
(ii) G is a torus if and only if it is connected;
(iii) G is a torus if and only if $X^*(G)$ is a free abelian group.

First observe that $\mathcal{G}(\mathbf{Z}^n) \simeq \mathbf{D}_n$, as follows from 3.2.2 and 3.2.6 (iii). Now $X^*(G)$ is isomorphic to a direct sum $\mathbf{Z}^n \oplus M$, where M is finite. By (4), G is isomorphic to the product of $\mathbf{D}_n$ and $\mathcal{G}(M)$. The latter group is finite, as follows for example from the argument of the proof of 3.2.6 (i). We have proved (i). (ii) is a consequence of (i) and (iii) also follows. $\square$

3.2.8. Proposition. [rigidity of diagonalizable groups] *Let G and H be diagonalizable groups and let V be a connected affine variety. Assume given a morphism of varieties $\phi : V \times G \to H$ such that for any $v \in V$ the map $x \mapsto \phi(v, x)$ defines a homomorphism of algebraic groups $G \to H$. Then $\phi(v, x)$ is independent of v.*

If $\psi \in X^*(H)$ then $\psi(\phi(v, x))$ can be written in the form

$$\psi(\phi(v, x)) = \sum_{\chi \in X^*(G)} f_{\chi,\psi}(v)\chi(x),$$

with $f_{\chi,\psi} \in k[V]$. For fixed v the right-hand side is a character of G. By Dedekind's theorem $f_{\chi,\psi}(v) = 1$ for one χ and 0 for the others, whence $f_{\chi,\psi}^2 = f_{\chi,\psi}$. The connectedness of V implies that $f_{\chi,\psi} = 1$ for one χ and 0 for the others (use 1.2.8 (2)). $\square$

We give an application of the proposition. If G is an arbitrary linear algebraic group and H a closed subgroup, we denote by $Z_G(H)$ and $N_G(H)$ the centralizer and normalizer of H in G, i.e.

$$Z_G(H) = \{x \in G \mid xyx^{-1} = y \text{ for all } y \in H\},$$

$$N_G(H) = \{x \in G \mid xHx^{-1} = H\}.$$

These are closed subgroups of G (check this) and $Z_G(H)$ is a normal subgroup of $N_G(H)$.

3.2.9. Corollary. *If H is a diagonalizable subgroup of G then $N_G(H)^0 = Z_G(H)^0$ and $N_G(H)/Z_G(H)$ is finite.*

The first point follows from the proposition, with $V = N_G(H)^0$, ϕ being the morphism $(x, y) \mapsto xyx^{-1}$ of $V \times H$ to H. The last point follows from 2.2.1 (i).   □

3.2.10. Exercises. p is the characteristic exponent of k. G is a diagonalizable group with character group X.

(1) Make diagonalizable groups and abelian groups without p-torsion into categories and describe an anti-equivalence between these categories.

(2) Let $\phi : G \to H$ be a homomorphism of diagonalizable groups. Denote by ϕ^* the induced homomorphism $X^*(H) \to X^*(G)$. If ϕ is injective (surjective) then ϕ^* is surjective (respectively injective).

(3) Describe a canonical isomorphism of abelian groups $G \simeq \mathrm{Hom}(X, k^*)$.

(4) For a closed subgroup H of G and a subgroup Y of X define

$$H^\perp = \{\chi \in X \mid \chi(H) = \{1\}\},$$

$$Y^\perp = \{x \in G \mid \chi(x) = 1 \text{ for all } \chi \in Y\}.$$

Then $(H^\perp)^\perp = H$, and $(Y^\perp)^\perp = Y$ if X/Y has no p-torsion.

(5) For a positive integer n prime to p, denote by G_n the subgroup of elements of G of order dividing n.

   (a) $(G_n)^\perp = nX$.

   (b) The subgroup of elements of finite order is dense in G.

(6) The group of automorphisms of an n-dimensional torus is isomorphic to the group $\mathbf{GL}_n(\mathbf{Z})$ of integral $n \times n$-matrices with an integral inverse.

We conclude this section with some material on tori. Let T be a torus. Put $X = X^*(T), Y = X_*(T)$ (3.2.1). For $\chi \in X, \lambda \in Y, a \in k^*$ the map $a \mapsto \chi(\lambda(a))$ defines a character of the multiplicative group. By 3.2.2 (with $n = 1$) there is an integer $\langle \chi, \lambda \rangle$ such that $\chi(\lambda(a)) = a^{\langle \chi, \lambda \rangle}$.

3.2.11. Lemma. *(i) $\langle \, , \, \rangle$ defines a perfect pairing between X and Y, i.e. any homomorphism $X \to \mathbf{Z}$ is of the form $\chi \mapsto \langle \chi, \lambda \rangle$ for some $\lambda \in Y$, and similarly for Y.*

In particular, Y is a free abelian group;
(ii) The map $a \otimes \lambda \mapsto \lambda(a)$ defines a canonical isomorphism of abelian groups $k^* \otimes Y \simeq T$.

It suffices to prove this in the case $T = \mathbf{D}_n$. Then (i) follows from the results of 3.2.2 (check this). The proof of (ii) uses the freeness of Y. $\qquad\square$

Let F be a subfield of k. An *F-torus* is an F-group which is a torus. An F-torus T which is F-isomorphic to some $\mathbf{D}_n$ is *F-split*. The study of non-split F-tori, which requires Galois theory, is deferred to Chapter 13.

3.2.12. Proposition. *(i) An F-torus T is F-split if and only if all its characters are defined over F. If this is so the characters form a basis of the algebra $F[T]$;*
(ii) Any rational representation over F of an F-split torus is a direct sum of one dimensional representations over F.

The proof of (i) is straightforward. (ii) is proved as 3.2.3 (c). $\qquad\square$

3.2.13. Let T, X and Y be as before. Let V be an affine T-space. We have a locally finite representation s of T in $k[V]$, as in 2.3.5. For $\chi \in X$ put

$$k[V]_\chi = \{f \in k[V] \mid s(t).f = \chi(t)f \text{ for all } t \in T\}.$$

It follows from 3.2.3 (c) that the subspaces $k[V]_\chi$ define an X-grading of the algebra $k[V]$, i.e.,

$$k[V] = \bigoplus_{\chi \in X} k[V]_\chi, \quad k[V]_\chi k[V]_\psi \subset k[V]_{\chi+\psi} \ (\chi, \psi \in X).$$

If $T = \mathbf{G}_m$ then $X = \mathbf{Z}$ and we have a structure of graded algebra on $k[V]$ in the usual sense.
If ϕ is a morphism of varieties $\mathbf{G}_m \to Z$ we shall write $\lim_{a \to 0} \phi(a) = z$ if ϕ extends to a morphism $\bar\phi : \mathbf{A}^1 \to Z$ with $\bar\phi(0) = z$. If $\phi'(a) = \phi(a^{-1})$ we define $\lim_{a \to \infty} \phi(a) = \lim_{a \to 0} \phi'(a)$.
If V is a T-space and $\lambda \in Y$, we denote by $V(\lambda)$ the set of $v \in V$ such that $\lim_{a \to 0} \lambda(a).v$ exists. Then $V(-\lambda)$ is the set of v such that $\lim_{a \to \infty} \lambda(a).v$ exists.

3.2.14. Lemma. *Assume that V is affine.*
(i) $V(\lambda)$ is a closed subset of V;
(ii) $V(\lambda) \cap V(-\lambda)$ is the set of fixed points of $\mathrm{Im}\,\lambda$ i.e. $\{v \in V \mid \lambda(k^*).v = \{v\}\}$.

Let $f = \sum_\chi f_\chi$ with $f_\chi \in k[V]_\chi$ be an element of $k[V]$. Then

$$s(\lambda(a)).f = \sum a^{\langle \chi, \lambda \rangle} f_\chi,$$

from which we see that $\lim_{a\to 0}\lambda(a).v$ exists if and only if v annihilates all functions of the V_χ with $\langle\chi,\lambda\rangle < 0$. This proves (i). Then $V(\lambda)\cap V(-\lambda)$ is the set v annihilating all V_χ with $\langle\chi,\lambda\rangle \neq 0$, which is also the set of v with $f(\lambda(a).v) = f(v)$ for all $f \in k[G]$, $a \in k^*$, i.e. the set of fixed points. $\square$

3.2.15. Example. Let G be an arbitrary linear algebraic group and $\lambda : \mathbf{G}_m \to G$ a cocharacter. We let $\mathbf{G}_m$ act on G by $a.x = \lambda(a)x\lambda(a)^{-1}$ ($a \in k^*, x \in G$). We denote by $P(\lambda)$ the set of $x \in G$ such that $\lim_{a\to 0} a.x$ exists. It is immediate that this is a subgroup, which by 3.2.14 (i) is closed. By 3.2.14 (ii) the intersection $P(\lambda)\cap P(-\lambda)$ is the centralizer of Im λ.

3.2.16. Exercises. (1) The category of $\mathbf{G}_m$-modules is equivalent to the category of finite dimensional graded vector spaces.
(2) Let $A = \oplus_{n\in\mathbf{Z}}A_n$ be a graded affine k-algebra without zero divisors. Assume $A \neq A_0$. Let $d\mathbf{Z}$ be the subgroup of $\mathbf{Z}$ generated by the n with $A_n \neq \{0\}$. Choose non-zero elements f, g and integers i, j with $f \in A_i, g \in A_j, i - j = d$ and let $B = A_{fg}$ (notation of 1.4.6). The grading of A induces one of B. Define an isomorphism $B_0 \otimes k[fg^{-1}, gf^{-1}] \simeq B$ and show that B_0 is an affine algebra.
(3) Deduce from the previous exercise the following properties of a $\mathbf{G}_m$-action on an affine variety V: There is a decomposition $V = \coprod_{i=0}^{N} V_i$ into disjoint irreducible locally closed pieces with the following properties:
(a) V_0 is the set of fixed points,
(b) For $i > 0$ there is an affine variety V_i', an isomorphism $\phi_i : V_i' \times k^* \to V_i$ and an integer d_i such that for $x \in V_i', t, u \in k^*$ we have $\phi_i(x, t^{d_i}u) = t.\phi_i(x, u)$,
(c) The closure of a V_i is a union of some V_j.
(4) In the situation of 3.2.15 let $G = GL(V)$. In that case there is the following description of a group $P(\lambda)$: there is a flag in V, i.e. a sequence of distinct subspaces $V = V_0 \supset V_1 \supset V_2 \supset ...$ of V such that $P(\lambda)$ is the group of all invertible maps of V stabilizing all V_i. (*Hint:* consider the case that $G = \mathbf{GL}_n$, $\lambda(a) = \mathrm{diag}(a^{h_1}, ... , a^{h_n})$ with $h_1 \geq h_2 \geq ... \geq h_n$).
(5) An *affine embedding* of T is an irreducible affine T-space V containing T as an open subvariety such that the action $T \times V \to V$ extends the product map $T \times T \to T$. Then V is an *affine toric variety*.
(a) In that case there is a finitely generated sub-semigroup S of the group X which generates X, such that $k[V]$ is isomorphic to the semigroup algebra $k[S]$. (*Hint:* view $k[V]$ as a T-stable subspace of $k[T]$.)
(b) Conversely, for every S with the properties of (a) there exists an equivariant affine embedding V with $k[V] \simeq k[S]$. It is unique up to isomorphism of T-spaces. (For more about toric varieties see [Od].)

3.3. Additive functions

3.3.1. Additive functions. An *additive function* on the linear algebraic group G is a homomorphism of algebraic groups $f : G \to \mathbf{G}_a$. The additive functions form a subspace $\mathcal{A} = \mathcal{A}(G)$ of the algebra $k[G]$. If F is a subfield of k and G is an F-group, we write $\mathcal{A}(F) = \mathcal{A}(G)(F)$ for the F-vector space of additive functions that are defined over F. Notice that, if $p = \operatorname{char} k$ is non-zero, then the p-th power of an additive function is again one. This fact is the reason for the introduction of a ring over which $\mathcal{A}$ will be a module.

First assume that $p > 0$. Then $\phi x = x^p$ defines an isomorphism of F onto a subfield F^p. Recall that F is *perfect* if $F^p = F$. We define a ring $R = R(F)$ as follows. The underlying additive group is the space of polynomials $F[T]$ and the multiplication is defined by

$$(\sum a_i T^i)(\sum b_j T^j) = \sum a_i (\phi^i b_j) T^{i+j}.$$

Then R is an associative, non-commutative ring (this is the case for any isomorphism ϕ of F onto a subfield). Notice that the subfield F of R does not lie in the center. The degree function deg on R is as in the case of the polynomial ring $F[T]$ and has the usual properties. They imply that R has no zero divisors. If $p = 0$ then F is perfect (by convention). We now define $R = R(F) = F$.

3.3.2. Lemma. *Assume that $p > 0$. Let $a, b \in R$ and assume that $\deg a > 0$.*
(i) There exist unique elements $c, d \in R$ such that $\deg d < \deg a$ and $b = ca + d$;
(ii) If F is perfect there also exist unique elements c, d with $\deg d < \deg a$ and $b = ac + d$.

The proof is like that of the well-known division algorithm in the polynomial ring $F[T]$. The proof of (ii) requires that one can extract p-th roots in F, whence the perfectness assumption. □

3.3.3. Lemma. *(i) Left ideals in R are principal. If F is perfect the same holds for right ideals;*
(ii) R is left noetherian. If F is perfect R is also right noetherian;
(iii) If F is perfect any finitely generated left R-module M is a direct sum of cyclic modules. If, moreover, M has no torsion then it is free.

The assertions are trivial if $p = 0$, so assume $p > 0$. Then (i) follows from 3.3.2, as in the case of $F[T]$ and (ii) is a consequence of (i). In the case of (iii) let $(m_i)_{1 \le i \le s}$ be a set of generators of M. We have a surjective homomorphism $R^s \to M$ sending the canonical basis element e_i to m_i. Let K be the kernel. It follows from (ii) that K is finitely generated, say by elements $\sum_{1 \le i \le t} a_{ji} e_j$. By multiplying the matrix $A = (a_{ij})$ on the left and right by suitable invertible square matrices, one reduces to the case that $a_{ij} = 0$ for $i \ne j$, in which case (iii) is obvious. The argument is the same as that in the case of $F[T]$, which can be found in [Jac4, p.177-178]. Because in the case of R

both the left and right division algorithms are needed, we must require in (iii) F to be perfect.   □

3.3.4. As in 3.3.1 let $\mathcal{A}(F)$ be the set of additive functions of the F-group G which are defined over F. If $p > 0$ we can define a structure of left R-module on $\mathcal{A}(F)$ by

$$\left(\sum a_i T^i\right).f = \sum a_i f^{p^i}.$$

If $p = 0$ then $R = F$ and it is trivial that $\mathcal{A}(F)$ is an R-module.

As an example, take $G = \mathbf{G}_a^n$. Then $F[G] = F[T_1, \dots, T_n]$. An additive function in $F[G]$ is now an *additive polynomial*, i.e. an element $f \in F[T_1, \dots T_n]$ satisfying

$$f(T_1 + U_1, \dots, T_n + U_n) = f(T_1, \dots, T_n) + f(U_1, \dots, U_n), \qquad (5)$$

where the T_i and U_j are indeterminates. The set of additive polynomials is a left R-module $\mathcal{A}(\mathbf{G}_a^n)(F)$.

3.3.5. Lemma. $\mathcal{A}(\mathbf{G}_a^n)(F)$ *is a free R-module with basis* $(T_i)_{1 \leq i \leq n}$.

The assertion means that an additive polynomial only involves monomials of the form $T_i^{p^j}$, where p is the characteristic exponent. Let D_i be partial derivation in $F[T_1, \dots, T_n]$ with respect to T_i. If f is an additive polynomial, it follows from (5) that $D_i f$ is a constant c_i for all i. Then $f - \sum c_i T_i$ is another additive polynomial g and all derivatives $D_i g$ are zero. If $p > 0$ this means that g involves only the p-th powers of the variables, i.e there is a polynomial h with $g = h(T_1^p, \dots, T_n^p)$. But then h is an additive polynomial of lower degree and we may assume by an induction that it is expressible in the $T_i^{p^j}$. Hence f is as asserted. The case $p = 0$ is left to the reader.   □

Now let G be an arbitrary F-group.

3.3.6. Lemma. (i) *If G is connected the R-module $\mathcal{A}(G)(F)$ is torsion free;*
(ii) *If $f_1, \dots, f_s$ are elements of $\mathcal{A}(G)(F)$ that are algebraically dependent over k, then they are linearly dependent over R.*

If $\mathcal{A}(G)(F)$ had torsion there would be a non-constant $f \in k[G]$ satisfying a relation

$$f^{p^h} + a_1 f^{p^{h-1}} + \dots + a_h f = 0,$$

with coefficients in k. Such an f would take only finitely many values, which is impossible if G is connected. This proves (i).

In the situation of (ii) there is a non-zero polynomial $H \in k[T_1, \dots, T_s]$ with $H(f_1, \dots, f_s) = 0$. Assume that H is such a polynomial with minimal degree. If

$x \in G$ then the polynomial $H(T_1 + f_1(x), \dots, T_s + f_s(x)) - H(T_1, \dots, T_s)$ also gives an algebraic dependence between the f_i, but has degree strictly smaller than that of H. Hence it is zero. Let $\tilde{H}(T_1, \dots T_n)$ be the coefficient of some monomial in the indeterminates U_i in

$$H(T_1 + U_1, \dots, T_s + U_s) - H(T_1, \dots, T_s) - H(U_1, \dots, U_s).$$

Then $\tilde{H}$ has degree smaller than H and $\tilde{H}(f_1, \dots f_s) = 0$. It follows that $\tilde{H} = 0$, which means that H is an additive polynomial. Write $H = \sum c_i H_i$, where the H_i are additive polynomials with coefficients in F and the c_i lie in k and are linearly independent over F. Then $H_i(f_1, \dots f_s) = 0$ for all i, and (ii) follows. $\qquad\square$

3.4. Elementary unipotent groups

3.4.1. We say that the unipotent linear algebraic group G is *elementary* if it is abelian and, moreover, if $p > 0$ its elements have order dividing p. G is a *vector group* if it is isomorphic to some $\mathbf{G}_a^n$. We first establish some auxiliary results, which will be needed in the discussion of the structural properties of elementary unipotent groups.

Assume that $p > 0$. If $n \in \mathbf{N}$ is a natural number we denote by

$$n = \sum_i n_i p^i$$

its p-adic expansion, where the integers n_i are uniquely determined by the requirement $0 \le n_i < p$. They are 0 for almost all $i \in \mathbf{N}$. If $m, n \in \mathbf{N}$ we write $n \le_p m$ if $n_i \le m_i$ for all $i \in \mathbf{N}$. If $m, n \in \mathbf{N}$ we write (m, n) for the binomial coefficient $m!(n!(m - n)!)^{-1}$, with the convention that it is zero if $n > m$.

3.4.2. Lemma. *(i)* $(m, n) \equiv \prod_i (m_i, n_i) \pmod{p}$;
(ii) $(m, n) \not\equiv 0 \pmod{p}$ *if and only if* $n \le_p m$.

Over a field of characteristic $p > 0$ we have $(T + 1)^m = \prod (T^{p^i} + 1)^{m_i}$. Then (i) follows by expanding the powers of $T + 1$ and equating coefficients of the powers of T. Now (ii) follows from (i). $\qquad\square$

3.4.3. We next establish a result about polynomial 2-cocycles. If $p > 0$ we define

$$c(T, U) = \sum_{i=1}^{p-1} p^{-1}(p, i) \, T^{p-i} U^i.$$

Notice that $p^{-1}(p, i)$ is an integer for $i \ne 0, p$.

Let F be a perfect field and assume that $f \in F[T, U]$ satisfies

$$f(T + U, V) + f(T, U) = f(U + V, T) + f(U, V). \qquad (6)$$

In (6) T, U, V are indeterminates.

3.4.4. Lemma. *(i) If $p = 0$ there is $g \in F[T]$ with*

$$f(T, U) = g(T + U) - g(T) - g(U);$$

(ii) If $p > 0$ there is $g \in F[T]$ such that $f(T, U) - g(T + U) + g(T) + g(U)$ is a linear combination l of polynomials c^{p^i};
(iii) If for $p > 0$ we have, moreover, $\sum_{1 \le i \le p-1} f(T, iT) = 0$, then the polynomial l of (ii) is 0.

If f satisfies (6) the same is true for its homogeneous components. It follows that we may assume f to be homogeneous of degree d. We use induction on d. If $d = 0$ the assertion is trivial. So assume $d > 0$. Putting $T = U = 0$ in (6) we find that $f(V, 0) = 0$. Putting $U = V = 0$ we obtain $f(0, T) = 0$. Write

$$f(T, U) = \sum_{h=0}^{d} c_h T^h U^{d-h}.$$

We have $c_0 = c_d = 0$. Equating coefficients of $T^h U^i V^j$ in both sides of the equality (6) we obtain

$$(h + i, h)c_{h+i} + \delta_{j,0}c_h = (i + j, j)c_{i+j} + \delta_{h,0}c_j, \tag{7}$$

where $h, i, j \in \mathbf{N}$, $h + i + j = d$. For $h = 0$ or $j = 0$ we find from (7) that

$$c_h = c_{d-h}. \tag{8}$$

Now assume $0 < h, j < d$. Then (7) and (8) imply

$$(d - h, j)c_h = (d - j, h)c_j. \tag{9}$$

If $p = 0$ we can rewrite this as

$$(d, j)c_h = (d, h)c_j,$$

and (i) readily follows.

From now on suppose that $p > 0$. From (9) with $j = 1$ and (8) we obtain

$$hc_h = (d - 1, d - h)c_1. \tag{10}$$

First assume that d is prime to p. Put

$$f_1(T, U) = f(T, U) - d^{-1}c_1((T + U)^d - T^d - U^d).$$

It follows from (10) that $\frac{\partial f_1}{\partial T} = 0$. Similarly $\frac{\partial f_1}{\partial U} = 0$. Hence f_1 is a polynomial in T^p and U^p. Since d is prime to p we have $f_1 = 0$, proving (ii) if p does not divide d.

Now assume that p divides d and that $c_h \neq 0$, with p not dividing h. If p divides j we have by 3.4.2 (ii) that p divides $(d - j, h)$ and (9) shows that p must divide $(d - h, j)$. If $d - h \geq p$ we have by 3.4.2 that $(d - h, p) \not\equiv 0 \pmod{p}$ and (9) with $j = p$ would lead to a contradiction. It follows that $d - h < p$ and by (8) also $h < p$. We conclude that $d = p$. Now (10) implies that f is a multiple of c.

There remains the case that all h with $c_h \neq 0$ are divisible by p. In that case f is the p-th power of a polynomial that also satisfies (6) (here we use the perfectness of F). Then (ii) follows by induction. We have proved (ii).

(iii) follows from (ii), observing that $\sum_{1 \leq i \leq p-1} c(T, iT)$ is a non-zero multiple of T^p. $\qquad\square$

3.4.5. We need a multi-variable generalization of 3.4.4. We now consider polynomials in $2n$ variables $\mathbf{T} = (T_1, \ldots, T_n)$, $\mathbf{U} = (U_1, \ldots, U_n)$. Write $F[\mathbf{T}, \mathbf{U}]$ for the polynomial algebra $F[T_1, \ldots, T_n, U_1, \ldots, U_n]$. Define $c_h \in F[\mathbf{T}, \mathbf{U}]$ to be $c(T_h, U_h)$, where c is as before. Assume now that $f \in F[\mathbf{T}, \mathbf{U}]$ satisfies

$$f(\mathbf{T} + \mathbf{U}, \mathbf{V}) + f(\mathbf{T}, \mathbf{U}) = f(\mathbf{U} + \mathbf{V}, \mathbf{T}) + f(\mathbf{U}, \mathbf{V}). \qquad (11)$$

3.4.6. Lemma. *(i) If $p = 0$ there is $g \in F[\mathbf{T}]$ such that*

$$f(\mathbf{T}, \mathbf{U}) = g(\mathbf{T} + \mathbf{U}) - g(\mathbf{T}) - g(\mathbf{U});$$

(ii) If $p > 0$ there is $g \in F[\mathbf{T}]$ such that $f(\mathbf{T}, \mathbf{U}) - g(\mathbf{T} + \mathbf{U}) + g(\mathbf{T}) + g(\mathbf{U})$ equals a linear combination l of powers $c_h^{p^i}$;
(iii) If for $p > 0$ we have, moreover, $\sum_{1 \leq i \leq p-1} f(\mathbf{T}, i\mathbf{T}) = 0$ then the polynomial l of (ii) equals 0.

(i) is proved as 3.4.4 (i), using a multi-variable binomial formula

$$(\mathbf{T} + \mathbf{U})^{\mathbf{m}} = \sum (\mathbf{m}, \mathbf{h}) \mathbf{T}^{\mathbf{h}} \mathbf{U}^{\mathbf{m} - \mathbf{h}}.$$

We leave it to the reader to work out the details.

We deduce (ii) and (iii) from the corresponding assertions of 3.4.4 by a trick. If $G \in F[\mathbf{T}]$ let $d(G)$ be the maximum of the degrees of G as a polynomial in one of the individual variables T_h. Let q be an integer $> d(G)$. It follows from the properties of q-adic expansions that $T_1^{a_1} T_2^{a_2} \ldots T_n^{a_n} \mapsto T^{a_1 + a_2 q + \cdots a_n q^{n-1}}$ defines a linear bijection of the space of $G \in F[\mathbf{T}]$ with $d(G) < q$ onto the subspace of $F[T]$ of polynomials of degree $< q^n$. We denote this map by $f \mapsto \tilde{f}$. We have a similar map, sending polynomials in two sets of variables $\mathbf{T}, \mathbf{U}$ of degree $< q$ in all the variables individually to polynomials in two variables T, U, of degree $< q^n$ in both T and U.
Now let f be as in (11) and assume $p > 0$. Choose a p-power q that is strictly larger than $d(f)$. If $G \in F[\mathbf{T}]$ with $d(G) < q$ and $H(\mathbf{T}, \mathbf{U}) = G(\mathbf{T} + \mathbf{U}) - G(\mathbf{T}) - G(\mathbf{U})$, then $\tilde{H}(T, U) = \tilde{G}(T + U) - \tilde{G}(T) - \tilde{G}(U)$, as follows by using 3.4.2 (i). Also, $\tilde{c}_h = c^{q^{h-1}}$. The polynomial $\tilde{f} \in F[T, U]$ satisfies (6). Apply 3.4.4 (ii) to $\tilde{f}$. The polynomials whose existence is asserted in 3.4.4 (ii) have degree $< q^n$ and hence can

be written in the form $\tilde{g}$ and $\tilde{l}$, with polynomials $g \in F[\mathbf{T}]$, $l \in F[\mathbf{T}, \mathbf{U}]$. These have
the properties of 3.4.6 (ii). In a similar manner, (iii) follows from 3.4.4 (iii).   □

After these preparations we come to one of the main results of this section. The
ring R and the R-module $\mathcal{A}(G)$ are as before.

3.4.7. Theorem. *The following properties of a linear algebraic group G are equivalent :*
(a) G is elementary unipotent;
(b) $\mathcal{A}(G)$ is an R-module of finite type. Its elements generate the algebra $k[G]$;
*(c) G is a vector group if $p = 0$ and a product of a vector group and a finite elemen-
tary abelian p-group if $p > 0$.*

Recall that an elementary abelian p-group is a product of cyclic groups of order
p.

Assume that G is elementary unipotent and connected. By 2.4.12 we may assume
that G is a closed subgroup of some group $\mathbf{U}_m$ of upper triangular unipotent matrices.
Denote by $f_{ij} \in k[G]$ the (i, j)-th matrix element function $(1 \le i < j \le m)$. These
functions generate the algebra $k[G]$. We prove (b) for this case by induction on m.
The case $m = 1$ being trivial, we may assume that $m > 1$ and that (b) is known for
connected elementary subgroups of $\mathbf{U}_{m-1}$.

There are two homomorphisms of algebraic groups ϕ_1, $\phi_2 : \mathbf{U}_m \to \mathbf{U}_{m-1}$, the first
one being obtained by erasing the first row and column of a matrix and the second
one by erasing the last row and column. Then $k[\phi_1 G]$ is generated by the f_{ij} with
$i > 1$ and $k[\phi_2 G]$ by the f_{ij} with $j < m$. Observing that an additive function for
$\phi_1 G$ or $\phi_2 G$ is also one for G, we conclude from our induction assumption that there
are additive functions $a_1, \dots, a_n \in k[G]$ such that the f_{ij} with $(i, j) \ne (1, m)$ all lie
in the subalgebra $k[a_1, \dots, a_n]$. Using 3.3.6 (ii) and 3.3.3 (iii) we see that we may
assume the a_h to be algebraically independent. By the multiplication rule of matrices
(using the notations of 3.4.5) there is $f \in k[\mathbf{T}, \mathbf{U}]$ such that for $x, y \in G$ we have

$$f_{1m}(xy) - f_{1m}(x) - f_{1m}(y) = f(a_1(x), \dots, a_n(x), a_1(y), \dots, a_n(y)).$$

It follows that f satisfies (11). If $p > 0$ the fact that the elements of G have order dividing p implies the property of 3.4.6 (iii). We conclude that there exists
$h \in k[a_1, \dots, a_n]$ such that $f_{1m} - h$ is an additive function. This shows that $k[G]$
is generated by a finite number of additive functions, which can be taken to be algebraically independent. Then G is a vector group (check this). We have established the
implications $(a) \Rightarrow (b)$ and $(b) \Rightarrow (c)$, if G is connected.

Now let G be an arbitrary elementary unipotent group. If $p > 0$ choose an element
in each coset of the identity component G^0. These representatives form an elementary
abelian p-group A and it is immediate that G is the direct product of A and the vector
group G^0. This proves (c). The verification of (b) is left to the reader.

If $p = 0$, G does not contain any elements of finite order > 1. On the other hand, it follows from the fact that G^0 is a vector group that each coset of G^0 is represented by an element of finite order. This can only be if G is connected. We have proved (b) and (c) in this case.

Since the implication (c) $\Rightarrow$ (a) is obvious the proof of 3.4.6 is complete.   □

3.4.8. Corollary. *Let G be an F-group. Then G is elementary unipotent if and only if one of the following equivalent conditions holds :*
(a) $\mathcal{A}(G)(F)$ generates $F[G]$;
(b) G is F-isomorphic to a closed F-subgroup of some $\mathbf{G}_a^n$.

Here $\mathcal{A}(G)(F)$ is as in 3.3.1. (a) follows from 3.4.7 (b), using that an additive function in $k[G]$ is a linear combination of additive functions in $F[G]$. To see this, observe that $f \in k[G]$ is additive if and only if $\Delta f = f \otimes 1 + 1 \otimes f$, where Δ denotes comultiplication (2.1.2). This shows that $\mathcal{A}(G)$ is the kernel of a linear map $k[G] \to k[G] \otimes k[G]$ which is defined over F. It then has a basis in $F[G]$.

We skip the proof of the equivalence of (a) and (b).   □

When F is perfect a connected elementary unipotent F-group is F-isomorphic to some $\mathbf{G}_a^n$, but this is not generally true (see 3.4.10 (3), (4)). We shall return to these matters in 14.3.

We can now deal with the classification of connected one dimensional groups.

3.4.9. Theorem. *Let G be a connected linear algebraic group of dimension one. Then G is isomorphic to either $\mathbf{G}_a$ or $\mathbf{G}_m$.*

We have already seen in 3.1.3 that G is commutative and either consists of semi-simple elements or is elementary unipotent. In the first case G is diagonalizable by 2.4.2 (ii), and then 3.2.7 (ii) gives that $G \simeq \mathbf{G}_m$. In the second case 3.4.7 (c) implies that $G \simeq \mathbf{G}_a$.   □

3.4.10. Exercises. F is a subfield of k.
(1) Let $R = R(k)$ be as in 3.3.1. Elementary unipotent groups over k form a category, which is anti-equivalent to the category of left R-modules of finite type. (For further results along these lines see 14.3.6).
(2) ($p > 0$) Let c be as in 3.4.3. Define a structure of algebraic group on k^2 with product

$$(x, x').(y, y') = (x + x', y + y' + c(x, x'))\ (x, x', y, y' \in k).$$

This group is connected, unipotent, commutative, of dimension two. Show that it is not isomorphic to $\mathbf{G}_a^2$.
(3) Assume F to be perfect and let G be a connected elementary unipotent F-group

that is F-isomorphic to a closed subgroup of a triangular unipotent group $\mathbf{U}_m$. Then G is F-isomorphic to some $\mathbf{G}_a^n$. (Remark: the triangulizability condition is redundant, see 14.1.2).

(4) F is a non-perfect field of characteristic p and $a \in F - F^p$. Then $G = \{(x, y) \in \mathbf{G}_a^2 \mid x^p - x = ay^p\}$ is an F-group isomorphic to $\mathbf{G}_a$ which is not F-isomorphic to $\mathbf{G}_a$ (*Hint*: use 2.1.5 (5)).

Notes

3.1.1 is due to Kolchin [Kol2, §3]. The name 'torus' for a connected diagonalizable group was coined by Borel in [Bo1]. He realized the important role played by these groups, similar to the role played by compact tori in the theory of compact Lie groups.

3.2 contains standard results on tori. The proof of the rigidity theorem 3.2.8 gives a stronger result: the affine variety V of the statement of that result may be replaced by any connected scheme over k. This implies that a diagonalizable group has no 'infinitesimal automorphisms.'

The theory of elementary unipotent groups bears some resemblance to the theory of tori, the character group being replaced by the $R(k)$-module $\mathcal{A}$ of 3.3.1. The use of the ring $R(k)$ seems to go back to [DG, Ch. IV, 3.6]. In [loc. cit., Ch. V, 3.4] one finds more general results, for arbitrary commutative unipotent groups. These are described by 'Dieudonné modules.'

One of the main results of this chapter is the classification theorem 3.4.9. The first published proof seems to be the one given by Grothendieck in [Ch4, Exp. 7]. In [Bo3, Ch. III, §10] a proof is given that uses the fact that an irreducible smooth projective curve with an infinite group of automorphisms fixing a point is isomorphic to $\mathbf{P}^1$. The proof given here is more elementary. We use the classification of elementary unipotent groups. We also need the result on polynomial cocycles of 3.4.4 (due to M. Lazard [Laz, lemme 3]). Another proof of the classification theorem (also using additive polynomials) can be found in [Hu1, no.20].

Chapter 4

Derivations, Differentials, Lie Algebras

We first discuss tangent spaces of algebraic varieties and related algebraic matters. In the second part of the chapter, Lie algebras of linear algebraic groups are introduced and their basic properties are established.

4.1. Derivations and tangent spaces

4.1.1. We recall the definition of a derivation. Let R be a commutative ring, A an R-algebra and M a left A-module. An *R-derivation* of A in M is an R-linear map $D : A \to M$ such that for $a, b \in A$

$$D(ab) = a.Db + b.Da.$$

It is immediate from the definitions that $D1 = 0$, whence $D(r.1) = 0$ for all $r \in R$.

The set $\mathrm{Der}_R(A, M)$ of these derivations is a left A-module, the module structure being defined by $(D + D')a = Da + D'a$ and $(b.D)(a) = b.Da$, if $D, D' \in \mathrm{Der}_R(A, M)$, $a, b \in A$.

The elements of $\mathrm{Der}_R(A, A)$ are the derivations of the R-algebra A. If $\phi : A \to B$ is a homomorphism of R-algebras and N is a B-module, then N is an A-module in an obvious way. If $D \in \mathrm{Der}_R(B, N)$ then $D \circ \phi$ is a derivation of A in N and the map $D \mapsto D \circ \phi$ defines a homomorphism of A-modules

$$\phi_0 : \mathrm{Der}_R(B, N) \to \mathrm{Der}_R(A, N),$$

whose kernel is $\mathrm{Der}_A(B, N)$. Thus we obtain an exact sequence of A-modules

$$0 \to \mathrm{Der}_A(B, N) \to \mathrm{Der}_R(B, N) \to \mathrm{Der}_R(A, N). \tag{12}$$

4.1.2. Tangent spaces, heuristic introduction. We use the notations of the first chapter. Let X be a closed subvariety of $\mathbf{A}^n$. We identify its algebra of regular functions $k[X]$ with $k[T_1, \ldots, T_n]/I$, where I is the ideal of polynomial functions vanishing on X. Assume that I is generated by the polynomials $f_1, \ldots, f_s$. Let $x \in X$ and let L be a line in $\mathbf{A}^n$ through x. Its points can be written as $x + tv$, where $v = (v_1, \ldots, v_n)$ is a direction vector, t running through k. The t-values of the points of the intersection $L \cap X$ are found by solving the set of equations

$$f_i(x + tv) = 0, \ 1 \le i \le s. \tag{13}$$

Clearly, $t = 0$ is a solution.

Let D_i be partial derivation in $k[T]$ with respect to T_i. Then

$$f_i(x + tv) = t \sum_{j=1}^{n} v_j (D_j f_i)(x) + t^2(\ldots),$$

and we see that $t = 0$ is a 'multiple root' of the set of equations (13) if and only if

$$\sum_{j=1}^{n} v_j (D_j f_i)(x) = 0, \ 1 \le i \le s.$$

If this is so we call L a tangent line and v a tangent vector of X in x.

Write $D' = \sum_{j=1}^{n} v_j D_j$, this is a k-derivation of $k[T]$. The last set of equations then says that $D' f_i(x) = 0$ for $1 \le i \le s$. Denoting by M_x the maximal ideal in $k[T]$ of functions vanishing at x, it follows that $D'I \subset M_x$. The linear map $f \mapsto (D'f)(x)$ factors through I and gives a linear map $D : k[X] \to k = k[X]/M_x$. Viewing k as a $k[X]$-module k_x via the homomorphism $f \mapsto f(x)$, we see that D is a k-derivation of $k[X]$ in k_x.

Conversely, any element of $\mathrm{Der}_k(k[X], k_x)$ can be obtained in this manner from a derivation D' of $k[T]$ with $D'I \subset M_x$. We conclude that there is a bijection of the set of tangent vectors v, such that (13) has a 'multiple root' $t = 0$, onto $\mathrm{Der}_k(k[X], k_x)$.

4.1.3. Tangent spaces. The heuristic description of tangent vectors of 4.1.2 suggests a formal definition of the tangent spaces of an algebraic variety. First let X be an affine variety. If $x \in X$ we define the *tangent space* $T_x X$ of X at x to be the k-vector space $\mathrm{Der}_k(k[X], k_x)$, where k_x is as in 4.1.2. Let $\phi : X \to Y$ be a morphism of affine varieties with corresponding algebra homomorphism $\phi^* : k[Y] \to k[X]$ (1.4.7). The induced linear map ϕ_0^* (see 4.1.1) is a linear map of tangent spaces

$$d\phi_x : T_x X \to T_{\phi x} Y,$$

the *differential* of ϕ at x, or the tangent map at x.

If $\psi : Y \to Z$ is another morphism of affine varieties then we have the chain rule

$$d(\psi \circ \phi)_x = d\psi_{\phi x} \circ d\phi_x.$$

If ϕ is an isomorphism then so is $d\phi_x$ and the differential of an identity morphism is an identity map.

We give two alternative descriptions of the tangent space $T_x X$. Let $M_x \subset k[X]$ be the maximal ideal of functions vanishing in x. If $D \in T_x X$ then D maps the elements of M_x^2 to 0. Hence D defines a linear function $\lambda(D) : M_x/M_x^2 \to k$.

4.1.4. Lemma. *λ is an isomorphism of $T_x X$ onto the dual of M_x/M_x^2.*

Let l be a linear function on M_x/M_x^2. Define a linear map $\mu(l) : k[X] \to k$ by $\mu(l)f = l(f - f(x) + M_x^2)$. Then $\mu(l) \in T_x X$, and μ is the inverse of λ. We skip the easy proof. $\square$

Another description of the tangent space uses the ring $\mathcal{O}_x$ of functions regular in x (see 1.4.3). It is a k-algebra which has a unique maximal ideal $\mathcal{M}_x$, consisting of the functions vanishing in x (1.4.4 (1)) and $\mathcal{O}_x/\mathcal{M}_x \simeq k$. We view k as an $\mathcal{O}_x$-module. There is an obvious algebra homomorphism $\alpha : k[X] \to \mathcal{O}_x$, inducing a linear map $\alpha_0 : \mathrm{Der}_k(\mathcal{O}_x, k) \to \mathrm{Der}_k(k[X], k_x)$.

4.1.5. Lemma. α_0 *is bijective.*

We have a linear map $\beta : \mathrm{Der}_k(k[X], k_x) \to \mathrm{Der}_k(\mathcal{O}_x, k)$, which comes from the formula for differentiating a quotient. Let $f \in \mathcal{O}_x$ and let $g, h \in k[X]$ be such that $h(x) \neq 0$ and that $hf - g$ vanishes in a neighborhood of x (see 1.4.1). If $D \in \mathrm{Der}_k(k[X], k_x)$ then

$$(\beta D)f = h(x)^{-2}(h(x)Dg - g(x)Dh)$$

defines an element of $\mathrm{Der}(\mathcal{O}_x, k)$ and it is immediate that α_0 and β are inverses. $\qquad\square$

4.1.6. Lemma. *Let ϕ be an isomorphism of X onto an affine open subvariety of Y. Then $d\phi_x$ is an isomorphism of $T_x X$ onto $T_{\phi x} Y$.*

ϕ induces an isomorphism $\mathcal{O}_{Y,\phi x} \simeq \mathcal{O}_{X,x}$. The assertion follows from the previous lemma (check the details). $\qquad\square$

4.1.7. We can now define the tangent space in a point x of an arbitrary algebraic variety X (see 1.6). It follows from 4.1.6 that if U and V are open affine neighborhoods of x in X with $V \subset U$ there is a canonical identification $T_x U \simeq T_x V$. This allows us to define the tangent space $T_x X$ to be $T_x U$, for U as above. The formal definition is

$$T_x X = \mathrm{limproj}\, T_x U,$$

a projective limit relative to the set of open affine neigborhoods of x, ordered by inclusion.

It is clear how to define for a morphism of varieties $\phi : X \to Y$ the tangent map

$$d\phi_x : T_x X \to T_{\phi x} Y.$$

We say that x is a *simple point* of X, or that X is *smooth* in x or that X is *non-singular* in x if $\dim T_x X = \dim X$ (the dimension of X, defined in 1.8.1). X is *smooth* or *non-singular* if all its points are simple.

4.1.8. Let X be an F-variety, where F is a subfield of k and let $x \in X(F)$ (1.6.14). First assume that X is an affine F-variety. The point x defines an algebra homomorphism $F[X] \to F$, which makes F into an $F[X]$-module F_x. Define

$$T_x X(F) = \mathrm{Der}_F(F[X], F_x);$$

this is a vector space over F. We have a canonical isomorphism

$$k \otimes_F T_x X(F) \simeq T_x X.$$

We call $T_x X(F)$ the space of F-rational points of $T_x X$ and we view it as an F-subspace of T_x. If ϕ is a morphism of affine F-varieties then for $x \in X(F)$ the tangent map $d\phi_x$ maps $T_x X(F)$ to $T_{\phi x} Y(F)$. If X is an arbitrary F-variety, the definition of $T_x X(F)$ is similar to that of $T_x X$, given in 4.1.7.

4.1.9. Exercises. (1) Using 4.1.4, describe $T_x X$ in the following cases:
(a) X is a point,
(b) $X = \mathbf{A}^n$,
(c) $X = \{(a, b) \in \mathbf{A}^2 \mid ab = 0\}$, $x = (0, 0)$,
(d) (char $k \neq 2, 3$) $X = \{(a, b) \in \mathbf{A}^2 \mid a^2 = b^3\}$, $x = (0, 0)$.
(2) Let X and Y be algebraic varieties, let $x \in X$, $y \in Y$. Then $T_{(x,y)}(X \times Y) \simeq T_x X \oplus T_y Y$.
(3) Let X be an affine algebraic variety. Let $k[\tau] = k[T]/(T^2)$ be the k-algebra of dual numbers (τ being the image of T). Show that there is a bijection of $T_x X$ onto the set of k-homomorphisms $\phi : k[X] \to k[\tau]$ such that $\phi(f) - f(x) \in k\tau$. (These are the '$k[\tau]$-valued points of X lying over x'.)
(4) If X is a closed subvariety of Y and $\phi : X \to Y$ is the injection morphism then $d\phi_x$ is injective for all $x \in X$.
(5) Complete the details in 4.1.8.

4.2. Differentials, separability

We shall need a number of results about derivations, in particular about derivations of fields. To deal with them we introduce differentials.

4.2.1. Let R be a commutative ring and A a commutative R-algebra. Denote by

$$m : A \otimes_R A \to A$$

the product homomorphism (so $m(a \otimes b) = ab$) and let $I = \operatorname{Ker} m$. This ideal of $A \otimes A$ is generated by the elements $a \otimes 1 - 1 \otimes a$ ($a \in A$). The quotient algebra $A \otimes A/I$ is isomorphic to A. We define the *module of differentials* $\Omega_{A/R}$ of the R-algebra A by

$$\Omega_{A/R} = I/I^2.$$

This is an $A \otimes A$-module, but since it is annihilated by I we may and shall view it as an A-module.

Denote by da or $d_{A/R}a$ the image of $a \otimes 1 - 1 \otimes a$ in $\Omega_{A/R}$. One checks that d is an R-derivation of A in $\Omega_{A/R}$ and that the da generate the A-module $\Omega_{A/R}$. The following result shows that $\Omega_{A/R}$ is a 'universal module for R-derivations of A.'

4.2.2. Theorem. *(i) For every A-module M the map* $\Phi : \mathrm{Hom}_A(\Omega_{A/R}, M) \to \mathrm{Der}_R(A, M)$ *sending ϕ to $\phi \circ d$ is an isomorphism of A-modules;*
(ii) A pair $(\Omega_{A/R}, d)$ of an A-module together with an R-derivation of A in $\Omega_{A/R}$ with the property of (i) is unique up to isomorphism.

Φ is a homomorphism of A-modules, which is injective since the da generate $\Omega_{A/R}$. Now let $D \in \mathrm{Der}(A, M)$. Define an R-linear map $\psi : A \otimes A \to M$ by $\psi(a \otimes b) = bDa$. Then

$$\psi(xy) = m(x)\psi(y) + m(y)\psi(x).$$

It follows that ψ vanishes on I^2, hence defines an R-linear map $\phi : \Omega_{A/R} \to M$, which in fact is A-linear. Since $\psi(a \otimes 1 - 1 \otimes a) = Da$ we have $\Phi(\phi) = D$. Hence Φ is surjective, proving (i). The proof of (ii) is standard. $\square$

4.2.3. If $\phi : A \to B$ is a homomorphism of R-algebras, there is a unique homomorphism of A-modules

$$\phi^0 : \Omega_{A/R} \to \Omega_{B/R},$$

with $\phi^0 \circ d_{A/R} = d_{B/R} \circ \phi$. If N is a B-module and if ϕ_0 is as in 4.1.1, there is a commutative diagram of A-modules:

$$
\begin{array}{ccc}
\mathrm{Hom}_B(\Omega_{B/R}, N) & \longrightarrow & \mathrm{Der}_R(B, N) \\
\downarrow & & \downarrow{\scriptstyle \phi_0} \\
\mathrm{Hom}_A(\Omega_{A/R}, N) & \longrightarrow & \mathrm{Der}_R(A, N).
\end{array}
$$

The horizontal arrows are isomorphisms, they are as in 4.2.2. The left-hand vertical arrow is induced by ϕ^0.

Now let A be an R-algebra of the form $A = R[T_1, \dots, T_m]/(f_1, \dots, f_n)$. Let t_i be the image of T_i in A and put $t = (t_1, \dots, t_m)$. Denote by D_i partial derivation in $R[T_1, \dots, T_m]$ with respect to T_i $(1 \le i \le m)$.

4.2.4. Lemma. *The dt_i $(1 \le i \le m)$ generate the A-module $\Omega_{A/R}$. The kernel of the A-homomorphism $\phi : A^m \to \Omega_{A/R}$ with $\phi e_i = dt_i$ is the submodule generated by the elements $\sum_{i=1}^m (D_i f_j(t)) e_i$ $(1 \le j \le n)$.*

Here (e_i) is the standard basis of A^n.

Let D be an R-derivation of A in an A-module M. If $f \in R[T_1, \dots, T_m]$ then

$$D(f(t)) = \sum_{i=1}^m (D_i f)(t).Dt_i, \tag{14}$$

whence

$$\sum_{i=1}^{m} (D_i f_j)(t).Dt_i = 0 \ (1 \leq j \leq n).$$

Let K be the submodule of A^n described in the lemma. A straightforward check shows that the A-module A^n/K (together with a derivation of A given by (14)) has the universal property of 4.2.2, hence is isomorphic to $\Omega_{A/R}$ by 4.2.2 (ii).   □

4.2.5. Exercises. (1) If $A = R[T_1, \ldots, T_m]$ then $\Omega_{A/R}$ is a free A-module with basis $(dT_i)_{1 \leq i \leq m}$.
(2) In the case of 4.2.4 with $m = n = 1$, give a necessary and sufficient condition on f_1 under which $\Omega_{A/R} = 0$. Consider the case when R is a field.
(3) Let A be an R-algebra which is an integral domain and let F be the quotient field of A. Then $\Omega_{F/R} \simeq F \otimes_A \Omega_{A/R}$.
(4) Let F be a field and let $E = F(x_1, \ldots, x_m)$ be an extension field of finite type. Then $\Omega_{E/F}$ is a finite dimensional vector space over E spanned by the dx_i.
(5) Let $A = k[T, U]/(T^2 - U^3)$. Show that $\Omega_{A/k}$ is not a free A-module.
(6) Let A and B be R-algebras. There is an isomorphism of $A \otimes_R B$-modules

$$\Omega_{A \otimes_R B/R} \simeq (\Omega_{A/R} \otimes_R B) \oplus (A \otimes_R \Omega_{B/R})$$

under which $d_{A \otimes_R B/R}$ corresponds to $(d_{A/R} \otimes_R \mathrm{id}_B) \oplus (\mathrm{id}_A \otimes_R d_{B/R})$.

4.2.6. We next discuss the case of fields. Let F be a field and let E, E' be two extensions of F of finite type, with $E' \subset E$. By (12) we have an exact sequence of groups

$$0 \to \mathrm{Der}_{E'}(E, E) \to \mathrm{Der}_F(E, E) \to \mathrm{Der}_F(E', E),$$

which is also an exact sequence of vector spaces over E (the vector space structures coming from the second arguments). Using 4.2.2 (i) we obtain an exact sequence of E-vector spaces

$$0 \to \mathrm{Hom}_E(\Omega_{E/E'}, E) \to \mathrm{Hom}_E(\Omega_{E/F}, E) \to \mathrm{Hom}_{E'}(\Omega_{E'/F}, E).$$

Since the map $u \mapsto 1 \otimes u$ of $\Omega_{E'/F}$ to $E \otimes_{E'} \Omega_{E'/F}$ induces an isomorphism of E-vector spaces

$$\mathrm{Hom}_E(E \otimes_{E'} \Omega_{E'/F}, E) \simeq \mathrm{Hom}_{E'}(\Omega_{E'/F}, E),$$

we get an exact sequence of E-vector spaces

$$0 \to \mathrm{Hom}_E(\Omega_{E/E'}, E) \to \mathrm{Hom}_E(\Omega_{E/F}, E) \to \mathrm{Hom}_E(E \otimes_{E'} \Omega_{E'/F}, E).$$

These vector spaces are finite dimensional by 4.2.5 (4). Dualizing we obtain an exact sequence of finite dimensional E-vector spaces

$$E \otimes_{E'} \Omega_{E'/F} \overset{\alpha}{\to} \Omega_{E/F} \to \Omega_{E/E'} \to 0, \tag{15}$$

which is basic in what follows. Notice that $\alpha(1 \otimes d_{E'/F}x) = d_{E/F}x$ $(x \in E')$.

Recall (see for example [La2, Ch. VII, §4]) that E is a *separable algebraic extension* of F or is *separably algebraic* over F if for each $x \in E$ there is a polynomial $f \in F[T]$ without multiple roots such that $f(x) = 0$. We may assume f to be irreducible, in which case the derivative f' is a non-zero polynomial. If char $F = 0$ every algebraic extension of F is separable.

4.2.7. Lemma. *If E is separably algebraic over E' then α is injective.*

From the discussion in 4.2.6 we see that the injectivity of α is equivalent to the surjectivity of the homomorphism of 4.1.1

$$\mathrm{Der}_F(E, E) \to \mathrm{Der}_F(E', E).$$

An equivalent property is: any F-derivation of E' in E can be extended to an F-derivation of E in E. To prove this it suffices to deal with the case of a simple extension $E = E'(x) \simeq E'[T]/(f)$, where f is an irreducible polynomial with $f'(x) \neq 0$. Let $D \in \mathrm{Der}_F(E', E)$. If $g = \sum_{i \geq 0} a_i T^i \in E'[T]$ define $Dg \in E[T]$ by $Dg = \sum(Da_i)T^i$. Then D is extendible to an F-derivation D' of E in E with $D'x = a$ if and only if $f'(x)a + (Df)(x) = 0$. Since $f'(x) \neq 0$ this equation has a unique solution and the lemma follows. $\square$

4.2.8. Lemma. *Let $E = F(x)$. Then $\dim_E \Omega_{E/F} \leq 1$. We have $\Omega_{E/F} = 0$ if and only if E is separably algebraic over F.*

If x is transcendental over F this follows from 4.2.5 (1), (3). If x is algebraic we are in the situation of 4.2.5 (2). $\square$

We denote by $\mathrm{trdeg}_F E$ the transcendence degree of E over F. If $E = F(x_1, ..., x_m)$ this is the maximal number of x_i that are algebraically independent over F. Recall that E is *purely transcendental* over F if the x_i can be taken such that the transcendence degree equals m. We say that E is *separably generated* over F if there is a purely transcendental extension E' of F, contained in E, such that E is separably algebraic over E'. Let $p = \mathrm{char}\, F$. If $p = 0$ then E is always separably generated.

We can now state the main result about fields.

4.2.9. Theorem. *(i) $\dim_E \Omega_{E/F} \geq \mathrm{trdeg}_F E$;*
(ii) Equality holds in (i) if and only if E is separably generated over F.

We prove (i) and (ii) together, by induction on $d = \dim_E \Omega_{E/F}$.
Let $E = F(x_1, \dots, x_m)$.

First let $d = 0$. If $m = 1$ we have (i) and (ii) by 4.2.8. If $m > 1$ then (15) with $E' = F(x_1)$ shows that $\Omega_{E/F(x_1)} = 0$. By induction on m we may assume that E is separably algebraic over $F(x_1)$. Using 4.2.7 we conclude from (15) that $\Omega_{F(x_1)/F} = 0$, and x_1 is separable over F by 4.2.8. It follows that E is separably algebraic over F, proving (i) and (ii) in the case $d = 0$. By induction on m one also shows that $\Omega_{E/F} = 0$ if E is separably algebraic over F.

Now let $d > 0$ and assume that (i) and (ii) are known for smaller values. By what we have already proved there is $x \in E$ with $d_{E/F}x \neq 0$. We use (15) with $E' = F(x)$. Since $\alpha(1 \otimes d_{F(x)/F}x) = d_{E/F}x \neq 0$ we have $\Omega_{F(x)/F} \neq 0$. Using 4.2.8 we conclude that $\dim_{F(x)} \Omega_{F(x)/F} = 1$ and that α is injective. Consequently, $\dim_E \Omega_{E/F} = \dim_E \Omega_{E/F(x)} + 1$. By induction we have $\dim_E \Omega_{E/F} \geq \operatorname{trdeg}_{F(x)} E + 1$. Since $\operatorname{trdeg}_F E = \operatorname{trdeg}_{F(x)} E + \operatorname{trdeg}_F F(x)$ (see [La2, Ch. X, 8.5]) and $\operatorname{trdeg}_F F(x) \leq 1$, (i) follows. If we have equality in (i) then x is transcendental over F, and by induction E is separably generated over $F(x)$, hence also over F.

To finish the proof we have to show that if E is separably generated over F we have equality in (i). Now apply (15) for E' a purely transcendental extension over which E is separably algebraic. We have already seen that $\Omega_{E/E'} = 0$. Using 4.2.7 we find, using 4.2.5 (1),(3), that $\dim_E \Omega_{E/F} = \dim_{E'} \Omega_{E'/F} = \operatorname{trdeg}_F E' = \operatorname{trdeg}_F E$, finishing the proof.   □

We say that E is *separable* over F if either $p = 0$ or if $p > 0$ and the following holds: Let $x_1, \dots, x_s$ be elements of E which are linearly independent over F. Then so are $x_1^p, \dots, x_s^p$. It is immediate that if F is perfect any extension E is separable. Recall that F is perfect if either $p = 0$ or if $p > 0$ and every element of F is a pth power. In particular, an algebraically closed field is perfect.

4.2.10. Proposition. *Assume that E is separable over F. Then E is separably generated over F.*

Let $E = F(x_1, \dots, x_m)$ and assume that $x_1, \dots, x_t$ are algebraically independent over F, with $t = \operatorname{trdeg}_F E$. Let E be separable over F and let $p > 0$ (for $p = 0$ the result is trivial). We may also assume that $0 \leq t < m$. If $t < m - 1$ then, by induction on m, there are algebraically independent elements $y_1, \dots, y_t$ in $F(x_1, \dots, x_{m-1})$ such that $F(x_1, \dots, x_{m-1})$ is separably algebraic over $F(y_1, \dots, y_t)$. Likewise, there are algebraically independent elements $z_1, \dots, z_t$ in $F(y_1, \dots, y_t, x_m)$ such that this field is separably algebraic over $F(z_1, \dots, z_t)$. Using a transitivity property of separably algebraic extensions (see [La2, Ch. VII, §4]) we conclude that E is separably algebraic over $F(z_1, \dots, z_t)$.

This reduces the proof to the case that $t = m - 1$. Since E is separable over F

there is a non-zero polynomial $f \in F[T_1, \ldots, T_m]$ such that $f(x_1, \ldots, x_m) = 0$ and that not all exponents of the powers of the indeterminates occurring in f are divisible by p. Using 4.2.4 (with $R = F$, $A = F[x_1, \ldots, x_m]$) and 4.2.5 (3) it follows that $\dim_E \Omega_{E/F} \leq m - 1$ (notice that the kernel of the homomorphism ϕ of 4.2.4 is non-zero). Now the assertion follows from 4.2.9. $\square$

The converse of 4.2.10 is also true (as a consequence of 4.2.12 (5)).

Let E, E', F be as in 4.2.6.

4.2.11. Corollary. *Assume F to be perfect. Either of the following conditions is necessary and sufficient for E to be separably generated over E':*
(a) $\alpha : E \otimes_{E'} \Omega_{E'/F} \to \Omega_{E/F}$ *is injective;*
(b) $\mathrm{Der}_F(E, E) \to \mathrm{Der}_F(E', E)$ *is surjective.*

The equivalence of (a) and (b) follows from 4.2.6. To obtain the criterion (a) observe that by 4.2.9 (ii) and 4.2.10 we have $\dim_{E'} \Omega_{E'/F} = \mathrm{trdeg}_F E'$, $\dim_E \Omega_{E/F} = \mathrm{trdeg}_F E$. It follows that $\dim_E \Omega_{E/E'} = \mathrm{trdeg}_{E'} E$ if and only if the map α of (15) is injective. Then 4.2.11 follows from 4.2.9 (ii). $\square$

4.2.12. Exercises. (1) If in the case of 4.2.11 E is separably algebraic over E', then α is an isomorphism.
(2) Let $p = 0$. If $x \in E$, $d_{E/F}x = 0$ then x is algebraic over F.
(3) Let $p > 0$. Show that $\Omega_{E/F} = 0$ if and only if $E = F(E^p)$. (*Hint:* use (15) and 4.2.8.)
(4) Assume that E has the following property: for any algebraic extension K of F the algebra $K \otimes E$ is reduced. Then E is separably generated over F.
(5) (a) If E is separable over E' and E' is separable over F then E is separable over F.

(b) If E is separably algebraic over F then E is separable over F.

4.2.13. In applying the preceding results to geometric questions we need some auxiliary results, pertaining to linear algebra.

Let R be an integral domain with quotient field F. If $f \in R$, $f \neq 0$ denote by R_f the ring $R[T]/(1 - fT)$ (see 1.4.6). We may and shall view it as the subring $R[f^{-1}]$ of F, i.e., the ring of elements of F of the form $f^{-n}a$ $(a \in R, n \geq 0)$. If M is an R-module we denote by M_f the R_f-module $R_f \otimes_R M$.

Let $A = (a_{ij})$ be an $m \times n$-matrix with entries in R. Denote by r the rank of A, viewed as a matrix with entries in F. Define the R-module $\mathcal{M}(A)$ or $\mathcal{M}_R(A)$ to be the quotient of R^n by the submodule generated by the elements $\sum_{j=1}^{n} a_{ij} e_j$ $(1 \leq i \leq m)$, where (e_j) is the canonical basis. We denote by $GL_m(R)$ the group of those $m \times m$-matrices with entries in R that have an inverse with entries in R.

4.2.14. Lemma. *(i) If $B \in GL_m(R)$ then $\mathcal{M}(BA) = \mathcal{M}(A)$,; if $C \in GL_n(R)$*

then $\mathcal{M}(AC) \simeq \mathcal{M}(A)$;
(ii) There exist $f \in R$, $f \neq 0$ and $B \in GL_m(R)$, $C \in GL_n(R)$ such that

$$A = B \begin{pmatrix} I_r & 0 \\ 0 & 0 \end{pmatrix} C.$$

Here I_r is an identity matrix. (i) is easy. The assertion of (ii) is true if R is a field, by linear algebra. Take $B \in GL_m(F)$, $C \in GL_n(F)$ with the required property and choose $f \in R$ such that B, C, B^{-1}, C^{-1} have entries in R_f. Then we have (ii).   □

4.2.15. Lemma. *There is $f \in R$, $f \neq 0$ such that $\mathcal{M}(A)_f$ is a free R_f-module of rank $n - r$. We may choose f such that $n - r$ of the images e'_i of the elements $1 \otimes e_i$ of $R_f \otimes R^n$ form a basis of $\mathcal{M}(A)_f$.*

The first point follows from 4.2.14 (ii). Let $(f_1, \dots, f_{n-r})$ be a basis of $\mathcal{M}(A)_f$ and assume that the elements $e'_{r+1}, \dots, e'_n$ are linearly independent over F. Write

$$e'_{r+i} = \sum_{j=1}^{n-r} c_{ij} f_j \ (1 \leq i \leq n - r),$$

with $c_{ij} \in R_f$, $\det(c_{ij}) \neq 0$. By modifying f we may assume that the inverse matrix $(c_{ij})^{-1}$ has entries in R_f. Then $e'_{r+1}, \dots, e'_n$ are as required.   □

4.3. Simple points

4.3.1. Let X be an irreducible affine variety over k. If $x \in X$ let again M_x be the maximal ideal in $k[X]$ of functions vanishing in x. If M is an R-module, put $M(x) = M/M_x M$; this is a vector space over k.

Let A be as in 4.2.13, with $R = k[X]$. Since the entries of A are functions on X the matrix $A(x)$ with entries in k is defined. It is clear that

$$\mathcal{M}_{k[X]}(A)(x) = \mathcal{M}_k(A(x))$$

(notations of 4.2.13). As before r is the rank of A, as a matrix with entries in the field $k(X)$.

4.3.2. Lemma. *(i) $\dim_{k(X)} \mathcal{M}_{k(X)}(A) = n - r$;*
(ii) The set of $x \in X$ such that $\dim_k \mathcal{M}(A)(x) = n - r$ is open and non-empty;
(iii) If $x \in X$ and $\dim_k \mathcal{M}(A)(x) = n - r$ there is $f \in k[X]$ with $f(x) \neq 0$ such that $\mathcal{M}(A)_f$ is a free $k[X]_f$-module of rank $n - r$.

One knows from linear algebra that $\dim_k \mathcal{M}(A)(x) = n - r$ if and only if r is the maximal size of a square submatrix of $A(x)$ with non-zero determinant. By the same result, r equals this maximal size for the matrix A. (i) and (ii) follow from these facts.

Let x be as in (iii). We may assume that $\det(a_{ij}(x)_{1 \le i, j \le r}) \ne 0$. Put $f = \det(a_{ij})_{1 \le i, j \le r}$. If $e'_i \in \mathcal{M}(A)_f$ is as in 4.2.15 then

$$\sum_{j=1}^{n} a_{ij} e'_j = 0 \ (1 \le i \le m).$$

Our assumption implies that we can express the e'_i with $1 \le i \le r$ as linear combinations with coefficients in $k[X]_f$ of $e'_{r+1}, \ldots, e'_n$. Using (i) we conclude that the latter elements form a basis of $\mathcal{M}(A)_f$, whence (iii). $\qquad\square$

We put $\Omega_X = \Omega_{k[X]/k}$. If $x \in X$ the tangent space $T_x X$ is isomorphic to $\mathrm{Hom}(\Omega_X, k_x)$, by 4.2.2 (i). Since for any $k[X]$-module M we have

$$\mathrm{Hom}_{k[X]}(M, k_x) \simeq \mathrm{Hom}_k(M(x), k)$$

it follows that

$$T_x X \simeq \mathrm{Hom}_k(\Omega_X(x), k).$$

The dual vector space of the tangent space $T_x X$ is the *cotangent space* $(T_x X)^*$. It can be identified with $\Omega_X(x)$.

We come now to the basic results on simple points (defined in 4.1.7).

4.3.3. Theorem. *Let X be an irreducible variety of dimension e.*
(i) If x is a simple point of X there is an affine open neighborhood U of x such that Ω_U is a free $k[U]$-module with a basis $(dg_1, \ldots, dg_e)$, for suitable $g_i \in k[U]$;
(ii) The simple points of X form a non-empty open subset of X;
(iii) For any $x \in X$ we have $\dim_k T_x X \ge e$.

We may assume that X is affine and that $k[X] = k[T_1, \ldots, T_m]/(f_1, \ldots f_n)$. With the notations of 4.2.4 we have $\Omega_X \simeq \mathcal{M}(A)$, where A is the $m \times n$-matrix $(D_j f_i(t))$. From 4.2.9 (ii) and 4.2.10 we see that $\dim_{k(X)} \Omega_{k(X)/k} = e$. If r is as before it follows from 4.3.2 (i) that $e = n - r$. Now (i) follows from 4.3.2 (iii) and 4.2.15, and (ii) is a consequence of 4.3.2 (ii). For any $x \in X$ the dimension of $T_x X$ is $n - s$, where s is the rank of $(D_j f_i(x))$. It is clear that $r \ge s$, whence (iii). $\qquad\square$

4.3.4. Exercises. (1) The functions $g_1, \ldots, g_e$ of 4.3.3 (i) are algebraically independent.
(2) Let X be an affine variety. If Ω_X is a free $k[X]$-module then X is smooth.

4.3.5. A morphism $\phi : X \to Y$ of irreducible varieties is called *dominant* if ϕX is dense in Y. It follows from 1.9.1 (ii) that, if ϕ is dominant, there is an injection of quotient fields $k(Y) \to k(X)$. So we can view $k(X)$ as an extension of $k(Y)$. We say that ϕ is *separable* if this extension is separably generated.

Assume, moreover, that X and Y are affine. The homomorphism

$$\phi^* : k[Y] \to k[X]$$

defining ϕ induces by 4.2.3 a homomorphism of $k[Y]$-modules

$$(\phi^*)^0 : \Omega_Y \to \Omega_X.$$

Let $x \in X$ and let k_x be as in 4.1.2. Viewed as a $k[Y]$-module (via ϕ^*) it is $k_{\phi x}$. Consider the linear map of 4.1.3

$$d\phi_x : T_x X \to T_{\phi x} Y.$$

From the preceding remarks, using the diagram of 4.2.3, we see that we can view $d\phi_x$ as the homomorphism

$$\mathrm{Hom}_{k[X]}(\Omega_X, k_x) \to \mathrm{Hom}_{k[Y]}(\Omega_Y, k_{\phi x})$$

deduced from $(\phi^*)^0$. It can also be viewed as a linear map

$$\mathrm{Hom}(\Omega_X(x), k) \to \mathrm{Hom}(\Omega_Y(\phi x), k).$$

4.3.6. Theorem. *Let $\phi : X \to Y$ be a morphism of irreducible varieties.*
(i) Assume that x is a simple point of X such that ϕx is a simple point of Y and that $d\phi_x$ is surjective. Then ϕ is dominant and separable;
(ii) Assume that ϕ is dominant and separable. Then the points $x \in X$ with the property of (i) form a non-empty open subset of X.

It follows from 4.3.3 that it suffices to consider the case that X and Y are affine and smooth, and that Ω_X and Ω_Y are free modules over $k[X]$ resp. $k[Y]$ of rank $d = \dim X$ and $e = \dim Y$.

The homomorphism $(\phi^*)^0$ of 4.3.5 leads to a homomorphism of free $k[X]$-modules

$$\psi : k[X] \otimes_{k[Y]} \Omega_Y \to \Omega_X.$$

Fixing bases of these modules, ψ is described by a $d \times e$-matrix A with entries in $k[X]$. Let $x \in X$ be such that $d\phi_x$ is surjective. Then by the remarks of 4.3.5, the matrix $A(x)$ has rank e. An argument involving determinants, as in the proof of 4.3.3, shows that the rank of A (as a matrix with entries in $k(X)$) is at least e. Since this rank is at most e it must equal e. It follows that ψ is injective. Then the same holds for $(\phi^*)^0$. As Ω_X and Ω_Y are free modules, the homomorphism $\phi^* : k[Y] \to k[X]$ is also injective, which means that ϕ is dominant. By 4.2.5 (3) it also follows that the homomorphism α of (15) in 4.2.6, with $E = k(X)$, $E' = k(Y)$, $F = k$, is injective (on suitable bases it is given by the matrix A). The separability of ϕ now follows from 4.2.11. We have proved (i).

If ϕ is dominant and separable, the rank of A (as a matrix with entries in $k(X)$) equals e. The set of $x \in X$ such that $A(x)$ has rank e is then non-empty and open, whence (ii). □

To conclude this section we give some consequences of the preceding results for homogeneous spaces (defined in 2.3.1).

4.3.7. Theorem. *Let G be a connected algebraic group.*
(i) Let X be a homogeneous space for G. Then X is irreducible and smooth. In particular, G is smooth;
(ii) Let $\phi : X \to Y$ be a G-morphism of homogeneous spaces. Then ϕ is separable if and only if the tangent map $d\phi_x$ is surjective for some $x \in X$. If this is so then $d\phi_x$ is surjective for all $x \in X$;
(iii) Let $\phi : G \to G'$ be a surjective homomorphism of algebraic groups. Then ϕ is separable if and only if $d\phi_e$ is surjective.

If X is as in (i) and $x \in X$, the morphism $G \to X$ sending g to $g.x$ is surjective. Hence X is irreducible by 1.2.3 (ii). Also, for fixed g, the map $x \mapsto g.x$ is an isomorphism of X. Hence x is simple if and only if $g.x$ is simple. Now (i) follows from 4.3.3 (ii) and (ii) from 4.3.6. Finally, (iii) is a consequence of (ii) (view G and G' as homogeneous spaces for G). □

4.4. The Lie algebra of a linear algebraic group

4.4.1. Let G be a linear algebraic group. Denote by λ and ρ the representation of G by left and right translations in $A = k[G]$ (2.3). We view $A \otimes_k A$ as the algebra of regular functions $k[G \times G]$. If $m : A \otimes A \to A$ is the multiplication map, then for $F \in k[G \times G]$ we have $(mF)(x) = F(x, x)$. So $I = \mathrm{Ker}\ m$ is the ideal of functions in $k[G \times G]$ vanishing on the diagonal (see 1.6.5). It is clear that for $x \in G$ the automorphisms $\lambda(x) \otimes \lambda(x)$ and $\rho(x) \otimes \rho(x)$ of $k[G \times G]$ stabilize I and I^2, hence induce automorphisms of $\Omega_G = I/I^2$, also denoted by $\lambda(x)$ and $\rho(x)$. We thus have representations λ and ρ of G in Ω_G, which are locally finite (by 2.3.6(i)). The derivation $d : A \to \Omega_G$ of 4.2.1 commutes with all $\lambda(x)$ and $\rho(x)$.

For $x \in G$ the map $\mathrm{Int}(x) : y \mapsto xyx^{-1}$ is an automorphism of the algebraic group G, fixing e. It induces linear automorphisms $\mathrm{Ad}\ x$ of the tangent space T_eG and $(\mathrm{Ad}\ x)^*$ of the cotangent space $(T_eG)^*$. Thus, for $u \in (T_eG)^*$ we have

$$((\mathrm{Ad}\ x)^*u)X = u(\mathrm{Ad}(x^{-1})X) \ (x \in G,\ X \in T_eG)$$

Let $M_e \subset A$ be the maximal ideal of functions vanishing in e. By 4.1.4, $(T_eG)^*$ can be identified with M_e/M_e^2. If $f \in A$ we denote by δf the element $f - f(e) + M_e^2$ of $(T_eG)^*$. For $X \in T_eG = \mathrm{Der}_k(A, k_e)$ we have $(\delta f)(X) = Xf$, as follows from the proof of 4.1.4.

4.4.2. Proposition. *There is an isomorphism of $k[G]$-modules*

$$\Phi : \Omega_G \to k[G] \otimes_k (T_e G)^*,$$

(the module structure of the right-hand side coming from the first factor), such that
(a) $\Phi \circ \lambda(x) \circ \Phi^{-1} = \lambda(x) \otimes \mathrm{id}$, $\Phi \circ \rho(x) \circ \Phi^{-1} = \rho(x) \otimes (\mathrm{Ad}\, x)^*$ ($x \in G$);
(b) if $f \in k[G]$ and $\Delta f = \sum_i f_i \otimes g_i$, then

$$\Phi(df) = -\sum_i f_i \otimes \delta g_i.$$

In (b) $\Delta : A \to A \otimes A$ is the comultiplication of 2.1.2 (so $(\Delta f)(x, y) = f(xy)$).

The map sending (x, y) to (x, xy) is an automorphism of the algebraic variety $G \times G$. The corresponding algebra automorphism ψ of $A \otimes A$ is given by

$$(\psi F)(x, y) = F(x, xy) \quad (x, y \in G).$$

It follows that ψI is the ideal of functions vanishing on $G \times \{e\}$, which is $k[G] \otimes M_e$. Then $\psi I^2 = k[G] \otimes M_e^2$ and it follows that ψ induces a bijection of Ω_G onto $k[G] \otimes M_e/M_e^2$. Let Φ be the composite of this bijection and the isomorphism coming from 4.1.4. From the definition of ψ it follows that for $x \in G$

$$(\lambda(x) \otimes \mathrm{id}) \circ \psi = \psi \circ (\lambda(x) \otimes \lambda(x)),$$

$$(\rho(x) \otimes \mathrm{Int}(x)) \circ \psi = \psi \circ (\rho(x) \otimes \rho(x)).$$

These formulas imply that Φ satisfies (a).
   With the notations of (b) we have

$$\psi(f \otimes 1 - 1 \otimes f)(x, y) = \sum_i f_i(x)(g_i(e) - g_i(y)),$$

from which we see that Φ satisfies (b).   □

4.4.3. We assume that the reader is familiar with the basic facts about Lie algebras (which can be found in [Bou2, Ch. 1] or [Jac1]).
   If A is an arbitrary commutative k-algebra, the space $\mathcal{D} = \mathrm{Der}_k(A, A)$ has a Lie algebra structure, the Lie product being given by $[D, D'] = D \circ D' - D' \circ D$ ($D, D' \in \mathcal{D}$).
   If $p = \mathrm{char}\, k > 0$ then by Leibniz's formula

$$D^p(ab) = \sum_{i=0}^{p} (p, i)(D^i a)(D^{p-i} b) = a(D^p b) + (D^p a)b \quad (a, b \in A, D \in \mathcal{D})$$

(where (p, i) is a binomial coefficient). So D^p is again a derivation. In this case $\mathcal{D}$ is an example of a *restricted Lie algebra* (or *p-Lie algebra*) . This means that the Lie algebra has a p-operation $D \mapsto D^{[p]}$ (which in the case of derivations is the ordinary p^{th} power) such that the following holds for $a \in k, D, D' \in \mathcal{D}$ (with $(\mathrm{ad}\ D)D' = [D, D']$)

(a) $(aD)^{[p]} = a^p D^{[p]}$,

(b) $\mathrm{ad}(D^{[p]}) = (\mathrm{ad}\ D)^p$,

(c) (Jacobson's formula) $(D + D')^{[p]} = D^{[p]} + D'^{[p]} + \sum_{i=1}^{p-1} i^{-1} s_i(D, D')$,

where $s_i(D, D')$ is the coefficient of a^i in $\mathrm{ad}(aD + D')^{p-1}(D')$.

   For a further discussion see [Bou2, Ch. I, p. 105-106].

   Now let G and A be as in 4.4.1. If necessary we write $\mathcal{D} = \mathcal{D}_G$. Then λ and ρ define representations of G in $\mathcal{D}$, denoted by the same symbols. So

$$\lambda(x)D = \lambda(x) \circ D \circ \lambda(x)^{-1} \ (x \in G, D \in \mathcal{D}),$$

and similarly for ρ. The *Lie algebra* $L(G)$ of G is the set of $D \in \mathcal{D}$ commuting with all $\lambda(x)$ $(x \in G)$. It is immediate that $L(G)$ is a subalgebra of $\mathcal{D}$, stable under the p-operation if $p > 0$. Since left and right translations commute, all $\rho(x)$ stabilize $L(G)$. We denote the induced linear maps also by $\rho(x)$.

4.4.4. Corollary. *There is an isomorphism of $k[G]$-modules*

$$\Psi : \mathcal{D}_G \to k[G] \otimes_k T_e G,$$

(the module structure of the right-hand side coming from the first factor), such that
(a) $\Psi \circ \lambda(x) \circ \Psi^{-1} = \lambda(x) \otimes \mathrm{id}, \Psi \circ \rho(x) \circ \Psi^{-1} = \rho(x) \otimes \mathrm{Ad}\ x \ (x \in G)$;
(b) (notations of 4.4.2 (b))

$$\Psi^{-1}(1 \otimes X)(f) = -\sum f_i(Xg_i) \ (X \in T_e G).$$

   This is a consequence of 4.4.2 and 4.2.2 (i). These results give an isomorphism of $\mathcal{D}_G$ onto $\mathrm{Hom}_{k[G]}(k[G] \otimes_k (T_e G)^*, k[G])$, which is a module isomorphic to $k[G] \otimes_k T_e G$. There is an isomorphism of the latter module onto the former which sends $f \otimes X$ to the homomorphism ϕ with $\phi(g \otimes u) = u(X)fg$ $(f, g \in k[G], X \in T_e G, u \in (T_e G)^*)$. The assertions of the corollary are readily checked. For the last one observe that $Xg_i = (\delta g_i)(X)$.   □

Let $\alpha_G = \alpha : \mathcal{D}_G \to T_e G$ be the linear map with $(\alpha_G D)f = (Df)(e)$.

4.4.5. Proposition. *(i) α induces an isomorphism of vector spaces $L(G) \simeq T_e G$. We have for $x \in G$*

$$\alpha \circ \rho(x) \circ \alpha^{-1} = \mathrm{Ad}\ x;$$

(ii) Ad is a rational representation of G in $T_e G$ (the adjoint representation).

Let Ψ be as in 4.4.4. It follows from 4.4.4 that $\Psi(L(G)) = 1 \otimes T_e G$. Moreover, 4.4.4 (b) implies that (with the previous notations)

$$(\alpha \circ \Psi^{-1})(1 \otimes X)(f) = -\sum_i f_i(e)(X g_i) = -X f,$$

since $f = \sum_i f_i(e) g_i$. Now (i) and (ii) readily follow.   □

4.4.6. Corollary. $\dim_k L(G) = \dim\ G$.

This is a consequence of (i) and 4.3.7 (i) (applied to the identity component G^0).

□

Next let H be a closed subgroup of G. Denote by $J \subset k[G]$ the ideal of functions vanishing on H, so $k[H] = k[G]/J$. Put

$$\mathcal{D}_{G,H} = \{ D \in \mathcal{D}_G \mid DJ \subset J \}.$$

Then $\mathcal{D}_{G,H}$ is a subalgebra of the Lie algebra $\mathcal{D}_G$ and there is an obvious homomorphism of Lie algebras $\phi : \mathcal{D}_{G,H} \to \mathcal{D}_H$. Notice that

$$T_e H = \{ X \in T_e G \mid X J = 0 \}.$$

4.4.7. Lemma. ϕ defines an isomorphism of $\mathcal{D}_{G,H} \cap L(G)$ onto $L(H)$.

It follows from the definitions that $\alpha_H \circ \phi$ is the restriction of α_G to $\mathcal{D}_{G,H}$. The injectivity of ϕ on $L(G)$ then follows from 4.4.5 (i). To finish the proof of the lemma we show that, if $X \in T_e H$, we have $D = \Psi^{-1}(1 \otimes X) \in \mathcal{D}_{G,H}$, where Ψ is as in 4.4.4. If $f \in J$ and $\Delta f = \sum f_i \otimes g_i$ then we may assume that for each i one of the elements f_i or g_i lies in J. Then 4.4.4 (b) shows that $Df \in J$, whence the lemma.   □

4.4.8. Henceforth we identify the Lie algebra $L(G)$ and the tangent space $T_e G$ via α_G. We thus obtain a Lie algebra structure on the latter space. We shall denote the Lie algebra of linear algebraic groups $G, H, \ldots$ either by $L(G)$, $L(H), \ldots$ or by the corresponding gothic letters $\mathfrak{g}$, $\mathfrak{h} \ldots$.

If $\phi : G \to G'$ is a homomorphism of linear algebraic groups, we write $d\phi$ for the tangent map $d\phi_e : \mathfrak{g} \to \mathfrak{g}'$. We call $d\phi$ the *differential* of ϕ.

If F is a subfield of k and G is an F-group, we denote the F-vector space $T_e G(F)$ of 4.1.8 by $L(G)(F)$ or $\mathfrak{g}(F)$. This is the set of F-rational points of $\mathfrak{g}$. It is a Lie algebra over F. If $\phi : G \to G'$ is a homomorphism of F-groups then $d\phi$ induces an F-linear map $\mathfrak{g}(F) \to \mathfrak{g}'(F)$.

4.4.9. Proposition. *Let* $\phi : G \to H$ *be a homomorphism of linear algebraic groups. Then* $d\phi$ *is a homomorphism of Lie algebras, which is compatible with the* p-*operation if* $p > 0$.

Using the factorization of ϕ

$$G \xrightarrow{\rho} G \times H \xrightarrow{\sigma} H,$$

where $\rho x = (x, \phi x)$ $(x \in G)$ and σ is a projection, we see that it suffices to prove the proposition in the cases that ϕ is an injection of a closed subgroup or a projection like σ. We leave the second case to the reader.

Let α_G be as in 4.4.5. If ϕ is an injection we have for $X \in \mathfrak{g}$, $f \in k[H]$

$$(\alpha_G^{-1}(X))(f \circ \phi) = (\alpha_H^{-1}(d\phi(X)))(f).$$

This formula implies the assertions. □

4.4.10. Examples. (1) $G = \mathbf{G}_a$. Then $k[G] = k[T]$. The derivations of $k[G]$ commuting with all translations $T \mapsto T + a$ $(a \in k)$ are the multiples of $X = \frac{d}{dT}$. If $p > 0$ we have $X^p = 0$. So $\mathfrak{g}$ is the one dimensional Lie algebra kX with $[X, X] = 0$ and $X^p = 0$ (if $p > 0$).
(2) $G = \mathbf{G}_m$. We have $k[G] = k[T, T^{-1}]$. The derivations of $k[G]$ commuting with the translations $T \mapsto aT$ $(a \in k^*)$ are the multiples of $T\frac{d}{dT}$. If $p > 0$ we have $X^p = X$. $\mathfrak{g}$ is as in the previous example, but the p-operation is different (if $p > 0$).
(3) $G = \mathbf{GL}_n$. Now $k[G] = k[T_{ij}, D^{-1}]_{1 \le i,j \le n}$, where $D = \det(T_{ij})$ (see 2.1.4 (3)). Denote by $\mathfrak{gl}_n$ the Lie algebra of all $n \times n$-matrices over k, with product $[X, Y] = XY - YX$, and the usual p^{th} power as p-operation if $p > 0$. If $X = (x_{ij}) \in \mathfrak{gl}_n$ then

$$D_X T_{ij} = -\sum_{h=1}^{n} T_{ih} x_{hj}$$

defines a derivation of $k[G]$ commuting with all left translations, hence lies in $L(G)$. Since the map $X \mapsto D_X$ is injective, it follows from 4.4.6 (since $\dim G = n^2$) that $L(G)$ consists of the D_X. We conclude that we can identify $\mathfrak{g}$ and $\mathfrak{gl}_n$ (with the p-power p-operation). For $x \in \mathbf{GL}_n$, $X \in \mathfrak{gl}_n$ we have $\mathrm{Ad}(x)X = xXx^{-1}$. Also, if H is a closed subgroup of $\mathbf{GL}_n$ we can view $\mathfrak{h}$ as a subalgebra of $\mathfrak{gl}_n$.

4.4.11. Exercises. (1) The Lie algebra of $\mathbf{SL}_n$ is the subalgebra $\mathfrak{sl}_n$ of $\mathfrak{gl}_n$ of matrices with trace zero. (*Hint:* use 4.4.7.)
(2) Determine the Lie algebras of the groups $\mathbf{D}_n$, $\mathbf{T}_n$, $\mathbf{U}_n$ of 2.1.4.
(3) Let $\phi : \mathbf{SL}_2 \to \mathbf{PSL}_2$ be the homomorphism of 2.1.5 (3). Show that $d\phi$ is bijective if and only if $p \ne 2$. Describe $d\phi$ if $p = 2$.
(4) Let T be a torus. There is a canonical isomorphism $L(T) \to k \otimes_{\mathbf{Z}} X_*(T)$ (where $X_*(T)$ is as in 3.2.1).
(5) $L(G) = L(G^0)$.
(6) Show that $\mathrm{Ad}\, x$ is an automorphism of the Lie algebra $\mathfrak{g}$ $(x \in G)$.
(7) Let $\phi : G \to H$ be a homomorphism of linear algebraic groups. Show that $(d\phi)((\mathrm{Ad}x)(X)) = \mathrm{Ad}(\phi(x))(d\phi(X))$ $(x \in G, X \in \mathfrak{g})$.

Next we give some differentiation formulas, to be used in the sequel. As before, G is a linear algebraic group. We denote by $\mu : G \times G \to G$ and $i : G \to G$ multiplication and inversion (2.1.1). We identify the vector spaces $L(G \times G)$ and $\mathfrak{g} \oplus \mathfrak{g}$ (see 4.1.9 (2)). In fact, the Lie algebras $L(G \times G)$ and $\mathfrak{g} \oplus \mathfrak{g}$ are isomorphic (we leave it to the reader to check this).

4.4.12. Lemma. $(d\mu)_{(e,e)} : \mathfrak{g} \oplus \mathfrak{g} \to \mathfrak{g}$ *is the map* $(X, Y) \mapsto X + Y$ *and* $(di)_e = -\mathrm{id}.$

μ defines a k-linear map

$$\bar{\mu} = (\mu^*)^0 : \Omega_G \to \Omega_{G \times G} = (\Omega_G \otimes k[G]) \oplus (k[G] \otimes \Omega_G)$$

(see 4.2.3 and 4.2.5 (6)). If $f \in k[G]$, $\Delta f = \sum f_i \otimes g_i$ then

$$\bar{\mu}(df) = \sum (df_i \otimes g_i + f_i \otimes dg_i).$$

Since $f = \sum f_i(e) g_i = \sum g_i(e) f_i$ we have

$$\bar{\mu}(df) - df \otimes 1 - 1 \otimes df \in M_{(e,e)} \Omega_{G \times G}.$$

Hence the linear map of $\Omega_G(e)$ to $\Omega_{G \times G}(e, e) = \Omega_G(e) \oplus \Omega_G(e)$ induced by $\bar{\mu}$ sends u to (u, u). As $(d\mu)_{(e,e)}$ is the dual of this map, the first assertion follows. The second one follows from the fact that $\mu \circ (\mathrm{id}, i)$ is the trivial map $G \to \{e\}$. $\square$

4.4.13. Lemma. *(i) Let* $\sigma : G \to G$ *be a morphism of varieties and put* $\phi(x) = (\sigma x) x^{-1}$. *Then* $d\phi_e = d\sigma_e - 1$;
(ii) Let $a \in G$. *If* $\psi(x) = axa^{-1}x^{-1}$ *then* $d\psi_e = \mathrm{Ad}\, a - 1$.

The morphism ϕ of (i) is the composite of the morphism $x \mapsto (\sigma(x), x)$ of G to $G \times G$ and $\mu \circ (\mathrm{id}, i)$. To prove the formula of (i) use the chain rule of 4.1.3 and 4.4.12 (and observe that the tangent map of the first morphism at x is $(d\sigma_x, \mathrm{id})$). The proof of (ii) is similar. $\square$

If V is a finite dimensional vector space over k, we write $\mathfrak{gl}(V)$ for the Lie algebra of endomorphisms of V. Then $\mathfrak{gl}(V) \simeq \mathfrak{gl}_{\dim V}$. If $\phi : G \to GL(V)$ is a rational representation (2.3.2 (3)) its differential $d\phi$ is a Lie algebra homomorphism $\mathfrak{g} \to \mathfrak{gl}(V)$, i.e., a representation of $\mathfrak{g}$ in V.

Now let G_1 and G_2 be two linear algebraic groups and let $\phi : G_i \to GL(V_i)$ be a rational representation ($i = 1, 2$). Let $\phi_1 \oplus \phi_2$ and $\phi_1 \otimes \phi_2$ be the direct sum and tensor product representations of $G_1 \times G_2$ in $V_1 \oplus V_2$ and $V_1 \otimes V_2$, respectively. We identify the vector spaces $L(G_1 \times G_2)$ and $\mathfrak{g}_1 \oplus \mathfrak{g}_2$.

4.4.14. Lemma. *(i)* $d(\phi_1 \oplus \phi_2) = d\phi_1 \oplus d\phi_2$;
(ii) $(d(\phi_1 \otimes \phi_2)(X_1, X_2))(v_1, v_2) = ((d\phi_1)(X_1))v_1 \otimes v_2 + v_1 \otimes ((d\phi_2)(X_2))v_2$.

We only prove (ii); the proof of (i) is similar. It suffices to deal with the case that $X_2 = 0$. Then observe that the left hand side of (ii) is the tangent map at e of the morphism $x \mapsto \phi_1(x)v_1 \otimes v_2$ of G to $V_1 \otimes V_2$, evaluated at X_1.   □

4.4.15. Exercises. (1) Let $\phi : G \to GL_n$ be a rational representation and write $\phi(x) = (f_{ij}(x))$, where $f_{ij} \in k[G]$. For $X \in \mathfrak{g}$ we have $d\phi(X) = (Xf_{ij})$.
(2) Let $V \subset k[G]$ be a finite dimensional subspace of $k[G]$ which is stable under all left multiplications $\lambda(x)$, $x \in G$. Let $\phi : G \to GL(V)$ be the rational representation defined by λ. For $X \in \mathfrak{g}$, $f \in V$ we have $d\phi(X)(f) = Xf$, where X is viewed as an element of $L(G) \subset \mathcal{D}_G$. Give a similar result for right translations. (*Hint*: use that $\rho(x) = \iota \circ \lambda(x^{-1}) \circ \iota$, where ι is the isomorphism of $k[G]$ induced by inversion).
(3) The differential of the adjoint representation Ad (4.4.5) is given by

$$d\mathrm{Ad}(X)(Y) = [X, Y] \ (X, Y \in \mathfrak{g}).$$

(*Hint*: first deal with the case of GL_n).
(4) (a) The Lie algebra of the commutator group (G, G) (see 2.2.8) is a subalgebra $\mathfrak{g}$ containing all elements $(\mathrm{Ad}(x) - 1)X$ and all $[X, Y]$ $(x \in G, \ X, Y \in \mathfrak{g})$.
   (b) If G is commutative (solvable) the same holds for $\mathfrak{g}$.
(5) Let $\phi : G \to GL(V)$ be a rational representation. Define the representation $\bigwedge^h \phi$ of G in the exterior power $\bigwedge^h V$ by $(\bigwedge^h \phi)(x)(v_1 \wedge \dots \wedge v_h) = \phi(x)v_1 \wedge \dots \wedge \phi(x)v_h$. Then $\bigwedge^h \phi$ is a rational representation and

$$d(\bigwedge^h \phi)(X)(v_1 \wedge \dots \wedge v_h) = \sum_{i=1}^{h} v_1 \wedge \dots \wedge (d\phi)(X)v_i \wedge \dots \wedge v_h.$$

(6) Let $s \in M_n$ and let $G = \{g \in GL_n \mid gs({}^t g) = s\}$. Then G is a closed subgroup of GL_n. Its Lie algebra is contained in $\{X \in \mathfrak{gl}_n \mid Xs + s({}^t X) = 0\}$.

4.4.16. As an application of the results of this chapter we shall prove a fundamental result about algebraic groups over finite fields. Let $F = \mathbf{F}_q$ be a finite field with q elements and assume k to be an algebraic closure of F. Notice that

$$F = \{a \in k \mid a^q = a\}.$$

If X is an affine F-variety then $f \mapsto f^q$ defines an algebra endomorphism of $F[X]$, whence an F-morphism $\sigma : X \to X$, depending functorially on X. This is the *Frobenius morphism* of X. If X is a closed F-subvariety of $\mathbf{A}^n$ and $x = (x_1, \dots, x_n) \in X$ then $\sigma x = (x_1^q, \dots, x_n^q)$. It is immediate from the definitions that the homomorphism $(\sigma^*)^0$ (see 4.3.5) is the zero homomorphism, from which it follows that $d\sigma_x = 0$ for all $x \in X$. It is also clear that the fixed point set $X^\sigma = \{x \in X \mid \sigma x = x\}$ is finite. Similar results hold for arbitrary F-varieties.

Now let G be an F-group, not necessarily affine. Its Frobenius morphism is an endomorphism of the algebraic group G.

4.4.17. Theorem. [Lang's theorem] *Assume G to be connected. Then $\Lambda x = (\sigma x)x^{-1}$ defines a surjective morphism of G.*

It follows from 4.4.13 (i) that $d\Lambda_e$ is bijective. Let $a \in G$. Denote by $\rho(x)$: $y \mapsto yx$ right translation by x in G. It is an isomorphism of the variety G. Put $\sigma' = \rho(\Lambda(a)) \circ \sigma$ and put $\Lambda'(x) = \sigma'(x)x^{-1}$. Then $\Lambda' = \Lambda \circ \rho(a)$ and $(d\Lambda')_e = d\Lambda_a \circ d\rho(a)_e$. As before, $(d\Lambda')_e$ is bijective. It follows that $d\Lambda_a$ is bijective for all a.

Let X be the closure of ΛG; this is an irreducible closed subvariety of G. It follows from 1.9.5 and 4.3.3 (ii) that there is $a \in G$ such that Λa is a simple point of X. We conclude that $\dim X = \dim T_{\Lambda a} X = \dim T_a G = \dim G$, whence $X = G$. By 1.9.5, ΛG contains a non-empty open subset U of G.

A similar argument shows that for $a \in G$ there is a non-empty open subset V of G consisting of elements of the form $\sigma(x)ax^{-1}$. Since G is irreducible, the intersection $U \cap V$ is non-empty. This means that $a \in \Lambda G$, proving Lang's theorem. $\square$

If G is as in the theorem, the fixed point set G^σ is a finite group. Many interesting finite groups are of this kind. Lang's theorem is a basic tool in the study of such groups, see e.g. [Ca]. We give an instance of an application in the next exercise.

4.4.18. Exercises. (1) Let G and G^σ be as before. Let $a \in G^\sigma$ and let

$$Z(a) = \{x \in G \mid xa = ax\}$$

be its centralizer. This is a closed subgroup, which is σ-stable. Assume that $Z(a)$ is connected. If $b \in G^\sigma$ is conjugate to a in G then b is conjugate to a in the finite group G^σ (more details about such matters can be found in [Bo2, Part E]).

(2) Let G be a connected linear algebraic group and $\tau : G \to G$ an endomorphism of algebraic groups such that $d\tau : \mathfrak{g} \to \mathfrak{g}$ is a nilpotent linear map. Then $x \mapsto (\tau x)x^{-1}$ is surjective.

4.4.19. Jordan decomposition in the Lie algebra. Let k be algebraically closed and let G be a linear algebraic group over k. We view the elements of the Lie algebra $\mathfrak{g}$ as derivations of $k[G]$, in particular they are linear maps of the vector space $k[G]$. We have a Jordan decomposition for $\mathfrak{g}$, which is similar to that of 2.4.8

Let $X \in \mathfrak{g}$. It follows from 2.3.6 and 4.4.15 (2) that X is a locally finite linear map of $k[G]$. Let $X = X_s + X_n$ be its additive Jordan decomposition (2.4.7).

4.4.20. Theorem. *(i) X_s and X_n lie in $\mathfrak{g}$ and $[X_s, X_n] = 0$;*

*(ii) If $\phi : G \to G'$ is a homomorphism of algebraic groups, then $((d\phi)X)_s = d\phi(X_s)$,
$((d\phi)X)_n = d\phi(X_n)$;
(iii) If $G = \mathbf{GL}_n$ then X_s and X_n are the semi-simple and nilpotent parts of the matrix
$X \in \mathbf{M}_n$ (see 4.4.10 (3)).*

The proof is similar to the proof of 2.4.8 and is left to the reader. X_s and X_n are
the *semi-simple* and *nilpotent* parts of X.

4.4.21. Exercises. (1) If G is a torus then all elements of $\mathfrak{g}$ are semi-simple. If G
is unipotent then all elements of $\mathfrak{g}$ are nilpotent.
(2) ($p = \text{char } k > 0$). If X is as above then $X_s^{[p]}$ and $X_n^{[p]}$ are the semi-simple and
nilpotent parts of $X^{[p]}$.

Notes

4.1 and 4.3 contain standard material on tangent spaces and simple points of algebraic varieties. The basic geometric results are 4.3.3 and 4.3.6.

The treatment given here makes use of modules of differentials. The discussion
in 4.2 of their formal properties has been kept brief. A more extensive discussion can
be found in [EGA, Ch. 0, §20] or in [Ma, Ch. 10]. We have also included the relevant
algebraic results about separably generated field extensions.

In the discussion of the Lie algebra of a linear algebraic group G and its properties,
use is also made of differentials. The basic properties of the Lie algebra of G (such as
4.4.5) are deduced from the properties of the differential module Ω_G given in 4.4.2.

Another approach to the Lie algebras uses the description of tangent spaces via
dual numbers (see 4.1.9 (3)). This is the approach followed in [Bo3].

Lang's theorem, basic in the study of 'finite groups of Lie type', was first proved
in [La1], with a view to applications to abelian varieties over finite fields.

Chapter 5

Topological Properties of Morphisms, Applications

The first part of the chapter deals with general results about morphisms of algebraic varieties. Then these results are applied in the theory of algebraic groups. One of the main items of the chapter is the construction in 5.5 of the quotient of a linear algebraic group by a closed subgroup.

5.1. Topological properties of morphisms

5.1.1. X and Y are two irreducible algebraic varieties over the algebraically closed field k and $\phi : X \to Y$ is a dominant morphism (4.3.5). We shall establish a number of general facts about the topological behavior of ϕ.

View the quotient field $k(X)$ as an extension of the field $k(Y)$ (see 1.8.1 and 1.9.1 (ii)). The transcendence degree $\operatorname{trdeg}_{k(Y)} k(X)$ equals $\dim X - \dim Y$. In 5.1.6 (ii) we shall give a geometric interpretation of this integer.

Let F be a field and E a finite algebraic extension. We denote by $[E : F]$ its degree. The elements of E that are separable over F form a subfield E_s, which is a separable algebraic extension of F. Its degree $[E : F]_s$ is the *separable degree* of the extension E/F. Let p be the characteristic. If $p = 0$ we have $E = E_s$. If $p > 0$ then E is a purely inseparable extension of E_s, i.e., for any $x \in E$ a power x^{p^e} lies in E_s (see [La2, Ch. VII, §4] for these facts). If $\phi : X \to Y$ is as before and $\dim X = \dim Y$ then $k(X)$ is an algebraic extension of $k(Y)$. In 5.1.6 (iii) we shall give a geometric interpretation of $[k(X) : k(Y)]_s$.

If $k(X) = k(Y)$ then ϕ is said to be *birational*.

5.1.2. Lemma. *ϕ is birational if and only if there is a non-empty open subset U of X such that ϕU is open and ϕ induces an isomorphism of varieties $U \simeq \phi U$.*

It follows from the definition of quotient fields (1.8.1) that ϕ is birational if the condition of the lemma is satisfied. Assume that ϕ is birational. We may assume X and Y to be affine. Then $k[X] = k[Y][f_1, \dots, f_r]$, where all f_i lie in $k(Y)$. Take $f \in k[Y]$ such that $f \neq 0$ and the ff_i lie in $k[Y]$. Then ϕ induces an isomorphism $k[Y]_f \simeq k[X]_f$, and $U = D_X(f)$ (see 1.3.5) is as required. □

The main result of this section is 5.1.6. We first deal with some special cases. Assume now, moreover, that X and Y are affine and that there is $f \in k[X]$ with $k[X] = k[Y][f]$ (ϕ being defined by the inclusion map $k[Y] \to k[X]$).

5.1.3. Lemma. *Assume that f is transcendental over $k(Y)$.*
(i) ϕ is an open morphism;

(ii) If Y' is an irreducible closed subvariety of Y then $\phi^{-1}Y'$ is an irreducible closed subvariety of X, of dimension dim $Y' + 1$.

Recall that an open (closed) map of topological spaces is a continuous map such that the image of an open set is open (respectively the image of a closed set is closed). We say that the morphism ϕ is open (closed) if it defines an open map (respectively a closed map).

We may assume that $X = Y \times \mathbf{A}^1$ and that ϕ is projection on the first factor. Let $g = \sum_{i=0}^{r} g_i T^i \in k[X] = k[Y][T]$. Then

$$\phi(D_X(g)) = \bigcup_{i=0}^{r} D_Y(g_i),$$

whence (i).

If Q is the prime ideal in $k[Y]$ defined by the irreducible closed subvariety Y', then $\phi^{-1}Y'$ is the set of points of X in which the functions of the ideal $P = Qk[X] = \{\sum_{i \geq 0} f_i T^i \mid f_i \in Q\}$ vanish. Then $k[X]/P \simeq (k[Y]/Q)[T]$. Since the last ring is an integral domain, P is a prime ideal and $\phi^{-1}Y'$ is irreducible. The last point of (ii) is clear. $\qquad\square$

5.1.4. Lemma. *Assume that f is separably algebraic over $k(Y)$. There is a non-empty open subset U of X with the following properties:*
(i) The restriction of ϕ to U defines an open morphism $U \to Y$;
(ii) If Y' is an irreducible closed subvariety of Y and X' is an irreducible component of $\phi^{-1}Y'$ that intersects U, then dim $X' = $ dim Y';
(iii) For $x \in U$ the fiber $\phi^{-1}(\phi(x))$ is a finite set with $[k(X) : k(Y)]$ elements.

We have $k[X] = k[Y][T]/I$ where I is the ideal of polynomials vanishing in f. Let F be the minimum polynomial of f over $k(Y)$ (the irreducible polynomial in $k(Y)[T]$ with leading coefficient one, with root f). Choose $a \in k[Y]$ such that all coefficients of F lie in $k[Y]_a$.

Let $f_1, \ldots, f_n$ be the roots of F, in some extension field of $k(Y)$. Since f is separable over $k(Y)$ the roots are distinct and the discriminant $d = \Pi_{i<j}(f_i - f_j)^2$ is a non-zero element of $k(Y)$, which can be expressed polynomially in the coefficients of F (see [Jac4, p. 250-251]). It follows that there is $b \in k[Y]$ and $m \geq 0$ such that $a^m d = b$.

We may replace X and Y by $D_X(ab)$, respectively $D_Y(ab)$. We are then reduced to proving the lemma when, moreover, the following holds:
(a) I contains the minimum polynomial F. From this it follows, using the division algorithm, that I is the ideal generated by F. It also follows that $k[X]$ is a free $k[Y]$-module.

(b) If $F(T) = \sum_{i=0}^{n} h_i T^i$ then for all $y \in Y$ the polynomial

$$F(y)(T) = \sum_{i=0}^{n} h_i(y) T^i$$

has distinct roots. We shall show that in this situation the statements of the lemma hold, with $U = X$. We may assume that

$$X = \{(y, t) \in Y \times \mathbf{A}^1 \mid F(y)(t) = 0\},$$

ϕ being the first projection. Let $G \in k[Y][T]$ and denote by g its image in $k[X]$. Then

$$D_X(g) = \{(y, t) \in X \mid G(y)(t) \neq 0\}.$$

Write $G = QF + R$, where $R = \sum_{i=0}^{n-1} r_i T^i$ is a polynomial in T of degree $< n = \deg F$. Then $\phi D_X(g)$ is the set of $y \in Y$ such that not all roots of $F(y)(T)$ are roots of $R(y)(T)$. Since the first polynomial has n distinct roots, this implies that

$$\phi D_X(g) = \bigcup_{i=0}^{n-1} D_Y(r_i),$$

whence (i).

Next let Y' be as in (ii) and let Q be the corresponding prime ideal in $k[Y]$. Then $\phi^{-1}Y'$ is the closed set defined by the ideal $Qk[X]$. Let $A = k[Y]/Q$ and denote by $\bar{F}$ the image of F in $A[T]$. We claim that $Qk[X]$ is a radical ideal, i.e., $A[T]/(\bar{F})$ is reduced (1.3.1). Let $H \in A[T]$ and assume that H^m is divisible by $\bar{F}$. We may assume that $\deg H < n$. It follows from property (b) that $\bar{F}$ has distinct roots and that H is divisible by $\bar{F}$, as polynomials with coefficients in the quotient field of A. But since H has lower degree than $\bar{F}$, this can only be if $H = 0$, which implies the claim.

By 1.2.6 (2) we know that $Qk[X]$ is an intersection of prime ideals of $k[X]$, say $Qk[X] = \bigcap_{i=1}^{r} P_i$. We may assume that there are no inclusions among the P_i. The irreducible components of the $\phi^{-1}Y'$ are the sets $V_X(P_i)$ (notation of 1.3.2). We show that $P_i \cap k[Y] = Q$ ($1 \leq i \leq r$). If this is not so we have, say, $P_1 \cap k[Y] \neq Q$. Take $x_1 \in P_1 \cap k[Y] - Q$ and $x_i \in P_i - P_1$ ($2 \leq i \leq r$). Then $x_1 x_2, ..., x_r \in Qk[X]$. Since $k[X]$ is free over $k[Y]$, it follows that $x_2, ..., x_r \in Qk[X] \subset P_1$, which is impossible if $r > 1$. If $r = 1$ we have a contradiction, since $Qk[X] \cap k[Y] = Q$.

It follows that the quotient field of $k[X]/P_i$ is an algebraic extension of the quotient field of A, which proves (ii).

If Y' is a point then Q is a maximal ideal of $k[Y]$ and $A = k$. The preceding analysis shows that now $\phi^{-1}Y'$ is the zero dimensional variety defined by the k-algebra $k[T]/(\bar{F})$. Since $\bar{F}$ is a polynomial of degree n with distinct roots, (iii) follows. $\square$

5.1.5. Lemma. *Let $p = \mathrm{char}\, k > 0$ and assume that $f^p \in k(Y)$. There is a non-empty open subset U of X with the following properties:*

(i) *The restriction of ϕ to U is an open morphism $U \to Y$ which induces a homeomorphism $U \simeq \phi U$;*
(ii) *If Y' is an irreducible closed subvariety of Y, there is at most one irreducible component X' of $\phi^{-1}Y'$ that intersects U. If X' exists, we have $\dim X' = \dim Y'$.*

Let $f^p = g$. Replacing X and Y by $D_X(a)$ and $D_Y(a)$ with suitable $a \in k[Y]$, we may assume that $g \in k[Y]$ and that $k[X]$ is free over $k[Y]$. We shall prove that we may then take $U = X$.

We view X as the set of points $(y, g(y)^{1/p})$ in $Y \times \mathbf{A}^1$, ϕ being the first projection. Then $k[X] = k[Y][T]/(T^p - g)$. The proof that ϕ is open is like that of 5.1.4 (i). Since ϕ is bijective (i) follows.

Let Y' be as in (ii) and let Q and A be as in the proof of 5.1.4. Let $\bar{g}$ be the image of g in A. Now $k[X]/Qk[X] \simeq A[T]/(T^p - \bar{g})$. If $T^p - \bar{g}$ is irreducible over the quotient field of A, then $Qk[X]$ is a prime ideal and (ii) follows. If $T^p - \bar{g}$ is reducible then in $A[T]/(T^p - \bar{g})$ the elements b with $b^p = 0$ form a prime ideal. It follows that $P = \{h \in k[X] \mid h^p \in Qk[X]\}$ is a prime ideal, and is the radical of $Qk[X]$. If $h \in P \cap k[Y]$ then $h^p \in Qk[X] \cap k[Y] = Q$. Hence $P \cap k[Y] = Q$ and (ii) follows, as before.   □

The following theorem is the main result of this section. Its content is that a morphism ϕ as in 5.1.1 behaves well on an open set. The exercises (3) and (4) of 5.1.8 show that one cannot expect good behavior everywhere.

5.1.6. Theorem. *Let X and Y be irreducible varieties and let $\phi : X \to Y$ be a dominant morphism. Put $r = \dim X - \dim Y$. There is a non-empty open subset U of X with the following properties:*
(i) *The restriction of ϕ to U is an open morphism $U \to Y$;*
(ii) *If Y' is an irreducible closed subvariety of Y and X' an irreducible component of $\phi^{-1}Y'$ that intersects U, then $\dim X' = \dim Y' + r$. In particular, if $y \in Y$, any irreducible component of $\phi^{-1}y$ that intersects U has dimension r;*
(iii) *If $k(X)$ is algebraic over $k(Y)$, then for all $x \in U$ the number of points of the fiber $\phi^{-1}(\phi x)$ equals $[k(X) : k(Y)]_s$.*

Assume we have a factorization $\phi = \phi' \circ \psi$, where $\psi : X \to Z$, $\phi' : Z \to Y$ are dominant morphisms, Z being irreducible. If (i) and (ii) hold for ϕ' and ψ they also hold for ϕ. To prove the theorem we may assume that X and Y are affine. Since $k[X]$ is a $k[Y]$-algebra of finite type, we can find a factorization of ϕ

$$ X = X_r \xrightarrow{\phi_r} X_{r-1} \xrightarrow{\phi_{r-1}} \dots \xrightarrow{\phi_2} X_1 \xrightarrow{\phi_1} X_0 = Y, $$

where the ϕ_i are cases covered by one of the three preceding lemmas. (i) and (ii) then follow by application of these lemmas.

In the case of (iii) let $k[X] = k[Y][f_1, \ldots, f_s]$ and let Z be the variety with

$$k[Z] = k[Y][f_1^{p^e}, \ldots, f_s^{p^e}],$$

e being such that all $f_i^{p^e}$ are separable over $k(Y)$. We obtain a factorization as in the beginning. Refining it we obtain a factorization of ϕ to which we can apply the last two lemmas.   □

The proof gives the following useful corollary.

5.1.7. Corollary. *In the case of 5.1.6, we may replace (i) by the following stronger property,*
(i)′ For any variety Z the restriction of ϕ to U defines an open morphism $U \times Z \to Y \times Z$.

It suffices to prove this for Z affine. Observe that if (i)′ holds for Z, and if Z' is a closed subvariety of Z, then (i)′ also holds for Z'. Hence it suffices to establish (i)′ for $Z = \mathbf{A}^m$. This will follow if we prove the corresponding result in the cases of 5.1.3, 5.1.4, 5.1.5. The first case is trivial. In the others (i)′ follows by observing that the minimum polynomial of an element of $k(X)$ over $k(Y)(T_1, \ldots, T_m)$ coincides with the minimum polynomial over $k(Y)$.   □

5.1.8. Exercises. (1) Using 5.1.6 (iii) show that an isomorphism $\phi : \mathbf{A}^1 \to \mathbf{A}^1$ is of the form $\phi t = at + b$ ($a \in k^*, b \in k$). Deduce that an isomorphism of the projective line $\mathbf{P}^1$ (see 1.6.13 (2) or 1.7.1) is induced by an element of $\mathbf{GL}_2$.
(2) Let $X = \{(x, y) \in \mathbf{A}^2 \mid x^2 = y^3\}$. Define $\phi : X \to \mathbf{A}^1$ by $\phi(x, y) = x^{-1}y$ if $(x, y) \neq (0, 0)$ and $\phi(0, 0) = 0$. Show that ϕ is a morphism of irreducible varieties that is birational and bijective, but is not an isomorphism of varieties.
(3) Consider the morphism $\phi : \mathbf{A}^2 \to \mathbf{A}^2$ with $\phi(x, y) = (x, xy)$. Show that it is birational, but not open. Determine the components of the fibers $\phi^{-1}z$, $z \in \mathbf{A}^2$.
(4) Define $\phi : \mathbf{A}^3 \to \mathbf{A}^3$ by $\phi(x, y, z) = (x, xy, z)$. Let $X = \{(x, y, z) \in \mathbf{A}^3 \mid y^2 = 1 + x\}$. Show that X and $Y = \phi X$ are irreducible closed subvarieties of $\mathbf{A}^3$ of dimension two. Put

$$Y' = \{(x, y, z) \in Y \mid y = xz, \ z^2 = 1 + x\}.$$

Show that Y' is irreducible, closed of dimension one and that if char $k \neq 2$, $\phi^{-1}Y' \cap X$ has a component of dimension zero.

5.2. Finite morphisms, normality

5.2.1. Let A be a ring and B an A-algebra. We say that B is *finite* over A if B is an A-module of finite type. We say that $b \in B$ is *integral over* A if it satisfies an equation

$$b^n + a_1 b^{n-1} + \ldots + a_n = 0$$

with coefficients in A.

5.2.2. Lemma. *Let B be an A-algebra of finite type. Then B is finite over A if and only if every element of B is integral over A.*

Assume that B is finite over A. There are $b_i \in B$ such that $B = Ab_1 + \ldots + Ab_m$. Let $b \in B$. There exist a_{ij} in A such that

$$bb_i = \sum_{j=1}^{m} a_{ij} b_j, \ 1 \le i \le m.$$

It follows that $\det(\delta_{ij} b - a_{ij}) = 0$, showing that b is integral over A. The proof of the converse statement is straightforward. $\square$

It follows from 5.2.2 that if, moreover, C is a B-algebra that is finite over B, then C is finite over A. It also follows from 5.2.2 that if B is any A-algebra, the elements of B that are integral over A form a subalgebra.

Now let $\phi : X \to Y$ be a morphism of affine varieties. The algebra homomorphism $\phi^* : k[Y] \to k[X]$ makes $k[X]$ into a $k[Y]$-algebra. We say that ϕ is *finite* if $k[X]$ is finite over $k[Y]$. Then $\phi^{-1}(y)$ is finite for all $y \in Y$.

5.2.3. Lemma. *A finite morphism is closed.*

Asume that ϕ is finite. Let $B = k[X]$, $A = k[Y]$. We have $B = (\phi^* A)[b_1, \ldots, b_h]$ and an easy argument (see the proof of 5.1.6) shows that we may assume $h = 1$. Applying 1.9.3 to B and its subring $\phi^* A$, we see that Im ϕ is closed. Applying this to a closed subvariety X' of X and the induced morphism $X' \to Y$, we conclude that $\phi(X')$ is closed. Hence ϕ is closed. $\square$

We say that ϕ is *locally finite* in a point $x \in X$ if there exists a finite morphism $\mu : Y' \to Y$ and an isomorphism ν of an open neighborhood U of x onto an open set in Y' such that $\mu \circ \nu$ is the restriction of ϕ to U.

Let $\psi : Y \to Z$ be another morphism of affine varieties.

5.2.4. Lemma. *If ϕ is locally finite in x and ψ in $\phi(x)$ then $\psi \circ \phi$ is locally finite in x.*

We may assume that $Y = D_{Z'}(f)$, where Z' is finite over Z, with $f \in k[Z']$. If Y' is finite over Y then $k[Y'] = B_f$, where B is integral over $k[Z']$. Hence B is integral over $k[Z]$. It follows that $Y' \simeq D_V(g)$, where V is finite over Z, with $g \in k[V]$. ☐

From now on we assume that X and Y are irreducible and that ϕ is dominant. We view $A = k[Y]$ as a subring of $B = k[X]$.

5.2.5. Lemma. *Assume that there is $b \in B$ with $B = A[b]$. Let $x \in X$. We have the following alternatives:*
(a) $\phi^{-1}(\phi x)$ is finite and ϕ is locally finite in x,
(b) $\phi^{-1}(\phi x) \simeq \mathbf{A}^1$.

We have $B = A[T]/I$, where I is the ideal of the polynomials $f \in A[T]$ with $f(b) = 0$. Let $\epsilon : A \to k$ be the homomorphism defining ϕx. Extend ϵ to a homomorphism $A[T] \to k[T]$ in the obvious manner. If $\epsilon I = \{0\}$ then $k[\phi^{-1}(\phi x)] \simeq k[T]$, whence $\phi^{-1}(\phi x) \simeq \mathbf{A}^1$.

If $\epsilon I \neq \{0\}$ the polynomials in ϵI vanish in $b(x)$; hence ϵI contains non-constant polynomials and no non-zero constants. This implies that $\phi^{-1}(\phi x)$ is finite. It also follows that there is $f \in I$ of the form

$$f_n T^n + \ldots + f_m T^m + \ldots + f_0,$$

where $\epsilon(f_n) = \ldots = \epsilon(f_{m+1}) = 0, \epsilon f_m \neq 0, m > 0$. Put $s = f_n b^{n-m} + \ldots + f_m$. Then $s \neq 0$ and

$$sb^m + f_{m-1} b^{m-1} + \ldots + f_0 = 0.$$

We see that sb is integral over $A[s]$ and that b is integral over the subring $A[s^{-1}]$ of the quotient field of A. But since $s \in A[b]$, it follows that s is integral over $A[s^{-1}]$, i.e., that s is integral over A. Now the assertion of (a) follows by observing that $B_s = A[sb, s]_s$. ☐

5.2.6. Proposition. *Let $x \in X$. If the fiber $\phi^{-1}(\phi x)$ is finite then ϕ is locally finite in x.*

We have $B = A[b_1, \ldots, b_h]$. If $h = 1$ the assertion is true by 5.2.5. We have a factorization of $\phi : X \overset{\psi}{\to} X' \overset{\phi'}{\to} Y$, where $k[X'] = A[b_1]$. Clearly $\psi^{-1}(\psi x)$ is finite. By induction on h we may assume that ψ is locally finite in x. We may then assume that there is a finite morphism of affine varieties $\psi' : X'' \to X'$ such that X is an affine open subset of X'' and that ψ is induced by ψ'.
Put $F = (\phi')^{-1}(\phi x)$. Assume that F is infinite. By 5.2.5 it is isomorphic to $\mathbf{A}^1$. Let C be a component of $(\psi')^{-1}(F)$ of dimension ≥ 1 passing through x. Now $X \cap C$ is an open subset of C containing x, hence must be infinite. But $X \cap C$ lies in the finite set $\phi^{-1}(\phi x)$ and we get a contradiction. Hence the components of $(\psi')^{-1}(F)$ of dimension ≥ 1 do not contain x. Replacing X by a suitable open neigborhood of x

we may assume that no such component exists. Then F is finite. The theorem follows by using 5.2.5 and 5.2.4.   □

5.2.7. Corollary. *In the situation of 5.2.6 we have* dim X = dim Y.

An integral domain A is *normal* if every element of its quotient field that is integral over A lies already in A. A point x of an irreducible variety X is normal if there exists an affine open neighborhood U of x such that $k[U]$ is normal. X is normal if all its points are normal.

The next result is (a version of) *Zariski's main theorem*.

5.2.8. Theorem. *Let $\phi : X \to Y$ be a morphism of irreducible varieties that is bijective and birational. Assume Y to be normal. Then ϕ is an isomorphism.*

Let $x \in X$. Replace X and Y by affine open neighborhoods U of x, respectively V of ϕx. We deduce from 5.2.6 that we may assume U is isomorphic to an affine open subset of an affine variety V' which is finite over V. But our birationality assumption implies that $k(V') \simeq k(V)$. Now the normality of Y implies that that finite morphism $V' \to V$ is in fact an isomorphism. This shows that ϕ is an isomorphism of ringed spaces, hence an isomorphism of varieties (see 1.4.7).   □

5.2.9. Exercises. (1) Let $\phi : X \to Y$ be a finite morphism of affine varieties. Show that for any variety Z the morphism $(\phi, \text{id}) : X \times Z \to Y \times Z$ is finite and closed.
(2) For any field F the polynomial algebra $F[T_1, \ldots , T_n]$ is normal.
(3) An irreducible affine variety X is normal if and only if $k[X]$ is normal.
(4) Let $\phi : X \to Y$ be a surjective morphism of irreducible affine varieties. If ϕ is locally finite in all points of X, then ϕ is finite. (*Hint*: For $f \in k[X]$ there are non-zero $F \in k[Y][T]$ with $F(f) = 0$. Consider the leading coefficients of such F).
(5) In 5.2.8 the normality assumption cannot be omitted.

5.2.10. Lemma. *Let A be a normal integral domain with quotient field F. Let B be an integral domain containing A, which is an A-algebra of finite type. Assume that the quotient field E of B is a separable algebraic extension of F. There is a non-zero element $a \in A$ such that B_a is normal.*

Let $B = A[b_1, \ldots , b_h]$. Using that any non-zero element of B divides a non-zero element of A one sees that it suffices to deal with the case $h = 1$. So assume $B = A[b]$. We may assume that b is integral over A. Then $(1, b, \ldots , b^{n-1})$ is a basis of E over F, where $n = [E : F]$.

Assume that $x = \sum_{i=0}^{n-1} a_i b^i$ $(a_i \in F)$ is integral over A. Then for $0 \leq j < n$

$$Tr(xb^j) = \sum_{i=0}^{n-1} a_i Tr(b^{i+j})$$

(where $Tr : E \to F$ is the trace function) is an element of F that is integral over A, since it is a sum of conjugates of xb^j (see [Jac4, p. 409]), which is integral by the remarks after 5.2.2. One also knows that

$$a = \det(Tr(b^{i+j}))_{0 \le i, j \le n-1} = \Pi_{i<j}(b_i - b_j)^2,$$

where the b_i are the conjugates of b, in a suitable extension of E (see [loc.cit., p. 250]). Clearly, $a \in A$. We have a set of linear equations for the a_i. By Cramer's rule we obtain that $aa_i \in A$. This shows that $x \in B_a$ and it follows that B_a is normal. $\qquad \square$

5.2.11. Proposition. *Let X be an irreducible variety. The set of its normal points is non-empty and open.*

That this set is open is clear from the definitions. Let $E = k(X)$. Using 4.2.10 we see that there is an affine open subset U of X such that (a) $k[U]$ is integral over a subalgebra A that is isomorphic to a polynomial algebra, (b) the quotient field E of B is a separable algebraic extension of the quotient field of A. Application of 5.2.10 (using 5.2.9 (2)) shows that X has normal points. $\qquad \square$

Remark. A result slightly weaker than 5.2.11, which would suffice for our purposes, follows from 4.3.3 (ii) by using that simple points are normal. We did not prove this fact (for a local algebra proof, see [Ma, p. 121]).

5.3. Homogeneous spaces

In the rest of this chapter we shall apply the algebro-geometric results of the preceding sections in cases involving algebraic groups. Let G be an algebraic group (which need not be linear) and let X be a homogeneous space for G (2.3.1). As usual, G^0 is the identity component of G.

5.3.1. Lemma. *(i) Each irreducible component of X is a homogeneous space for G^0;*
(ii) The components of X are open and closed and X is their disjoint union.

Let X' be an orbit of G^0 in X. Since G acts transitively on X, it follows from 2.2.1 (i) that X is the disjoint union of finitely many translates $g.X'$. Each of them is a G^0-orbit and is irreducible. It follows from 2.3.3 (ii) that all G^0-orbits are closed. Now (i) and (ii) readily follow. $\qquad \square$

5.3.2. Theorem. *Let G be an algebraic group and let $\phi : X \to Y$ be an equivariant homomorphism of homogeneous spaces for G. Put $r = \dim X - \dim Y$.*
(i) For any variety Z the morphism $(\phi, \mathrm{id}) : X \times Z \to Y \times Z$ is open;

(ii) If Y' is an irreducible closed subvariety of Y and X' an irreducible component of
$\phi^{-1}Y'$*, then* $\dim X' = \dim Y' + r$*. In particular, if* $y \in Y$ *then all irreducible compo-*
nents of $\phi^{-1}y$ *have dimension* r*;*
(iii) ϕ *is an isomorphism if and only if it is bijective and if for some* $x \in X$ *the tangent*
map $d\phi_x : T_x X \to T_{\phi x} Y$ *is bijective.*

Using 5.3.1 we reduce the proof to the case that G is connected and X, Y are
irreducible. Then ϕ is surjective, hence dominant. Let $U \subset X$ be an open subset with
the properties of 5.1.6 and 5.1.7. Then all translates $g.U$ enjoy the same properties.
Since these cover X we have (i) and (ii).

If ϕ is bijective we conclude from 5.1.6 (iii) that $k(X)$ is a purely inseparable
extension of $k(Y)$. If $d\phi_x$ is surjective for some x, we see from 4.3.7 (ii) that this
extension is also separable. Hence $k(X) = k(Y)$ and ϕ is birational. Using 5.1.2 and
a covering argument, we conclude that ϕ is an isomorphism, proving (iii).   □

5.3.3. Corollary. *Let* $\phi : G \to G'$ *be a surjective homomorphism of algebraic*
groups.
(i) $\dim G = \dim G' + \dim \operatorname{Ker} \phi$*;*
(ii) ϕ *is an isomorphism if and only if* ϕ *and the tangent map* $d\phi_e$ *are bijective.*

View the groups G and G' as homogeneous spaces for G, the first one via left
translations and the second via the action $g.g' = \phi(g)g'$. Now apply the theorem.   □

If G and G' are linear algebraic groups, the condition on the tangent map of (ii)
can be rephrased as: the Lie algebra homomorphism $d\phi : \mathfrak{g} \to \mathfrak{g}'$ is bijective. In 2.1.4
(2) we already encountered an example where ϕ is bijective but is not an isomorphism.

5.3.4. Lemma. *The components of a homogeneous space are normal.*

This follows from 5.2.11 and 5.3.1.   □

5.3.5. Exercises. (1) If char $k = 0$, a homomorphism of algebraic groups is an
isomorphism if and only if it is an isomorphism of abstract groups. Similarly for homogeneous spaces.
(2) (a) Let $\phi : G \to G'$ be a homomorphism of algebraic groups. Then $\dim \operatorname{Ker} \phi +$
$\dim \operatorname{Im} \phi = \dim G$;
   (b) Assume, moreover, that ϕ is surjective, G is connected and $\dim G' = \dim G$.
Then $\operatorname{Ker} \phi$ is a finite subgroup of the center of G. (*Hint*: for $g \in \operatorname{Ker} \phi$ consider the
morphism $x \mapsto xgx^{-1}g^{-1}$).
(3) Let $\phi : G \to G'$ be a homomorphism of connected linear algebraic groups such
that the Lie algebra homomorphism $d\phi : \mathfrak{g} \to \mathfrak{g}'$ is bijective. Show that we have the
case of 2 (b).

5.4. Semi-simple automorphisms

5.4.1. Let G be a connected linear algebraic group and let σ be an automorphism of the algebraic group G. We put

$$G_\sigma = \{x \in G \mid \sigma x = x\};$$

this is a closed subgroup of G. We denote by χ the morphism $G \to G$ with $\chi x = (\sigma x)x^{-1}$. The differential $d\sigma$ is an automorphism of the Lie algebra $\mathfrak{g}$. We put

$$\mathfrak{g}_\sigma = \{X \in \mathfrak{g} \mid d\sigma(X) = X\}.$$

Since $\chi(G_\sigma) = \{e\}$ we have $d\chi(L(G_\sigma)) = \{0\}$. Using 4.4.13 (i) we see that $L(G_\sigma) \subset \mathfrak{g}_\sigma = \operatorname{Ker} d\chi$. Equality need not hold (for an example see 5.4.9 (1)). Let G act on itself by $g.x = (\sigma g)xg^{-1}$ ($g, x \in G$). Then χ defines a dominant morphism $\psi : G \to \overline{\chi G}$ which is G-equivariant, G acting on G by left translations and on $\overline{\chi G}$ in the manner just defined. Also, χG is an orbit in $\overline{\chi G}$ for this action, so is open in $\overline{\chi G}$ by 2.3.3 (i). Using 4.3.3 (ii) we find that e is a simple point of $\overline{\chi G}$. By 5.3.2 (ii) we have $\dim \overline{\chi G} = \dim G - \dim G_\sigma$.

5.4.2. Lemma. *We have $L(G_\sigma) = \mathfrak{g}_\sigma$ if and only if ψ is separable.*

Using 4.4.13 (i) we obtain

$$\dim \operatorname{Im}(d\psi_e) = \dim \mathfrak{g} - \dim \mathfrak{g}_\sigma \le \dim \mathfrak{g} - \dim L(G_\sigma) =$$

$$= \dim G - \dim G_\sigma = \dim \overline{\chi G}.$$

From 4.3.6 we see that ψ is separable if and only if the outer members are equal, and the lemma follows.

$\square$

We say that σ is *semi-simple* if the induced automorphism σ^* of $k[G]$ is semi-simple in the sense of 2.4.7. Such automorphisms can be characterized in another manner.

5.4.3. Lemma. *σ is semi-simple if and only if there is an isomorphism ϕ of G onto a closed subgroup of some $\mathbf{GL}_n$ and a semi-simple element $s \in \mathbf{GL}_n$ normalizing ϕG such that $\phi(\sigma x) = s\phi(x)s^{-1}$ ($x \in G$).*

Let σ be semi-simple. Let ϕ be as in the proof of 2.3.7 (i). We may assume that σ^* stabilizes the space V of that proof. Take for s the automorphism defined by the restriction of σ^* to V; then ϕ and s are as required. Conversely, if ϕ and s are as stated, then σ is semi-simple. (Use that $k[G]$ is a quotient of $k[\mathbf{GL}_n]$ and that the inner automorphism of $\mathbf{GL}_n$ defined by s induces a semi-simple automorphism of $k[\mathbf{GL}_n]$).

5.4.4. Theorem. *Assume σ to be semi-simple.*
(i) χG is closed and the morphism $\psi : G \to \chi G$ is separable;
(ii) $L(G_\sigma) = \mathfrak{g}_\sigma$.

By 5.4.3 we may assume that G is a closed subgroup of $\mathbf{GL}_n$ and that $\sigma x = sxs^{-1}$ $(x \in G)$, where s is a semi-simple element of $\mathbf{GL}_n$. We may assume that s is a diagonal matrix. We first prove the separability of ψ or, equivalently, property (ii) (see 5.4.2). If $G = \mathbf{GL}_n$ then it is easy to check that (ii) holds. If G is arbitrary, extend σ to an automorphism of $\mathbf{GL}_n$, viz. the inner automorphism $\mathrm{Int}(s)$ defined by s and extend χ similarly. Let $X \in T_e \overline{\chi G}$. Since $T_e \overline{\chi G} \subset T_e \chi(\mathbf{GL}_n)$ and since we already know (ii) for $\mathbf{GL}_n$, there is $Y \in \mathfrak{gl}_n$ with $X = d\sigma(Y) - Y$. The semi-simplicity of s implies that $d\sigma$ is a semi-simple automorphism of $\mathfrak{gl}_n$, stabilizing the subspace $\mathfrak{g}$. Because of the semi-simplicity of $d\sigma$, this subspace has a $d\sigma$-stable complement. This implies that we may take Y in $\mathfrak{g}$ and that $d\psi_e$ is surjective, whence the separability of ψ by 4.3.6 (i). It remains to be proved that χG is closed.

Put $m(T) = \prod_{i=1}^{r}(T - a_i)$, the a_i being the distinct eigenvalues of s. Let $S \subset \mathbf{GL}_n$ be the set of elements x with the following properties: (a) x normalizes G, (b) $m(x) = 0$, (c) the characteristic polynomial of the restriction of $\mathrm{Ad}\,x$ to $\mathfrak{g}$ equals that of $d\sigma$. Then S is a closed subset, containing s. Since m has distinct roots, all elements of S are semi-simple.

For $x \in S$, put $G_x = \{g \in G \mid gxg^{-1} = x\}$ and $\mathfrak{g}_x = \{X \in \mathfrak{g} \mid \mathrm{Ad}(x)X = X\}$. By (ii) we have $\dim G_x = \dim \mathfrak{g}_x$. But $\dim \mathfrak{g}_x$ equals the multiplicity of the eigenvalue 1 of the restriction of $\mathrm{Ad}\,x$ to $\mathfrak{g}$, which equals $\dim \mathfrak{g}_\sigma$ by condition (c). It follows that $\dim G_x = \dim G_\sigma$ for all $x \in X$. Now G operates on S by inner automorphisms and by 5.3.2 (ii) (applied to a morphism $g \mapsto gxg^{-1}$ of G to an orbit) all orbits have dimension $\dim G - \dim G_\sigma$. It follows from 2.3.3 (i) and 1.8.2 that all orbits must be closed. Since χG is an orbit we have proved (i). $\qquad\square$

5.4.5. Corollary. *Let $s \in G$ be semi-simple.*
(i) The conjugacy class $C = \{xsx^{-1} \mid x \in G\}$ is closed. The morphism $x \mapsto xsx^{-1}$ is separable;
(ii) Let $Z = \{x \in G \mid xsx^{-1} = s\}$ be the centralizer of s. Then $\mathfrak{g} = (\mathrm{Ad}(s) - 1)\mathfrak{g} \oplus L(Z)$.

Let $\sigma x = s^{-1}xs$. Then σ is a semi-simple automorphism of G. If χ is as before, then $C = s.\mathrm{Im}\chi$ and $Z = G_\sigma$. The assertions follow from 5.4.4. For the last point notice that $\mathrm{Ad}\,s$ is a semi-simple endomorphism of $\mathfrak{g}$. $\qquad\square$

Conjugacy classes in G need not be closed. For an example see 5.4.9 (6).

5.4.6. Assume that D is a diagonalizable linear algebraic group (3.2.1) which acts as a group of automorphisms on our connected linear algebraic group G. This means that G is a D-space (2.3.1) and that for all $d \in D$ the morphism $g \mapsto d.g$ is an automorphism of G. It follows from 3.2.3 that these are semi-simple automorphisms. Let

$$Z_G(D) = \{g \in G \mid d.g = g \text{ for all } d \in D\}.$$

D also acts as a group of automorphisms on the Lie algebra $\mathfrak{g}$. Let

$$\mathfrak{z}_{\mathfrak{g}}(D) = \{X \in \mathfrak{g} \mid d.X = X \text{ for all } d \in D\}.$$

5.4.7. Corollary. $L(Z_G(D)) = \mathfrak{z}_{\mathfrak{g}}(D)$.

We proceed by induction on $\dim G$, starting with the trivial group. If D acts trivially on $\mathfrak{g}$, the assertion follows from 5.4.4 (ii). Otherwise choose $d \in D$ such that the fixed point set of d in $\mathfrak{g}$ is a proper subspace of $\mathfrak{g}$. By 5.4.4 (ii) this subspace is the Lie algebra of the fixed point group G_d of d acting in G. This is a subgroup of smaller dimension. Since D is commutative we have $\dim Z_G(D) = \dim Z_{G_d^0}(D)$. We can now apply induction. $\square$

The corollary applies, in particular, if D is a diagonalizable subgroup of G, acting by inner automorphisms.

Finally we give another application of 5.4.4 (ii).

5.4.8. Corollary. *Assume G to be a connected, nilpotent, linear algebraic group. The set G_s of semi-simple elements is a subgroup of the center of G.*

For $x, y \in G$ write (x, y) for their commutator $xyx^{-1}y^{-1}$. Since G is nilpotent, there is $n > 0$ such that all iterated commutators $(x_1(...(x_n, x_{n+1})...))$ equal e. Let $s \in G$ be semi-simple and let σ be the inner automorphism $\mathrm{Int}(s)$. Then, χ being as in 5.4.1, we have $\chi x = (s, x)$. It follows that $\chi^n G = \{e\}$. Using 4.4.13 (ii) we obtain that $\mathrm{Ad}(s) - 1$ is a linear map of $\mathfrak{g}$ which is both semi-simple and nilpotent, hence it is the zero map. From 5.4.4 (ii) we conclude that σ is trivial, i.e., that s lies in the center of G. The assertion now follows from 2.4.3 (i). $\square$

A more precise result will be given in 6.3.2. It should be noticed that the connectedness assumption in 5.4.8 is essential (see 5.4.9 (5)).

5.4.9. Exercises. (1) Assume char $k = 2$ and let $G = \mathrm{SL}_2$. Denote by σ the inner automorphism of G defined by

$$\begin{pmatrix} 1 & 1 \\ 0 & 1 \end{pmatrix}.$$

Show that $L(G_\sigma) \neq \mathfrak{g}_\sigma$.

(2) Let $G = \mathrm{GL}_n$ and let σ be any inner automorphism of G. Show that $L(G_\sigma) = \mathfrak{g}_\sigma$.

(3) Let G be a connected closed subgroup of $\mathbf{GL}_n$. Assume that the subspace $\mathfrak{g}$ of $\mathfrak{gl}_n$ has a complement that is stable under Ad G.

(a) If σ is an inner automorphism of G then $L(G_\sigma) = \mathfrak{g}_\sigma$. (*Hint*: proceed as in the first paragraph of the proof of 5.4.4.)

(b) Let $x \in G$ and let C be the conjugacy class of x in $\mathbf{GL}_n$. Show that $C \cap G$ consists of finitely many conjugacy classes of G. (*Hint*: let X be a component of $C \cap G$ containing x and let D be the conjugacy class of x in G; using (a) show that $T_x X = T_x D$ and $X = D$).

(4) (a) Show that the number of unipotent conjugacy classes in $\mathbf{GL}_n$ is finite. (Use Jordan normal forms, see [La2, Ch. XV, §3].)

(b) Let σ be an automorphism of $\mathbf{GL}_n$ of finite order prime to $p = \text{char } k$ (if $p > 0$). Using (3) show that the number of unipotent conjugacy classes of G_σ^0 is finite. (The same argument, due to Richardson [Ri], can be used to prove finiteness of the number of unipotent conjugacy classes in other situations, see [St5, p. 106]).

(5) Give an example of an irreducible, nilpotent, finite group of complex 2×2-matrices (showing that 5.4.8 is false for non-connected groups).

(6) (a) Show that (for arbitrary k) the conjugacy class in $\mathbf{GL}_2$ of the matrix of (1) is not closed.

(b) More generally, let $x = 1 + X$ be a unipotent element of $\mathbf{GL}_n$, where X is a nilpotent matrix. Using Jordan normal forms, define a cocharacter $\lambda : \mathbf{G}_m \to \mathbf{GL}_n$ (3.2.1) such that $\lambda(x)X\lambda(x)^{-1} = x^2 X$ $(x \in k^*)$. Show that 1 lies in the closure of the conjugacy class of x.

5.5. Quotients

In this section G denotes a linear algebraic group and H a closed subgroup, with respective Lie algebras $\mathfrak{g}$ and $\mathfrak{h}$. Let F be a subfield of k such that G is an F-group and H is an F-subgroup. We shall establish the existence of a quotient variety G/H. We begin with some auxiliary results.

5.5.1. Lemma. *There exists a finite dimensional subspace V of $k[G]$ together with a subspace W of V such that*
(a) V is stable under all right translations $\rho(x)$ $(x \in G)$;
(b) we have

$$H = \{x \in G \mid \rho(x)W = W\},$$

$$\mathfrak{h} = \{X \in \mathfrak{g} \mid X.W \subset W\};$$

(c) V is defined over F and W is an F-subspace of V.

Let $I \subset k[G]$ be the ideal of functions vanishing on H and let V be a finite dimensional $\rho(G)$-stable subspace of $k[G]$ that is defined over F and contains a set of generators $(f_1, \ldots, f_r)$ of I which lie in $F[G]$ (see 2.3.6). Take $W = V \cap I$. We shall show that the requirements of the lemma are met.

If $x \in H$ then $\rho(x)W = W$ by 2.3.8. Conversely, if this is so, then $\rho(x)f_i \in I$ $(1 \le i \le r)$, whence $\rho(x)I \subset I$. By 2.3.8 we have $x \in H$. The proof of the corresponding Lie algebra property (b) is similar, using 4.4.7 instead of 2.3.8. Observe that, if ϕ is the rational representation of G in V defined by ρ, we have for $f \in V$ and $X \in \mathfrak{g}$ that $d\phi(X).f = X.f$ (see 4.4.15 (2)). (c) is clear from the definitions. $\square$

Now let V be an arbitrary finite dimensional vector space and W a subspace of dimension d. The dth exterior power $\bigwedge^d V$ contains the one dimensional subspace $L = \bigwedge^d W$. Let ϕ be the canonical representation of $GL(V)$ in $\bigwedge^d V$. The differential of ϕ is described in 4.4.15 (5).

5.5.2. Lemma. *(i) Let $x \in GL(V)$. We have $x.W = W$ if and only if $(\phi x)(L) = L$; (ii) Let $X \in \mathfrak{gl}(V)$. We have $X.W \subset W$ if and only if $(d\phi X)(L) \subset L$.*

The 'only if' parts are clear. Choose a basis $(v_1, \ldots, v_n)$ of V such that $(v_1, \ldots, v_d)$ is one of W. The exterior products $v_{i_1} \wedge \ldots \wedge v_{i_d}$ with $i_1 < \ldots < i_d$ form a basis of $\bigwedge^d V$ and $v_1 \wedge \ldots \wedge v_d$ is a basis vector of the one dimensional space L. Let $x \in GL(V)$. We may choose our basis such that, moreover, $v_{l+1}, \ldots, v_{l+d}$ is a basis of $x.W$, for some l. Put $e = v_1 \wedge \ldots \wedge v_d$, $f = v_{l+1} \wedge \ldots \wedge v_{l+d}$. Then $(\phi x)e$ is a multiple of f. If $l > 0$ then e and f are linearly independent and x does not stabilize L. This implies (i).

If $X \in \mathfrak{gl}(V)$ we have by 4.4.15 (5)

$$(d\phi X)e = \sum_{i=1}^{d} v_1 \wedge \ldots \wedge X v_i \wedge \ldots \wedge v_d.$$

Writing $X v_i = \sum a_{ij} v_j$ it follows that

$$(d\phi X)e = \sum_{i=1}^{d} \sum_{j} a_{ij} v_1 \wedge \ldots \wedge v_j \wedge \ldots \wedge v_d,$$

from which we see that, if $a_{ij} \ne 0$ for $i \le d$ and $j > d$, the subspace L is not mapped into itself by $(d\phi X)$, proving (ii). $\square$

The lemmas imply the following result. G and H are as before.

5.5.3. Theorem. *There exists a rational representation $\phi : G \to GL(V)$ over F and a non-zero $v \in V(F)$ such that*

$$H = \{x \in G \mid (\phi x)v \in kv\},$$
$$\mathfrak{h} = \{X \in \mathfrak{g} \mid (d\phi X)v \in kv\}.$$

The following corollary is important for the construction of G/H.

5.5.4. Corollary. *There is a quasi-projective homogeneous space X for G, together with a point $x \in X$ such that:*
(a) the isotropy group of x in G is H,
(b) the morphism $\psi : g \mapsto g.x$ of G to X defines a separable morphism $G^0 \to \psi G^0$;
(c) the fibers of ψ are the cosets gH $(g \in G)$.

Recall that a quasi-projective variety is an open subvariety of a projective variety (1.7.1).

Let V and v be as in the theorem and let x be the point in the projective space $\mathbf{P}(V)$ defined by the line kv (see 1.7.2 (2)). Denote by $\pi : V - \{0\} \to \mathbf{P}(V)$ the map sending a vector to the line passing through it. Now G acts on $\mathbf{P}(V)$ by $g.\pi(v) = \pi(\phi(g)v)$. Denote by X the G-orbit of x. By 2.3.3 (i) it is a quasi-projective variety. Now (a) and (c) follow from 5.5.3. The property (b) is a consequence of the Lie algebra assertion of 5.5.3 and 4.3.7 (ii). □

A *quotient of G by H over F* is a pair $(G/H, a)$ of a homogeneous space G/H for G over F together with a point $a \in (G/H)(F)$ such that the following universal property holds:

for any pair (Y, b) of a G-space Y for G over F and a point $b \in Y(F)$ whose isotropy group contains H, there exists a unique equivariant F-morphism $\phi : G/H \to Y$ such that $\phi a = b$..

In the next theorem we prove the existence of a quotient over k, i.e., we ignore the ground field F. The existence of quotients over F will be established in 12.2, as an application of the criteria for ground fields of 11.2.

5.5.5. Theorem. *A quotient $(G/H, a)$ over k exists and is unique up to a G-isomorphism. In fact, if X and x are as in 5.5.4 then (X, x) is such a quotient.*

The uniqueness part of the theorem is trivial. To prove the existence we first define $(G/H, a)$ in the category of ringed spaces and then show it to be isomorphic as a ringed space to the pair (X, x) of 5.5.4. The points of our ringed space G/H are the cosets gH $(g \in G)$ and a is the coset H. Let $\pi : G \to G/H$ be the canonical map. We define $U \subset G/H$ to be open if $\pi^{-1}U$ is open in G. This defines a topology on G/H such that π is an open map. We define a sheaf $\mathcal{O}$ of k-valued functions on G/H (see 1.4.2) as follows: if $U \subset G/H$ is open then $\mathcal{O}(U)$ is the ring of functions f on U such that $f \circ \pi$ is regular on $\pi^{-1}U$ (check that this defines indeed a sheaf of functions).

G acts transitively on G/H by left translations (set-theoretically). If $x \in G$ the map $gH \mapsto xgH$ defines an isomorphism of the ringed space $(G/H, \mathcal{O})$. It is a straightforward matter to verify that, if (Y, b) is as in the definition of a quotient, there is a unique G-morphism of ringed spaces $\phi : G/H \to Y$ with $\phi a = b$. (We have $\phi(gH) = g.b.$).

Let X, x and ψ be as in 5.5.4. We have, in particular, a G-morphism of ringed spaces $\phi : G/H \to X$ with $\phi(gH) = g.x$. We shall prove that this is an isomorphism of ringed spaces, which will imply that the ringed space G/H is an algebraic variety with the required properties.

First observe that by property (c) of 5.5.4 ϕ is a continuous bijection. If $U \subset G/H$ is open then $\phi U = \psi(\pi^{-1}U)$ is open by 5.3.2 (i). It follows that ϕ is a homeomorphism of topological spaces.

To prove that ϕ is an isomorphism of ringed spaces, the following has to be established: if U is open in X, the homomorphism of k-algebras $\mathcal{O}_X(U) \to \mathcal{O}(\phi^{-1}(U))$ defined by ϕ is an isomorphism. By the definition of $\mathcal{O}$ this means that, for any regular function f on $V = \psi^{-1}U$ such that $f(gh) = f(g)$ ($g \in V, h \in H$), there is a unique regular function F on U such that $F(\psi g) = f(g)$, if $g \in V$. As a consequence of 5.3.1 (ii) we may assume G to be connected.

Let $\Gamma = \{(g, f(g)) \mid g \in V\} \subset V \times \mathbf{A}^1$ be the graph of f and put $\Gamma' = (\psi, \mathrm{id})\Gamma$, so $\Gamma' \subset U \times \mathbf{A}^1$. Since Γ is closed in $V \times \mathbf{A}^1$ (1.6.11 (i)), application of 5.3.2 (i) shows that

$$(\psi, \mathrm{id})(V \times \mathbf{A}^1 - \Gamma) = U \times \mathbf{A}^1 - \Gamma'$$

is open in $U \times \mathbf{A}^1$. Hence Γ' is closed in $U \times \mathbf{A}^1$. Let $\lambda : \Gamma' \to U$ be the morphism induced by the first projection. It follows from the definitions that λ is bijective. From property (b) of 5.5.4 we see that λ is separable. By 5.1.6 (iii) λ is birational. U is normal by 5.3.4. Now Zariski's main theorem (5.2.8) shows that λ is an isomorphism. This implies that there is a regular function F on U such that $\Gamma' = \{(u, F(u) \mid u \in U\}$, which is what we wanted to prove. This finishes the proof of 5.5.5. $\square$

5.5.6. Corollary. *(i) G/H is a quasi-projective variety of dimension* $\dim G - \dim H$; *(ii) If G is connected, the morphism $g \mapsto g.a$ of G to G/H is separable.*

The quasi-projectivity of G/H and the assertion about separability follow from 5.5.4. The dimension formula is a consequence of 5.3.2 (ii). $\square$

5.5.7. Comments. Assume that H is a linear algebraic group and X an H-variety. We let H act on the right. For simplicity assume X to be irreducible. As in the proof of 5.5.5, we can construct in this situation a ringed space $(X/H, \mathcal{O})$, together with a continuous map of ringed spaces $\phi : X \to X/H$. We say that *the quotient X/H exists* if (a) ϕ is open and separable, (b) the ringed space X/H is an algebraic variety. Easy examples (e.g. $\mathbf{G}_m$ acting linearly in $\mathbf{A}^1$) show that in general these properties do not hold. See also [Bo3, p. 95]. A situation in which a quotient exists is given in the next lemma. Let G and H be as before and let $\pi : G \to G/H$ be the canonical morphism. Let U be an open subset of G/H. A *section* (of π) on U is a morphism $\sigma : U \to G$ such that $\pi \circ \sigma = \mathrm{id}_U$. In that case $(x, h) \mapsto \sigma(x)h$ defines an isomorphism of varieties $U \times H \simeq \pi^{-1}U$. We say that π *has local sections* if G/H is covered by open sets on each of which σ has a section. Assume that X is a variety with a left H-action. Let H act on $G \times X$ by $(g, x).h = (gh, h^{-1}x)$.

5.5.8. Lemma. *Assume that π has local sections. Then the quotient $(G \times X)/H$ exists.*

One obtains the quotient by glueing together varieties $U \times X$, where U runs through a finite open covering of G/H by open sets on which π has a section. We leave the details to the reader.   □

The quotient is denoted by $G \times^H X$. One has a morphism $G \times^H X \to G/H$ with local sections, whose fibers are isomorphic to X. The quotient is the *fibre bundle over G/H associated to X*.

5.5.9. Exercises. G is a linear algebraic group and H a closed subgroup.
(1) The following properties are equivalent: (a) G/H is irreducible, (b) H meets all components of G, (c) $G = G^0 H$.
(2) An open subvariety of an affine algebraic variety is called a *quasi-affine variety*. If all homomorphisms of algebraic groups $H \to \mathbf{G}_m$ are trivial, then G/H is quasi-affine.
(3) Let $G = \mathbf{SL}_2$.
   (a) If H is the subgroup of upper triangular matrices then G/H is isomorphic to the projective line $\mathbf{P}^1$.
   (b) If H is the subgroup of upper triangular unipotent matrices then G/H is isomorphic to $\mathbf{A}^2 - \{0\}$ (a quasi-affine variety that is not affine, see 1.6.13 (3)).
(4) Let H_i be a closed subgroup of the linear algebraic group G_i, $i = 1, 2$. Show that there is an isomorphism of $G_1 \times G_2$-spaces $G_1 \times G_2/H_1 \times H_2 \simeq G_1/H_1 \times G_2/H_2$.
(5) Let H and K be connected closed subgroups of the linear algebraic group G.
   (a) Show that HK is an irreducible quasi-affine subvariety of G of dimension

$$\dim H + \dim K - \dim(H \cap K).$$

(*Hint*: let $H \times K$ act on G by $(x, y).g = xgy^{-1}$.)
   (b) Let $\pi : G \to G/K$ be the canonical morphism. Show that the restriction of π to H is a separable morphism if and only if $\dim(L(H) \cap L(K)) = \dim(H \cap K)$.
   (c) If char $k = 0$ then $L(H \cap K) = L(H) \cap L(K)$.
   (d) Let char $k = 2$, $G = \mathbf{SL}_2$, with H the subgroup of diagonal matrices and $K \neq H$ a conjugate of H. Show that $L(H \cap K) \neq L(H) \cap L(K)$.

(6) Assume that char $k = 0$. Let $\mathfrak{a}$ be a Lie subalgebra of $L(G)$. Show there is a unique smallest closed subgroup of G whose Lie algebra contains $\mathfrak{a}$. (More about these matters can be found in [Bo3, Ch. II, §7]).

(7) (a) Let char $k \neq 2$. Define a bijective morphism of $\mathbf{GL}_n/\mathbf{O}_n$ onto the space of symmetric $n \times n$-matrices. Show that $\dim \mathbf{O}_n = \frac{1}{2}n(n-1)$.

(b) Prove by a similar argument that $\dim \mathbf{Sp}_{2n} = n(2n+1)$.

(One needs some results about alternate and symmetric matrices, see [Jac4, p. 334, p. 338]).

(8) Let $A = \{f \in k[G] \mid f(gh) = f(g)$ for all $g \in G, h \in H\}$.

(a) If G/H is affine then $k[G/H] = A$.

(b) For G/H to be affine it is necessary and sufficient that A be a k-algebra of finite type that separates the cosets of H. This means that if xH and yH are distinct cosets there is $f \in A$ such that the restrictions of f to these cosets, are different constants. (*Hint*: if A has these properties, let X be an affine algebraic variety with $k[X] = A$. Show that left translations induce a structure of homogeneous space on X that has the properties of 5.5.5.)

We next discuss the quotient by a normal subgroup.

5.5.10. Proposition. *Let G be a linear algebraic group and H a closed, normal subgroup.*

(i) G/H is an affine variety;

(ii) Provided with the usual group structure, G/H is a linear algebraic group.

Make G/H into a homogeneous space for $G \times G$ by $(x, y).gH = xgy^{-1}H$. The isotropy group of the coset H contains $H \times H$. Using the universal property of quotients and 5.5.9 (4), we see that $(xH, yH) \mapsto xy^{-1}H$ defines a morphism of varieties $G/H \times G/H \to G/H$, which makes G/H into an algebraic group. It remains to be proved that G/H is affine. For this we use 5.5.3. Let V, ϕ and v be as in 5.5.3. For any character χ of H put

$$V_\chi = \{x \in V \mid \phi(h)x = \chi(h)x \text{ for all } h \in H\}.$$

From the linear independence of characters, it follows that the subspace V' of V spanned by the V_χ is the direct sum of the non-zero V_χ. Since H is a normal subgroup, G permutes the V_χ, hence stabilizes V'. We may assume that $V = V'$. Let W be the vector space of linear maps of V stabilizing each V_χ. Define a morphism $\psi : G \to GL(W)$ by

$$\psi(x)f = \phi(x)f\phi(x)^{-1} \quad (x \in G, f \in W).$$

If $\psi(x) = \mathrm{id}$, then $\phi(x)$ commutes with all $f \in W$, which implies that $\phi(x)$ acts as a scalar multiplication in each V_χ. This implies that $x \in H$. Hence ψ induces an injective map $\lambda : G/H \to GL(W)$ that is, in fact, a homomorphism of algebraic groups. By 2.2.5 (ii) the image of λ is a closed subgroup of $GL(W)$. We shall show that λ is an isomorphism of G/H onto that image, which will finish the proof.

Using 5.3.3 (ii) one sees that it suffices to prove: if $X \in L(G)$ and $d\psi_e(X) = 0$, then $X \in L(H)$. Now it follows from 4.4.15 (1) that $d\psi_e(X)(f) = d\phi_e(X).f - f.d\phi_e(X)$. Consequently, if $d\psi_e(X) = 0$, we have that $d\phi_e(X)$ commutes with all $f \in W$, and it follows that $X \in L(H)$. $\qquad\qquad\qquad\qquad\qquad\qquad\qquad\qquad\qquad\qquad\qquad$ $\square$

5.5.11. Exercises. G is a linear algebraic group.
(1) Let H and N be closed subgroups of G, with $N \subset H$. If N is normal then $G/H \simeq (G/N)/(H/N)$.
(2) Show that the character group $X^*(G)$ of G (3.2.1) is a finitely generated abelian group. (*Hint*: reduce to the case that G is connected and consider $G/(G, G)$.)

Notes

In 5.1 and 5.2 we establish by elementary means a number of algebro-geometric results, mainly dealing with 'generic' properties, which hold on dense open sets. We did not need to go into the theory of local rings.

The material of these sections is needed in the theory of algebraic groups, for example in the construction of quotients in 5.5. But we have kept to the strictly necessary. For example, in the situation of 5.1.6 we did not give a proof of the inequality $\dim \phi^{-1}y \geq r$ for all $y \in Y$. We only prove the weaker result 5.2.7. A thorough treatment of these matters can be found in [EGA], see e.g. [loc.cit., Ch.IV, 13.1]. This requires a considerable amount of commutative algebra.

There is a connection between the 'equidimensionality' property of 5.1.6 (ii) and the 'universal openness' property of 5.1.7. If Y is normal they are equivalent (theorem of Chevalley, see [loc.cit.,14.4.1]).

The examples of 5.1.8 (3),(4) are taken from [Bou1, ex.13, p.80]. The argument of the second part of the proof of 5.2.5 is due to Chevalley [Ch3, p.177].

The discussion of normality in 5.2 is kept as brief as possible. We need Zariski's main theorem in the construction of quotients in 5.5. The argument of the proof of 5.4.4 goes back to [BoT1, §10].

The construction of the quotient of an algebraic group G by a closed subgroup is a–rather delicate–essential part of the theory. The idea to use 5.5.3 in the case that G is linear is due to Borel (see[Bo3, Ch. II, §2]). 5.5.3 is, essentially, due to Chevalley. See [Ch2, Ch. II, §2] and [Ch4, exp. 4, p. 03]. The quotient of any (not necessarily linear) algebraic group by a closed subgroup always exists, also over a ground field F. See for example [We2].

Chapter 6

Parabolic Subgroups, Borel Subgroups, Solvable Groups

In this chapter basic ingredients of the theory of linear algebraic groups are introduced: maximal tori, Borel groups, parabolic subgroups. Fundamental results are the conjugacy theorems for Borel groups and maximal tori (6.2.7 and 6.4.1). The structure theory of connected solvable groups is also treated. The chapter begins with a brief discussion of complete varieties.

6.1. Complete varieties

In this section $X, Y, \ldots$ are algebraic varieties over the algebraically closed field k.

6.1.1. The algebraic variety X is said to be *complete* if for any variety Y the projection morphism $X \times Y \to Y$ is closed, i.e. maps closed sets onto closed sets. One might view the notion of completeness as an analogue, in the category of algebraic varieties, of the notion of compactness in the category of locally compact topological spaces (see exercise 6.1.7 (1)). The example in 1.9.1 shows that the affine line $\mathbf{A}^1$ is not complete. We shall see presently that projective varieties (introduced in 1.7.1) are complete.

6.1.2. Proposition. *Let X be complete.*
(i) A closed subvariety of X is complete;
(ii) If Y is complete then so is $X \times Y$;
(iii) If $\phi : X \to Y$ is a morphism then ϕX is closed and complete;
(iv) If X is a subvariety of Y then X is closed;
(v) If X is irreducible then any regular function on X is constant;
(vi) If X is affine then X is finite.

(i) and (ii) follow from the definition of completeness. To prove (iii) let $\Gamma = \{(x, \phi x) \in X \times Y \mid x \in X\}$ be the graph of ϕ. It is a closed subset of $X \times Y$ (1.6.11 (i)) that is isomorphic to X, hence is complete. ϕX is closed because it is the image of Γ under the second projection. Its completeness follows from the completeness of Γ.

(iv) is a consequence of (iii), applied to the injection $X \to Y$. A regular function on X is a morphism $X \to \mathbf{A}^1$ that defines a morphism $\phi : X \to \mathbf{P}^1$. If X is irreducible and ϕ non-constant then ϕX is a non-empty dense subset which by (iv) must be $\mathbf{P}^1$. This is impossible, whence (v). Finally, (vi) is a consequence of (v). $\qquad\square$

6.1.3. Theorem. *A projective variety is complete.*

By 6.1.2 (i) it suffices to prove that projective n-space $\mathbf{P}^n$ is complete, i.e. that for any variety Y the projection morphism $\pi : \mathbf{P}^n \times Y \to Y$ is closed. It suffices to deal with the case that Y is affine and irreducible. Assume this. Put $A = k[Y]$, $S = A[T_0, \ldots, T_n]$. We can view S as an algebra of functions on $k^{n+1} \times Y$. If I is a proper homogeneous ideal in S (defined as in 1.7.3) put

$$ \mathcal{V}^*(I) = \{(x^*, y) \mid f(x, y) = 0 \text{ for all } f \in I\}. $$

Here x^* denotes, as in 1.7.1, the point of $\mathbf{P}^n$ defined by $x \in k^{n+1} - \{0\}$. We have results like those of 1.7.4 and 1.7.5 (2): the $\mathcal{V}^*(I)$ are the closed subsets of $\mathbf{P}^n \times Y$, $\mathcal{V}^*(I) = \emptyset$ if and only if there is $h > 0$ such that $T_i^h \in I$ ($0 \le i \le n$) and $\mathcal{V}^*$ is irreducible if and only if $\sqrt{I}$ is a prime ideal.

We have to show that all sets $\pi \mathcal{V}^*(I)$ are closed. We may assume I to be a proper prime ideal (use 1.2.4). We may also assume that the restriction of π to $\mathcal{V}^*(I)$ is dominant, which means that $A \cap I = 0$. Under these assumptions we have to prove the following: if $y \in Y$ there is $x^* \in \mathbf{P}^n$ with $(x^*, y) \in \mathcal{V}^*(I)$. Let M be the maximal ideal in A of functions vanishing in y. Then $J = MS + I$ is a proper homogeneous ideal in S. What we have to prove is that $\mathcal{V}^*(J) \ne \emptyset$. Assume that $\mathcal{V}^*(J) = \emptyset$. Then there is $h > 0$ such that $T_i^h \in J$ for all i, or equivalently, there is $l > 0$ such that the set $S_l \subset S$ of homogeneous polynomials of degree l lies in J. Put $N = S_l/S_l \cap I$. This is an A-module of finite type; let $(n_1, ..., n_r)$ be a set of generators. Our assumptions imply that $N = MN$. Hence there are $m_{ij} \in M$ such that

$$ n_i = \sum_{i=1}^{r} m_{ij} n_j \ (1 \le i \le r). $$

Put $a = \det(\delta_{ij} - m_{ij})$. Then $a n_i = 0$ for all i, so $aN = 0$ and $aS_l \subset I$. We have $a - 1 \in M$. It follows that $a \notin I$ (otherwise we had $J = S$). Since I is a prime ideal we conclude that $S_l \subset I$, i.e. that $N = 0$. But this means that $\mathcal{V}^*(I) = \emptyset$, which is absurd. (The statement that $MN = N$ implies $N = 0$ is a version of Nakayama's lemma, see [La2, Ch.IX,§1].) It follows that $\mathcal{V}^*(J) \ne 0$, which we had to prove. $\square$

A *curve* is an algebraic variety of dimension one. Let C be an irreducible smooth curve. Another property of complete varieties is that a morphism of a non-empty open set of C to a complete variety can be extended to all of C. In the next lemma we prove this in a particular case. (For a proof of the general statement, which requires facts from curve theory, see [Sha, p.265]). As usual, F is a subfield of k.

6.1.4. Lemma. *Let U be a non-empty F-open subset of $\mathbf{P}^1$, X a projective F-variety and $\phi : U \to X$ an F-morphism. Then ϕ extends to an F-morphism $\mathbf{P}^1 \to X$.*

For F-structures on projective varieties see 1.7.2 (3) and 1.7.5 (3). It suffices to prove the lemma for $X = \mathbf{P}^n$. It then follows that ϕ can be described as follows. For

$x \in U \cap \mathbf{A}^1$ we have

$$\phi(x) = (\phi_o(x), \dots , \phi_n(x))^*,$$

where the ϕ_i $(0 \le i \le n)$ are rational functions in $F(T)$, not all zero, having poles outside U. Now if $x \in k$ we can write, by multiplying the ϕ with a suitable rational function

$$\phi(x) = (\psi_0(x), \dots , \psi_n(x))^* \ (x \in U \cap \mathbf{A}^1),$$

where the ψ_i are polynomials not all vanishing in x. This shows that ϕ is regular in all points of $\mathbf{A}^1$. Similarly, ϕ is regular in the remaining point ∞ of $\mathbf{P}^1$ (see 1.6.13 (2)). $\square$

The last results of this section will not be needed in the sequel.

6.1.5. Lemma. *Let X, Y, Z be irreducible varieties, with X complete. Assume that $\phi : X \times Y \to Z$ is a morphism and put $\phi_y(x) = \phi(x, y)$ $(x \in X, y \in Y)$. If there is $a \in Y$ such that ϕ_a is constant, then ϕ_y is constant for all $y \in Y$.*

Let Γ be the graph of ϕ, it is a closed subset of $X \times Y \times Z$. Since X is complete the projection of Γ on $Y \times Z$, i.e. the set

$$A = \{(y, \phi(x, y)) \mid x \in X, y \in Y\}$$

is closed. It is also irreducible. Let $\pi : A \to Y$ be the projection. Then $\pi^{-1}a$ consists of one point, whence by 5.2.7 $\dim A = \dim Y$. If $x \in X$ the subset $A_x = \{(y, \phi(x, y) \mid y \in Y\}$ is closed, being the graph of a morphism $Y \to Z$. It is isomorphic to Y, hence is irreducible. Its dimension equals $\dim A$. This implies that $A_x = A$ for $x \in X$, which means that ϕ_y is constant for all $y \in Y$. $\square$

6.1.6. Theorem. *Let G be a connected algebraic group that is complete as a variety. Then G is commutative.*

Apply 6.1.5 with $X = Y = Z = G$, $\phi(x, y) = xyx^{-1}$ and $a = e$. $\square$

The groups of the theorem are the *abelian varieties*. An example was given in 2.1.4 (5). For more about abelian varieties we refer to [Mu1].

6.1.7. Exercises. (1) Let X be a locally compact Hausdorff space. Show that X is compact if and only if for any locally compact space Y the projection morphism $X \times Y \to Y$ is closed. (*Hint:* use a one point compactification of X, i.e. a compact space $\hat{X}$ containing X as a subspace such that $\hat{X} - X$ consists of one point).
(2) Let G be a connected algebraic group.
(a) A closed, connected, complete subgroup of G lies in the center.
(b) There exists a unique maximal subgroup with the properties of (a).

6.2. Parabolic subgroups and Borel subgroups

In this section G is a linear algebraic group over k.

6.2.1. Lemma. *Let X and Y be homogeneous spaces for G and $\phi : X \to Y$ a bijective G-morphism. Then X is complete if and only Y is complete.*

It follows from 5.3.2 (i) that for any variety Z the map $(\phi, \mathrm{id}) : X \times Z \to Y \times Z$ is a homeomorphism of topological spaces. The lemma is a consequence of this fact. $\square$

A closed subgroup P of G is a *parabolic subgroup,* or is parabolic, if the quotient variety G/P is complete.

6.2.2. Lemma. *If P is a parabolic subgroup of G then G/P is a projective variety.*

This follows from 6.1.2 (iv), recalling that by 5.5.6 (i) G/P is a quasi-projective variety, i.e. an open subvariety of a projective variety. $\square$

6.2.3. Lemma. *Let P be parabolic in G and Q parabolic in P. Then Q is parabolic in G.*

We have to show that, for any variety X, the projection map $G/Q \times X \to X$ is closed. Using 5.3.2 (i) one sees that this is tantamount to proving that, for any closed set $A \subset G \times X$ such that $(g, x) \in A$ implies $(gQ, x) \subset A$, the projection A' of A in X is closed. Consider the morphism $\alpha : P \times G \times X \to G \times X$ with $\alpha(p, g, x) = (gp, x)$. If A is as above then $\alpha^{-1}A = \{(p, g, x) \mid (gp, x) \in A\}$, which is closed in $P \times G \times X$. The completeness of P/Q implies that the projection of this set in $G \times X$, i.e. the set $\bigcup_{(g,x) \in A} (gP, x)$, is closed. By the completeness of G/P the projection of this set in X is closed. Since this projection is A' the lemma is proved. $\square$

6.2.4. Lemma. *(i) Let P be a parabolic subgroup of G. If Q is a closed subgroup of G containing P then Q is parabolic;*
(ii) P is parabolic in G, if and only if P^0 is parabolic in G^0.

(i) follows from 6.1.2 (iii), as G/Q is the image of G/P under a morphism. To prove (ii) first observe that G^0 is parabolic in G. If P is parabolic in G, then P^0 is parabolic in G by the previous lemma and also in G^0, since G^0/P^0 is closed in G/P^0. If, conversely, P^0 is parabolic in G^0 then it is parabolic in G by the previous lemma and P is parabolic in G by (i). $\square$

6.2.5. Proposition. *A connected group G contains proper parabolic subgroups if and only if G is non-solvable.*

Assume that G is a closed subgroup of some $GL(V)$. Then G acts on the projective space $\mathbf{P}(V)$. Let X be a closed orbit for this action (2.3.3 (ii)). Then X is a projective variety, which is complete by 6.1.3. Take $x \in X$ and let P be its isotropy group. Then $gP \mapsto g.x$ defines a bijective morphism of homogeneous spaces $G/P \to X$ and 6.2.1 shows that P is a parabolic subgroup. If $P = G$ put $V_1 = V/x$. Then G acts on $\mathbf{P}(V_1)$. There is a closed orbit for this action, whence a parabolic subgroup P_1. If $P_1 = G$ continue in this manner. We either obtain a proper parabolic subgroup or we have that G is isomorphic to a group of upper triangular matrices, hence is solvable (2.1.5 (4)).

To finish the proof we have to show that, if G is connected and solvable, it has no proper parabolic subgroups. Assume that P is one of maximal dimension. By 6.2.4 (ii) we may assume that P is connected. The commutator subgroup (G, G) is a closed connected subgroup (2.2.8 (i)) and $Q = P.(G, G)$ is a connected parabolic subgroup containing P. By our assumptions we have either $Q = G$ or $Q = P$. In the first case we have a bijection of homogeneous spaces for (G, G)

$$(G, G)/(G, G) \cap P \to G/P.$$

By 6.2.1 it follows that $(G, G) \cap P$ is parabolic in (G, G). By induction on $\dim G$ we may assume that $(G, G) \cap P = (G, G)$, i.e. that $(G, G) \subset P$, which contradicts the assumption that $Q = G$. In the case that $Q = P$ we have again $(G, G) \subset P$. But then P is a normal subgroup of G, and G/P is affine by 5.5.10 (i). Using 6.1.2 (vi) we obtain a contradiction. $\square$

6.2.6. Theorem. *(Borel's fixed point theorem) Let G be a connected solvable linear algebraic group and X a complete G-variety. There exists a point in X that is fixed by all elements of G.*

By 2.3.3 (ii) G has a closed orbit in X. The isotropy group of a point of that orbit is parabolic. By 6.2.5 this group must be all of G, whence the theorem. $\square$

A *Borel subgroup* of G is a closed, connected, solvable, subgroup of G, which is maximal for these properties. Such subgroups exist (take one of maximal dimension).

6.2.7. Theorem. *(i) A closed subgroup of G is parabolic if and only if it contains a Borel subgroup;*
(ii) A Borel subgroup is parabolic;
(iii) Two Borel subgroups of G are conjugate.

By 6.2.4 (ii) we may assume G to be connected. Let B be a Borel subgroup and P any parabolic subgroup. Applying 6.2.6 to B and the complete variety G/P we see that P contains a conjugate of B, which is also a Borel group. To finish the proof of (i) it suffices by 6.2.4 (i) to prove (ii). We may assume G to be non-solvable. By 6.2.5 there exists a proper parabolic subgroup P. By what we already proved we may assume that $B \subset P$. Clearly, B is a Borel subgroup of P. By induction on $\dim G$ we may assume that B is parabolic in P. Now (ii) follows from 6.2.3.

If B and B' are two Borel subgroups then B' is conjugate to a subgroup of B and B to a subgroup of B'. Hence $\dim B = \dim B'$, which implies (iii). $\square$

6.2.8. Corollary. *Let $\phi : G \to G'$ be a surjective homomorphism of linear algebraic groups. Let P be a parabolic subgroup (respectively: a Borel subgroup) of G. Then ϕP is a subgroup of G' of the same type.*

By 6.2.7 (i) it suffices to deal with the case of a Borel group. Then ϕP is closed, connected, solvable. Moreover, the morphism $G/P \to G'/\phi P$ induced by ϕ is surjective. By 6.1.2 (iii) $G'/\phi P$ is complete, so ϕP contains a Borel group of G' by 6.2.7 (i). It follows that ϕP is a Borel group. $\square$

Let B be a Borel group of G. Denote by $C(G)$ the center of G. It is a closed subgroup.

6.2.9. Corollary. *If G is connected then $C(G)^0 \subset C(B) \subset C(G)$.*

$C(G)^0$ is closed, connected and commutative, so lies in a Borel group. By 6.2.7 (iii) it lies in all Borel groups, whence the first inclusion. $\square$
    If $g \in C(B)$ the morphism $x \mapsto gxg^{-1}x^{-1}$ induces a morphism $G/B \to G$, which must be constant by 6.1.2 (vi), whence the second inclusion.

6.2.10. Corollary. *If B is nilpotent then $G^0 = B$.*

A connected nilpotent group G contains a non-trivial closed, connected group H in its center, viz. the subgroup generated by the non-trivial iterated commutators of maximal length (H is closed and connected by 2.2.8 (i)). It follows from 6.2.9 that H is a normal subgroup of G. Passing to G/H the corollary follows by induction on $\dim G$.
                              $\square$

6.2.11. Exercises. G is a connected linear algebraic group.
(1) Let $G = \mathbf{GL}_n$.
(a) The subgroup $\mathbf{T}_n$ of upper triangular matrices is a Borel subgroup of G.
(b) Let G act on $V = k^n$. A *flag* in V of length s is a sequence of distinct subspaces

V_i $(1 \le i \le s)$ of V with $\{0\} \ne V_1 \subset V_2 \subset \ldots \subset V_s$. The flag is *complete* if $s = n - 1$ (in which case dim $V_i = i$ for all i). The Borel subgroups of G are the subgroups that fix a given complete flag. (*Hint:* for (a): Let G act on $\mathbf{P}^{n-1}$. Using that the isotropy group of a point of $\mathbf{P}^{n-1}$ is a parabolic subgroup, prove by induction on n that $G/\mathbf{T}^n$ is complete).

(2) Assume that char $k \ne 2$. Let $V = k^n$ and define a non-degenerate symmetric bilinear form $(\,,\,)$ on V by

$$((x_1, \ldots, x_n), (y_1, \ldots y_n)) = \sum_{i=1}^{n} x_i y_i.$$

Then the orthogonal group $G = \mathbf{O}_n$ (2.1.4 (4)) is the set of $g \in \mathbf{GL}_n$ such that $(gx, gy) = (x, y)$ for $x, y \in V$. A subspace W of V is *isotropic* if the restriction of the form to W is zero. A *complete isotropic flag* is a flag whose subspaces are all isotropic, of maximal length (which is $[n/2]$). The Borel subgroups of G are the subgroups fixing a given complete isotropic flag. (*Hint:* proceed as in the preceding exercise. One has to use some results from the theory of quadratic forms, see [Jac4, 6.5]).

(3) Let $V = k^{2n}$ and define a non-degenerate alternating bilinear form $\langle\,,\,\rangle$ on V by

$$\langle(x_1, \ldots, x_{2n}), (y_1, \ldots, y_{2n})\rangle = \sum_{i=1}^{n}(x_i y_{n+i} - x_{n+i} y_i).$$

Then the symplectic group $G = \mathbf{Sp}_{2n}$ (2.1.4 (4)) is the set of $g \in \mathbf{GL}_{2n}$ such that $\langle gx, gy \rangle = \langle x, y \rangle$ for $x, y \in V$. Give a description of the Borel groups similar to the one of the preceding exercise (for the required results about alternating forms see [Jac4, 6.9]).

(4) Let B be a Borel subgroup of G. If σ is an automorphism of the algebraic group G that fixes all elements of B then $\sigma = \mathrm{id}$.

(5) Let P and Q be parabolic subgroups of G with $P \subset Q$. Let X be a closed subset of G such that $XP = X$. Then XQ is closed.

6.3. Connected solvable groups

In this section we deal with the structure theory of connected solvable linear algebraic groups. G is such a group.

6.3.1. Theorem. [Theorem of Lie-Kolchin] *Assume that G is a closed subgroup of $\mathbf{GL}_n$. There is $x \in \mathbf{GL}_n$ such that $xGx^{-1} \subset \mathbf{T}_n$.*

As before $\mathbf{T}_n$ is the group of upper triangular matrices. Using induction on n it suffices to prove that the elements of G have a non-zero common eigenvector. This follows from Borel's fixed point theorem 6.2.6, applied to G acting on $\mathbf{P}^{n-1}$.

But there is a more elementary proof. By induction on $\dim G$ we may assume that there is a common eigenvector for the elements of the commutator group (G, G) (which is closed and connected by 2.2.8 (i)). If χ is a character of (G, G) let

$$V_\chi = \{v \in V \mid g.v = \chi(g)v \text{ for } g \in (G, G)\}.$$

Then G permutes the distinct non-zero spaces V_χ. Since G is connected it must stabilize all V_χ. We may replace G by its restriction to one of the non-zero V_χ, i.e. we may assume that there is χ with $V = V_\chi$. Then the elements of (G, G) act as scalar multiplications. Since a commutator $xyx^{-1}y^{-1}$ $(x, y \in GL(V))$ has determinant 1, these scalar multiplications also have determinant 1. This implies that (G, G) is finite. Being connected, (G, G) is trivial. So G is commutative and there is a common eigenvector by 2.4.2 (i). □

6.3.2. Corollary. *Assume, moreover, that G is nilpotent.*
(i) The sets G_s, G_u of semi-simple resp. unipotent elements are closed, connected subgroups. G_s is a central torus of G;
(ii) The product map $G_s \times G_u \to G$ is an isomorphism of algebraic groups.

G_s lies in the center of G by 5.4.8. Using 2.4.2 (ii) we obtain a decomposition of k^n into a direct sum of G-invariant subspaces, in each of which the elements of G_s act as scalar multiplications. By the theorem we can bring the restriction of G to each summand in triangular form. Now proceed as in the proof of 3.1.1. □

6.3.3. Corollary. *(i) The commutator subgroup (G, G) is a closed, connected, unipotent, normal, subgroup;*
(ii) The set G_u of unipotent elements is a closed, connected, nilpotent, normal, subgroup of G. The quotient group G/G_u is a torus.

(G, G) is closed and connected. By 6.3.1 we may assume that G is a closed subgroup of $\mathbf{T}_n$. It is then obvious that (G, G) is unipotent.

Since $G_u = G \cap \mathbf{U}_n$ (where $\mathbf{U}_n$ is the group of upper triangular unipotent matrices) we see that G_u is a closed normal subgroup, which is nilpotent because $\mathbf{U}_n$ is nilpotent (2.4.13). We have an injective homomorphism of the algebraic group G/G_u into the torus $\mathbf{T}_n/\mathbf{U}_n$. It follows that G/G_u is commutative and that all its elements are semi-simple (by 2.4.8 (ii)). Being connected it must be a torus (use 3.1.1 and 3.2.7 (ii)).

It remains to be proved that G_u is connected. Its identity component G_u^0 is a normal subgroup of G. Passing to G/G_u^0 we are reduced to showing: if G_u is finite it is trivial. In that case G_u lies in the center of G (2.2.2 (4)), and G is nilpotent. But it follows from 6.3.2 that G_u is connected, hence must be trivial. □

6.3.4. Lemma. *Assume that G is not a torus. There exists a closed normal subgroup N of G that is isomorphic to $\mathbf{G}_a$ and lies in the center of G_u.*

Let H be a non-trivial closed, connected, normal subgroup of G that lies in the center of G_u. Such a group exists, see the proof of 6.2.10. If char $k = p > 0$ we may also assume that $H^p = \{e\}$ (replace H by its image under a suitable map $x \mapsto x^{p^e}$). By 3.4.7, H is isomorphic to a vector group $\mathbf{G}_a^m$. If $m = 1$ we are done. Otherwise let $\mathcal{A} \subset k[H]$ be the space of additive functions on H (3.3.1). The torus $T = G/G_u$ acts on H via conjugation. We obtain a representation of T in $k[H]$ with invariant subspace $\mathcal{A}$. Using 2.3.6 (i) and 3.2.3 (c) we find $f \in \mathcal{A}$ which is a simultaneous eigenvector for the elements of T. Then $(\mathrm{Ker}\ f)^0$ is a subgroup with the same properties as H, but of lower dimension. The lemma follows by induction. □

We come now to the main result of this section. A *maximal torus* of G is a subtorus that has the same dimension as $S = G/G_u$. It will follow from 6.3.6 (i) that a maximal torus is also a torus that is maximal in the set-theoretical sense.

6.3.5. Theorem. *(i) Let $s \in G$ be semi-simple. Then s lies in a maximal torus. In particular: maximal tori exist;*
(ii) The centralizer $Z_G(s)$ of a semi-simple element $s \in G$ is connected;
(iii) Two maximal tori of G are conjugate;
(iv) If T is a maximal torus, the product map $\pi : T \times G_u \to G$ is an isomorphism of varieties.

We first prove (iv). It is clear that $G = T.G_u$. Now G is a homogeneous space of the group $T \times G_u$, for the action $(t, u).x = txu^{-1}$. The isotropy group of e is trivial, since $T \cap G_u = \{e\}$. It follows from 4.4.12 that the tangent map $d\pi_{(e,e)}$ is the map $(X, Y) \mapsto X - Y$ of $L(T) \times L(G_u)$ to $L(G)$. Since $L(T) \cap L(G_u) = \{0\}$ (as a consequence of 4.4.21 (1)) the tangent map is injective, hence bijective (because of dimensions). Now 5.3.2 (iii) shows that π is an isomorphism of varieties.

Next we prove the other assertions in the case that $\dim G_u = 1$. Since G_u is connected we know by 3.4.9 that it is isomorphic to $\mathbf{G}_a$. Fix an isomorphism $\phi : \mathbf{G}_a \to G_u$ and let $\psi : G \to S = G/G_u$ be the canonical map. We have $\dim S = \dim G - 1$. There is a character α of S such that

$$g\phi(a)g^{-1} = \phi(\alpha(\psi g)a) \ (g \in G, a \in k). \tag{16}$$

If α is trivial then G is commutative, and everything follows from 3.1.1. So assume α to be non-trivial.

Let $s \in G$ be semi-simple and put $Z = Z_G(s)$. By 5.4.5 (ii) we have a direct sum decomposition $\mathfrak{g} = (\mathrm{Ad}(s)-1)\mathfrak{g} \oplus \mathfrak{z}$. Since $\psi(sxs^{-1}) = \psi(x)$ we have $d\psi \circ (\mathrm{Ad}(s) - 1) = 0$, whence $(\mathrm{Ad}(s) - 1)\mathfrak{g} \subset \mathrm{Ker}\ d\psi = L(G_u)$ (the last equality coming from 5.5.6 (ii)). It follows that $\dim(\mathrm{Ad}(s) - 1)\mathfrak{g} \leq 1$ and $\dim Z^0 = \dim \mathfrak{z} \geq \dim G - 1$. Now assume that $\alpha(\psi s) \neq 1$. Such s exist, for example the semisimple part of a $g \in G$ with $\alpha(\psi g) \neq 1$. Then (16) shows that $Z \cap G_u = \{e\}$. It follows that Z^0 is a closed, connected subgroup of G, of dimension $\dim G - 1$, with $Z_u^0 = \{e\}$. By

6.3.3 (ii) it is a torus. It is a maximal torus and by (iv) we know that $G = Z^0 G_u$.
If $g = xy$ ($x \in Z^0, y \in G_u$) commutes with s then y commutes with s, whence
by (16) $y = e$. So $Z = Z^0$. We have shown that the centralizer of a semi-simple
element with $\alpha(\psi s) \neq 1$ is connected. If $\alpha(\psi s) = 1$ then $L(G_u) \subset \mathfrak{z}$ and we can now
conclude that $\mathrm{Ad}(s) = \mathrm{id}$. By 5.4.5 (ii) this means that s lies in the center of G. It
then lies in a maximal torus, for example the centralizer of a semi-simple element s'
with $\alpha(\psi s') \neq 1$.

It remains to prove (iii) (in the case $\dim G_u = 1$). If T is a maximal torus there
is $t \in T$ with $\alpha(\psi t) \neq 1$ and then $T = Z_G(t)$. Let T' be another maximal torus and
let $t' \in T'$ be an element with $\alpha(\psi t') \neq 1$. Then $T = Z_G(t)$, $T' = Z_G(t')$. Write
$t' = t\phi(a)$, where $t \in T$, $a \in k$. From (16) we find for $b \in k$

$$\phi(b) t' \phi(b)^{-1} = t\phi(a + (\alpha(\psi t')^{-1} - 1)b).$$

We can take b such that the right-hand side equals t. Then $\phi(b) T' \phi(b)^{-1} = T$,
proving (iii).

Now consider the general case. Assume $\dim G_u > 1$. Let N be as in 6.3.4. Put
$\bar{G} = G/N$. Then $\dim G/G_u = \dim \bar{G}/\bar{G}_u$. Let s be semi-simple in G and let $\bar{s}$ be its
image in $\bar{G}$. By induction on $\dim G_u$ we may assume that $\bar{s}$ lies in a maximal torus $\bar{T}$
of $\bar{G}$. The inverse image of $\bar{T}$ in G is a closed, connected subgroup H of G containing
s, and $\dim H_u \leq 1$. We know that s lies in a maximal torus of H, which is also one of
G. This proves (i). The proof of (iii) is similar, and can be left to the reader.

Let $G_1 = \{g \in G \mid sgs^{-1}g^{-1} \in N\}$. This is a closed subgroup containing
$Z = Z_G(s)$ and N, and

$$G_1/N \simeq Z_{\bar{G}}(\bar{s}).$$

We may assume that $Z_{\bar{G}}(\bar{s})$ is connected. From 5.5.9 (1) we then conclude that G_1 is
connected. If $G_1 \neq G$ we have by induction on $\dim G$ that Z is connected. Assume
now that $G_1 = G$. We may also assume that s is non-central. Then an argument
similar to the one used to prove (ii) in the case $\dim G_u = 1$ shows that $G = Z^0.N$,
$Z^0 \cap N = \{e\}$, whence $Z = Z^0$. This completes the proof of the theorem. □

6.3.6. Corollary. *Let $H \subset G$ be a subgroup of G whose elements are semi-simple.*
(i) H is contained in a maximal torus of G. In particular: a subtorus of G is con-
tained in a maximal torus;
(ii) The centralizer $Z_G(H)$ is connected and coincides with the normalizer $N_G(H)$.

H is commutative, since the restriction to H of the canonical homomorphism
$G \to G/G_u$ is bijective. If H lies in the center of G the assertions are obvious.
Otherwise, take a non-central element s of H. By 6.3.5 (ii) the centralizer $Z_G(s)$
is connected. It contains H. Now (i) and the connectedness of $Z_G(H)$ follow by

induction on dim G. Finally, if $x \in N_G(H)$ then for $h \in H$

$$xhx^{-1}h^{-1} \in H \cap (G, G) \subset H \cap G_u = \{e\},$$

whence $N_G(H) = Z_G(H)$. $\square$

We shall not discuss in this section questions about solvable groups involving a ground field F. They will be taken up in chapter 14.

6.3.7. Exercises. (1) Give an example of a finite solvable subgroup of $\mathbf{SL}_2(\mathbf{C})$ that is not conjugate to a group of triangular matrices (showing that 6.3.1 is false for non-connected groups).
(2) Let G be a connected linear algebraic group whose elements are semi-simple. Show that G is a torus. (*Hint:* consider a Borel subgroup of G).
In the next exercises G is a connected solvable linear algebraic group.
(3) There is a sequence $\{e\} = G_0 \subset G_1 \subset ... \subset G_{n-1} \subset G_n = G$ of closed, connected, normal, subgroups of G such that the quotients G_i/G_{i-1} are isomorphic to either $\mathbf{G}_a$ or $\mathbf{G}_m$.
(4) Let G be unipotent and let H be a proper closed, connected, subgroup. Show that $\dim N_G H > \dim H$. (*Hint:* consider $Z(G).H$).
(5) Let H be a closed, connected, nilpotent, subgroup of G that coincides with its normalizer. Show that H is the centralizer of a maximal torus. (*Hint:* By 6.3.2 (ii) we have $H = H_s \times H_u$. Use 6.3.6 (i) to show that H_s is a maximal torus).

6.4. Maximal tori, further properties of Borel groups

In this section G is a connected linear algebraic group.

A *maximal torus* of G is a subtorus of G that is not strictly contained in another subtorus. A *Cartan subgroup* of G is the identity component of the centralizer of a maximal torus. (In fact, such a centralizer is connected, as we shall see in 6.4.7).

6.4.1. Theorem. *Two maximal tori of G are conjugate.*

Fix a Borel subgroup B of G. A maximal torus T, being connected and solvable, lies in some Borel group. By 6.2.7 (iii) T is conjugate to a subtorus of B, which must be a maximal torus of B. The theorem now follows from 6.3.5 (iii). $\square$

6.4.2. Proposition. *Let T be a maximal torus of G and $C = Z_G(T)^0$ the corresponding Cartan subgroup.*
(i) C is nilpotent and T is its maximal torus;
(ii) There exist elements $t \in T$ lying in only finitely many conjugates of C.

Clearly, C contains T as a central subgroup. A Borel group B of C containing T then also has T as a central subgroup and must be nilpotent (since $B/T \simeq B_u$ is nilpotent). By 6.2.10 we have $C = B$ and by 6.3.2 (i) T is the only maximal torus of C. This proves (i). For the proof of (ii) we use the following lemma.

6.4.3. Lemma. *Let S be a subtorus of G. There exists $s \in S$ with $Z_G(s) = Z_G(S)$.*

On account of 2.3.7 (i) it suffices to prove this for $G = \mathbf{GL}_n$. We may then also assume by 3.2.3 (c) that S consists of diagonal matrices. The diagonal entries define characters of S; let $\chi_1, \ldots, \chi_m$ be the distinct characters of S so obtained. The elements $s \in S$ with $\chi_i(s) \neq \chi_j(s)$ if $i \neq j$ have the required property. They form a dense open subset of S. $\qquad\square$

Now choose $t \in T$ such that $Z_G(t) = Z_G(T)$. If t lies in a conjugate gCg^{-1}, then $g^{-1}tg \in T$ and $T \subset Z_G(g^{-1}tg) = g^{-1}Tg$. Since T is a maximal torus it follows that $g \in N_G(T)$. We know from 3.2.9 that C has finite index in the last group and (iii) follows. $\qquad\square$

6.4.4. Lemma. *Let H be a closed subgroup of G and denote by X the union of the conjugates xHx^{-1} $(x \in G)$.*
(i) X contains a non-empty open subset of its closure $\bar{X}$. If H is parabolic then X is closed;
(ii) Assume that H has finite index in its normalizer N and that there exist elements of H that lie in only finitely many conjugates of H. Then $\bar{X} = G$.

We may assume that H is connected. Then $Y = \{(x, y) \in G \times G \mid x^{-1}yx \in H\}$ is a closed subset of $G \times G$, which is isomorphic to $G \times H$, hence is irreducible. Also, if $(x, y) \in Y$ then $(xH, y) \subset Y$. It follows from 5.3.2 (i) and 1.2.3 (ii) that $Y_1 = \{(xH, y) \mid x^{-1}yx \in H\}$ is an irreducible closed subset of $G/H \times G$. Since $X = \pi Y_1$, where π is the second projection, (i) follows by 1.9.5 and the definition of parabolic subgroups. Since the fibers of the projection morphism $Y_1 \to G/H$ all have dimension $\dim H$ it follows from 5.1.6 (ii) that $\dim Y_1 = \dim G$. Now let $x \in H$ lie in finitely many conjugates of H. Since H has finite index in N it follows that $\pi^{-1}x$ is finite. Then 5.2.7 shows that $\dim \bar{X} = \dim Y_1 = \dim G$, and (ii) follows. $\qquad\square$

6.4.5. Theorem. *(i) Every element of G lies in a Borel subgroup;*
(ii) Every semi-simple element of G lies in a maximal torus;
(iii) The union of the Cartan subgroups of G contains a dense open subset.

Let T be a maximal torus, $C = Z_G(T)^0$ the corresponding Cartan subgroup and B a Borel subgroup containing C (which exists because C is connected and nilpotent). Apply 6.4.4 with $H = C$. It follows from 6.4.2 (i) that $N_G(C) = N_G(T)$. By 3.2.9

we know that C has finite index in its normalizer. By 6.4.2 (ii) the conditions of 6.4.4 (ii) are satisfied, and (iii) follows. Next apply 6.4.4 (i) with $H = B$. Then (i) follows from (iii). Finally, (ii) follows from (i) and 6.3.5 (i). $\square$

6.4.6. Corollary. *Let B be a Borel subgroup of G. Then $C(B) = C(G)$.*

An element in $C(G)$ lies in a Borel subgroup by 6.4.5 (i), hence lies in all of them by the conjugacy of Borel groups. So $C(G) \subset C(B)$. The reverse inclusion was already proved in 6.2.9. $\square$

6.4.7. Theorem. *Let S be a subtorus of G.*
(i) The centralizer $Z_G(S)$ is connected;
(ii) If B is a Borel subgroup containing S then $Z_G(S) \cap B$ is a Borel subgroup of $Z_G(S)$. All Borel subgroups of $Z_G(S)$ are obtained in this way.

Put $Z = Z_G(S)$. Take $g \in Z$ and let B be a Borel subgroup containing g. Put

$$X = \{xB \in G/B \mid x^{-1}gx \in B\}.$$

Then X is a closed subvariety of G/B (it is a fiber of the projection $Y_1 \to G$ of the proof of 6.4.4, with $H = B$). By 6.1.2 (i) X is complete. Now S acts on X via left multiplication, and by the fixed point theorem 6.2.6 there exists $xB \in X$ with $x^{-1}Sx \subset B$. This means that there exists a Borel subgroup containing both g and S. It follows from 6.3.5 (ii) and 6.3.6 (ii) that g lies in the identity component Z^0, whence (i).

Let B be as in (ii). Then $Z \cap B$ is connected (6.3.6 (ii)) and solvable. To prove the first part of (ii) it suffices to show that $Z/Z \cap B$ is complete. There is a bijective morphism of homogeneous spaces (for Z) of $Z/Z \cap B$ onto the image of $Z.B$ in G/B. Using 6.2.1 and the openness of $G \to G/B$, we conclude that it suffices to prove that $Y = Z.B$ is closed in G. Being the image of the irreducible variety $Z \times B$ under a morphism, Y is irreducible. The closure $\bar{Y}$ is also irreducible, hence connected (1.2.3 (i), 1.2.7).

If $y \in Y$ we have $y^{-1}Sy \subset B$. This also holds when $y \in \bar{Y}$. Consider the morphism $\phi : \bar{Y} \times S \to B/B_u$ sending (y, s) to $y^{-1}syB_u$. Applying 3.2.8 to ϕ (with $V = \bar{Y}, G = S, H = B/B_u$) we conclude that for $y \in \bar{Y}$ we have $y^{-1}sy \in sB_u$. Then $y^{-1}Sy$ is a maximal torus of SB_u. By the conjugacy of maximal tori of that group there is $z \in B_u$ with $y^{-1}Sy = z^{-1}Sz$. It follows that $y \in Z.B = Y$. Hence Y is closed, as asserted.

The last point of (ii) follows by the conjugacy of Borel subgroups. $\square$

6.4.8. Corollary. *Let T be a maximal torus of G.*
(i) $C = Z_G(T)$ is a Cartan subgroup of G;
(ii) If B is a Borel subgroup containing T then B contains C.

Apply the theorem with $S = T$, recalling that Cartan subgroups are nilpotent (6.4.2 (i)). $\square$

6.4.9. Theorem. *Let B be a Borel subgroup of G. Then $N_G(B) = B$.*

We prove the theorem by induction on dim G. The assertion is trivial if G is solvable. Put $H = N_G(B)$ and let $x \in H$. Fix a maximal torus T of B. Then xTx^{-1} is also a maximal torus of B. Using the conjugacy of maximal tori of B we see that we may assume that $xTx^{-1} = T$. Consider the homomorphism $\psi : t \mapsto xtx^{-1}t^{-1}$ of T into itself. The image is a closed, connected subgroup of T. There are two cases:
(a) Im ψ is a proper subgroup of T. Then $S = (\text{Ker } \psi)^0$ is a non-trivial torus. Moreover, x lies in $Z = Z_G(S)$ and normalizes the Borel group $Z \cap B$ of Z. If $Z \neq G$ we have $x \in B$ by induction. If $Z = G$ then S lies in the center of G. Passing to G/S we can again use induction to obtain that $x \in B$.
(b) ψ is surjective. Let ϕ, V and v be as in 5.5.3 (with $F = k$). Then $\phi(B_u)$ and $\phi(T)$ fix v, because B_u is unipotent and T lies in the commutator group (H, H). But this implies that ϕ induces a morphism of the complete variety G/B to the affine variety V. By 6.1.2 (vi) it follows that G fixes v, i.e. that $H = G$. This means that B is a normal subgroup of G. But then G/B is a unipotent group and G is solvable, whence $H = G = B$. $\square$

6.4.10. Corollary. *Let P be a parabolic subgroup of G. Then P is connected and $N_G(P) = P$.*

By 6.2.7 (i) P contains a Borel subgroup B, which lies in P^0. If $x \in N_G(P)$ then xBx^{-1} is also a Borel group of P^0. By the conjugacy of Borel subgroups there is $y \in P^0$ with $xBx^{-1} = yBy^{-1}$. By 6.4.9, $y^{-1}x$ lies in B, whence $x \in P^0$. $\square$

6.4.11. Corollary. *Let P and Q be two conjugate parabolic subgroups of G whose intersection contains a Borel subgroup B. Then $P = Q$.*

Let $P = xQx^{-1}$. Then B and xBx^{-1} are two Borel subgroups of P, which are conjugate in P. As in the proof of 6.4.10 we find that $x \in P$, whence $P = Q$. $\square$

6.4.12. Corollary. *Let T be a maximal torus of G and B a Borel subgroup containing T. The map $x \mapsto xBx^{-1}$ induces a bijection of $N_G(T)/Z_G(T)$ onto the set of Borel subgroups containing T.*

Surjectivity of the map follows from the conjugacy of maximal tori of B. Injectivity follows from 6.4.9 and 6.3.6 (ii). $\square$

6.4.13. Let $\mathcal{B}$ be the set of Borel subgroups of G. Fix $B \in \mathcal{B}$. By 6.2.7 (iii) and

6.4.9 the map $xB \mapsto xBx^{-1}$ defines a bijection of G/B onto $\mathcal{B}$. Via this bijection we can define a structure of projective variety on $\mathcal{B}$. It is independent of the choice of B. We have thus defined the *variety of Borel subgroups* of G. More generally, fixing a parabolic subgroup $P \supset B$ we can define a structure of projective variety $\mathcal{P}$ on the set of conjugates of P. We have a surjective morphism of homogeneous spaces $\mathcal{B} \to \mathcal{P}$. By 6.4.11 it is the map sending a Borel group to the unique conjugate of P that contains it.

6.4.14. If N and N' are normal subgroups of G then $N.N'$ is also one. Using this fact and 2.2.7 (i), it follows that there is a maximal closed, connected, normal, solvable, subgroup of G, viz. a group with these properties of maximal dimension. This is the *radical $R(G)$* of G. Similarly, there is a maximal closed, connected, unipotent, subgroup of G, the *unipotent radical $R_u(G)$*. We have $R_u(G) = R(G)_u$. G is *semisimple* if $R(G) = \{e\}$ and *reductive* if $R_u(G) = \{e\}$. It follows from 6.2.7 (iii) that $R(G)$ is the identity component of the intersection of all Borel subgroups.

6.4.15. Exercises. G is a connected linear algebraic group.
(1) Let H be a closed subgroup of G containing a maximal torus T. Then $N_G(H) \subset H^0.N_G(T)$.
(2) Call $x \in G$ *regular* if the multiplicity of the root 1 of the characteristic polynomial of the linear map Ad x of $\mathfrak{g}$ is minimal.

    (a) The regular elements form a non-empty open subset of G.

    (b) $x \in G$ is regular if and only if its semi-simple part x_s is regular.

    (c) A semi-simple element is regular if and only if its centralizer has minimal dimension. (*Hint*: use 5.4.4 (ii)).

    (d) A semi-simple element x is regular if and only if $Z_G(x)^0$ is a Cartan subgroup. (*Hint*: use 6.4.3 and 6.4.5 (ii)).

(3) (a) A maximal nilpotent, closed subgroup C of G such that $C = N_G(C)^0$ is a Cartan subgroup.

    (b) Let C be a maximal nilpotent subgroup of G with the property that each subgroup of finite index of C has finite index in its normalizer. Then C is a Cartan subgroup. (This is a group-theoretical characterization of Cartan subgroups. *Hint* for the proof: show that the closure of a nilpotent subgroup is nilpotent and deduce that C is closed and satisfies the conditions of (a)).

(4) Let $x = x_s x_u$ be the Jordan decomposition of $x \in G$. Show that $x \in Z_G(x_s)^0$.

(5) Let char $K \neq 2$, $G = \mathrm{SO}_n$ ($n \geq 3$). There exist semi-simple elements in G whose centralizer is not connected. (*Hint*: consider elements of order two.)

(6) Let T be a maximal torus of G, with corresponding Cartan subgroup $C = Z_G(T)$, and let B be a Borel subgroup containing C. Assume that $\sigma : G \to G$ is a surjective homomorphism of algebraic groups with $\sigma B = B$.

    (a) Define a morphism $\phi_b : G \times B \to G$ by $\phi_b(x, c) = (\sigma x)b^{-1}cx^{-1}b$ ($b, c \in B, x \in G$). Then $\phi_b(e, e) = e$. Show that the subspace Im $(d\phi_b)_{(e,e)}$ of $\mathfrak{g}$ contains $\mathfrak{b}$

and $(\text{Ad}(b)d\sigma - 1)\mathfrak{g}$. (*Hint*: use 4.4.13.)

(b) Show that there is $b \in B$ such that $\sigma T = b^{-1}Tb$ and that the linear map of $\mathfrak{g}/\mathfrak{b}$ induced by $\text{Ad}(b)d\sigma$ has no eigenvalue 1. (*Hint*: first take b such that the first condition is satisfied and modify it by an element of T).

(c) If $b \in B$ is as in (b) then ϕ_b is dominant.

(d) Show that ϕ_e is surjective, i.e. that any element of G can be written in the form $(\sigma x)bx^{-1}$ with $b \in B$, $x \in G$. (*Hint*: adapt the argument used to prove 6.4.5 (i)).

(e) Deduce from (c) that an arbitrary surjective homomorphism of algebraic groups $\sigma : G \to G$ fixes a Borel subgroup.

Notes

The elementary proof of the completeness of projective varieties in 6.1.3 is taken from [EGA, Ch.II, 5.5.3].

The elegant result 6.1.6. is due to Chevalley. We have followed the proof given in [Sha, p. 152-153].

Most of the results about solvable groups and Borel subgroups are to be found in Borel's fundamental paper [Bo1]. In the proof of 6.2.7 we have avoided the use of flag varieties, which are used in Borel's proof. Instead we use 6.2.3 and 6.2.5.

Theorem 6.3.1 is due to Kolchin [Kol1, Ch. 1]. It is an analogue of Lie's theorem for Lie algebras over **C**.

The name 'parabolic subgroups' seems to appear for the first time in [BoT1].

The proof of the main result 6.3.5 of 6.3 is somewhat different from Borel's original proof. We have exploited the auxiliary result 6.3.4.

The fundamental result 6.4.9 is due to Chevalley. The proof given here, which is simpler than Chevalley's proof (in [Ch4, Exp. 9]) is due to Borel.

The results of 6.4.15 (6) are due to Steinberg [St3, no. 7].

Chapter 7

Weyl Group, Roots, Root Datum

In this chapter we introduce combinatorial data associated to a linear algebraic group: Weyl group, root system and root datum. Important results are the classification of semi-simple groups of rank one (7.2.4) and the characterization in 7.6 of the unipotent radical. In this chapter G denotes a connected linear algebraic group and T a maximal torus of G. The character group of T is denoted by X. We shall not discuss questions involving ground fields. These will be taken up in the later chapters.

7.1. The Weyl group

7.1.1. If S is a torus and $r : S \to GL(V)$, a rational representation of S (2.3.2 (3)), then is by 3.2.3 (c) V is a direct sum of one dimensional subspaces, in each of which an element $s \in S$ acts as multiplication by $\chi(s)$, where χ is a character of S. The characters so obtained are the *weights* (of S in V). The non-zero subspaces

$$V_\chi = \{v \in V \mid r(s)v = \chi(s)v \text{ for all } s \in S\}$$

are the *weight spaces*. A non-zero vector in a weight space is a *weight vector*. We denote by P the set of non-zero weights of T, acting via the adjoint representation Ad in the Lie algebra $\mathfrak{g}$ of G (4.4.5 (ii)). It is a finite subset of X.

7.1.2. Lemma. *Let S be a subtorus of T. Then $Z_G(S) = Z_G(T)$ if and only if S is not contained in any of the subgroups* Ker α *of T, where $\alpha \in P$.*

$Z_G(S)$ is a connected subgroup of G (6.4.7 (i)) that contains $Z_G(T)$. It follows from 5.4.7 that $Z_G(S) \neq Z_G(T)$ if and only if Ad(S) fixes a non-zero element of a weight space $\mathfrak{g}_\alpha$ where $\alpha \in P$. The lemma follows from this observation. $\square$

For $\alpha \in P$ we denote by G_α the centralizer of the subtorus (Ker $\alpha)^0$ of T. It is a closed connected subgroup. It follows from 7.1.2, that if S is a subtorus of T with $Z_G(S) \neq Z_G(T)$, then there is $\alpha \in P$ such that $Z_G(S)$ contains G_α.

7.1.3. Lemma. *(i) The G_α ($\alpha \in P$) generate G;*
(ii) If all G_α are solvable then G is solvable.

By 2.2.7 (i) the G_α ($\alpha \in P$) generate a closed, connected subgroup H. By 5.4.7 the Lie algebra of G_α contains the Lie algebra $\mathfrak{c}$ of the centralizer of T and the weight space $\mathfrak{g}_\alpha$. Since $\mathfrak{c}$ and these weight spaces span $\mathfrak{g}$, and since the Lie algebra of G_α is contained in $\mathfrak{h}$, we must have $\mathfrak{h} = \mathfrak{g}$, whence $H = G$. This proves (i).

Let B be a Borel subgroup of G containing T. If G_α is solvable then it is contained in B, by 6.4.7 (ii). It follows from (i) that if all G_α are solvable we must have $G = B$, hence G is solvable. $\square$

7.1.4. We denote by P' the set of $\alpha \in P$ such that G_α is non-solvable. By 7.1.3 (ii) P' is empty if and only if G is solvable. By 3.2.9 the group $W = W(G, T) = N_G(T)/Z_G(T)$ is finite. This is the *Weyl group* of (G, T). It follows from the definitions that W acts faithfully as a group of automorphisms of X (a free abelian group of finite rank, see 3.2.7 (iii)). We identify W with this group. W permutes the elements of P and P'. It is clear that, if S is a subtorus of G, the Weyl group $W(Z_G(S), T)$ is a subgroup of $W(G, T)$. Moreover, if S lies in the center of G the canonical homomorphism $G \to G/S$ induces an isomorphism $W(G, T) \simeq W(G/S, T/S)$. By 6.4.12 there is a bijection of W onto the set of Borel subgroups containing T. If B is such a group, there is also a bijection of this set on the set of fixed points of T in G/B (T acting via left translations).

Fix $\alpha \in P'$. The group G_α contains the torus $S = (\text{Ker } \alpha)^0$ in its center and the Weyl group W_α of (G_α, T) is isomorphic to that of $(G_\alpha/S, T/S)$. Since T/S is isomorphic to $\mathbf{G}_m$ it follows that W_α has order ≤ 2 (by 3.2.10 (6) with $n = 1$).

7.1.5. Proposition. *Assume that G is non-solvable and that* $\dim T = 1$.
(i) W has order two;
(ii) If B is a Borel subgroup of G then $\dim G/B = 1$.

Fix an isomorphism $\lambda : \mathbf{G}_m \to T$. Let B be a Borel group containing T. Let $\phi : G \to GL(V)$ be a representation with the properties of 5.5.3 for the subgroup B (with $F = k$). We may assume that V is spanned by the images $(\phi x)v$ (where v is as in 5.5.3). Then ϕ defines an isomorphism of G/B onto a closed subvariety of $\mathbf{P}(V)$ (see 5.5.4). We identify G/B with this variety. Choose a basis $(e_1, ..., e_n)$ of V consisting of weight vectors for the representation $\rho = \phi \circ \lambda$ of $\mathbf{G}_m$. We may assume that $\rho(a)e_i = a^{m_i} e_i$ ($a \in k^*, 1 \leq i \leq n$) with $m_1 \geq m_2 \geq ... \geq m_n$. If $x \in V - \{0\}$ write $x^* = kx$; this is a point in $\mathbf{P}(V)$. An easy argument shows that, if the last coordinate of $x \in \mathbf{P}(V)$ is non-zero, then $x_0^* = \lim_{a \to 0} \rho(a)x^*$ exists, i.e. the morphism $a \mapsto \rho(a)x^*$ of $\mathbf{G}_m$ to $\mathbf{P}(V)$ extends to a morphism $\mathbf{A}^1 \to \mathbf{P}(V)$ (see 3.2.13). Similarly, if the first coordinate of y is non-zero, then $y_\infty^* = \lim_{a \to \infty} \rho(a)y^*$ exists. It is clear that the points x_0^* and y_∞^* are fixed points for the action of $\phi(T)$ on $\mathbf{P}(V)$. Moreover, two such points are distinct, otherwise all points of $\mathbf{P}(V)$ would be fixed points for $\phi(T)$. Then T would be contained in all Borel groups of G. This is impossible as T lies in at most two Borel groups (see 7.1.4).

From our assumption that $\phi(G)v$ spans V, it follows that there exist points x and y as above such that $x^*, y^* \in G/B$. The corresponding fixed points of T, say p_0 and p_∞, then lie in G/B. It follows that T has at least two fixed points in G/B, i.e that T lies in at least two Borel groups. Now (i) follows.

Let l be the first coordinate function on V, it is a linear function. The points z^* of G/B with $l(z) = 0$ form a closed T-stable subset Σ, which is non-empty (it contains p_∞). Its components are T-stable. If a component had dimension > 0, an argument like the one of the first part of the proof would show that it contained two distinct fixed points of T, at least one of which would be different from p_0 and p_∞, contradicting (i). Hence Σ is finite. Now l defines a regular function f on a suitable open neighborhood U of p_0 in G/B with $f(p_0) = 0$. Since Σ is finite, the fiber $f^{-1}(0)$ is finite. Then (ii) follows from 5.2.7. $\square$

7.1.6. Let $\alpha \in P'$. We know by 7.1.5 (i) that W_α has order two. Choose $n_\alpha \in N_{G_\alpha}(T) - Z_{G_\alpha}(T)$. Let s_α the image of n_α in W. We denote by $X^\vee = \mathrm{Hom}(X, \mathbf{Z})$ the dual of X and by $\langle \ , \ \rangle$ the pairing between X and $X^\vee$. We view $X^\vee$ as the group of cocharacters of T (see 3.2.11 (i)). We identify X, $X^\vee$ with subgroups of $V = \mathbf{R} \otimes X$, respectively $V^\vee = \mathbf{R} \otimes X^\vee$. The induced pairing between V and $V^\vee$ will also be denoted by $\langle \ , \ \rangle$.

7.1.7. It will be convenient to introduce a positive definite symmetric bilinear form $(\ , \ )$ on V that is invariant for the induced action of W. Such forms exist: take any positive definite symmetric bilinear form f on V and define for $x, y \in V$

$$(x, y) = \sum_{w \in W} f(w.x, w.y).$$

Then the s_α ($\alpha \in P'$) are Euclidean reflections (or symmetries), for the metric defined by $(\ , \ )$. It is well-known (see e.g. [Jac4, p. 345]) that

$$s_\alpha(x) = x - 2(\alpha, \alpha)^{-1}(x, \alpha)\alpha. \tag{17}$$

7.1.8. Lemma. *(i) There exists a unique $\alpha^\vee \in V^\vee$ with $\langle \alpha, \alpha^\vee \rangle = 2$ such that for $x \in X$*

$$s_\alpha(x) = x - \langle x, \alpha^\vee \rangle \alpha;$$

(ii) If $\beta \in P'$ and $G_\beta = G_\alpha$ then $s_\beta = s_\alpha$.

From (17) we see that the element $\alpha^\vee$ of (i) must satisfy

$$\langle x, \alpha^\vee \rangle = 2(\alpha, \alpha)^{-1}(x, \alpha) \ (x \in V).$$

This determines $\alpha^\vee$ uniquely, and (i) follows. If $G_\alpha = G_\beta$ then we may take $n_\alpha = n_\beta$, whence $s_\alpha = s_\beta$. $\square$

7.1.9. Theorem. *W is generated by the s_α ($\alpha \in P'$).*

We proceed by induction on $\dim G$. The argument is similar to the one of the proof of 6.4.9.

Let $w \in W$ be represented by $x \in N_G(T)$ and consider the homomorphism $\psi : t \mapsto x t x^{-1} t^{-1}$ of T into itself. There are two cases:

(a) Im ψ is a proper subgroup of T. We can then proceed as in the proof of 6.4.9.

(b) ψ is surjective. This implies that the linear map $w - 1$ of V is injective (see 3.2.10 (2)), hence is bijective. Let α be a root. There is $x \in V$ with $(w - 1)x = \alpha$. Then

$$(x, x) = (w.x, w.x) = (x + \alpha, x + \alpha) = (x, x) + 2(x, \alpha) + (\alpha, \alpha),$$

and it follows that $2(\alpha, \alpha)^{-1}(x, \alpha) = -1$. Consequently $s_\alpha.x = x + \alpha = w.x$. Hence $s_\alpha w$ has an eigenvalue 1 and falls in case (a). □

7.1.10. Exercises. The notations are as in 7.1.9.
(1) Let $\alpha \in P'$, $w \in W$. Then $s_{w.\alpha} = w s_\alpha w^{-1}$.
(2) If $w \in W$ fixes $x \in V$ it is a product of reflections s_α ($\alpha \in P'$) that fix x.
(3) An element of W is a product of at most dim T reflections s_α ($\alpha \in P'$).

7.2. Semi-simple groups of rank one

7.2.1. The integer dim T is called the *rank* of G. By the conjugacy of maximal tori it is an invariant of G. The *semi-simple rank* of G is the rank of $G/R(G)$, where $R(G)$ is the radical of G (6.4.14). In this section we assume that G is *of rank one and non-solvable*. Let B be a Borel subgroup containing T and put $U = B_u$. Fix $n \in N_G(T) - Z_G(T)$ representing the non-trivial element of W (see 7.1.5 (i)). Then $ntn^{-1} = t^{-1}$ ($t \in T$) and $n^2 \in Z_G(T)$.

7.2.2. Lemma. *(i) G is the disjoint union of B and UnB;*
(ii) $R(G) = (U \cap nUn^{-1})^0$;
(iii) dim $U/U \cap nUn^{-1} = 1$.

Let $x \in G/B$ be the coset B. Then $n.x \neq x$ and x, $n.x$ are the two fixed points of T in G/B (see 7.1.4 and 7.1.5 (i)). Since $n^{-1}Bn \neq B$ we have $Un.x \neq \{n.x\}$. We know by 7.1.5 (ii) that dim $G/B = 1$. Using 1.9.5 we see that the complement of $Un.x$ is a finite set S. Since the torus T normalizes U, it must permute the points of S, hence must fix these points. This means that $S \subset \{x, n.x\}$. As $x \notin Un.x$, $n.x \in Un.x$ we conclude that $Un.x = G/B - \{x\}$, which is equivalent to (i). Since $U \cap nUn^{-1}$ is the isotropy group of $n.x$ in U, (iii) follows from (i) and 5.3.2 (ii), using that dim $G/B = 1$. From 6.3.7 (4) we conclude that $(U \cap nUn^{-1})^0$ is normal in U. Since this group is also normalized by T and n, it follows from (i) that it is a normal subgroup of G. Now (ii) follows by observing that the radical $R(G)$ cannot contain a torus. □

7.2.3. Lemma. *Assume that G is semi-simple, of rank one.*
(i) dim $U = 1$, $Z_G(T) = T$ and $U \cap nUn^{-1} = \{e\}$;
(ii) There is a unique weight α of T in $\mathfrak{g}$ such that $\mathfrak{g}$ is the direct sum of $\mathfrak{t}$ and two one dimensional weight spaces $\mathfrak{g}_\alpha$, $\mathfrak{g}_{-\alpha}$, with $L(U) = \mathfrak{g}_\alpha$, $L(nUn^{-1}) = \mathfrak{g}_{-\alpha}$;
(iii) The product map $(u, b) \mapsto unb$ is an isomorphism of varieties $U \times B \to UnB = G - B$.

From the preceding lemma we see that $\dim U = 1$, whence $\dim B = 2$. We also see that $U \cap nUn^{-1}$ is finite and unipotent. Since this group is normalized by T, it lies in the centralizer of T, which is connected and contained in B by 6.4.8 (ii). As $\dim B = 2$, we have $Z_G(T) = T$ or $Z_G(T) = B$. The second case is impossible (otherwise B was nilpotent and G would be solvable by 6.2.10). It follows that $Z_G(T) = T$ and $U \cap nUn^{-1} = \{e\}$. This proves (i).

By 3.4.9 we know that U is isomorphic to $\mathbf{G}_a$. Let $u : \mathbf{G}_a \to U$ be an isomorphism. There is a character α of T such that

$$tu(a)t^{-1} = u(\alpha(t)a) \quad (a \in k, t \in T),$$

and α is non-trivial since $Z_G(T) = T$. If $X \in L(U)$ is a non-zero element in the image of the differential du, then for $t \in T$

$$\mathrm{Ad}(t)X = \alpha(t)X.$$

Also, $\mathrm{Ad}(n)X \in L(nUn^{-1})$ and

$$\mathrm{Ad}(t)\mathrm{Ad}(n)X = \alpha(t)^{-1}\mathrm{Ad}(n)X.$$

From 7.2.2 (i) it follows that $\dim G \leq 3$. On the other hand $\mathfrak{t} \oplus kX \oplus k\mathrm{Ad}(n)X$ is a three dimensional subspace of $\mathfrak{g}$, so must be all of $\mathfrak{g}$. Now (ii) follows.

We finally prove that $(v, b) \mapsto vb$ is an isomorphism of the group $nUn^{-1} \times B$ onto $G - nB$, which is a statement equivalent to (iii). We can view this as an equivariant map of homogeneous spaces for $nUn^{-1} \times B$. Using (ii) and 4.4.12 it follows that the tangent map at (e, e) is bijective. Then apply 5.3.2 (iii). $\square$

7.2.4. Theorem. *Assume that G is connected, semi-simple, of rank one. Then G is isomorphic to* $\mathbf{SL}_2$ *or* $\mathbf{PSL}_2$.

The notations are as before. We have seen that $\dim U = 1$. Let u and α be as in the proof of 7.2.3. Fix an isomorphism $t : \mathbf{G}_m \to T$. There is an integer m such that $t(x)u(y)t(x)^{-1} = u(x^m y)$ $(x \in k^*, y \in k)$. Then $\alpha(t(x)) = x^m$. It follows from 7.2.3 (i) that $m \neq 0$. We may assume that $m > 0$. Also, $nt(x)n^{-1} = t(x^{-1})$. Put $n^2 = t(\epsilon)$. Then $t(\epsilon) = nt(\epsilon)n^{-1} = t(\epsilon^{-1})$, whence $\epsilon^2 = 1$.

It follows from 6.3.5 (iv) and 7.2.3 (iii) that $(x, y, z) \mapsto u(x)nt(y)u(z)$ defines an isomorphism of varieties $\mathbf{G}_a \times \mathbf{G}_m \times \mathbf{G}_a \to G - B$. If $y \neq 0$ then $nu(y)n^{-1} \notin B$, from which it follows that there are rational functions $f, g \neq 0, h \in k(\mathbf{G}_a)$ with poles at most in 0 and ∞, such that for $y \neq 0$

$$nu(y)n^{-1} = u(f(y))nt(g(y))u(h(y)). \tag{18}$$

Conjugating both sides of (18) by $t(z)$ $(z \neq 0)$ we obtain

$$nu(z^{-m}y)n^{-1} = u(z^m f(y))nt(z^{-2}g(y))u(z^m h(y)),$$

from which we conclude that

$$f(z^{-m}y) = z^m f(y), \quad g(z^{-m}y) = z^{-2}g(y), \quad h(z^{-m}y) = z^m h(y).$$

It follows that $g(y^m) = y^2 g(1)$ and $f(y) = ay^{-1}$, $h(y) = by^{-1}$, with $a, b \in k$. We conclude that $m = 1$ or 2. Taking inverses of both sides of (18) we find that $a = b$ and $g(-y) = \epsilon g(y)$.

If $a = b = 0$ we would have $n \in B$, contradicting 7.2.2 (i). So a and b are non-zero. By modifying n we may assume that $a = b = -1$. It also follows that if $m = 1$ we have $g(y) = cy^2$ and $\epsilon = 1$, whereas if $m = 2$ we have $g(y) = cy$ and $\epsilon = -1$, where $c \in k^*$. Now (18) becomes

$$nu(y)n^{-1} = u(-y^{-1})nt(g(y))u(-y^{-1}) \ (y \neq 0).$$

Let $y \neq 0, -1$ and apply this formula for y, $y+1$ and 1, respectively. The right-hand side of

$$nu(y+1)n^{-1} = nu(y)n^{-1}.nu(1)n^{-1}$$

then gives another expression for $u(f(y+1))$. By straightforward computation we find

$$-(y+1)^{-1} = -y^{-1} + g(y)^{-m}(y^{-1}+1)^{-1},$$

whence $g(y)^m = y^2$. If $m = 2$ we have $g(y) = \eta y$, with $\eta = \pm 1$. Replacing n by $nt(\eta)$ (which does not affect the normalization $a = b = -1$) we may assume that $g(y) = y$. We conclude that we may assume that

$$nu(y)n^{-1} = u(-y^{-1})nt(y^{m'})u(-y^{-1}) \ (y \neq 0), \quad n^2 = t((-1)^{m'}), \qquad (19)$$

where $mm' = 2$. We also have the relations

$$u(x+y) = u(x)u(y), \quad t(zw) = t(z)t(w), \quad t(z)u(x)t(z)^{-1} = u(z^m x), \qquad (20)$$

where $x, y \in k$, $z, w \in k^*$.

It follows from 7.2.2 (i) that the group structure of G is determined by (19) and (20). Let $G_1 = SL_2$, denote by B_1 and T_1 the subgroups of upper triangular resp. diagonal matrices and put

$$n_1 = \begin{pmatrix} 0 & 1 \\ -1 & 0 \end{pmatrix}.$$

Then G_1 is the disjoint union of B_1 and $U_1 n_1 B_1$. In fact, $U_1 n_1 B_1$ is the set V of matrices

$$X = \begin{pmatrix} x & y \\ z & t \end{pmatrix}$$

in $\mathbf{SL}_2$ with $z \neq 0$, which is an open subset of $\mathbf{SL}_2$. Also put

$$u_1(x) = \begin{pmatrix} 1 & x \\ 0 & 1 \end{pmatrix}, \quad t_1(y) = \begin{pmatrix} y & 0 \\ 0 & y^{-1} \end{pmatrix} \quad (x \in k, \ y \in k^*).$$

The map $(x, y, z) \mapsto u_1(x)n_1t_1(y)u_1(z)$ defines an isomorphism of $\mathbf{G}_a \times \mathbf{G}_m \times \mathbf{G}_a$ onto V. In fact, if X is as before we have

$$X = u_1(z^{-1}x)n_1t_1(-z)u_1(z^{-1}t). \tag{21}$$

For t_1, u_1 and n_1 we have the multiplication rules of (19) and (20) with $m = 2$ and these rules describe the group structure of $G_1 = \mathbf{SL}_2$. It then follows that there exists a surjective homomorphism of abstract groups $\phi : G_1 \to G$ with $\phi(u_1(x)) = u(x)$, $\phi(n_1) = n$, $\phi(t_1(y)) = t(y^{m'})$ $(x \in k, \ y \in k^*)$. From (21) we conclude that the restriction of ϕ to V is a morphism of V onto UnB. Also, the restriction of ϕ to a translate $g_1.V$ is a morphism. It follows that ϕ is a homomorphism of algebraic groups $G_1 \to G$. If $m = 2$ the restrictions of ϕ to V and its translates are isomorphisms and it follows that then G is isomorphic to $\mathbf{SL}_2$.

We have $k[\mathbf{SL}_2] = k[T_1, T_2, T_3, T_4]/(T_1T_4 - T_2T_3 - 1) = k[t_1, t_2, t_3, t_4]$ and $k[\mathbf{PSL}_2]$ is the subalgebra generated by the elements t_it_j $(1 \leq i, j \leq 4)$, whence a morphism $\pi : \mathbf{SL}_2 \to \mathbf{PSL}_2$, which is a homomorphism of algebraic groups (see 2.1.5 (3)). Then $k[V] = k[\mathbf{SL}_2]_{t_3} = k[t_1, t_4, t_3, t_3^{-1}]$ (notation of 1.4.6). Now assume that $m = 1$. It follows from (21) that the subalgebra $\phi^* k[UnB]$ of $k[V]$ is $k[t_3^{-1}t_1, t_3^{-1}t_4, t_3^2, t_3^{-2}]$, which coincides with $k[\mathbf{PSL}_2]_{t_3^2}$. This shows that there is an isomorphism of varieties $\pi(V)$ onto UnB. Using translations we conclude that there is an isomorphism of varieties $\psi : \mathbf{PSL}_2 \to G$ such that $\phi = \psi \circ \pi$. Then ψ is an isomorphism of algebraic groups. This completes the proof of the theorem. $\qquad \square$

7.2.5. Exercises. (1) Let G be as in 7.2.4. Show that G has no proper normal, closed subgroups of dimension > 0. Deduce that $G = (G, G)$.

(2) Let r be a non-trivial rational representation of $\mathbf{SL}_2$. Then Im r is isomorphic to $\mathbf{SL}_2$ or $\mathbf{PSL}_2$.

(3) A connected linear algebraic group of dimension two is solvable.

7.3. Reductive groups of semi-simple rank one

We begin with a general result that will be needed presently.

7.3.1. Proposition. *Let G be a connected, reductive, linear algebraic group.*
(i) The radical $R(G)$ is a central torus. It is the identity component of the center of G;
(ii) $R(G) \cap (G, G)$ is finite.

From the definitions of 6.4.14 and 6.3.5 (iv) it follows that the radical is a torus. That it lies in the center follows from 3.2.9. The last point of (i) follows from 3.1.1.

Assume that G is a closed subgroup of $GL(V)$. Decompose V into a direct sum of weight spaces for the radical (7.1.1), say

$$V = \oplus V_\chi.$$

G permutes the weight spaces. Being connected, it stabilizes each weight space. Then an argument as in the (second) proof of 6.3.1 proves (ii). □

7.3.2. We assume in the rest of this section that G is connected, reductive and of semi-simple rank one. Let C be the radical of G. Then G/C is semi-simple of rank one, hence isomorphic to $\mathbf{SL}_2$ or $\mathbf{PSL}_2$ by 7.2.4. Using 7.3.1 (ii) we see that (G, G) is connected, semi-simple, of rank one. Let T_1 be a maximal torus in (G, G) and T a maximal torus of G containing T_1. It follows from 7.2.3 (ii) and 7.1.3 (ii) that the set P has two elements $\pm\alpha$. We have

$$\mathfrak{g} = \mathfrak{t} \oplus \mathfrak{g}_\alpha \oplus \mathfrak{g}_{-\alpha},$$

as in 7.2.3 (ii). The subspaces $\mathfrak{g}_{\pm\alpha}$ are one dimensional.

7.3.3. Lemma. *(i) There exists a homomorphism of algebraic groups $u_\alpha : \mathbf{G}_a \to G$ such that $t u_\alpha(x) t^{-1} = u_\alpha(\alpha(t)x)$ $(x \in k, \ t \in T)$ and* $\mathrm{Im}\, du_\alpha = \mathfrak{g}_\alpha$. *If u'_α is a homomorphism with the same properties, there is a unique $a \in k^*$ with $u'_\alpha(x) = u_\alpha(ax)$; (ii) T and $\mathrm{Im}\, u_\alpha$ generate a Borel subgroup of G, whose Lie algebra is $\mathfrak{t} \oplus \mathfrak{g}_\alpha$.*

If u_α is as in (i) we have

$$t u_\alpha(x) t^{-1} u_\alpha(x)^{-1} = u_\alpha((\alpha(t) - 1)x),$$

which shows that $\mathrm{Im}\, u_\alpha \subset (G, G)$. It follows that we may assume that G is semi-simple. In this case we can take for u_α the composite of an isomorphism u of $\mathbf{G}_a$ onto the unipotent part U of a Borel group containing T_1 (as in the proof of 7.2.4) and the inclusion $U \to (G, G)$. The uniqueness statement of (i) follows from the fact that the automorphisms of $\mathbf{G}_a$ are scalar multiplications (2.1.5 (5)). (ii) follows from 7.2.3 (ii). □

7.3.4. Let $\lambda : \mathbf{G}_m \to T_1$ be an isomorphism. With the notations of 7.1.6, view λ as an element of the group of cocharacters $X^\vee$. The first formula of 7.3.3 (i) shows that $\pm\langle\alpha, \lambda\rangle$ is an integer m as in the first paragraph of the proof of 7.2.4, which was proved to equal 1 or 2 (according as $(G, G) \simeq \mathbf{PSL}_2$ or $\mathbf{SL}_2$). The Weyl group $W((G, G), T_1)$ has order two (7.1.5 (i)). Let $n \in N_{(G,G)}(T_1)$ represent its non-trivial element and let s_α be the reflection in $V = \mathbf{R} \otimes X$, which it defines. Let $\alpha^\vee \in V^\vee$ be as in 7.1.8. Then

$$s_\alpha(x) = x - \langle x, \alpha^\vee\rangle\alpha.$$

7.3.5. Lemma. *(i)* $\alpha^\vee \in X^\vee$. *We have* $\operatorname{Im} \alpha^\vee = T_1$;
(ii) $n^2 = \alpha^\vee(-1)$.

Denote by $s_\alpha^\vee$ the contragredient of s_α (an automorphism of $X^\vee$). Then

$$s_\alpha^\vee(y) = y - \langle \alpha, y \rangle \alpha^\vee \quad (y \in X^\vee).$$

Since $s_\alpha(t) = t^{-1}$ for $t \in T_1$ we have $s_\alpha^\vee(\lambda) = -\lambda$. It follows that

$$2\lambda = \langle \alpha, \lambda \rangle \alpha^\vee.$$

We conclude that $\alpha^\vee = \pm\lambda$ if $(G, G) \simeq \mathbf{SL}_2$ and $\alpha^\vee = \pm 2\lambda$ if $(G, G) \simeq \mathbf{PSL}_2$. This implies (i). It follows from the second formula (19) of 7.2 that (ii) holds, for a particular choice of n. Now observe that for all $t \in T_1$ we have $(nt)^2 = n^2$. □

Let α be as before and let B be the Borel subgroup of G containing T whose Lie algebra is $\mathfrak{t} \oplus \mathfrak{g}_\alpha$ (see 7.3.3 (ii)). Let $\chi \in X$. The composite of χ and the homomorphism $B \to B/B_u \to T$ defines a character of B, also denoted by χ.

7.3.6. Proposition. *Let* $f \in k[G]$ *be a regular function on* G *whose restriction to* (G, G) *is non-constant. Assume that for* $g \in G$, $b \in B$ *we have* $f(gb) = \chi(b) f(g)$. *Then* $\langle \chi, \alpha^\vee \rangle > 0$.

The restriction of f to (G, G) has the same property as f, relative to a Borel subgroup of (G, G). Since $\operatorname{Im} \alpha^\vee \subset (G, G)$ by 7.3.5 (i) we may work in (G, G), i.e. we may assume that G is semi-simple. By 7.2.4, G is isomorphic to $\mathbf{SL}_2$ or $\mathbf{PSL}_2$. Since $k[\mathbf{PSL}_2] \subset k[\mathbf{SL}_2]$ it suffices to deal with the case that $G = \mathbf{SL}_2$. Take B to be the upper triangular subgroup and T to be the diagonal subgroup. Then it is readily checked that for $x \in k^*$ we have

$$\alpha(\operatorname{diag}(x, x^{-1})) = x^2, \quad \alpha^\vee(x) = \operatorname{diag}(x, x^{-1}).$$

Put $\langle \chi, \alpha^\vee \rangle = a$. From (21) we see that there is a polynomial in one indeterminate g such that for $z \neq 0$

$$f \begin{pmatrix} 1 & 0 \\ z & 1 \end{pmatrix} = z^a g(z^{-1}).$$

But this function of $z \in k^*$ must be regular for $z = 0$, which can only be if $a \geq 0$. If $a = 0$ then g must be constant and it follows from (21) that f is constant. □

7.3.7. Exercises. $G = \mathbf{SL}_2$. The notations are as in the proof of 7.3.6.
(1) (a) Let $\chi(\operatorname{diag}(x, x^{-1})) = x^a$, with $a \geq 0$. Let $V_a \subset k[G]$ the set of functions with the property of 7.3.6. Show that V_a is a subspace of dimension $a + 1$ that is invariant under left translations.

(b) Let ρ_a be the rational representation of G in V_a by left translations. There exists a basis $(e_0, ..., e_a)$ of V_a such that

$$\rho_a \begin{pmatrix} x & 0 \\ 0 & x^{-1} \end{pmatrix} e_i = x^{a-2i} e_i \ (x \in k^*),$$

$$\rho_a \begin{pmatrix} 1 & x \\ 0 & 1 \end{pmatrix} e_i = \sum_{j=0}^{a} (-1)^{i-j} (i, j) x^{i-j} e_j \ (x \in k),$$

where (i, j) is a binomial coefficient.

(c) Let $p = \text{char } k$. If $p = 0$ or $p > a$ the representation ρ_a is irreducible.

(2) Let $r : G \to GL(V)$ be a finite dimensional rational representation of G and let $r^\vee : G \to GL(V^\vee)$ be the dual representation. So $V^\vee$ is the dual of V and for $v \in V$, $u \in V^\vee$ we have $(r^\vee(g)u)(v) = u(r(g)^{-1}v))$.

(a) There exist $\chi \in X$ and a non-zero $v \in V$ such that for $b \in B$ we have $r(b)v = \chi(b)v$. We have $a = \langle \chi, \alpha^\vee \rangle \geq 0$. Define a linear map $\phi : V^\vee \to k[G]$ by $(\phi u)(g) = u(r(g)v)$.

(b) If r is irreducible then ϕ is injective.

(c) Any irreducible representation is isomorphic to a quotient representation and also to a subrepresentation of some ρ_a.

(d) If $p = 0$ an irreducible rational representation of G is isomorphic to a unique ρ_a.

(3) Notations of the preceding exercises.

(a) $V_a^\vee$ has a basis $(f_0, ..., f_a)$ with the following properties

$$\rho_a^\vee \begin{pmatrix} x & 0 \\ 0 & x^{-1} \end{pmatrix} f_i = x^{-a+2i} f_i \ (x \in k^*),$$

$$\rho_a^\vee \begin{pmatrix} 1 & x \\ 0 & 1 \end{pmatrix} f_0 = \sum_{i=0}^{a} x^i f_i \ (x \in k).$$

If (f_i') is another basis with these properties there exists $a \in k^*$ with $f_i' = a f_i$.

(b) Let $r : G \to GL(V)$ be a rational representation with the following properties: $\dim V = a + 1$, there exists $v \in V - \{0\}$ with

$$\rho \begin{pmatrix} x & y \\ 0 & x^{-1} \end{pmatrix} v = x^a v \ (x \in k^*, y \in k)$$

and the vectors $(\rho(g) - 1)v \ (g \in G)$ span V. Then ρ is isomorphic to $\rho_a^\vee$. (*Hint* : use the map of ϕ of the preceding exercise.)

7.4. Root data

7.4.1. A *root datum* is a quadruple $\Psi = (X, R, X^\vee, R^\vee)$, where
(a) X and $X^\vee$ are free abelian groups of finite rank, in duality by a pairing $X \times X^\vee \to$ $\mathbf{Z}$, denoted by $\langle\ ,\ \rangle$;
(b) R and $R^\vee$ are finite subsets of X and $X^\vee$, and we are given a bijection $\alpha \mapsto \alpha^\vee$ of R onto $R^\vee$.

For $\alpha \in R$ we define endomorphisms s_α and $s_\alpha^\vee$ of X and $X^\vee$ by

$$s_\alpha(x) = x - \langle x, \alpha^\vee \rangle \alpha, \quad s_\alpha^\vee(y) = y - \langle \alpha, y \rangle \alpha^\vee \quad (x \in X, \ y \in X^\vee).$$

The following axioms are imposed.
(RD 1) *If* $\alpha \in R$ *then* $\langle \alpha, \alpha^\vee \rangle = 2$;
(RD 2) *If* $\alpha \in R$ *then* $s_\alpha R = R$, $s_\alpha^\vee(R^\vee) = R^\vee$.

From (RD 1) it follows that $s_\alpha^2 = 1$ and $s_\alpha \alpha = -\alpha$. The *Weyl group* $W = W(\Psi)$ of Ψ is the group of automorphisms of X generated by the s_α ($\alpha \in R$). Notice the symmetry in the definition between X and $X^\vee$. We see that $\Psi^\vee = (X^\vee, R^\vee, X, R)$ also defines a root datum, which is the *dual root datum*. R is the set of *roots* of Ψ and $R^\vee$ the set of *coroots*. Denote by Q the subgroup of X generated by R and put $V' = \mathbf{R} \otimes Q$. If $R \neq \emptyset$ then R is a *root system* in V' in the sense of [Bou2, Ch.VI, § 1]. This means that the following axioms are satisfied :
(RS 1) *R is finite and generates V', and $0 \notin R$;*
(RS 2) *If* $\alpha \in R$ *there is* $\alpha^\vee$ *in the dual of V' such that (with the previous notations)* $\langle \alpha, \alpha^\vee \rangle = 2$ *and the endomorphism s_α stabilizes R;*
(RS 3) *If* $\alpha \in R$ *then* $\alpha^\vee(R) \subset \mathbf{Z}$.

Likewise, $R^\vee$ is a root system. W is isomorphic to the Weyl group $W(R)$ of [loc.cit.].

7.4.2. Exercise. Let $\Psi = (X, R, X^\vee, R^\vee)$ be a root datum. Define a homomorphism $f : X \to X^\vee$ by $f(x) = \sum_{\alpha \in R} \langle x, \alpha^\vee \rangle \alpha^\vee$.
(a) For $\alpha \in R$ we have $f(\alpha) = \frac{1}{2} \langle \alpha, f(\alpha) \rangle \alpha^\vee$. Show that

$$X_0 = \operatorname{Ker} f = \{x \in X \mid \langle x, \alpha^\vee \rangle = 0 \text{ for all } \alpha \in R\}.$$

(b) Show that $Q \cap X_0 = \{0\}$ and that $Q + X_0$ has finite index in X.
(c) Show that the Weyl group W is finite.

7.4.3. Now let G be an arbitrary connected linear algebraic group. We use the notations introduced in 7.1. Let $\beta \in P'$ and consider the group G_β. Apply the observations of 7.3.2 to the reductive group $H = G_\beta / R_u(G_\beta)$. We find two non-trivial characters $\pm \alpha'$ of the image of T in H. Since this image is isomorphic to T we find two corresponding characters $\pm\alpha$ of T. We have $(\text{Ker }\alpha)^0 = (\text{Ker }\beta)^0$, which implies that α is a rational multiple of β. It follows that $\alpha \in P'$. The characters α so obtained, if β runs through P', are the *roots* of G relative to T (or of (G, T)). The set of roots is denoted by R or $R(G, T)$. It is empty if and only if G is solvable (by 7.1.3 (ii)).

Using 7.3.5 we obtain a map $\alpha \mapsto \alpha^\vee$ of R onto a subset $R^\vee$ of $X^\vee$. We claim that it is bijective. It suffices to prove injectivity. Now if α, $\beta \in R$ and $\alpha^\vee = \beta^\vee$, then

$$s_\alpha s_\beta(x) = x + \langle x, \alpha^\vee \rangle (\alpha - \beta) \ (x \in X).$$

Since $\langle \alpha - \beta, \alpha^\vee \rangle = 0$, all eigenvalues of the linear map $s_\alpha s_\beta$ of V are 1. But this linear map has finite order (because the Weyl group is finite). Hence $s_\alpha s_\beta = 1$ and $\alpha = \beta$. The elements of $R^\vee$ are the *coroots* (of G relative to T). We have defined the ingredients of a root datum. The axioms (RD 1) and (RD 2) hold (see 7.1.8 (i) and 7.1.4). So we can associate to a connected linear algebraic group G and a maximal torus T of G a root datum $\Psi = \Psi(G, T)$. Since maximal tori are conjugate, Ψ is uniquely determined by G, up to isomorphism. The same holds for its root system $R = R(G, T)$. This root system is *reduced*, i.e. it has the property of the following lemma.

7.4.4. Lemma. *If $\alpha \in R$, $c \in \mathbf{Q}$ and $c\alpha \in R$ then $c = \pm 1$.*

We have $G_\alpha = G_{c\alpha}$. The lemma follows from the observation that the pair of roots $\{\pm\alpha\}$ is uniquely determined by G_α. □

7.4.5. Let $(X, R, X^\vee R^\vee)$ be a root datum with Weyl group W. Fix a W-invariant positive definite symmetric bilinear form on V (see 7.1.7). A subset R^+ of R is a *system of positive roots* if there exists $x \in V$ with $(\alpha, x) \neq 0$ for all $\alpha \in R$ such that

$$R^+ = \{\alpha \in R \mid (\alpha, x) > 0\}.$$

An equivalent definition is: there is $\lambda \in X^\vee$ with $\langle \alpha, \lambda \rangle \neq 0$ for all $\alpha \in R$ such that

$$R^+ = \{\alpha \in R \mid \langle \alpha, \lambda \rangle > 0\}$$

(check this). It follows that R^+ has the following properties:
(a) the convex hull of R^+ in V does not contain 0,
(b) R is the disjoint union of R^+ and $-R^+$.
(a) and (b) imply:
(c) If α, $\beta \in R^+$ and $\alpha + \beta \in R$ then $\alpha + \beta \in R^+$. It also follows that $(R^+)^\vee$ is a system of positive roots in $R^\vee$ (the definition of this notion is clear).

Let B be a Borel subgroup of G containing T and let $\alpha \in R(G, T)$. By 6.4.7 (ii), $G_\alpha \cap B$ is a Borel group of G_α. Then $B' = G_\alpha \cap B / R_u(G_\alpha) \cap B$ is a Borel subgroup of the reductive group $G' = G_\alpha / R_u(G_\alpha)$, containing the image T' of T. Let $\pm\alpha'$ be the characters of T' corresponding to $\pm\alpha$. We see from 7.3.3 (ii) that $L(B')$ is the direct sum of $L(T')$ and a one dimensional weight space, whose weight is either α' or $-\alpha'$. It follows that B picks out one root from each pair of roots $\pm\alpha$. Let $R^+(B)$ be the set of roots so obtained, when α runs through $R(G, T)$.

7.4.6. Proposition. $R^+(B)$ *is a system of positive roots.*

Choose a rational representation $\phi : G \to GL(A)$ with a non-zero vector $a \in A$ such that B is the stabilizer of the line ka (5.5.3). With the notation of 7.3.6 there is a character χ of T such that $(\phi b).a = \chi(b)a$. Let l be a linear function on A and put $F(g) = l((\phi g)a)$, $(g \in G)$. Then $F \in k[G]$ and if $b \in B$ we have $F(gb) = \chi(b)F(g)$. Let α be a root and consider the restriction of F to G_α. Since the unipotent radical $R_u(G_\alpha)$ fixes a, the function F is the pull-back of a function $F' \in k[G']$ (see the definition of the sheaf of functions on a quotient in the proof of 5.5.5). It follows that F' has the property of 7.3.6 relative to the Borel group B'. Application of 7.3.6 shows that $\langle \chi, \alpha^\vee \rangle \geq 0$. If we had equality, then the restriction of F to G_α would be constant for all l, from which it would follow that G_α stabilized a, which is impossible. It follows that $(R^+(B))^\vee$ is a system of positive roots in $R^\vee$. Hence $R^+(B)$ is a system of positive roots in R. $\square$

7.4.7. Exercises. (1) $G = \mathbf{GL}_n$ and T is the subgroup of diagonal matrices.
   (a) Define characters α_{ij} of T by $\alpha_{ij}(\mathrm{diag}(x_1, \ldots, x_n)) = x_i x_j^{-1}$ ($1 \leq i, j \leq n$, $i \neq j$). Then α_{ij} is a root of (G, T).
   (b) $\mathfrak{g}$ is the direct sum of $\mathfrak{t}$ and the weight spaces $\mathfrak{g}_{\alpha_{ij}}$. Show that G is reductive.
   (c) The root datum $\Psi(G, T)$ is isomorphic to $(X, R, X^\vee, R^\vee)$ with $X = X^\vee = \mathbf{Z}^n$ (the pairing being the standard one), $R = R^\vee = \{\epsilon_i - \epsilon_j \mid i \neq j\}$, where (ϵ_i) is the canonical basis.
   (d) Let $B \subset G$ be the Borel subgroup of upper triangular matrices (6.2.11 (1)). Show that $R^+(B)$ is the set of α_{ij} with $i < j$.
(2) (a) $G_1 = \mathbf{SL}_n$ and T_1 is the subgroup of diagonal matrices, which is a maximal torus. With the notations of (1), the root datum $\Psi(G_1, T_1)$ is isomorphic to $(X_1, R_1, X_1^\vee, R_1^\vee)$, where $X_1 = X/\mathbf{Z}(\epsilon_1 + \ldots + \epsilon_n)$, $X_1^\vee = \{(x_1, \ldots, x_n) \in X^\vee \mid \sum_i x_i = 0\}$, with the obvious pairing. If $\pi : X \to X_1$ is the canonical map then $R_1 = \pi R$, $R_1^\vee = R^\vee$.
   (b) Let $Z \subset \mathbf{GL}_n$ be the subgroup of scalar matrices, which is the center of $\mathbf{GL}_n$. Put $G_2 = G/Z$, $T_2 = T/Z$. Then T_2 is a maximal torus in G_2. Show that $\Psi(G_2, T_2)$ is isomorphic with the dual $(X_1^\vee, R_1^\vee, X_1, R_1)$ of the root datum of (a) (defined in 7.4.1).
(3) (char $k \neq 2$) Let $V = k^{2n+1}$ ($n \geq 1$) and let Q be the quadratic form on V defined by

$$Q((\xi_0, \ldots, \xi_{2n+1})) = \xi_0^2 + \sum_{i=1}^n \xi_i \xi_{n+i}.$$

Write $(v, w) = Q(v+w) - Q(v) - Q(w)$, this is a symmetric bilinear form on V. Let

G be the identity component of the group of $t \in GL(V)$ with $Q(tv) = Q(v)$ $(v \in V)$.

(a) G is isomorphic to $\mathbf{SO}_{2n+1}$ (recall that $\mathbf{SO}_{2n+1}$ is connected by 2.2.2 (2) or 2.2.9 (1)).

(b) The Lie algebra $\mathfrak{g}$ is the set of all $t \in End(V)$ such that $(tv, w) + (v, tw) = 0$ $(v, w \in V)$ (*Hint*: First show that $\mathfrak{g}$ is contained in the set of these t and then use 5.5.9 (7))).

(c) The subgroup T of maps

$$(\xi_0, \ldots, \xi_{2n+1}) \mapsto (\xi_0, x_1\xi_1, \ldots, x_n\xi_n, x_1^{-1}\xi_{n+1}, \ldots, x_n^{-1}\xi_{2n+1}),$$

with $x_i \in k^*$, is a maximal torus of G. The maps $(x_1, ..., x_n) \mapsto x_i^\epsilon$ and $(x_1, ..., x_n) \mapsto x_i^\epsilon x_j^\eta$, where $i \neq j$ and $\epsilon, \eta = \pm 1$ define roots of (G, T).

(d) $\mathfrak{g}$ is the direct sum of $\mathfrak{t}$ and the weight spaces $\mathfrak{g}_\alpha$, where α runs through the roots of (b). Show that G is semi-simple.

(e) The root datum $\Psi(G, T)$ is isomorphic to $(X, R, X^\vee, R^\vee)$, where $X = X^\vee = \mathbf{Z}^n$ (standard pairing), $R = \{\pm\epsilon_i, \pm\epsilon_i \pm \epsilon_j \mid i \neq j\}$ and $R^\vee = \{\pm 2\epsilon_i, \pm\epsilon_i \pm \epsilon_j \mid i \neq j\}$.

(4) (char $k \neq 2$) Let $V = k^{2n}$ $(n \geq 2)$ and let Q be the quadratic form on V with

$$Q((\xi_1, \ldots, \xi_{2n})) = \sum_{i=1}^n \xi_i\xi_{n+i}.$$

Define $(,)$ and G as in (3) and let T be the subgroup of maps

$$(\xi_1, \ldots, \xi_{2n}) \mapsto (x_1\xi_1, \ldots, x_n\xi_n, x_1^{-1}\xi_{n+1}, \ldots, x_n^{-1}\xi_{2n}).$$

Then G is semi-simple, isomorphic to $\mathbf{SO}_{2n}$, and T is a maximal torus. The root datum $\Psi(G, T)$ is isomorphic to $(X, R, X^\vee, R^\vee)$, with $X = X^\vee = \mathbf{Z}^n$ and $R = R^\vee = \{\pm\epsilon_i \pm \epsilon_j \mid i \neq j\}$.

(5) Let $V = k^{2n}$ $(n \geq 1)$ and let $(,)$ be the alternating bilinear form with

$$((\xi_1, ..., \xi_{2n}), (\eta_1, ..., \eta_{2n})) = \sum_{i=1}^n (\xi_i\eta_{n+i} - \xi_{n+i}\eta_i).$$

Define T as in (4) and let G be the group of $t \in GL(V)$ with $(tv, tw) = (v, w)$ $(v, w \in V)$. Then G is connected, semi-simple, isomorphic to $\mathbf{Sp}_{2n}$, and T is a maximal torus of G. The root datum of (G, T) is isomorphic to the dual of the root datum of (3).

(6) (char $k = 2$). Define G as in (3).

(a) All elements of G fix $e_0 = (1, 0, ..., 0)$, hence there is a homomorphism $\phi : G \to GL(V/ke_0)$.

(b) Ker ϕ is trivial and Im $\phi \simeq \mathbf{Sp}_{2n}$.

(c) Show the assertions (c),(d),(e) of (3) remain true.

(d) The Lie algebra of G is the set of all $t \in End(V)$ such that $te_0 = 0$ and $(v, tv) = 0$ $(v \in V)$.

(7) (char $k = 2$). The notations are as in (4).

(a) G is a closed subgroup of the group of (6), and $\dim G = n(2n - 1)$.

(b) The Lie algebra of G is the set of all $t \in End(V)$ such that $(v, tv) = 0$ ($v \in V$).

(c) Show that the statements of (4) remain true.

The root systems introduced in (1),(3),(4),(5) are the root systems of respective types A_{n-1}, B_n, D_n, C_n.

7.5. Two roots

Let $\Psi = (X, R, X^\vee, R^\vee)$ be a root datum. Let W be the Weyl group of Ψ. It is a finite group (7.4.2 (c)). Let α, $\beta \in R$ be linearly independent roots. By 7.4.4 we have $\alpha \neq \pm\beta$.

7.5.1. Lemma. *(i) $\langle \alpha, \beta^\vee \rangle$ has one of the values $0, 1, 2, 3$. If $|\langle \alpha, \beta^\vee \rangle| > 1$ then $|\langle \beta, \alpha^\vee \rangle| = 1$;*
(ii) In the four cases of (i) the order of $s_\alpha s_\beta$ is, respectively, $2, 3, 4, 6$;
(iii) If $\langle \alpha, \beta^\vee \rangle = 0$ then $\langle \beta, \alpha^\vee \rangle = 0$.

s_α and s_β stabilize the two dimensional subspace of $V = \mathbf{R} \otimes X$ spanned by α and β. On the basis (α, β) of that space $s_\alpha s_\beta$ is represented by the matrix

$$M_{\alpha\beta} = \begin{pmatrix} \langle \alpha, \beta^\vee \rangle \langle \beta, \alpha^\vee \rangle - 1 & \langle \beta, \alpha^\vee \rangle \\ -\langle \alpha, \beta^\vee \rangle & -1 \end{pmatrix}.$$

Since W is finite $s_\alpha s_\beta$ has finite order. Hence the eigenvalues of $M_{\alpha\beta}$ are two conjugate roots of unity and the absolute value of the trace of $M_{\alpha\beta}$ is at most 2. As $M_{\alpha\beta}$ cannot be the identity matrix, the eigenvalues cannot both be 1. Now (i) and (ii) readily follow.

If $\langle \alpha, \beta^\vee \rangle = 0$ then $M_{\alpha,\beta}$ is a triangular matrix which can only have finite order if it is diagonal. This implies (iii). $\square$

Fix a system of positive roots R^+ in R. If $\alpha \in R$ we write $\alpha > 0$ if $\alpha \in R^+$ and $\alpha < 0$ if $-\alpha \in R^+$.

7.5.2. Lemma. *There exists $w \in W$ such that $w\alpha > 0$, $w\beta > 0$.*

We may reduce the proof to the case that X is spanned by α and β. Put $a = \langle \alpha, \beta^\vee \rangle \langle \beta, \alpha^\vee \rangle$, then $a = 0, 1, 2, 3$ by 7.5.1 (i). If $a = 0$ then $\langle \alpha, \beta^\vee \rangle = \langle \beta, \alpha^\vee \rangle = 0$ by 7.5.1 (iii). If $a = 0$ then $-s_\alpha \alpha = s_\beta \alpha = \alpha$ and similarly for β. The assertion readily follows.

Now assume $a > 0$. We may assume that $\alpha < 0$. If also $\beta < 0$ we may take $w = s_\alpha s_\beta s_\alpha$ if $a = 1$, for then $w\alpha = -\beta$, $w\beta = -\alpha$. If $a > 1$ then $s_\alpha s_\beta$ has even order (7.5.1 (ii)) and a power of that element maps α and β onto their negatives. It follows that we may assume $\alpha < 0$ and $\beta > 0$. If $\langle \beta, \alpha^\vee \rangle > 0$ we may take $w = s_\alpha$. So assume $\langle \beta, \alpha^\vee \rangle < 0$. Let $a = 1$. If $\alpha + \beta > 0$ we can again take $w = s_\alpha$. If $\alpha + \beta < 0$ the formula for the matrix $M_{\alpha\beta}$ of the proof of 7.5.1 shows that we may take $w = s_\alpha s_\beta$. If $a > 1$ we may assume, replacing if necessary α and β by $-\beta$ and $-\alpha$, that $\langle \alpha, \beta^\vee \rangle = -1$.

Let $a = 2$. Then

$$s_\alpha\alpha = -\alpha, \qquad s_\beta\alpha = \alpha + \beta, \quad s_\alpha s_\beta\alpha = \alpha + \beta,$$
$$s_\alpha\beta = 2\alpha + \beta, \quad s_\beta\beta = -\beta, \qquad s_\alpha s_\beta\beta = -2\alpha - \beta.$$

If $2\alpha + \beta > 0$ we took $w = s_\alpha$. If $\alpha + \beta < 0$ take $w = -s_\beta$. If $2\alpha + \beta < 0$, $\alpha + \beta > 0$ take $w = s_\alpha s_\beta$.

Finally let $a = 3$. Then

$$s_\alpha\alpha = -\alpha, \quad s_\beta\alpha = \alpha + \beta, \quad s_\alpha s_\beta\alpha = 2\alpha + \beta, \quad s_\beta s_\alpha\alpha = -\alpha - \beta,$$
$$s_\alpha\beta = 3\alpha + \beta, \quad s_\beta\beta = -\beta, \quad s_\alpha s_\beta\beta = -3\alpha - \beta, \quad s_\beta s_\alpha\beta = 3\alpha + 2\beta.$$

and

$$s_\alpha s_\beta s_\alpha\alpha = -2\alpha - \beta, \quad s_\alpha s_\beta s_\alpha\beta = 3\alpha + 2\beta.$$

If $3\alpha + \beta > 0$ we took $w = s_\alpha$. If $\alpha + \beta < 0$ take $w = -s_\beta$, if $3\alpha + \beta < 0$, $2\alpha + \beta > 0$ take $w = s_\alpha s_\beta$, if $\alpha + \beta > 0$, $3\alpha + 2\beta < 0$ take $w = -s_\beta s_\alpha$ and if $2\alpha + \beta < 0$, $3\alpha + 2\beta > 0$ take $w = s_\alpha s_\beta s_\alpha$. $\qquad\square$

Let $(\ ,\ )$ be a positive definite symmetric bilinear form with the property of 7.1.7. If $\alpha, \beta \in R$ and $\langle \alpha, \beta^\vee \rangle = -1$, $\langle \beta, \alpha^\vee \rangle = -a$ then α and β can be viewed as two vectors in $\mathbf{R}^2$ such that the ratio of their lengths is $\sqrt{a}$. If ϕ is the angle between them, we have $\cos\phi = -\frac{1}{2}\sqrt{a}$. One could also prove the lemma in a geometric way, using these facts.

7.5.3. Exercises. (1) Work out a geometric proof of 7.5.2.
(2) Let α and β be as in 7.5.2. If $a > 0$ assume that $\langle \alpha, \beta^\vee \rangle = -1$, $\langle \beta, \alpha^\vee \rangle = -a$. The set S described below is a subset of R that is stable under s_α and s_β.
$a = 0$, $S = \{\pm\alpha, \pm\beta\}$,
$a = 1$, $S = \{\pm\alpha, \pm\beta, \pm(\alpha + \beta)\}$,
$a = 2$, $S = \{\pm\alpha, \pm\beta, \pm(\alpha + \beta), \pm(2\alpha + \beta)\}$,
$a = 3$, $S = \{\pm\alpha, \pm\beta, \pm(\alpha + \beta), \pm(2\alpha + \beta), \pm(3\alpha + \beta), \pm(3\alpha + 2\beta)\}$.
(This exercise gives the classification of root systems of rank two, to be used in 9.1.)

7.6. The unipotent radical

7.6.1. G is again a connected linear algebraic group, and T a maximal torus. The main result of this section is the characterization of the unipotent radical $R_u G$ given in 7.6.3.

We denote by C the identity component of the intersection of the unipotent parts B_u of the Borel subgroups B of G that contain T. If α is a root of (G, T) we denote by C_α the identity component of the intersection of the B_u, where B runs over the Borel subgroups containing T with $\alpha \in R^+(B)$. Then C is a closed subgroup of C_α.

7.6.2. Lemma. $\dim C_\alpha / C \leq 1$.

Since T normalizes C_α it follows that the subalgebra $L(C_\alpha)$ of the Lie algebra $\mathfrak{g}$ is spanned by weight vectors of T. With the notations of 7.1, such a weight vector X lies in the Lie algebra of some G_γ with $\gamma \in P$. It follows from 6.4.7 (ii) that $R_u(G_\gamma)$ lies in all Borel subgroups containing T. So if $X \in L(R_u(G_\gamma))$ we have $X \in L(C)$. We conclude that $L(C_\alpha)$ is spanned by $L(C)$ and root vectors (i.e. weight vectors for roots). If X is a root vector for a root β then $\beta \in R^+(B)$ for all Borel groups $B \supset T$ with $\alpha \in R^+(B)$. Fix a Borel subgroup B with the latter property and put $R^+ = R^+(B)$. Using 6.4.12 we conclude that (with the notation of 7.5.2) we have $w\beta > 0$ for all elements w of the Weyl group W with $w\alpha > 0$. If α and β are linearly independent, application of 7.5.2 to α and $-\beta$ leads to a contradiction, and we must have $\beta = \alpha$. The assertion follows from the fact that the space $\mathfrak{g}_\alpha$ of 7.3.2 is one dimensional. □

7.6.3. Theorem. $R_u G = C$.

It is clear that $R_u G \subset C$. To prove the reverse inclusion it suffices to show that C is a normal subgroup of G. Using 7.1.3 (i) we see that it also suffices to prove that for each $\gamma \in P$ the group G_γ normalizes C. Now G_γ is generated by its Borel subgroups that contain T (this is trivial if G_γ is solvable and follows from a similar fact for $\mathbf{SL}_2$ if G_γ is non-solvable). Such a Borel group lies either in $T.C$, or in a subgroup $T.C_\alpha$, for some root α (by 6.4.7 (ii)).It follows from 7.6.2 and 6.3.7 (4) that C is a normal subgroup of C_α, which proves the theorem. □

7.6.4. Corollary. *Assume G to be reductive.*
(i) If S is a subtorus of G then $Z_G(S)$ is connected and reductive;
(ii) $Z_G(T) = T$, i.e. Cartan subgroups are maximal tori;
(iii) The center $C(G)$ of G lies in T.

To prove (i) we may assume that $S \subset T$. (i) follows from the theorem, using 6.4.7. Assertion (ii) follows from (i) with $S = T$, and 6.4.2 (i). For (iii) observe that $C(G) \subset Z_G(T)$. □

We have now assembled the ingredients needed for the theory of reductive groups, to be taken up in the next chapter.

Notes

The introduction of the combinatorial data of a linear algebraic group (roots...) is due to Chevalley. The main results of this chapter are all due to him; they are exposed in [Ch4].

Roots were introduced in [loc. cit., Exp. 12]. We have taken here as the fundamental combinatorial notion the notion of a root datum (following [SGA3, exp. XXI]). We have introduced this as soon as possible, for not necessarily reductive groups.

The proof of the classification theorem 7.2.4 exploits the 'Bruhat decomposition' 7.2.2 (i). The argument goes back to Zassenhaus (see [Gor, p. 393]). It is convenient to have 7.2.4 available at an early stage. A consequence is the integrality result 7.3.5 (i).

The important characterization 7.6.3 of the unipotent radical is due to Chevalley [Ch4, exp.12]. We have tried to stress the role played in his proof by roots and Weyl group.

Chapter 8

Reductive Groups

This chapter presents the fundamental results about reductive groups. G denotes a connected, reductive, linear algebraic group and T is a maximal torus of G.

8.1. Structural properties of a reductive group

Let $(X, R, X^\vee, R^\vee)$ be the root datum of (G, T) (see 7.4.3).

8.1.1. Proposition. *(i) For $\alpha \in R$ there exists an isomorphism u_α of $\mathbf{G}_a$ onto a unique closed subgroup U_α of G such that $t u_\alpha(x) t^{-1} = u_\alpha(\alpha(t)x)$ $(t \in T, \ x \in k)$. We have $\operatorname{Im} d u_\alpha = \mathfrak{g}_\alpha$, the weight space for the weight α of T;*
(ii) T and the U_α ($\alpha \in R$) generate G.

Weight spaces were defined in 7.1.1.
If $\alpha \in R$ the group G_α of 7.1 is reductive by 7.6.4 (i) and has semisimple rank one. The existence of u_α and the last point of (i) then follow from 7.3.3. If U_α is as in (i) then $U_\alpha \subset G_\alpha$ and the uniqueness of U_α follows from 7.3.3 (ii) (recall that there are two Borel subgroups of G_α containing T). (ii) follows from 7.1.3 (i), observing that a group G_α is generated by T, U_α and $U_{-\alpha}$ (which follows from the formula of 7.3.2). $\qquad\square$

8.1.2. Corollary. *The roots of R are the non-zero weights of T in $\mathfrak{g}$. For each $\alpha \in R$ the weight space $\dim \mathfrak{g}_\alpha$ has dimension one.*

If $\beta \in P$ (defined in 7.1.1) then G_β is reductive by 7.6.4 (i) and has semi-simple rank one. It must be a G_α with $\alpha \in R$. By the formula of 7.3.2 we have $\beta = \pm\alpha$. The last point follows from 8.1.1 (i). $\qquad\square$

8.1.3. Corollary. *Let B be a Borel subgroup of G containing T and let $\alpha \in R$.*
(i) The following properties are equivalent: (a) $\alpha \in R^+(B)$, (b) $U_\alpha \subset B$, (c) $\mathfrak{g}_\alpha \subset \mathfrak{b}$;
(ii) $\dim B = r + \frac{1}{2}|R|$, $\dim G = r + |R|$, where $r = \dim T$.

Here $R^+(B)$ is as in 7.4.6. (i) follows from the definition of $R^+(B)$, using 7.3.3 (ii). Using 8.1.2 one determines $\dim \mathfrak{b}$ and $\dim \mathfrak{g}$, giving (ii). $\qquad\square$

Let $W = N_G(T)/T$ be the Weyl group of (G, T) (see 7.1.4). We identify it with the Weyl group of the root datum of (G, T). For $\alpha \in R$ we have the reflection $s_\alpha \in W$ of 7.1.8. Then $s_{-\alpha} = s_\alpha$.

8.1.4. Lemma. *(i) The u_α may be chosen such that for all $\alpha \in R$*

$$n_\alpha = u_\alpha(1)u_{-\alpha}(-1)u_\alpha(1)$$

lies in $N_G(T)$ and has image s_α in W. For $x \in k^*$ we have

$$u_\alpha(x)u_{-\alpha}(-x^{-1})u_\alpha(x) = \alpha^\vee(x)n_\alpha; \tag{22}$$

(ii) $n_\alpha^2 = \alpha^\vee(-1)$ and $n_{-\alpha} = n_\alpha^{-1}$;
(iii) If $u \in U_\alpha - \{1\}$ there is a unique $u' \in U_{-\alpha} - \{1\}$ such that $uu'u \in N_G(T)$;
(iv) If $(u'_\alpha)_{\alpha \in R}$ is a second family with the properties of 8.1.1 (i) and 8.1.4 (i), there exist $c_\alpha \in k^*$ such that

$$u'_\alpha(x) = u_\alpha(c_\alpha x), \ c_\alpha c_{-\alpha} = 1 \ (\alpha \in R, \ x \in k).$$

We have $U_\alpha \subset (G_\alpha, G_\alpha)$ (by the formula in the proof of 7.3.3). The latter group is isomorphic to $\mathbf{SL}_2$ or $\mathbf{PSL}_2$ by 7.2.4. Using the homomorphism $\mathbf{SL}_2 \to \mathbf{PSL}_2$ of the proof of 7.2.4 and the description of $\alpha^\vee$ given in the proof of 7.3.5, we reduce the proof of (i) to the case that $G = \mathbf{SL}_2$, T being the diagonal torus. Define the character α of T by $\alpha(\mathrm{diag}(x, x^{-1})) = x^2$. A straightforward check shows that, with the notations of the proof of 7.2.4, we may take

$$u_\alpha = u_1, \ u_{-\alpha}(x) = n_1 u_\alpha(-x)n_1^{-1} = \begin{pmatrix} 1 & 0 \\ x & 1 \end{pmatrix} (x \in k).$$

Then $n_\alpha = n_1$ and 7.3.5 (ii) gives the first formula of (ii). The second formula follows by a check in $\mathbf{SL}_2$. The existence of u' with the property of (iii) follows from (i). The uniqueness is proved by a computation in $\mathbf{SL}_2$ which is left to the reader. Finally, (iv) is a consequence of 7.3.3 (i) and (iii). $\qquad\square$

We shall call a family $(u_\alpha)_{\alpha \in R}$ with the properties of 8.1.1 (i) and 8.1.4 (i) a *realization* of the root system $R = R(G, T)$ in G. Notice that by (22) a realization determines the coroots $\alpha^\vee$.

8.1.5. Theorem. *Assume that G is semi-simple.*
(i) The U_α $(\alpha \in R)$ generate G;
(ii) $G = (G, G)$;
(iii) Let $G_1 \neq G$ be a non-trivial connected, closed, normal subgroup of G. Then G_1 is semi-simple. There is a similar subgroup G_2 such that $(G_1, G_2) = \{e\}$, $G_1 \cap G_2$ is finite and $G = G_1 G_2$;
(iv) The number of minimal non-trivial connected, closed, normal, subgroups of G is finite. If $G_1, ..., G_r$ are these groups then $(G_i, G_j) = \{e\}$ if $i \neq j$ and $G_i \cap \Pi_{j \neq i} G_j$ is finite. Moreover $G = G_1...G_r$ and the G_i have no closed, normal subgroups of dimension > 0.

Semi-simple groups were defined in 6.4.14. The intersection of the kernels of all roots is a subgroup H of T that is centralized by all U_α, hence is normal in G because of 8.1.1 (ii). Since G is semi-simple H must be finite. This implies that the roots span a subgroup of finite index of X, and also that the groups Im $\alpha^\vee$ ($\alpha \in R$) generate T. Now (i) follows from 8.1.1 (ii) and (22). Since $U_\alpha \subset (G_\alpha, G_\alpha) \subset (G, G)$, (ii) also follows.

Next let G_1 be as in (iii). Since $R(G_1)$ is normal in G, it must be trivial. Hence G_1 is semi-simple. Let T_1 be a maximal torus of G_1. We may assume that $T_1 \subset T$. If $\alpha \in R$ and $U_\alpha \not\subset G_1$ it follows from the formula of the proof of 7.3.3 that $T_1 \subset \text{Ker } \alpha$, so that T_1 centralizes U_α. Conversely, if this is so then $U_\alpha \not\subset G_1$ by 7.6.4 (ii). Put $R_1 = \{\alpha \in R \mid U_\alpha \subset G_1\}$, $R_2 = R - R_1$. Then $R_1, R_2 \neq \emptyset$. Let $\alpha \in R_1$, $\beta \in R_2$. Put

$$u_y(x) = u_\beta(y)u_\alpha(x)u_\beta(-y) \ (x, \ y \in k).$$

Then $u_y(x) \in G_1$ and $tu_yt^{-1} = u_y(\alpha(t)x)$ for $t \in T_1$. It follows from 8.1.1 (i) that there is a morphism $f : \mathbf{A}^1 \to \mathbf{A}^1$ such that $u_y(x) = u_\alpha(f(y)x)$ and $f(y) \neq 0$ for all y. This can only be if f is the constant $f(0) = 1$. It follows that U_α and U_β commute. If G_2 is the subgroup generated by the U_β with $\beta \in R_2$, then $(G_1, G_2) = \{e\}$, $G = G_1G_2$. That $G_1 \cap G_2$ is finite follows from the fact that it is a closed normal subgroup not containing any U_α. This proves (iii) and (iv) is a consequence of (iii). $\square$

We drop the assumption that G be semi-simple.

8.1.6. Corollary. *(i)* $G = R(G).(G, G)$;
(ii) (G, G) *is semi-simple.*

(i) follows from 8.1.5 (ii), applied to the group $G/R(G)$, which is semi-simple. (ii) is a consequence of (i). $\square$

Recall that by 7.3.1, $R(G)$ is a central torus and that $R(G) \cap (G, G)$ is finite. If A is a subgroup of X we denote by

$$A^\perp = \{y \in X^\vee \mid \langle A, y \rangle = \{0\}\}$$

its annihilator, a subgroup of $X^\vee$. The group

$$\tilde{A} = \{x \in X \mid \mathbf{Z}.x \cap A \neq \{0\}\}$$

is the *rational closure* of A in X. Then $\tilde{A}/A$ is the torsion subgroup of X/A. We have similar notions for subgroups of $X^\vee$.

8.1.7. Lemma. *For any subgroup A of X we have* $\tilde{A} = (A^\perp)^\perp$.

In the case that A is a direct summand of X we have $\tilde{A} = A$, and an easy argument proves the lemma. We have this case if X/A has no torsion, in fact then X/A is free (see [Jac4, p. 184]), which implies that A is a direct summand. We conclude that the lemma holds for $\tilde{A}$. The general case follows by observing that $(\tilde{A})^{\perp} = A^{\perp}$. □

Let Q $(Q^{\vee})$ be the subgroup of X $(X^{\vee})$ spanned by R (respectively: $R^{\vee}$).

8.1.8. Proposition. *(i) The center $C(G)$ of G is the intersection of the kernels Ker α $(\alpha \in R)$;*
(ii) The radical $R(G)$ is the subgroup of T generated by the groups Im y, $y \in Q^{\perp}$. We have $X^{*}(R(G)) \simeq X/\tilde{Q}$, $X_{*}(R(G)) \simeq Q^{\perp}$;
(iii) The subtorus T_1 of T generated by the groups Im $\alpha^{\vee}$, $(\alpha \in R)$ is a maximal torus of (G, G). We have $X^{*}(T_1) \simeq X/(Q^{\vee})^{\perp}$, $X_{*}(T_1) \simeq \tilde{Q}^{\vee}$.

$C(G)$ lies in T by 7.6.4 (iii). Then (i) follows from 8.1.1. Let $y \in X_{*}(T)$. By (i), y lies in the subgroup $X_{*}(R(G))$ if and only if $y \in Q^{\perp}$. This proves the first and the last statements of (ii). We have $X^{*}(R(G)) = (Q^{\perp})^{\vee}$. By 3.2.10 (4), this is a quotient of X. It then follows that $(Q^{\perp})^{\vee} \simeq X/(Q^{\perp})^{\perp} = X/\tilde{Q}$, proving (ii).

We have $X_{*}(T_1) = Q^{\vee}$. Introduce a positive definite form $(\ ,\ )$ on $V = \mathbf{R} \otimes X$ as in 7.1.7. We may assume $X^{\vee} \subset V$, such that $\alpha^{\vee} = 2(\alpha, \alpha)^{-1}\alpha$ $(\alpha \in R)$. Then $\mathbf{R} \otimes Q^{\perp}$ is the orthogonal complement of $\mathbf{R} \otimes Q^{\vee}$. It follows that $T = T_1.R(G)$ and that $T_1 \cap R(G)$ is finite. Hence $\dim T_1 = \dim T/R(G)$, the dimension of a maximal torus of the semi-simple group $G/R(G)$, which by 8.1.6 is a quotient of (G, G) by a finite group. Then $\dim T_1$ must equal the dimension of a maximal torus of (G, G), which proves the first point of (iii). Using 3.2.10 (4) we conclude that the character group $X^{*}(T_1)$ is isomorphic to $X/(Q^{\vee})^{\perp}$. Then its dual $X_{*}(T_1)$ is isomorphic to $((Q^{\vee})^{\perp})^{\perp} = \tilde{Q}^{\vee}$. □

We also have that $\mathbf{R} \otimes Q$ is the orthogonal complement of $\mathbf{R} \otimes (Q^{\vee})^{\perp}$. This implies that R maps injectively into $X/(Q^{\vee})^{\perp}$. We identify R with its image.

8.1.9. Corollary. *The root datum of $((G, G), T_1)$ is $(X/(Q^{\vee})^{\perp}, R, \tilde{Q}^{\vee}, R^{\vee})$.* This follows from 8.1.8. □

We have the surjective product map $(G, G) \times R(G) \to G$. By 8.1.8 (ii) and 8.1.9 the root datum of the first group (relative to the maximal torus $T_1 \times R(G)$) is

$$(X/(Q^{\vee})^{\perp} \oplus X/\tilde{Q}, (R, \{0\}), \tilde{Q}^{\vee} \oplus Q^{\perp}, (R^{\vee}, \{0\})).$$

The product map induces an injective homomorphism of character groups

$$i : X \to X/(Q^{\vee})^{\perp} \oplus X/\tilde{Q}.$$

8.1.10. Lemma. $\operatorname{Im} i = \{(y + (Q^\vee)^\perp, z + \tilde{Q} \mid y - z \in (Q^\vee)^\perp \oplus \tilde{Q}\}.$

The proof is straightforward.   □

8.1.11. Assume that G is semi-simple. It follows from 8.1.8 (ii) that $Q^\perp = \{0\}$, which implies that $X = \tilde{Q}$ and that Q has finite index in X. The finite group $C^* = C^*(G) = X/Q$ is the *cocenter* of G. Put

$$ P = \{x \in V \mid \langle x, R^\vee \rangle \subset \mathbf{Z}\}, $$

where V is as before. Then P is a lattice in V with $Q \subset P$. If the root system R in V is given, there are only finitely many possibilities for X. Q is the *root lattice* of R and P the *weight lattice*. The semi-simple group G is *adjoint* if $X = Q$ and *simply connected* if $X = P$. The finite abelian group P/Q is the *fundamental group* of R.

8.1.12. Exercises. The notations are as before.
(1) Let H be a connected subgroup of G containing T, with Lie algebra $\mathfrak{h}$.
   (a) For $\alpha \in R$ we have $U_\alpha \subset H$ if and only if $\mathfrak{g}_\alpha \subset \mathfrak{h}$. (*Hint*: If $\mathfrak{g}_\alpha \subset \mathfrak{h}$ consider $Z_H((\operatorname{Ker}\alpha)^0)$).
   (b) Let R' be the set of roots $\alpha \in R$ with $U_\alpha \subset H$. Then $\mathfrak{h} = L(T) + \sum_{\alpha \in R'} \mathfrak{g}_\alpha$ and $\dim H = \dim T + |R'|$.
(2) Let w be an element of the Weyl group of (R, T), represented by $n \in N_G(T)$. If $\alpha \in R$ there is $c_{n,\alpha} \in k^*$ with $n u_\alpha(x) n^{-1} = u_{w.\alpha}(c_{n,\alpha}x)$ $(x \in k)$.
(3) (a) Let $t \in T$. The connected centralizer $Z_G(t)^0$ is generated by T and the U_α with $\alpha(t) = 1$. (*Hint*: use 5.4.7.)
   (b) The centralizer of a semi-simple element of G is reductive.
(4) (a) Define the notion of a direct sum of root systems.
   (b) A root system is *reducible* if it is a non-trivial direct sum and *irreducible* otherwise. Show that if G is semi-simple, the root system R is irreducible if and only if G has no proper connected, closed, normal, subgroup, in which case G is said to be *quasi-simple*. The root systems of the groups G_i of 8.1.5 (iv) are irreducible.
(5) (a) The cokernel of the homomorphism i of 8.1.10 is isomorphic to $X/((Q^\vee)^\perp \oplus \tilde{Q})$.
   (b) The cocenter $C^*((G, G))$ is isomorphic to $X/((Q^\vee)^\perp \oplus Q)$.
(6) (a) (G, G) is adjoint if and only if $X = Q \oplus (Q^\vee)^\perp$;
   (b) (G, G) is simply connected if and only if $\tilde{Q}^\vee = Q^\vee$.
(7) Determine root lattice, weight lattice and fundamental group for each of the root systems A_{n-1}, B_n, C_n, D_n introduced in 7.4.7.
(8) Assume that G is semi-simple. There is an injective homomorphism

$$ \operatorname{Hom}(C(G), k^*) \to C^*(G) $$

that is bijective if the characteristic does not divide the order of the center $C(G)$.

8.2. Borel subgroups and systems of positive roots

Let B be a Borel subgroup of G containing T. We denote its unipotent radical by B_u or U and write R^+ for the system of positive roots $R^+(B)$ defined by B (see 7.4.6).

8.2.1. Proposition. *Let $(\alpha_1, \alpha_2, ..., \alpha_m)$ be a numbering of the roots in R^+. The morphism $\phi : \mathbf{G}_a^m \to B_u$ with $\phi(x_1, \ldots, x_m) = u_{\alpha_1}(x_1)u_{\alpha_2}(x_2)\ldots u_{\alpha_m}(x_m)$ is an isomorphism of varieties. In particular, B_u is generated by the groups U_α with $\alpha \in R^+$.*

Taking into account 8.1.1 and 8.1.2, the proposition follows from the following result.

8.2.2. Lemma. *Let H be a connected, solvable linear algebraic group. Let S be a maximal torus of H. Assume that there is a set of isomorphisms v_i $(1 \le i \le n)$ of $\mathbf{G}_a$ onto closed subgroups of H such that*
(a) there exist non-trivial characters β_i of S, no two of which are linearly dependent, with $s v_i(x)s^{-1} = v_i(\beta_i(s)x)$ $(s \in S,\ x \in k,\ 1 \le i \le n)$,
(b) the weight spaces $\mathfrak{h}_{\beta_i}$ $(1 \le i \le n)$ are one dimensional and span $L(H_u)$.
Then the morphism $\psi : \mathbf{G}_a^n \to H_u$ with $\psi(x_1, \ldots, x_n) = v_1(x_1)\ldots v_n(x_n)$ is an isomorphism of varieties.

The proof of 8.2.2 is by induction on n. If $n = 1$ we have $H_u = \operatorname{Im} v_1$ (look at dimensions). If $n > 1$ let N be a normal subgroup in the center of H_u isomorphic to $\mathbf{G}_a$ (see 6.3.4). Then $L(N)$ is an S-stable one dimensional subspace of $L(H_u)$, which by our assumptions must be one of the weight spaces $\mathfrak{h}_{\beta_j}$. Application of 5.4.7 shows that $Z_H(\operatorname{Ker}(\beta_j)^0)$ is a group with the properties of H and a one dimensional unipotent radical (here one uses the linear independence of the β_i). By what was established in the case $n = 1$ we have $N = \operatorname{Im} v_j$.

For $i \ne j$ let $w_i : \mathbf{G}_a \to H/N$ be the homomorphism induced by v_i. We claim that H/N and the w_i satisfy the assumptions of the lemma, relative to the image of S in H/N. This is clear except for the fact that the w_i are isomorphisms. Since $\operatorname{Im}(v_i) \cap \operatorname{Im}(v_j) = \{e\}$ $(i \ne j)$ (check this), w_i is injective. From (b) it follows that the differential dw_i is also injective. By 5.3.3 (ii), w_i is an isomorphism onto a subgroup of H/N, and the claim follows.

By induction we may assume the result to be true for H/N. Since N is central it readily follows that ψ is bijective. Using 4.4.12 we see that the tangent map $d\psi_{(0,\ldots,0)}$ is bijective. By 4.3.6 (i) and 5.1.6 ψ is birational. It now follows from 5.3.4 and 5.2.8 that ψ is an isomorphism. □

As before, $(g, h) = ghg^{-1}h^{-1}$ $(g,\ h \in G)$. We fix a total order of R (i.e., a numbering of the roots).

8.2.3. Proposition. *Let α, $\beta \in R$, with $\alpha \neq \pm\beta$. There exist constants $c_{\alpha,\beta;i,j} \in k$ such that*

$$(u_\alpha(x), u_\beta(y)) = \prod_{i\alpha+j\beta\in R;\ i,j>0} u_{i\alpha+j\beta}(c_{\alpha,\beta;i,j}x^i y^j)\ (x,\ y \in k),$$

the order of the factors in the right-hand side being prescribed by the ordering of R.

Using 7.5.2 and 8.1.12 (2) we see that it is sufficient to prove this when α, $\beta \in R^+$. Then U_α, $U_\beta \subset B_u$ and

$$(u_\alpha(x), u_\beta(y)) = \prod_{\gamma\in R^+} u_\gamma(P_\gamma(x, y)),$$

where $P_\gamma \in k[T, U]$, the order of the factors being the prescribed one. Conjugating by $t \in T$ we obtain

$$P_\gamma(\alpha(t)x, \beta(t)y) = \gamma(t)P_\gamma(x, y).$$

Using the linear independence of characters we see that $P_\gamma \neq 0$ only if $\gamma = i\alpha + j\beta$, with $i,\ j \geq 0$. Since α and β are linearly independent (7.4.4), such i and j are unique. To finish the proof we have to show that i and j are non-zero. Suppose, for example, that in the formula of the proposition we had a non-trivial factor with $j = 0$. By 7.4.4 the corresponding i equals 1. Then the commutator $(u_\alpha(x), u_\beta(y))$ would equal $u_\alpha(cx)$ times a product like the one in the proposition, with $c \neq 0$. Putting $y = 0$ we arrive at a contradiction. $\square$

8.2.4. Proposition. *Let $\tilde{R}^+$ be an arbitrary system of positive roots in R (7.4.5).*
(i) T and the U_α with $\alpha \in \tilde{R}^+$ generate a Borel subgroup of G;
(ii) There is a unique $w \in W$ with $\tilde{R}^+ = w.R^+$.

As before, W is the Weyl group of (G, T). For $n \geq 1$ denote by $\tilde{R}_n^+$ the set of roots in $\tilde{R}^+$ that are integral linear combinations with strictly positive coefficients of at least n roots of $\tilde{R}^+$, and let U_n be the subgroup of G generated by the U_α with $\alpha \in \tilde{R}_n^+$. Using property (c) of 7.4.5 and the preceding proposition, one proves by descending induction on n that U_n is a closed, connected, unipotent, subgroup of G, normalized by T. Using 8.2.2 one sees that $\dim U_n = |\tilde{R}_n^+|$. It follows that $\tilde{B} = T.U_1$ is a closed, connected, solvable, subgroup of G, of dimension $\dim T + \frac{1}{2}|R|$. By 6.2.7 (iii) and 8.1.3 (ii) $\tilde{B}$ must be a Borel group containing T with $R^+(\tilde{B}) = \tilde{R}^+$. Now (ii) follows from 6.4.12. $\square$

Denote by $\mathcal{W}$ the set of Borel subgroups of G that contain T. By 6.4.12 the Weyl group W acts simply transitively on $\mathcal{W}$. We say that B, $B' \in \mathcal{W}$ are *adjacent* if

$$\dim(B \cap B') = \dim B - 1 = \dim B' - 1.$$

We say that two systems of positive roots R^+ and $\tilde{R}^+$ are adjacent if $R^+ \cap \tilde{R}^+$ has one element less than R^+. As in 7.1.7, let $(\ ,\ )$ be a positive definite symmetric bilinear form on $V = \mathbf{R} \otimes X$ that is W-invariant.

8.2.5. Proposition. (i) *If B, $B' \in W$ there is a family $B = B_0, B_1, \ldots, B_h = B'$ in W such that B_i and B_{i+1} are adjacent $(0 \le i \le h-1)$;*
(ii) *If R^+ and $\tilde{R}^+$ are two systems of positive roots, there is a family of systems of positive roots, $R^+ = R_0^+, R_1^+, \ldots, R_h^+ = \tilde{R}^+$ such that R_i^+ and R_{i+1}^+ are adjacent $(0 \le i \le h-1)$.*

By 8.2.4 it suffices to prove (ii). By the definition of systems of positive roots (7.4.5) there exist $x, y \in V$ such that R^+ $(\tilde{R}^+)$ is the set of $\alpha \in R$ with $(x, \alpha) > 0$ (respectively, $(y, \alpha) > 0$). The set of x with this property is an open subset of V (with its euclidean topology). Changing x a little if necessary we may assume that

$$(x, \alpha)^{-1}(x, \beta) \neq (y, \alpha)^{-1}(y, \beta),$$

for all pairs (α, β) of linearly independent roots. For $t \in [0, 1]$ put $x(t) = (1 - t)x + ty$. If $\alpha \in R$ and $(x(t), \alpha) = 0$ then $(x, \alpha) \neq (y, \alpha)$ and $t = t_\alpha = ((x, \alpha) - (y, \alpha))^{-1}(x, \alpha)$. We have $t_\alpha \neq t_\beta$ if $\alpha \neq \pm\beta$. It follows that there is a subdivision $0 = t_0 < t_1 < \ldots < t_h = 1$ of $[0, 1]$ such that for $0 < i < h$ each t_i is a t_α and that for $\alpha \in R^+$ the numbers $(x(t), \alpha)$ have the same sign if t lies in the interval $I_i = (t_i, t_{i+1})$. Moreover, the signs attached to I_i and I_{i+1} by the roots of R^+ differ for only one such root. Put $R_i^+ = \{\alpha \in R \mid (x(t), \alpha) > 0 \text{ for } t \in I_i\}$ $(0 \le i \le h-1)$. These are systems of positive roots with the required properties. $\square$

8.2.6. Lemma. *Let R^+ and $\tilde{R}^+$ be adjacent systems of positive roots. There is a unique $\alpha \in R^+$ with $\tilde{R}^+ = s_\alpha . R^+$.*

Let x and y be as in the proof of the preceding proposition and let $R^+ \cap \tilde{R}^+ = R^+ - \{\alpha\}$. If $\beta \in R^+ - \{\alpha\}$ and $\langle \beta, \alpha^\vee \rangle \ge 0$ then

$$(s_\alpha.\beta, y) = (\beta, y) - \langle \beta, \alpha^\vee \rangle (\alpha, y) > 0,$$

whence $s_\alpha.\beta \in \tilde{R}^+$. If $\langle \beta, \alpha^\vee \rangle < 0$ then $s_\alpha.\beta$ lies in R^+ and is different from α, so lies in $\tilde{R}^+$. We have shown that $s_\alpha.(R^+ - \{\alpha\}) \subset \tilde{R}^+$. Since $s_\alpha.\alpha = -\alpha$ and $(-\alpha, y) > 0$ the lemma follows. $\square$

Let R^+ be a system of positive roots and let $D = D(R^+)$ be the set of roots $\alpha \in R^+$ such that $s_\alpha.R^+$ and R^+ are adjacent. If B is the Borel subgroup of G containing T with $R^+ = R^+(B)$ (see 7.4.6 and 8.2.4 (i)) we also write $D = D(B)$. We call D the *basis* of R defined by R^+ (or by B). Its elements are the *simple roots* of R^+. Let $S = S(R^+)$ (or $S(B)$) be the set of reflections s_α with $\alpha \in D$. These are the *simple reflections* defined by R^+ (or B). For $w \in W$ we have $D(w.R^+) = w.D(R^+)$, $S(w.R^+) = wS(R^+)w^{-1}$.

8.2.7. Lemma. *(i) Let $\alpha \in D$. Then s_α permutes the elements of $R^+ - \{\alpha\}$;*
(ii) If α, $\beta \in D$, $\alpha \neq \beta$ then $\langle \alpha, \beta^\vee \rangle \leq 0$.

(i) was established in the proof of the preceding lemma.
If α and β are as in (ii) we have, with x as before,

$$(x, \langle \alpha, \beta^\vee \rangle s_\alpha \beta + s_\beta \alpha) = (1 - \langle \alpha, \beta^\vee \rangle \langle \beta, \alpha^\vee \rangle)(x, \alpha).$$

If we had $\langle \alpha, \beta^\vee \rangle > 0$, the left-hand side would be > 0, since $s_\alpha \beta$ and $s_\beta \alpha$ lie in R^+ by (i). But the right-hand side would be ≤ 0 by 7.5.1 (i). This contradiction proves (ii). $\square$

8.2.8. Theorem. *(i) S generates W;*
(ii) $R = W.D$;
(iii) The roots of D are linearly independent. A root in R^+ is a linear combination $\sum_{\alpha \in D} n_\alpha \alpha$ with $n_\alpha \in \mathbf{Z}$, $n_\alpha \geq 0$. These two properties characterize the subset D of R^+.

Let W' be the subgroup of W generated by the simple reflections. In the situation of 8.2.5 (ii), there exists by 8.2.6 a simple root α such that $R_1^+ = s_\alpha.R^+$. Then $s_\alpha R_1^+, ..., s_\alpha R_h^+$ is a family as in 8.2.5 (ii), with $h - 1$ elements. By induction on h we obtain that W' acts transitively on the set of systems of positive roots. We next prove linear independence of the simple roots.

Let $\sum m_\alpha \alpha = 0$ be a dependence relation, with real coefficients. Let $D_1 = \{\alpha \in D \mid m_\alpha > 0\}$. We may assume that $D_1 \neq \emptyset$. Rewrite the relation as

$$t = \sum_{\alpha \in D_1} m_\alpha \alpha = \sum_{\beta \in D - D_1} n_\beta \beta.$$

Then, x and $(\ ,\ )$ being as before $(x, t) = \sum_{D_1} m_\alpha (x, \alpha) > 0$. But

$$(t, t) = \sum_{\alpha \in D_1, \beta \in D - D_1} m_\alpha n_\beta (\alpha, \beta) \leq 0$$

by 8.2.7 (ii). It follows that $t = 0$, which leads to a contradiction. Hence the roots of D are linearly independent.

Let $\alpha \in R^+ - D$ and $\beta \in D$. Then $s_\beta \alpha \in R^+ - D$ (8.2.7 (i)) and $(x, s_\beta \alpha) = (x, \alpha) - \langle \alpha, \beta^\vee \rangle(x, \beta)$. If $\langle \alpha, \beta^\vee \rangle > 0$ we have $0 < (x, s_\beta \alpha) < (x, \alpha)$, and α is the sum of $s_\beta \alpha$ and a positive integral multiple of β. It follows that (ii) and the second assertion of (iii) hold if we establish the following: if $\alpha \in R^+ - D$ there is $\beta \in D$ with $\langle \alpha, \beta^\vee \rangle > 0$. We prove the equivalent statement: if $\alpha \in R$ and $\langle \alpha, \beta^\vee \rangle \leq 0$ for all $\beta \in D$ then $\alpha \in -R^+$.

Let α have this property and write

$$\alpha = \sum_D c_\beta \beta + t,$$

with real coefficients c_β and $t \in V = \mathbf{R} \otimes X$ in the subspace V_0 orthogonal to all simple roots. The elements of W' stabilize the subspace of V spanned by D and fix all vectors in V_0. Applying s_α to both members of the equality we conclude that $t = 0$. Now put $u = \alpha - \sum_{c_\beta < 0} c_\beta \beta$. Then

$$(u, u) = (\alpha - \sum_{c_\beta < 0} c_\beta \beta, \sum_{c_\gamma \geq 0} c_\gamma \gamma) \leq 0,$$

whence $u = 0$. It follows that $(x, \alpha) < 0$, whence $\alpha \in -R^+$, as was to be established.

The two properties of (iii) imply that D is the set of the roots in R^+ that cannot be written as a linear combination, with strictly positive integral coefficients, of at least two other roots in R^+, which shows that D is unique.

In the proof of (ii) we have actually shown that $R = W'.D$. Let $\beta \in R$ and take $w \in W'$, $\alpha \in D$ with $\beta = w.\alpha$. Then $s_\beta = ws_\alpha w^{-1} \in W'$. Since the s_β generate W it follows that $W' = W$, whence (i). $\square$

8.2.9 Remark. The fact that W acts transitively on the set of systems of positive roots also follows from 8.2.4 (ii). However, the proof of 8.2.8 is elementary, in the sense that it does not use the theory of algebraic groups. We shall use this later.

8.2.10. Corollary. *G is generated by T and the groups $U_{\pm\alpha}$ with $\alpha \in D$.*

Let G_1 be the subgroup generated by these groups. It is a closed, connected, subgroup (2.2.7), containing all subgroups G_α of 7.1 with $\alpha \in D$. It follows that the Weyl group W_1 of (G_1, T), which is a subgroup of W, contains S. By part (i) of the theorem we have $W_1 = W$. Using part (ii) and 8.1.12 (2) it follows that G_1 contains all U_α, hence coincides with G (8.1.1 (ii)). $\square$

8.2.11. Exercises. (1) The notations are as in 7.4.7. R is one of the root systems A_{n-1}, B_n, C_n, D_n.

(a) The set D described below is a basis of R.

$$A_{n-1} : \{\epsilon_1 - \epsilon_2, ..., \epsilon_{n-1} - \epsilon_n\},$$
$$B_n : \{\epsilon_1 - \epsilon_2, ..., \epsilon_{n-1} - \epsilon_n, \epsilon_n\},$$
$$C_n : \{\epsilon_1 - \epsilon_2, ..., \epsilon_{n-1} - \epsilon_n, 2\epsilon_n\},$$
$$D_n : \{\epsilon_1 - \epsilon_2, ..., \epsilon_{n-1} - \epsilon_n, \epsilon_{n-1} + \epsilon_n\}.$$

(b) R is irreducible (8.1.12 (4)).

(c) The Weyl group W is as follows.

A_{n-1} : $W = S_n$, acting as group of permutations of the basis $(\epsilon_i)_{1 \leq i \leq n}$,

B_n, C_n : W is the group of linear maps of $V = \mathbf{R}^n$ with $\epsilon_i \mapsto \eta_i \epsilon_{\sigma i}$, where $\sigma \in S_n$ and $\eta_i = \pm 1$,

D_n : W is the subgroup of the preceding group whose elements satisfy $\Pi_i \eta_i = 1$.

(2) A root system R is reducible if and only if there is a non-trivial decomposition $D = D_1 \coprod D_2$ such that $(\alpha, \beta) = 0$ for $\alpha \in D_1$, $\beta \in D_2$.

(3) Let G be semi-simple. Let P be the weight lattice (8.1.11).

(a) P has a basis of *fundamental weights* $(\varpi_\alpha)_{\alpha \in D}$, where $< \varpi_\alpha, \beta^\vee > = \delta_{\alpha\beta}$. For α, $\beta \in D$ we have $s_\alpha \varpi_\beta = \varpi_\beta - \alpha$.

(b) Put $\rho = \sum_{\alpha \in D} \varpi_\alpha$. Show that $\rho = \frac{1}{2} \sum_{\beta \in R^+} \beta$, where R^+ is the system of positive roots with basis D. (*Hint*: consider the action of simple reflections on both sides of the equality.)

(4) Let R be a root system with system of positive roots R^+. If $\alpha \in R^+$ there exists a chain α_1, α_2, ..., $\alpha_h = \alpha$ such that α_1 and all differences $\alpha_{i+1} - \alpha_i$ $(1 \le i \le h - 1)$ are simple roots.

8.3. The Bruhat decomposition

8.3.1. The notations are as before. For $w \in W$ put

$$R(w) = \{\alpha \in R^+ \mid w.\alpha \in -R^+\}.$$

By 8.2.7 (i) we have $R(s_\alpha) = \{\alpha\}$ if $\alpha \in D$. This implies that for $\alpha \in D$

$$
\begin{aligned}
R(ws_\alpha) &= s_\alpha.R(w) \cup \{\alpha\} && \text{if } w.\alpha \in R^+, \\
R(ws_\alpha) &= s_\alpha(R(w) - \{\alpha\}) && \text{if } w.\alpha \in -R^+.
\end{aligned}
\tag{23}
$$

Recall that by 8.2.8 (i) W is generated by S. If $w \in W$ the *length* $l(w)$ of w (relative to S) is the smallest integer $h \ge 0$ such that w is a product of h elements of S. A *reduced decomposition* of w is a sequence $\mathbf{s} = (s_1, \ldots, s_h)$ in S with $w = s_1...s_h$ and $h = l(w)$. Notice that $l(w) = l(w^{-1})$.

8.3.2. Lemma. *Let* $\mathbf{s} = (s_1, \ldots, s_h)$ *be a reduced decomposition of* $w \in W$, *with* $s_i = s_{\alpha_i}$ $(\alpha_i \in D)$.

(i) $R(w) = \{\alpha_h, s_h.\alpha_{h-1}, \ldots, s_h...s_2.\alpha_1\}$;

(ii) The number of elements of $R(w)$ *is* $l(w)$;

(iii) If $\alpha \in D$ *then* $l(ws_\alpha) = l(w) + 1$ *if* $w.\alpha \in R^+$ *and* $l(ws_\alpha) = l(w) - 1$ *if* $w.\alpha \in -R^+$, *and similarly for* $s_\alpha w$;

(iv) If $\alpha \in D$ *and* $w.\alpha \in -R^+$ *there is a reduced decomposition of* w *with last element* s_α;

(v) If $\mathbf{s}' = (s_1', \ldots, s_h')$ *is any reduced decomposition of* w, *there is* i *with* $1 \le i \le h$ *such that* $s_1...s_{i-1}s_{i+1}...s_h = s_1'...s_{h-1}'$.

(i) holds for $h = 1$. Let $h > 1$ and put $w' = s_1...s_{h-1}$. By induction we may assume that $R(w') = \{\alpha_{h-1}, s_{h-1}.\alpha_{h-2}, \ldots, s_{h-1}...s_2.\alpha_1\}$. Then (i) will follow from (23) if $w'\alpha_h \in R^+$. If this were not the case we would have $\alpha_h = s_{h-1}...s_{i+1}.\alpha_i$ for some i, whence $s_h = s_{h-1}...s_{i+1}s_is_{i+1}...s_{h-1}$ and $w = s_1...s_{i-1}s_{i+1}...s_{h-1}$, contradicting the minimality of h. We have proved (i).

(ii) is a consequence of (i) and (iii) follows from (ii) and (23). For the last point use $l(w) = l(w^{-1})$. If α is as in (iv), then using (i) we see that there is i with

$ws_\alpha = s_1...s_{i-1}s_{i+1}...s_h$, whence (iv). This also implies (v). □

If $s, t \in S$ and $s \neq t$ we denote by $m(s, t)$ the order of st in W. (From 7.5.1 (ii) we know that $m(s, t) \in \{2, 3, 4, 6\}$.) Notice that $m(s, t) = m(t, s)$.

8.3.3. Proposition. *Let μ be a map of S into a multiplicative monoid with the following property: if $s, t \in S$, $s \neq t$ then*

$$\mu(s)\mu(t)\mu(s)... = \mu(t)\mu(s)\mu(t)...,$$

where in both sides the number of factors is $m(s, t)$. Then there exists a unique extension of μ to W such that if $\mathbf{s} = (s_1, ..., s_h)$ is a reduced decomposition of $w \in W$ we have

$$\mu(w) = \mu(s_1)...\mu(s_h). \tag{24}$$

We have to show that the right-hand side $\mu(\mathbf{s})$ of (24) is independent of the choice of the reduced decomposition of w. Let $\mathbf{s}$ as above and $\mathbf{s}' = (s_1', ..., s_h')$ be two reduced decompositions of w. We proceed by induction on $h = l(w)$. We may assume that $h > 1$. Let i be as in 8.3.2 (v). If $i > 1$ then $s_i...s_h = s_{i+1}...s_h s_h'$ has length $h - i + 1 < h$. By induction we have

$$\mu(s_i)...\mu(s_h) = \mu(s_{i+1})...\mu(s_h)\mu(s_h')$$

and

$$\mu(\mathbf{s}') = \mu(s_1)...\mu(s_{i-1})\mu(s_{i+1})...\mu(s_h)\mu(s_h'),$$

from which we conclude that $\mu(\mathbf{s}) = \mu(\mathbf{s}')$. It follows that we may assume that the two reduced decompositions have the form $\mathbf{s} = (..., s, t)$, $\mathbf{s}' = (..., t, s)$, with distinct elements $s, t \in S$. Proceeding in the same manner with the two new reduced decompositions we further reduce to the case that $\mathbf{s} = (..., t, s, t)$, $\mathbf{s}' = (..., s, t, s)$. Continuing we end up with reduced decompositions covered by the assumptions. □

8.3.4. Theorem. *The set of generators S and relations $s^2 = 1$, $(st)^{m(s,t)} = 1$ $(s, t \in S, s \neq t)$ give a presentation of W.*

Let $\tilde{W}$ be the group defined by generators $\tilde{s}$ $(s \in S)$ and relations $\tilde{s}^2 = 1$, $(\tilde{s}\tilde{t})^{m(s,t)} = 1$ $(s \neq t)$. We have a homomorphism $\pi : \tilde{W} \to W$ sending $\tilde{s}$ to s. By 8.3.3 there is a map $\mu : W \to \tilde{W}$ with $\mu(w) = \tilde{s}_1...\tilde{s}_h$ if $\mathbf{s} = (s_1, ..., s_h)$ is a reduced decomposition of w. It is immediate that $\pi \circ \mu$ and $\mu \circ \pi$ are identity maps. Hence π is bijective. □

By 8.2.4 (ii) there is a unique element $w_0 \in W$ with $w_0.R^+ = -R^+$. By 8.3.2 (ii) it is the element of W with maximal length. We put $B_u = U$.

8.3.5. Lemma. *Let* $w \in W$.
(i) The groups U_α *with* $\alpha \in R(w)$ *generate a closed, connected subgroup* U_w *of* U *that is normalized by* T. *We have* $U_w = \prod_{\alpha \in R(w)} U_\alpha$ *(the product being taken in any order);*
(ii) The product morphism $U_w \times U_{w_0 w} \to U$ *is an isomorphism of varieties.*

U_w is closed and connected by 2.2.7 (i). It is normalized by T. Using 8.2.3 we see that the product of (i) is a group, hence must coincide with U_w. Now (ii) follows from 8.2.1. $\square$

Let $(\dot{w})_{w \in W}$ be a set of representatives in $N_G(T)$ of the elements of W. We denote by $C(w)$ the double coset $B\dot{w}B$. It is an orbit of $B \times B$ acting in G, hence is a locally closed subvariety of G, i.e. an open subvariety of a closed subvariety of G (by 2.3.3 (i)).

8.3.6. Lemma. *Let* $\mathbf{s} = (s_1, \dots, s_h)$ *be a reduced expression of* $w \in W$, *with* $s_i = s_{\alpha_i}$ $(\alpha_i \in D)$.
(i) The morphism $\phi : \mathbf{A}^h \times B \to G$ *with*

$$\phi(x_1, \dots, x_h; b) = u_{\alpha_1}(x_1)\dot{s}_1 u_{\alpha_2}(x_2)\dot{s}_2 \dots u_{\alpha_h}(x_h)\dot{s}_h b$$

defines an isomorphism $\mathbf{A}^h \times B \simeq C(w)$.
(ii) The map $(u, b) \mapsto u\dot{w}b$ *is an isomorphism of varieties* $U_{w^{-1}} \times B \simeq C(w)$.

We have $C(w) = B\dot{w}B = U\dot{w}B = U_{w^{-1}}U_{w_0 w^{-1}}\dot{w}B$ (8.3.5 (ii)). Since by 8.1.12 (2) we have $\dot{w}^{-1}U_{w_0 w^{-1}}\dot{w} \subset B$ it follows that $C(w) = U_{w^{-1}}\dot{w}B$. By 8.3.2 (i)

$$R(w^{-1}) = \{\alpha_1, s_1\alpha_2, \dots, s_1 \dots s_{h-1}\alpha_h\}.$$

Using 8.1.12 (2) we obtain

$$U_{w^{-1}} = U_{\alpha_1}\dot{s}_1(U_{\alpha_2} \dots U_{s_2 \dots s_{h-1}\alpha_h})\dot{s}_1^{-1},$$

whence

$$U_{w^{-1}} = U_{\alpha_1}(\dot{s}_1 U_{w^{-1}s_1}\dot{s}_1^{-1}),$$

and

$$C(w) = U_{w^{-1}}\dot{w}B = U_{\alpha_1}\dot{s}_1 U_{w^{-1}s_1}\dot{s}_1^{-1}\dot{w}B = U_{\alpha_1}\dot{s}_1 C(s_1 w).$$

By induction on h we may assume that the assertion of the lemma is true for $s_1 w = s_2 \dots s_h$. We can then conclude from the last formula that ϕ is surjective. ϕ is the composite of an isomorphism $\mathbf{A}^h \times B \to U_{w^{-1}} \times B$ (see 8.3.5 (i)) and the morphism of (ii). We have to prove that the latter morphism is an isomorphism or, equivalently, that $(u, b) \mapsto \dot{w}^{-1}u\dot{w}b$ defines an isomorphism of $U_{w^{-1}} \times B$ onto $\dot{w}^{-1}.C(w)$. This can be done by viewing both spaces as homogeneous spaces for $U_{w^{-1}} \times B$ and applying 5.3.2 (iii). We skip the details. $\square$

8.3.7. Lemma. *Let* $w \in W$, $s \in S$. *We have*

$$C(s).C(w) = C(sw) \qquad\qquad if\ l(sw) = l(w) + 1,$$
$$C(s).C(w) = C(w) \cup C(sw) \quad if\ l(sw) = l(w) - 1.$$

Let $s = s_\alpha$ $(\alpha \in D)$. By 8.3.6 (ii) we have $C(s) = U_\alpha \dot{s} B$, whence $C(s).C(w) = U_\alpha \dot{s} C(w)$. For $l(sw) = l(w) + 1$ the assertion follows from 8.3.6 (i). If $l(sw) = l(w) - 1$ we have $C(s).C(w) = C(s).C(s).C(sw)$. The lemma will follow if we show that

$$C(s).C(s) = C(e) \cup C(s). \tag{25}$$

Using 7.2.2 (i) we see that $C(s) \cup C(e)$ is the group G_α of 7.1.3. By 7.2.4 the quotient by its radical is isomorphic to $\mathbf{SL}_2$ or $\mathbf{PSL}_2$. It then suffices to prove (25) in the case that $G = \mathbf{SL}_2$. We leave the proof to the reader. $\square$

8.3.8. Theorem. [Bruhat's lemma] *G is the disjoint union of the double cosets* $C(w)$ $(w \in W)$.

Put $G_1 = \bigcup_{w \in W} C(w)$. From the preceding lemma we conclude that $C(s).G_1 = G_1$. It follows that G_1 is stable under multiplication by the subgroups T, U_α, $U_{-\alpha}$ $(\alpha \in D)$. As these groups generate G (8.2.10), we must have $G_1 = G$.

Let w, $w' \in W$ and assume that $C(w) \cap C(w') \neq \emptyset$. Then $C(w) = C(w')$. Since by 8.3.6 (i) we have $\dim C(w) = l(w) + \dim B$, it follows that $l(w) = l(w')$. We may assume that $l(w) > 0$. By 8.3.2 there is $s \in S$ with $l(sw) = l(w) - 1$ and by 8.3.7

$$C(sw) \subset C(s).C(w') \subset C(w') \cup C(sw'),$$

whence $C(sw) = C(w')$ or $C(sw) = C(sw')$. By induction on $l(w)$ we find that either $sw = w'$ or $sw = sw'$. The first case is impossible because $l(sw) \neq l(w')$, hence $w = w'$. $\square$

8.3.9. Corollary. [Bruhat decomposition] *An element of G can be written uniquely in the form* $u\dot{w}b$ *with* $w \in W$, $u \in U_{w^{-1}}$, $b \in B$.

This follows from the theorem and 8.3.5 (ii). $\square$

8.3.10. Corollary. *The intersection of two Borel subgroups of G contains a maximal torus.*

We may assume that the two Borel groups are B and $B' = gBg^{-1}$. Write $g = b\dot{w}b'$, with $w \in W$, $b, b' \in B$. Then $bTb^{-1} \subset B \cap B'$. $\square$

8.3.11. Corollary. *There is a unique open double coset, viz.* $C(w_0)$, *where* w_0 *is the longest element of* W.

$C(w_0)$ is the only double coset with dimension equal to $\dim G$ (as follows from 8.3.5 (i) and 8.1.3 (ii)). It is open by 2.3.3 (i). $\qquad\qquad\qquad\qquad\qquad\qquad\qquad\square$

8.3.12. Exercises. (1). Put $\mathcal{B} = G/B$. The group G acts on $\mathcal{B} \times \mathcal{B}$ by $g.(xB, yB) = (gxB, gyB)$. For $w \in W$ let $O(w)$ be the orbit of $(B, \dot{w}B)$. Show that the following statement is equivalent to 8.3.8: $\mathcal{B} \times \mathcal{B}$ is the disjoint union of the orbits $O(w)$.
(2) Let $G = \mathbf{GL}_n$. Now the points of $\mathcal{B}$ can be viewed as complete flags in $V = k^n$ (see 6.2.11 (1)). Let $(V_1, \dots, V_{n-1})$ and $(V_1', \dots, V_{n-1}')$ be two complete flags. There is a unique permutation $\sigma \in S_n$ together with a basis $(e_1, \dots, e_n)$ of V such that $(e_1, \dots, e_i)$ is a basis of V_i and $(e_{\sigma.1}, \dots, e_{\sigma.i})$ a basis of V_i' ($1 \le i \le n-1$). Using the preceding exercise, deduce another proof of Bruhat's lemma in this particular case.
(3) Assume that G is defined over the finite field F and let σ be the Frobenius morphism (4.4.16). Assume that $\sigma B = B$. Then σ acts on $\mathcal{B} = G/B$.
(a) For $w \in W$ put

$$Z(w) = \{x \in \mathcal{B} \mid (x, \sigma x) \in O(w)\},$$

where $O(w)$ is as in (1). Show that $Z(w)$ is a locally closed subset on which the finite group G^σ acts.
(b) $\overline{Z(w)}$ is a union of some $Z(x)$ (see also 8.5.4).
(c) Let $G = \mathbf{GL}_n$, $B = \mathbf{T}_n$, the group of upper triangular matrices. We have $W = S_n$ (8.2.11 (1)). Let w be the cyclic permutation $(12\dots n)$. Using the preceding exercise show that in this case $Z(w)$ is the set of complete flags $(V_1, \dots, V_{n-1})$ in $V = k^n$ such that there exists $v \in V$ with the property that $(v, \sigma v, \dots, \sigma^{i-1}v)$ is a basis of V. Define a bijection of $Z(w)$ onto the complement of the union of all rational hyperplanes in $\mathbf{P}^{n-1}$ (i.e., hyperplanes given by a linear equation with coefficients in F).
(4) Let $t \in T$. With the notations of 8.3.9 let $g = u\dot{w}b$ be an element of the centralizer $Z_G(t)$. Then u, $\dot{w}$ and b centralize t. Show that $Z_G(t)$ is generated by $Z_G(t)^0$ and the set of $\dot{w}$ where $w \in W$, $w.t = t$. (*Hint*: use 6.3.5 (ii).)

8.4. Parabolic subgroups

In this section we shall use Bruhat's lemma to describe the parabolic subgroups of G containing B (and hence, by the conjugacy of Borel groups, all parabolic subgroups).

8.4.1. If I is a subset of our basis D, denote by W_I the subgroup of W generated by the reflections s_α with $\alpha \in I$ and by $R_I \subset R$ the set of roots that are linear combinations of the roots in I. Put $S_I = (\bigcap_{\alpha \in I} \text{Ker } \alpha)^0$, $L_I = Z_G(S_I)$. Then L_I is a connected reductive subgroup of G with maximal torus T and Borel subgroup $B_I = B \cap L_I$ (6.4.7).

8.4.2. Lemma. *(i) The root system* $R(L_I, T)$ *and the Weyl group* $W(L_I, T)$ *are* R_I, *respectively* W_I;
(ii) The system of positive roots $R_I^+(B_I)$ *is* $R_I^+ = R^+ \cap R_I$ *and the corresponding basis of* R_I *is* I.

Let Q_I be the subgroup of X spanned by I. The cocharacter group $X_*(S_I)$ is the annihilator $(Q_I)^\perp$. From 8.1.2 and 5.4.7 we see that $R(L_I, T)$ is the set of $\alpha \in R = R(G, T)$ which lie in $((Q_I)^\perp)^\perp = \tilde{Q}_I$. From 8.2.8 (iii) we conclude that α lies in R_I. Now (i) follows. That $R_I^+(B_I) = R_I^+$ follows by using 8.1.3 (i). The proof that I is the corresponding basis is left to the reader. $\qquad\square$

8.4.3. Theorem. *(i)* $P_I = \bigcup_{w \in W_I} C(w)$ *is a parabolic subgroup of* G *containing* B *and* L_I;
(ii) The unipotent radical $R_u(P_I)$ *is generated by the* U_α *with* $\alpha \in R^+ - R_I$;
(iii) The product map $L_I \times R_u(P_I) \to P_I$ *is an isomorphism of varieties;*
(iv) If P *is a parabolic subgroup containing* B *there is a unique subset* I *of* D *such that* $P = P_I$.

It follows from 8.3.7 that P_I is the subgroup of G generated by $C(e) = B$ and the $C(s_\alpha)$ with $\alpha \in I$. Since $U_{\pm \alpha}$ and T lie in $C(e) \cup C(s_\alpha)$ we conclude from 8.2.10 that $L_I \subset P_I$. Let w_0' be the longest element of W_I. Then $\dim C(w_0') > \dim C(w)$ for $w \in W_I$, $w \neq w_0'$. Using 2.3.3 (i) we see that $C(w_0')$ contains a non-empty open subset of the closure $\overline{P_I}$. By 2.2.4 (ii), P_I is closed. This proves (i).

If $\alpha \in R^+ - R_I$ then $\alpha = \sum_{\beta \in D} n_\beta \beta$ with $n_\beta > 0$ for some $\beta \in D - I$. This implies that also $s_\gamma \alpha \in R^+ - R_I$ for $\gamma \in I$. Hence W_I stabilizes $R^+ - R_I$. It follows that $R(w_0') = R_I^+$ (notation of 8.3.1). Applying 8.3.5 (ii) with $w = w_0'$ we obtain a decomposition $U = U_1.U_2$, where U_1 (U_2) is generated by the α with $\alpha \in R_I^+$ (respectively: $\alpha \in R^+ - R_I^+$). Also, U_2 is normalized by the elements $\dot{w}$ for $w \in W_I$ (use 8.1.12 (2)). According to 7.6.3, $R_u(P_I)$ is the identity component of

$$\bigcap_{w \in W_I} \dot{w} U \dot{w}^{-1} = (\bigcap_{w \in W_I} \dot{w} U_1 \dot{w}^{-1}).U_2.$$

Since L_I is reductive the last intersection is $\{e\}$ and $R_u(P_I) = U_2$, which establishes (ii).

For $w \in W_I$ write $C'(w) = B_I \dot{w} B_I$ (assuming that $\dot{w}$ has been chosen in L_I). It follows from 8.3.5 (ii) that $C(w) = C'(w).R_u(P_I)$. Using Bruhat's lemma for L_I, we conclude that $P_I = L_I.R_u(P_I)$. The argument yields, more precisely, that the product map $L_I \times R_u(P_I) \to P_I$ is bijective. Now (iii) follows by an application of 5.3.3 (ii).

Let P be a closed subgroup containing B. The set R_1 of roots of (P, T) (see 7.4.3) is a subset of R and the set of positive roots $R_1^+(B)$ (7.4.6) intersects D in a subset I. If $\alpha \in I$ the subgroup G_α of 7.1 relative to P coincides with the similar group relative to G (since their Lie algebras coincide), and it follows that $U_{-\alpha} \subset P$. By 8.2.10 we conclude that $L_I \subset P$, moreover $R_u(P_I) \subset B \subset P$, whence (by (iii)) $P_I \subset P$. The root systems of L_I and P are isomorphic. It follows from 8.1.3 (ii) that

$\dim L_I = \dim P/R_u(P)$, which implies that

$$P \subset L_I.R_u(P) \subset L_I.B \subset P_I.$$

Hence $P = P_I$.   □

Let P be a parabolic subgroup of G. A *Levi subgroup* L of P is a closed subgroup of P such that the product map $L \times R_u(P) \to P$ is bijective.

8.4.4. Corollary. *A maximal torus of P lies in a unique Levi subgroup.*

We may assume that $P = P_I$, as in 8.4.3, the maximal torus being T. The statement of the corollary then asserts that L_I is independent of the choice of the Borel subgroup B (with $T \subset B \subset P$). This follows from the fact that two such Borel groups are conjugate by an element $\dot{w}$ with $w \in W_I$ (see 6.4.12).   □

Let λ be a cocharacter of G, i.e. a homomorphism of algebraic groups $\mathbf{G}_m \to G$. In 3.2.15 we have associated to λ a closed subgroup $P(\lambda)$ of G, viz.

$$P(\lambda) = \{x \in G \mid \lim_{a \to 0} \lambda(a)x\lambda(a)^{-1} \text{ exists}\}.$$

8.4.5. Proposition. $P(\lambda)$ *is a parabolic subgroup of G. Any parabolic subgroup of G is of this form.*

We may assume that λ is non-trivial and that $\operatorname{Im} \lambda \subset T$. So $\lambda \in X^{\vee}$. Let R^+ be a set of positive roots such that the roots $\alpha \in R$ with $\langle \alpha, \lambda \rangle > 0$ lie in R^+. (To obtain one, first choose $\mu \in V$ close to λ not orthogonal to any root and, with the usual notations, define $R^+ = \{\alpha \mid (\alpha, \mu) > 0\}$). If $\alpha \in R^+$ then $U_\alpha \subset P(\lambda)$. Hence $P(\lambda)$ contains the Borel group $B \supset T$ with $R^+(B) = R^+$ and $P(\lambda)$ is parabolic. In fact, we have (relative to this Borel group) $P(\lambda) = P_I$, where I is the set of simple roots orthogonal to λ. This also implies that every parabolic subgroup is a $P(\lambda)$. □

8.4.6. Exercises. (1) Let H be a closed connected subgroup of G containing T. Let R' be the set of roots $\alpha \in R$ with $U_\alpha \subset H$. Show that H is parabolic if and only if $R' \cup (-R') = R$. (*Hint*: if the condition holds, take a Borel subgroup B of G such that $B' = H \cap B$ is a Borel group of H and show that $B' = B$.)

(2) Let P and Q be parabolic subgroups of G. Then $(P \cap Q)R_u P$ is parabolic.

(3) Let I and J be subsets of D. Let A be a set of representatives in W of the double cosets $W_I w W_J$. Show that G is the disjoint union of the double cosets $P_I \dot{w} P_J$ ($w \in A$).

(4) $G = \mathbf{GL}_n$, acting in $V = k^n$. Let $\mathcal{F} = (V_1, ... V_s)$ be a flag in V (6.2.11 (1)). Take a basis $(e_1, ..., e_n)$ such that each V_h has a basis $(e_1, e_2, ..., e_{n_h})$, let the Borel group B

be the stabilizer of the complete flag $(ke_1, ke_1 + ke_2, ...)$ and let T be the maximal torus whose elements fix all subspaces ke_h.

(a) The stabilizer $P_{\mathcal{F}}$ of $\mathcal{F}$ is a parabolic subgroup of G.

(b) Let D be the basis of the root system $R(G, T)$ defined by B, described in 8.2.11 (1). By 8.4.3 (iv) there is a unique subset I of D with $P_{\mathcal{F}} = P_I$. Determine I.

(c) Any parabolic subgroup P of G is the stabilizer of some flag $\mathcal{F}$.

(d) Discuss similar results for the case of special orthogonal groups (char $k \neq 2$) and symplectic groups (see 6.2.11).

(5) The unipotent radical of the parabolic subgroup $P(\lambda)$ of 8.4.5 is the group $U(\lambda) = \{x \in G \mid \lim_{a \to 0} \lambda(a) x \lambda(a)^{-1} = e\}$.

(6) Let G be semi-simple and simply connected (8.1.11). Let P be a parabolic subgroup of G, with Levi group L. Then the commutator group (L, L) is semi-simple and simply connected. (*Hint*: use 8.1.8 (iii).)

8.5. Geometric questions related to the Bruhat decomposition

The notations are as in 8.3. The decomposition of 8.3.8 implies a similar result for the homogeneous space $\mathcal{B} = G/B$, which is called a *flag variety* (it is uniquely determined by G, up to isomorphism). Let $\pi : G \to \mathcal{B}$ be the canonical map. Put $X(w) = \pi C(w)$.

8.5.1. Proposition. (i) $\mathcal{B}$ *is the disjoint union of the locally closed subvarieties* $X(w)$ $(w \in W)$. *They are the B-orbits in* $\mathcal{B}$;

(ii) $X(w)$ $(w \in W)$ *is an affine variety isomorphic to* $\mathbf{A}^{l(w)}$;

(iii) $X(w)$ *contains a unique point* x_w *fixed by* T;

(iv) *There is a cocharacter* $\lambda \in X^\vee$ *such that for all* $x \in X(w)$ *we have* $\lim_{a \to 0} \lambda(a).x = x_w$.

(i) is a consequence of 8.3.8. By 8.3.6 (ii) and 8.3.5 (i) we have $\pi C(w) \simeq U_{w^{-1}} \simeq \mathbf{A}^{l(w)}$, whence (ii). The fixed point x_w of of (iii) is $\pi(\dot{w})$. For the cocharacter of (iv) take any $\lambda \in X^\vee$ such that $\langle \alpha, \lambda \rangle > 0$ for all $\alpha \in R^+$. Then $\lambda(a) u_\alpha(b) \lambda(a)^{-1} = u_\alpha(a^{\langle \alpha, \lambda \rangle} b)$ $(a \in k^*, b \in k)$, from which it follows (using 8.3.5 (i)) that for $u \in U_{w^{-1}}$ we have $\lim_{a \to 0} \lambda(a) u \lambda(a)^{-1} = e$. This implies (iv). $\qquad \square$

8.5.1 gives a stratification of $\mathcal{B}$, i.e. a decomposition of $\mathcal{B}$ into locally closed pieces called strata. The strata $X(w)$ are affine spaces, called *Bruhat cells*. A closure $S(w) = \overline{X(w)}$ is a *Schubert variety*. There is a unique open stratum $X(w_0)$, called the *big cell*. Its translates $g.X(w_0)$ cover $\mathcal{B}$. This observation has the following consequence.

8.5.2. Lemma. π *has local sections.*

Local sections were defined in 5.5.7. That π has a section on $X(w_0)$ follows from
8.3.6 (ii). Using translations we obtain the lemma. $\square$

8.5.3. It follows from 5.5.8 that for any B-variety Z we have an associated fiber
bundle $G \times^B Z$ over B. Interesting cases are : (a) Z is a vector space with a linear
action of B. In the particular case of a one dimensional vector space, $G \times^B Z$ is a line
bundle over B with G-action (see 8.5.7); (b) $Z = B$, the action being conjugation.
In this case the associated fibre bundle is of interest for the finer study of conjugacy
classes of G (see [Slo, no. 4]).

8.5.4. Let $w \in W$. It follows from 2.3.3 (i) that the closure $\overline{C(w)}$ is a union of
some $C(x)$. We define an order relation on W by $x \leq w$ if $C(x) \subset \overline{C(w)}$ (it is immediate that this relation defines an ordering). We could also have defined the relation in
a similar manner, using the Bruhat cells. We shall give a combinatorial description of
the order. Let $\mathbf{s} = (s_1, ...s_h)$ be a reduced decomposition of $w \in W$ (8.3.1) and denote
by $W(\mathbf{s})$ the set of $x \in W$ that can be written in the form $x = s_{i_1}...s_{i_m}$ with $m \geq 0$ and
$i_1 < ... < i_m$, i.e. $W(\mathbf{s})$ is the set of elements obtained by erasing some factors of the
product $s_1...s_h$.

8.5.5. Proposition. *Let* $w, x \in W$. *Then* $x \leq w$ *if and only if* $x \in W(\mathbf{s})$.

Write $P_i = C(e) \cup C(s_i)$ ($1 \leq i \leq h$). By 8.4.3 (i) this is a parabolic subgroup.
It follows from 6.2.11 (5) that $P_1...P_h$ is an irreducible closed subset of G. By 8.3.7 it
is the union of the double cosets $C(y)$ with $y \in W(\mathbf{s})$. Among these there is a unique
one of maximal dimension, namely $C(w)$, so $\overline{C(w)}$ is contained in $P_1...P_h$. Since
both sets are irreducible, closed, and have the same dimension, they must coincide. $\square$

8.5.6. Corollary. $W(\mathbf{s})$ *is independent of the choice of the reduced decomposition*
$\mathbf{s}$ *of* w.

The corollary can also be proved combinatorially (see [Hu2, 5.10]).

8.5.7. Equivariant line bundles on B.

With the usual notations, let $\chi \in X$ be a character of T and also denote by χ the
character of B that it defines (see 7.3.6). Let V be the one dimensional B-module
with underlying vector space k and action $b.a = \chi(b)^{-1}a$ ($a \in k$, $b \in B$). We write
$L(\chi) = G \times^B V$ (see 8.5.3 and 5.5.8). This is a G-variety, the action coming from
left translations in G. We have a G-morphism $\rho : L(\chi) \to B$ with local sections,
whose fibers are k. $L(\chi)$ is the *equivariant line bundle* on B defined by χ.

If $g \in G$, $a \in k$ denote by $g * a$ the image of (g, a) in $L(\chi)$. Let σ be a section
of ρ on B. There is a unique $f \in k[G]$ with $\sigma(gB) = g * f(g)$ and we have for

$g \in G,\ b \in B$

$$f(gb) = \chi(b)f(g). \qquad (26)$$

These functions form a vector space $\Gamma(\chi)$, the space of sections of the line bundle $L(\chi)$.

We denote by X^+ the set of $\chi \in X$ such that $\langle \chi, \alpha^\vee \rangle \geq 0$ for all $\alpha \in D$.

8.5.8. Theorem. $\Gamma(\chi)$ *is a finite dimensional vector space that is non-zero if and only if $\chi \in X^+$.*

Assume that $\Gamma(\chi) \neq \{0\}$. Using 7.3.6 one shows that $\chi \in X^+$.

Now assume that $\chi \in X^+$. Put $\Omega = C(w_0)$ and let $\Gamma(\Omega, \chi)$ be the vector space of regular functions f on Ω satisfying (26) for $g \in \Omega$, $b \in B$. We have a locally finite representation λ of B in $\Gamma(\Omega, \chi)$ by left translations. By 6.3.1 there is $h \in \Gamma(\Omega, \chi)$ which is a simultaneous eigenvector for the elements of $\lambda(B)$. Hence $h(u\dot{w}_0 b) = \chi(b)h(\dot{w}_0)$ $(u \in U,\ b \in B)$. We shall show that h extends to a regular function on G, corresponding to a section of $L(\chi)$. We first show that h extends to a regular function on the union of Ω and a translate $\dot{s}.\Omega$, where $s \in S$.

By 8.2.1 we can write an element $u \in U$ in the form

$$u = \prod_{\alpha \in R^+} u_\alpha(x_\alpha),$$

the product being determined by an ordering of the roots. It follows from 8.2.3 that if α is a simple root the coordinate x_α is uniquely determined, i.e. is independent of the choice of the ordering. For $\alpha \in D$ we write $\xi_\alpha(u) = x_\alpha$. By 8.3.6 (ii) we have $\Omega = TY$, where $Y = U\dot{w}_0 U$, a closed subset of Ω. If $x = tu\dot{w}_0 u' \in \Omega$ write $\xi_\alpha(x) = \xi_\alpha(u)$ and put $Y_\alpha = \{y \in Y \mid \xi_\alpha(y) \neq 0\}$ $(\alpha \in D)$. Let n_α be as in 8.1.4 (i).

8.5.9. Lemma. *(i) For $\alpha \in D$ there is a morphism $\phi_\alpha : Y_\alpha \to Y_\alpha$ such that for $t \in T$, $y \in Y_\alpha$*

$$n_\alpha(ty) = (s_\alpha t).(\alpha^\vee)(-\xi_\alpha(y))^{-1}.\phi_\alpha(y).$$

We have $\xi_\alpha(\phi_\alpha y) = -\xi_\alpha(y)$ and $\phi_\alpha^2 = \mathrm{id}$;
(ii) $\Omega \cap n_\alpha.\Omega = T.Y_\alpha$.

We have $U = U_\alpha U_1$, with $U_1 = \prod_{\beta \in R^+ - \{\alpha\}} U_\beta$. This is a normal subgroup of U that is also normalized by $U_{-\alpha}$. From formula (22) in 8.1.4 we obtain for $a \neq 0$

$$n_\alpha u_\alpha(a)U_1\dot{w}_0 U \subset \alpha^\vee(-a)^{-1}u_\alpha(-a)U_{-\alpha}U_1\dot{w}_0 U = \alpha^\vee(-a)^{-1}u_\alpha(-a)U_1\dot{w}_0 U.$$

This formula implies the assertions of (i). It also follows that $T.Y_\alpha \subset \Omega \cap n_\alpha.\Omega$. Since $n_\alpha.\Omega = T.Y_\alpha \subset C(s_\alpha w_0)$, (ii) follows from 8.3.8. $\qquad \square$

As a consequence of the lemma, we can evaluate h on $n_\alpha.\Omega \cap \Omega$ as follows:

$$h(n_\alpha u \dot{w_0} u't) = (-\xi_\alpha(u))^{-\langle \chi, w_0 \alpha^\vee \rangle} \chi(t) h(\dot{w_0}).$$

(Notice that $\xi_\alpha(u) \neq 0$.) Since $\chi \in X^+$, this formula defines a regular function on all of $n_\alpha.\Omega$. It follows that h can be extended to a regular function on $\Omega \cup n_\alpha.\Omega$.

The set Σ of $g \in G$ such that h can not be extended to a regular function in a neighborhood of g is closed and stable under left and right multiplication by B. Hence Σ is a union of double cosets $C(w)$, where $w \neq w_0$ and also $w \neq s_\alpha w_0$ for all simple roots α, because for a simple root α we have $C(s_\alpha w_0) \cap n_\alpha.C(w_0) \neq \emptyset$. Using 8.3.6 (ii) we conclude that $\dim \Sigma \leq \dim G - 2$. Consider a translate U of Ω. It is isomorphic to $\mathbf{A}^{2N} \times T$ (where $N = |R^+|$). h defines a regular function on an open subset of U, whose complement has dimension $\leq \dim U - 2$. Using the fact that $k[U]$ is a unique factorization domain and 1.8.4 (3), we conclude that this function is regular on all of U. Since the translates of Ω cover G we see that h extends to a regular function on G. We have shown that $\Gamma(\chi)$ is non-zero if $\chi \in X^+$.

It remains to be shown that $\Gamma(\chi)$ has finite dimension. In the proof we use the order relation on X defined as follows: $\mu \leq \nu$ if $\nu - \mu$ is a sum of positive roots. Notice that $\mu \leq \nu$ implies $w_0 \nu \leq w_0 \mu$. T acts semi-simply on $\Gamma(\chi)$ by left translations: $(\lambda(t))f(g) = f(t^{-1}g)$, so we can decompose $\Gamma(\chi)$ into weight spaces. The restriction of $f \in \Gamma(\chi)$ to Ω is given by $f(u\dot{w_0}u't) = f'(u)\chi(t)f(\dot{w_0})$ $(u, u' \in U, t \in T)$, where f' is a regular function on U. Introduce coordinates x_α $(\alpha \in R^+)$ on U as before. A straightforward calculation then shows that the weights of T in $\Gamma(\chi)$ are all $\leq -w_0 \chi$. Since the Weyl group permutes the weights, they are also all $\geq -\chi$. The finite dimensionality now is a consequence of the following fact, the proof of which is left to the reader. For fixed $\mu \in X$ the number of vectors with non-negative integral coordinates $(m_\alpha)_{\alpha \in R^+}$ satisfying $\sum_{\alpha \in R^+} m_\alpha \alpha \leq \mu$ is finite. $\square$

The rational representation of G in $\Gamma(\chi)$ are basic objects in the representation theory of G. If $\chi \in X^+$ there is a unique quotient of $\Gamma(\chi)$ that affords an irreducible representation of G, and any irreducible representation is isomorphic to one obtained in this manner. For these facts, which are not hard to prove, see for example [Hu1, no.31]. If char $k = 0$ the G-modules $\Gamma(\chi)$ are irreducible.

8.5.10. Exercises. (1) Notations of 8.5.1. Show that $\mathcal{B} = \bigcup_{w \in W} \dot{w}.X(w_0)$. (*Hint*: put $\mathcal{B}' = \bigcup \dot{w}.X(w_0)$. Show that the closed set $\mathcal{B} - \mathcal{B}'$ is empty by considering the T-action on it.) It follows that $G = \bigcup_{w \in W} \dot{w}.C(w_0)$.

(2) Notations of 8.5.5. Let $X(\mathbf{s}) = P_1 \times^B P_2 \times^B \dots \times^B P_h$, an iterated associated fiber space (5.5.8).

(a) $X(\mathbf{s})$ is a smooth variety.

(b) The product map of G induces a morphism $\rho : X(\mathbf{s}) \to G$. Show that the image of ρ is $\overline{C(w)}$ and that the restriction of ρ to the inverse image of $C(w)$ is an isomorphism.

(c) Establish similar results for the Schubert varieties $S(w)$.

One can show that ρ is a resolution of $C(w)$ (cf. [Dem, no.3]).

(3) Let w, $w' \in W$. The order relation on W is as in 8.5.4.

(a) If $X(w) \cap \dot{w}_0.X(w_0w) \neq \emptyset$ we have $w' \leq w$ (*Hint*: consider the T-action on this intersection.)

(b) Conversely, if $w' \leq w$ then $X(w) \cap \dot{w}_0.X(w_0w) \neq \emptyset$.

(c) $X(w) \cap \dot{w}_0.X(w_0w) = \dot{w}B$.

Notes

Most of the results of this chapter are due to Chevalley, and can be found in [Ch4]. Along the way, we have proved the results on Weyl groups and root systems that we need.

The first occurrence of a result like Bruhat's lemma seems to be in 1950 in a book in Russian by Gelfand and Neumark, for the case of $\mathbf{GL}_n(\mathbf{C})$ (see the German translation, [GN, §18]).For classical (orthogonal and symplectic) groups over $\mathbf{C}$, F. Bruhat established the result in 1954 (see [Bru]). For algebraic groups see [Ch4, exp. 13]. The theory of Bruhat decompositions in groups was axiomatized by Tits in his theory of 'BN-pairs', called 'Tits systems' in [Bou2]. The theory is exposed in [loc. cit., Ch. IV, §2]. The results of 8.3 provide the ingredients of a Tits system on G. We did not go into these matters. Nor did we mention in 8.3 or 8.4 the building associated to G. For the theory of buildings we refer to [Ron].

The varieties $Z(w)$ of 8.3.12 (3) are the 'Deligne–Lusztig varieties', introduced in [DelL]. They are fundamental for the representation theory of the finite group G^σ, see e.g. [Ca].

Parabolic subgroups are treated in [BoT1], they do not yet occur explicitly in [Ch4]. Several results of 8.4 follow from the theory of Tits systems. We have given independent proofs.

Subgroups of semi-simple Lie groups similar to the $U(\lambda)$ of 8.4.6 (5) occur in the older literature under the name 'horyspherical subgroups' (see e.g. [Gel, p. 78]).

The order on W of 8.5.4 is sometimes called the Bruhat order, but it was first introduced by Chevalley. The proof of 8.5.5 follows [BoT1, Compl., p. 267].

We have included in 8.5 a proof of the basic result 8.5.8 on equivariant line bundles on the flag manifold. In a somewhat different formulation this is already contained in [Ch4, exp. 15].

We have not gone into the representation theory of reductive groups. This is treated at length in [Jan1].

Chapter 9

The Isomorphism Theorem

G denotes a connected, reductive, linear algebraic group over k and T a maximal torus of G. The main result of this chapter is that the root datum $\Psi(G, T)$ introduced in 7.4.3 determines G up to isomorphism. In the proof of this uniqueness result we shall study in detail the way in which G is built up from T and the groups U_α ($\alpha \in R$) of 8.1.1 (i). We shall get involved with a number of technicalities about root systems.

9.1. Two dimensional root systems

In this section R is a root system in a real vector space V. We do not assume that it comes from an algebraic group.

9.1.1. Classification. Assume that V is two dimensional. Let (α, β) be a basis of R. It follows from 7.5.1 and 8.2.7 (ii) that we may assume that $\langle \beta, \alpha^\vee \rangle = -a$, where $a = 0, 1, 2, 3$, and that $\langle \alpha, \beta^\vee \rangle = 0$ or -1, if $a = 0$, respectively $a \neq 0$. Using 8.2.8 and 8.2.9 it follows that the root system R coincides with the set S of 7.5.3 (2), in the four possible cases. Hence there are (up to isomorphism) four possibilities for R. For $a = 1, 2$ we have the root systems A_2, respectively B_2 of 7.4.7. With the notations used there one can take $\alpha = \epsilon_1 - \epsilon_2$, $\beta = \epsilon_2 - \epsilon_3$, respectively $\alpha = \epsilon_1$, $\beta = \epsilon_2 - \epsilon_1$.

For $a = 0$ we have the root system $A_1 \times A_1$. For $a = 3$ we obtain the root system G_2. Its twelve roots can be realized in the euclidean plane by taking $\alpha = \epsilon_1$, $\beta = -\frac{3}{2}\epsilon_1 + \frac{1}{2}\sqrt{3}\,\epsilon_2$. The coroots are given in terms of the euclidean metric by the formula of the proof of 7.1.8. The root systems A_2, B_2, G_2 are irreducible. They are pictured below.

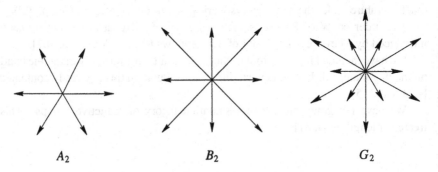

A_2 $\qquad\qquad\qquad\qquad$ B_2 $\qquad\qquad\qquad\qquad$ G_2

In 7.4.7 there also appeared the two dimensional root system C_2. This is isomorphic to the root system B_2.

We next review some facts about general root systems, involving two dimensional systems. Let R be a reduced root system and denote by W the Weyl group. Fix a system of positive roots R^+, whose basis is D. The *rank* of R is the number of elements of D. The bilinear form $(\ ,\ )$ is as in 7.1.7.

9.1.2. Lemma. *Let $\alpha, \beta \in R$ be linearly independent roots. There is a system of positive roots R^+ such that α is a simple root of R^+ and β is a positive linear combination of α and another simple root.*

Let $(e_1, \ldots, e_n)$ be a basis of V with $e_1 = \alpha, e_2 = \beta$. Define a total order on V as follows. If $x = x_1 e_1 + \ldots + x_i e_i$ with $x_i \neq 0$ then $x > 0$ if $x_i > 0$. For $x, y \in V$ define $x > y$ if $x - y > 0$. Then $R^+ = \{\alpha \in R \mid \alpha > 0\}$ is a system of positive roots in the sense of 7.4.5 (check this) and α is the smallest element of R^+ for our order. It follows from the characterization of simple roots mentioned at the end of the proof of 8.2.8 that α is a simple root of R^+. It also follows that β is a positive linear combination of simple roots $\leq \beta$. Such simple roots being linear combinations of α and β, there can only be two of them. The lemma follows. $\qquad\square$

9.1.3. Lemma. *Let $\alpha, \beta \in R$ be linearly independent roots. The set of integers i such that $\beta + i\alpha \in R$ is a segment $[-c, b]$ containing 0. We have $b - c = -\langle \beta, \alpha^\vee \rangle$, and $b + c \leq 3$.*

By the previous lemma we may assume that R is two dimensional and that there is a system of positive roots R^+ such that α is a simple root of R^+ and $\beta \in R^+$. The first statement of the lemma is readily checked in the four possible two dimensional root systems of 7.5.3 (2).

The sequence of roots $(\beta - c\alpha, \ldots, \beta + b\alpha)$ is the *α-string through β*, its *length* is $b + c + 1$. The reflection s_α permutes the roots of the string and interchanges its first and last root. This remark implies the formula for $b - c$. The inequality for $b + c$ follows from 7.5.1. $\qquad\square$

The set S of reflections s_α ($\alpha \in D$) generates the Weyl group W (8.2.8 (i)). The length function l on W is as in 8.3.1. If $\alpha, \beta \in D$, denote by $R_{\alpha,\beta}$ the set of roots that are linear combinations of α and β. If $\alpha \neq \beta$ this is a two dimensional root system with basis (α, β), in which $R^+_{\alpha,\beta} = R^+ \cap R_{\alpha,\beta}$ is a system of positive roots. We write $R_\alpha = R_{\alpha,\alpha} = \{\pm\alpha\}$. We denote the Weyl groups of these root systems by $W_{\alpha,\beta}$.

9.1.4. Lemma. *Let $\alpha, \beta \in D$ and $w \in W$ be such that $w.\alpha = \beta$. There exist $\alpha_1, \alpha_2, \ldots, \alpha_{s+1} \in D$ and $w_1, \ldots, w_s \in W$ with the following properties:*
(a) $w = w_s \ldots w_2 w_1$ and $l(w) = l(w_1) + \ldots + l(w_s)$,
(b) $\alpha_1 = \alpha$, $\alpha_{s+1} = \beta$ and $w_i.\alpha_i = \alpha_{i+1}$ ($1 \leq i \leq s$),
(c) If $\alpha_i \neq \alpha_{i+1}$ then $w_i \in W_{\alpha_i,\alpha_{i+1}}$ and if $\alpha_i = \alpha_{i+1}$ there is $\gamma \neq \alpha_i$ such that $w_i \in W_{\alpha_i,\gamma}$ ($1 \leq i \leq s$).

The lemma tells that the move from α to β defined by w can be built up from similar moves, each taking place inside a root system $R_{\gamma,\delta}$.

Assume that $w \neq 1$. By 8.3.2 (i) there is $\gamma \in D$ with $w.\gamma \in -R^+$. Then $\gamma \neq \alpha$. It follows from 8.2.4 (ii) that there is $w_1 \in W_{\alpha,\gamma}$ with $w_1^{-1}.R_{\alpha,\gamma}^+ = (w^{-1}.R^+) \cap R_{\alpha,\gamma}$. By 8.3.2 (ii), $l(w)$ is the number of $\delta \in R^+$ with $w.\delta \in -R^+$ and $l(w_1)$ is the number of $\delta \in R_{\alpha,\gamma}^+$ with this property. Using that w_1 stabilizes $R^+ - R_{\alpha,\gamma}$ we also see that $l(ww_1^{-1})$ equals the number of $\delta \in R^+ - R_{\alpha,\gamma}$ with the same property. It follows that $l(w) = l(w_1) + l(ww_1^{-1})$. We have that $\alpha = w^{-1}.\beta$ is a simple root for the system of positive roots $(w^{-1}.R^+) \cap R_{\alpha,\gamma} = w_1^{-1}.R_{\alpha,\gamma}^+$ in $R_{\alpha,\gamma}$. It follows that $\alpha \in w_1^{-1}.\{\alpha, \gamma\}$. Put $\alpha_2 = w_1.\alpha$. Then $ww_1^{-1}.\alpha_2 = \beta$. Since $l(ww_1^{-1}) < l(w)$ we may assume by induction that the assertion holds for α_2, β and ww_1^{-1}. The lemma follows.   □

9.2. The structure constants

Now R is again the root system of our reductive group G.

9.2.1. We fix a realization $(u_\alpha)_{\alpha \in R}$ of R in G. Recall that u_α ($\alpha \in R$) is an isomorphism of $\mathbf{G}_a$ onto a subgroup U_α of G that is normalized by T, and that we have the properties of 8.1.1 (i) and 8.1.4 (i). We fix a total order on R. If $\alpha, \beta \in R$ are linearly independent, it follows from 8.2.3 that there exist unique constants $c_{\alpha,\beta;i,j} \in k$, defined if $i, j > 0$, $i\alpha + j\beta \in R$, such that

$$u_\alpha(x)u_\beta(y)u_\alpha(x)^{-1} = u_\beta(y) \prod u_{i\alpha+j\beta}(c_{\alpha,\beta;i,j}x^i y^j), \qquad (27)$$

the product being over the pairs (i, j) as above. The order of the factors is prescribed by our total order. The $c_{\alpha,\beta;i,j}$ are the *structure constants*, relative to the realization and the order. If $(c'_{\alpha,\beta;i,j})$ is the set of structure constants for another realization (and the same order), we see from 8.1.4 (iv) that there exist $c_\alpha \in k^*$ ($\alpha \in R$) such that $c_\alpha c_{-\alpha} = 1$ and that

$$c'_{\alpha,\beta;i,j} = c_\alpha^{-i} c_\beta^{-j} c_{i\alpha+j\beta} c_{\alpha,\beta;i,j}.$$

We call such sets of structure constants $(c_{\alpha,\beta;i,j})$ and $(c'_{\alpha,\beta;i,j})$ *equivalent*. Put $c_{\alpha,\beta;0,1} = c_{\alpha,\beta;1,0} = 1$ and $c_{\alpha,\beta;i,j} = 0$ in the cases where we had not yet defined this constant. Now let $\alpha, \beta \in R$ be arbitrary. As in 8.1.4 (i), put $n_\alpha = u_\alpha(1)u_{-\alpha}(-1)u_\alpha(1)$. This element normalizes T and represents the reflection s_α. For $\beta \in R$, $x \in k$ define $u'_\beta(x) = n_\alpha u_{s_\alpha.\beta}(x)n_\alpha^{-1}$. Then (u'_β) is a realization of R. It follows that there exist $d_{\alpha,\beta} \in k^*$ such that

$$n_\alpha u_\beta(x)n_\alpha^{-1} = u_{s_\alpha.\beta}(d_{\alpha,\beta}x).$$

9.2.2.   Lemma.   *Let α and β be linearly independent roots and denote by* $(\beta - c\alpha, \dots, \beta + b\alpha)$ *the α-string through β (9.1.3).*

(i) $d_{\alpha,\beta} = \sum_{i=\max(0,c-b)}^{c} (-1)^i c_{-\alpha,\beta;i,1} c_{\alpha,\beta-i\alpha;i+b-c,1}$;

(ii) We have $d_{-\alpha,\beta} = (-1)^{\langle\beta,\alpha^\vee\rangle} d_{\alpha,\beta}$, $d_{\alpha,\beta} d_{-\alpha,-\beta} = d_{\alpha,\beta} d_{\alpha,s_\alpha.\beta} = (-1)^{\langle\beta,\alpha^\vee\rangle}$,
$d_{\alpha,\beta} d_{\alpha,-\beta} = 1$, $d_{\alpha,\alpha} = -1$;

(iii) $n_\alpha n_\beta n_\alpha^{-1} = (s_\alpha.\beta)^\vee (d_{\alpha,\beta}) n_{s_\alpha.\beta}$.

We first assume that $c = 0$ i.e., that $\beta - \alpha \notin R$. Using formulas like (27) one sees that $V = \prod_{i\alpha+j\beta\in R, j>0} U_{i\alpha+j\beta}$ is a subgroup of G and that $V_1 = \prod_{i\alpha+j\beta\in R, j>1} U_{i\alpha+j\beta}$ is a normal subgroup of V. Both V and V_1 are normalized by $U_{\pm\alpha}$. The map $\phi : (x_0, \dots, x_b) \mapsto \prod_{i=0}^{b} u_{\beta+i\alpha}(x_i)$ induces an isomorphism of groups $M = k^{b+1} \simeq V/V_1$. The closed subgroup $G_\alpha = Z_G((\mathrm{Ker}\ \alpha)^0)$ of 7.1.3 is connected, reductive (7.6.4 (i)) and its commutator subgroup H is connected, semi-simple (7.3.1). It follows from the proof of 7.2.4 that there is a surjective homomorphism of algebraic groups $\psi : \mathbf{SL}_2 \to H$ such that for $x \in k$

$$\psi \begin{pmatrix} 1 & x \\ 0 & 1 \end{pmatrix} = u_\alpha(x), \quad \psi \begin{pmatrix} 1 & 0 \\ x & 1 \end{pmatrix} = u_{-\alpha}(x).$$

Then

$$\psi \begin{pmatrix} 0 & 1 \\ -1 & 0 \end{pmatrix} = n_\alpha.$$

The group H is generated by U_α and $U_{-\alpha}$ (8.1.5 (i)). Since these groups normalize V and V_1, the same holds for H. Hence we can define an action ρ of $\mathbf{SL}_2$ on M by

$$\phi(\rho(g)m)V_1 = \psi(g)\phi(m)\psi(g)^{-1} V_1,$$

for $g \in \mathbf{SL}_2$, $m \in M$. Using (27) we see that

$$\left.\begin{aligned}
\rho \begin{pmatrix} 1 & x \\ 0 & 1 \end{pmatrix} e_i &= \sum_{h=0}^{b-i} c_{\alpha,\beta+i\alpha;h,1} x^h e_{i+h} \\
\rho \begin{pmatrix} 1 & 0 \\ x & 1 \end{pmatrix} e_i &= \sum_{h=0}^{i} c_{-\alpha,\beta+i\alpha;h,1} x^h e_{i-h},
\end{aligned}\right\} \tag{28}$$

where (e_i) is the canonical basis. (28) implies that ρ is a rational representation of $\mathbf{SL}_2$. For $x \in k^*$ we have

$$\rho \begin{pmatrix} x & 0 \\ 0 & x^{-1} \end{pmatrix} e_i = x^{\langle\beta+i\alpha,\alpha^\vee\rangle} e_i.$$

Put

$$n = \begin{pmatrix} 0 & 1 \\ -1 & 0 \end{pmatrix}.$$

Then

$$\rho(n)e_i = d_{\alpha,\beta+i\alpha} e_{b-i}.$$

Also,

$$\rho \begin{pmatrix} 1 & x \\ 0 & 1 \end{pmatrix} \rho \begin{pmatrix} 1 & 0 \\ -x^{-1} & 1 \end{pmatrix} e_i = \rho \begin{pmatrix} 0 & x \\ -x^{-1} & 0 \end{pmatrix} \rho \begin{pmatrix} 1 & -x \\ 0 & 1 \end{pmatrix} e_i \quad (x \in k^*).$$
(29)

Using (28) we can write the left-hand side of (29) in terms of the basis (e_h). Comparing coefficients of e_{b-i} in both sides we obtain a formula for $d_{\alpha,\beta+i\alpha}$, which is equivalent to the formula stated in (i).

The formula for $d_{-\alpha,\beta}$ of (ii) follows from $n_{-\alpha} = \alpha^{\vee}(-1)n_\alpha$ (see 8.1.4 (ii)). Moreover

$$n_\alpha n_\beta n_\alpha^{-1} = n_\alpha u_\beta(1) u_{-\beta}(-1) u_\beta(1) n_\alpha^{-1} = u_{s_\alpha.\beta}(d_{\alpha,\beta}) u_{-s_\alpha.\beta}(-d_{\alpha,-\beta}) u_{s_\alpha.\beta}(d_{\alpha,\beta}).$$

Using 8.1.4 (iii) we see that $d_{\alpha,-\beta} = d_{\alpha,\beta}^{-1}$. Then (iii) follows from 8.1.4 (i). The remaining assertions also readily follow.   □

9.2.3. Lemma. *Let $\alpha, \beta, \gamma \in R$ and assume that β and γ are linearly independent.*
(i) $c_{s_\alpha.\beta,s_\alpha.\gamma;i,j} - d_{\alpha,\beta}^{-i} d_{\alpha,\gamma}^{-j} d_{\alpha,i\beta+j\gamma} c_{\beta,\gamma;i,j}$ is uniquely determined by the $c_{\delta,\epsilon;i',j'}$ and the $d_{\alpha,\delta}$, where δ and ϵ are positive linear combinations of β and γ and $i' + j' < i + j$. In particular, $c_{s_\alpha.\beta,s_\alpha.\gamma;1,1} = d_{\alpha,\beta}^{-1} d_{\alpha,\gamma}^{-1} d_{\alpha,\beta+\gamma} c_{\beta,\gamma;1,1}$;
(ii) $c_{\gamma,\beta;i,j} - (-1)^i c_{\beta,\gamma;i,j}$ is uniquely determined by the $c_{\delta,\epsilon;i',j'}$ of (i). In particular, $c_{\gamma,\beta;1,1} = -c_{\beta,\gamma;1,1}$.

(i) is proved by conjugating both sides of (27) (for β and γ) by n_α and rearranging the product in the right-hand side in the correct order, using (27) repeatedly. (ii) also follows by a repeated application of (27). We skip the details.   □

9.2.4. We now assume that R has rank two and is irreducible, so is of one of the types A_2, B_2, G_2. Let (α, β) be a basis of R with the properties of 9.1.1 and let R^+ be the corresponding set of positive roots. The bilinear form $(\, , \,)$ on the ambient vector space V is as before. We identify the coroot $\gamma^{\vee}$ with $2(\gamma, \gamma)^{-1}\gamma$. In the case A_2 we normalize such that $(\gamma, \gamma) = 2$ for all roots γ. In the cases B_2 and G_2 we have two different root lengths, i.e. we have long and short roots. We normalize such that $(\gamma, \gamma) = 2$ if γ is short. Then $(\gamma, \gamma) = 4, 6$ for a long root γ of B_2, respectively G_2. The short roots of B_2 are $\pm\alpha$, $\pm(\alpha + \beta)$ and its long roots are $\pm\beta$, $\pm(2\alpha + \beta)$. The short roots of G_2 are $\pm\alpha$, $\pm(\alpha + \beta)$, $\pm(2\alpha + \beta)$ and the long roots $\pm\beta$, $\pm(3\alpha + \beta)$, $\pm(3\alpha + 2\beta)$. These facts follow from 7.5.3 (2), or can be read off from the figures of 9.1.1.

9.2.5. Proposition. *(i) We can normalize the u_γ such that all structure constants lie in $\mathbf{Z}.1$ and that the following holds:*
(a) $c_{\alpha,\beta,i,j} = 1$ whenever $i, j > 0$ and $i\alpha + j\beta \in R$,

(c) all $d_{\alpha,\gamma}$, $d_{\beta,\gamma}$ $(\gamma \in R^+)$ are ± 1;

(d) if R is of type G_2 we have $c_{\beta,3\alpha+\beta;1,1} = 1$;

(ii) The properties (a) and (d) (in type G_2) uniquely determine all structure constants.

We use the following steps.

(a) If γ, $\delta \in R$ are linearly independent, there exists by 9.1.2 and 8.2.4 (ii) an element w of the Weyl group W such that $w.\gamma = \alpha$ or β and $w.\delta \in R^+$. Apply 9.2.2 (iii) and 9.2.3 (i). Using that W is generated by the simple reflections s_γ $(\gamma \in D)$ (8.2.8 (i)) and also using induction on $i + j$, we see that $c_{\gamma,\delta,i,j}$ is determined by the values $c_{\alpha,\epsilon;i',j'}$, $c_{\beta,\epsilon;i',j'}$, $d_{\alpha,\epsilon}$, $d_{\beta,\epsilon}$ where $\epsilon \in R^+$. If γ and δ have different lengths, it follows from 9.2.3 (ii) that we may assume γ to be the short one. In that case we only have to consider the $c_{\alpha,\epsilon;i',j'}$ with ϵ long.

(b) 9.2.2 (i) can be used to express some of the d's in terms of structure constants and 9.2.2 (ii) gives relations between the d's. In particular, we obtain from 9.2.2 (i)

$$d_{\gamma,\delta} = 1 \text{ if } \gamma + \delta \notin R, \ -\gamma + \delta \notin R,$$

$$d_{\gamma,\delta} = c_{\gamma,\delta;1,1} \text{ if } \gamma + \delta \in R, \ \gamma - \delta \notin R, \ 2\gamma + \delta \notin R. \tag{30}$$

In the latter case we have, moreover,

$$d_{\delta,\gamma} = -c_{\gamma,\delta;1,1}. \tag{31}$$

(c) In the cases B_2 and G_2 we exploit the representation ρ of SL_2 in the space M introduced in the proof of 9.2.2.

We discuss the three possible cases.

A_2. Now $i = j = 1$ in all non-zero structure constants $c_{\gamma,\delta;i,j}$. By step (a) and (30),(31) all structure constants can be expressed in the $d_{\alpha,\gamma}$ and $d_{\beta,\gamma}$ with $\gamma \in R^+$. We have $d_{\alpha,\beta} = c_{\alpha,\beta;1,1} = -d_{\beta,\alpha}$. We also have to consider $d_{\alpha,\alpha+\beta}$, $d_{\beta,\alpha+\beta}$. From 9.2.2 (ii) we obtain

$$d_{\alpha,\alpha+\beta} = d_{\alpha,s_\alpha.\beta} = -d_{\alpha,\beta}^{-1}.$$

Also, $d_{\beta,\alpha} = -d_{\alpha,\beta}$, $d_{\beta,\alpha+\beta} = d_{\alpha,\beta}^{-1}$. We can normalize such that $d_{\alpha,\beta} = 1$. The assertions of the proposition follow.

B_2. Using steps (a) and (b) we see that the structure constants are determined by $c_{\alpha,\beta;1,1}$, $c_{\alpha,\beta;2,1}$, $c_{\alpha+\beta;1,1}$, $d_{\alpha,\beta}$, $d_{\alpha,\alpha+\beta}$, $d_{\beta,\alpha}$. By 9.2.3 (ii) and 9.2.2

$$c_{\alpha,\beta;1,1} = -c_{\beta,\alpha,1,1} = d_{-\beta,\alpha} = -d_{\beta,\alpha} \neq 0,$$

and

$$c_{\alpha,\beta;2,1} = d_{-\alpha,\beta} = d_{\alpha,\beta} \neq 0.$$

We now use step (c). Use the notations of the proof of 9.2.2. In the present case $b = 2$. We have $\rho(n)e_0 = xe_2$, $\rho(n)e_2 = x^{-1}e_0$, with $x = d_{\alpha,\beta}$. Since the determinant of $\rho(n)$ must be 1, we conclude that $\rho(n)e_1 = -e_1$, which means that $d_{\alpha,\alpha+\beta} = -1$. Using 9.2.3 (ii) we obtain

$$-1 = d_{-\alpha,\alpha+\beta} = 1 - c_{\alpha,\alpha+\beta;1,1}c_{-\alpha,2\alpha+\beta;1,1}.$$

Now $c_{-\alpha,2\alpha+\beta;1,1} = c_{s_\alpha.\alpha,s_\alpha.\beta;1,1}$. Using 9.2.3 (i) we obtain

$$c_{-\alpha,2\alpha+\beta;1,1} = d_{\alpha,\beta}^{-1}c_{\alpha,\beta,1,1} = -d_{\alpha,\beta}^{-1}d_{\beta,\alpha}.$$

Normalizing such that (a) holds, the assertions follow.

G_2. This is the most complicated case. By step (a) the structure constants are determined by following ones: $c_{\alpha,\beta;i,j}$ with $(i,j) = (1,1), (2,1), (3,1), (3,2)$, $c_{\alpha,\alpha+\beta;i,j}$ with $(i,j) = (1,1), (2,1), (1,2)$, $c_{\alpha,2\alpha+\beta;1,1}$, $c_{\beta,3\alpha+\beta;1,1}$, together with the $d_{\alpha,\epsilon}, d_{\beta,\delta}$ $(\epsilon, \delta \in R^+)$. Using 9.2.2 (i) we obtain $c_{\alpha,\beta;1,1} = -d_{\beta,\alpha} \neq 0$ and $c_{\alpha,\beta;3,1} = d_{\alpha,\beta} \neq 0$. Then, using 9.2.3 (i)

$$c_{-\alpha,3\alpha+\beta;1,1} = c_{s_\alpha.\alpha,s_\alpha.\beta;1,1} \neq 0.$$

We now use step (c). The notations are as in the proof of 9.2.2. Let M' be the subspace of the four dimensional vector space M generated by the vectors $(\rho(g) - 1)e_3$ $(g \in SL_2)$. It is stable under $\rho(SL_2)$. Since e_3 is an eigenvector for all diagonal matrices, M' contains e_3, and also $e_0 \in k^*\rho(n)e_3$. By (28)

$$\rho\begin{pmatrix} 1 & 0 \\ 1 & 1 \end{pmatrix}e_3 - e_3 \in c_{-\alpha,3\alpha+\beta;1,1}e_2 + ke_1 + ke_0.$$

Since $c_{-\alpha,3\alpha+\beta;1,1} \neq 0$ it follows that $e_2 \in M'$ and also $e_1 \in k^*\rho(n)e_2$. So $M' = M$. Applying part (b) of 7.3.7 (3) we find that ρ is isomorphic to the dual of the representation ρ_3 of SL_2. Put

$$u(x) = \rho\begin{pmatrix} 1 & x \\ 0 & 1 \end{pmatrix} \quad (x \in k).$$

Using part (a) of 7.3.7 (3) we see that there is a unique basis $(f_i)_{0 \le i \le 3}$ of M with $f_0 = e_0$ and such that

$$\left.\begin{aligned}
u(x)f_0 &= f_0 + xf_1 + x^2f_2 + x^3f_3 \\
u(x)f_1 &= f_1 + 2xf_2 + 3x^2f_2 \\
u(x)f_2 &= f_2 + 3xf_3 \\
u(x)f_3 &= f_3
\end{aligned}\right\}$$

The matrix coefficients of $u(x)$ on this basis are the $c_{\alpha,i\alpha+\beta;j,1}$, where the $u_{i\alpha+\beta}$ $(0 \le i \le 3)$ are normalized such that $c_{\alpha,\beta;i,1} = 1$ $(0 \le i \le 3)$. Assume this has been done. Then all structure constants $c_{\alpha,\gamma,i,j}$ and $c_{\beta,\gamma;i,j}$ with $\gamma \in R^+$ are known, except for

$a = c_{\alpha,\beta;3,2}$, $b = c_{\alpha,\alpha+\beta;1,2}$, $c = c_{\beta,3\alpha+\beta;1,1}$. Now $(3\alpha + \beta, \beta)$ is a basis of a root system of type A_2. Using what we already know, we conclude that $c \neq 0$. To deal with the remaining structure constants a, b, c we make (27) more explicit. It follows from (27) that, ϕ being as in the proof of 9.2.2, we have for $m, m' \in M$, $x \in k$

$$\phi(m)\phi(m') = \phi(m + m')u_{3\alpha+2\beta}(f(m, m')), \tag{32}$$

$$u_\alpha(x)\phi(m)u_\alpha(x)^{-1} = \phi(u(x)m)u_{3\alpha+2\beta}(g(x, m)). \tag{33}$$

Here f is the bilinear form on M given (with obvious notations) by

$$f(\sum x_i f_i, \sum x_i' f_i) = -dx_2 x_1' - cx_3 x_0',$$

with $d = c_{\alpha+\beta,2\alpha+\beta;1,1}$. Moreover,

$$g(x, \sum x_i f_i) = bxx_1^2 - dx^2 x_0 x_1 + ax^3 x_0^2.$$

We have the 1-cocycle relation

$$g(x + y, m) = g(x, u(y)m) + g(y, m) \ (x, y \in k, m \in M).$$

Inserting the formula for g we find after a straightforward calculation that $b = -d = -3a$. It follows from (32) and (33) that

$$f(u(x)m, u(x)m') - f(m, m') = g(x, m + m') - g(x, m) - g(x, m'), \tag{34}$$

from which we see that the alternating bilinear form $[\ ,\ ]$ on M with

$$[m, m'] = f(m, m') - f(m, m')$$

is invariant under all $u(x)$ $(x \in k)$. In particular,

$$0 = [f_0, f_1] = [f_0 + xf_1 + x^2 f_2 + x^3 f_3, f_1 + 2xf_2 + 3x^2 f_3] = 3c + d,$$

whence $d = -3c$. If $p = \text{char } k \neq 3$ we can conclude that $c = a$, and $b = -d = -3a$. If $p = 3$ we have $b = d = 0$. Using (34) a brief calculation shows that again $a = c$. Since $c \neq 0$ by (31), we can normalize $u_{3\alpha+2\beta}$ such that $a = 1$.
The assertions of (i) will follow if we prove part (c). We have already seen that

$$d_{\alpha,\beta} = c_{\alpha,\beta;3,1}, \quad -d_{\beta,\alpha} = c_{\alpha,\beta;1,1}.$$

By our normalization these equal 1. Also, all $d_{\alpha,i\alpha+\beta}$ are ± 1, since these are matrix elements of $\rho(n)$ (see the proof of 9.2.2). Moreover $d_{\alpha,3\alpha+2\beta} = 1$, since $(3\alpha+2\beta)\pm\alpha \notin R$. Similarly, $d_{\beta,2\alpha+\beta} = 1$. Using (31) we obtain that $d_{\beta,3\alpha+\beta} = 1$. We are left with $d_{\beta,\alpha+\beta}$. Since $\alpha + \beta = s_\alpha\beta$, this equals $d_{\beta,\alpha}^{-1}$ by 9.2.2 (ii). This finishes the proof of 9.2.5. $\qquad\qquad\square$

9.2.6. Exercise. Assume R to be of rank two. Give an explicit description of the structure of the unipotent group U generated by the U_α with $\alpha \in R^+$, by a formula like (32).

9.3. The elements n_α

9.3.1. R is again arbitrary. The notations are as before. We fix a realization $(u_\alpha)_{\alpha \in R}$ of R in G. For $\alpha \in R$ the element n_α of 8.1.4 (i) represents the reflection $s_\alpha \in W$. Put $t_\alpha = \alpha^\vee(-1)$. By 8.1.4 (ii) we have $n_\alpha^2 = t_\alpha$. This an element of order ≤ 2 of T. Let $\alpha, \beta \in R$ be linearly independent. We denote by $m(\alpha, \beta)$ the order of $s_\alpha s_\beta$. By 7.5.1 (ii) $m(\alpha, \beta)$ equals one of the integers 2, 3, 4, 6.

9.3.2. Proposition. *Assume that α and β are simple roots, relative to some system of positive roots. Then*

$$n_\alpha n_\beta n_\alpha \ldots = n_\beta n_\alpha n_\beta \ldots,$$

the number of factors on either side being $m(\alpha, \beta)$.

(α, β) is a basis of a two dimensional root system. We discuss the four cases of 9.1.1. The case $A_1 \times A_1$ is left to the reader.

A_2. It follows from 9.2.2 (iii) that

$$n_\alpha n_\beta n_\alpha = (\alpha + \beta)^\vee(d_{\alpha,\beta})n_{\alpha+\beta}t_\alpha = (\alpha + \beta)^\vee(d_{\alpha,\beta})t_\beta n_{\alpha+\beta}.$$

Likewise,

$$n_\beta n_\alpha n_\beta = (\alpha + \beta)^\vee(d_{\beta,\alpha})t_\alpha n_{\alpha+\beta}.$$

We have seen in the proof of 9.2.5 that $d_{\beta,\alpha} = -d_{\alpha,\beta}$. Also, $(\alpha + \beta)^\vee = \alpha^\vee + \beta^\vee$ (use the facts of 9.1.1). It follows that $n_\alpha n_\beta n_\alpha = n_\beta n_\alpha n_\beta$. Since $m(\alpha, \beta) = 3$ we have proved the assertion.

B_2. Now $m(\alpha, \beta) = 4$. The assertion to be proved is equivalent to

$$(n_\alpha n_\beta)^2 = (n_\beta n_\alpha)^2.$$

From 9.2.2 (iii) we find

$$n_\alpha n_\beta n_\alpha = (2\alpha + \beta)^\vee(d_{\alpha,\beta})t_{\alpha+\beta}n_{2\alpha+\beta},$$

whence

$$(n_\alpha n_\beta)^2 = (2\alpha + \beta)^\vee(d_{\alpha,\beta})t_{\alpha+\beta}n_{2\alpha+\beta}n_\beta,$$

$$(n_\beta n_\alpha)^2 = (2\alpha + \beta)^\vee(d_{\alpha,\beta})t_\alpha n_{2\alpha+\beta}n_\beta$$

(use that n_β and $n_{2\alpha+\beta}$ commute and that $s_\beta.(\alpha+\beta) = \alpha$, $s_\beta.(2\alpha+\beta) = 2\alpha+\beta$). We have $(\alpha+\beta)^\vee = \alpha^\vee + 2\beta^\vee$, which shows that $t_{\alpha+\beta} = t_\alpha$, and the assertion follows.

G_2. Now $m(\alpha,\beta) = 6$. Using 9.2.2 (iii) it follows that we have to prove that

$$n_\alpha n_\beta n_\alpha = (3\alpha+\beta)^\vee(d_{\alpha,\beta})t_{2\alpha+\beta}n_{3\alpha+\beta} = tn_{3\alpha+\beta}$$

and

$$n_\beta n_\alpha n_\beta = (\alpha+\beta)^\vee(d_{\beta,\alpha})t_{3\alpha+2\beta}n_{\alpha+\beta} = t'n_{\alpha+\beta}$$

commute. Since $n_{3\alpha+\beta}$ and $n_{\alpha+\beta}$ commute, we have to show that

$$t.s_{3\alpha+\beta}(t') = t'.s_{\alpha+\beta}(t).$$

Using that $s_{3\alpha+\beta}.(\alpha+\beta) = \alpha+\beta$, $s_{3\alpha+\beta}.(3\alpha+2\beta) = \beta$, $s_{\alpha+\beta}.(3\alpha+\beta) = 3\alpha+\beta$, $s_{\alpha+\beta}.(2\alpha+\beta) = \alpha$, the result to be proved simplifies to

$$t_{2\alpha+\beta}t_\beta = t_{3\alpha+2\beta}t_\alpha.$$

This follows from

$$(2\alpha+\beta)^\vee + \beta^\vee = 2\alpha^\vee + 4\beta^\vee, \quad (3\alpha+2\beta)^\vee + \alpha^\vee = 2\alpha^\vee + 2\beta^\vee.$$

$\square$

9.3.3. Fix a system of positive roots R^+ in R and let D be the corresponding basis of R. Let $\mathbf{s} = (s_{\alpha_1}, ..., s_{\alpha_h})$ (with $\alpha_1, ..., \alpha_h \in D$) be a reduced decomposition of $w \in W$ (8.3.1). It follows from 8.3.3 and 9.3.2 that the element $n_{\alpha_1}...n_{\alpha_h}$ of the normalizer $N_G(T)$ is independent of the choice of the reduced decomposition of w. We denote this element by $\phi(w)$. So $\phi(W)$ is a set of representatives in $N_G(T)$ of the elements of W, uniquely determined by R^+ and the realization of R.

9.3.4. Exercises. (1) If $w, w' \in W$ then $\phi(ww') = \phi(w)\phi(w')c(w,w')$, where $c(w,w')$ is an element of order ≤ 2 of T. If $l(ww') = l(w) + l(w')$ we have $c(w,w') = 1$ (l is the length function on W defined by R^+, see 8.3.1).
(2) Let $\overline{W}$ be the subgroup of G generated by the n_α, $\alpha \in D$. Then $\overline{W}$ is an extension of W by an elementary abelian 2-group. (More about $\overline{W}$ can be found in [Ti1].)

9.3.5. Proposition. *If $\gamma, \delta \in D$ and $w \in W$ are such that $w.\gamma = \delta$ then*

$$\phi(w)u_\gamma(x)\phi(w)^{-1} = u_\delta(x) \ (x \in k).$$

There is $a \in k^*$ with $\phi(w)u_\gamma(x)\phi(w)^{-1} = u_\delta(ax)$. We have to show that $a = 1$. Using 9.1.4 and 9.3.4 (1) we reduce the proof to the case that R is two dimensional.

Let α and β be as in 9.1.1. We again discuss the four possible cases. The case $A_1 \times A_1$ is skipped. The reader is advised to use a picture of the root system in the other cases.

A_2. We may assume that $\gamma = \alpha$, $\delta = \beta$. We then must have $w = s_\alpha s_\beta$. With the notations of 9.2.1 we have $a = d_{\alpha,\alpha+\beta} d_{\beta,\alpha}$. It follows from what we established in the proof of 9.2.5 that $a = 1$.

B_2. Since α and β have different lengths we must have $\gamma = \delta$. If $\gamma = \beta$ we have $w = s_{2\alpha+\beta}$ and the result follows from the fact that $n_{2\alpha+\beta}$ commutes with the elements of U_β. If $\gamma = \alpha$ then $w = s_{\alpha+\beta} = s_\beta s_\alpha s_\beta$. Using that $d_{\alpha,\alpha+\beta} = -1$ (which was established in the proof of 9.2.5) we obtain $a = -d_{\beta,\alpha+\beta} d_{\beta,\alpha}$. By 9.2.2 (ii) this equals $-(-1)^{\langle \alpha, \beta^\vee \rangle} = 1$.

G_2. As in the previous case we have $\gamma = \delta$. If $\gamma = \alpha$ then $w = s_{3\alpha+2\beta}$ and the assertion follows from the fact that $n_{3\alpha+2\beta}$ commutes with the elements of U_α. If $\gamma = \beta$ we have $w = s_{2\alpha+\beta}$ and

$$\phi(w) = n_\alpha n_\beta n_\alpha n_\beta n_\alpha.$$

By 8.1.4 (iv) we have $\phi(w) u_{-\beta}(x) \phi(w)^{-1} = u_{-\beta}(a^{-1}x)$. Formula (22) of 8.1.4 gives $\phi(w) n_\beta \phi(w)^{-1} = \beta^\vee(a) n_\beta$, whence $(n_\alpha n_\beta)^3 = \phi(w) n_\beta = \beta^\vee(a)(n_\beta n_\alpha)^3$. From 9.3.2 we conclude that $\beta^\vee(a) = 1$, whence $a = \alpha(\beta^\vee(a))^{-1} = 1$.   □

9.4. A presentation of G

9.4.1. In this section we describe G as an abstract group. The root datum of (G, T) is $\Psi = (X, R, X^\vee, R^\vee)$. We fix a system of positive roots R^+ in R, a total order on R and a realization of R in G. Denote by $(c_{\alpha,\beta;i,j})$ the corresponding set of structure constants (9.2.1).

Let D be the basis of R defined by R^+. If $\alpha, \beta \in D$ and $\alpha \neq \beta$, denote as in 9.1 by $R_{\alpha,\beta}$ the two dimensional root system with basis (α, β). Put $R_1 = \bigcup_{\alpha,\beta \in D, \alpha \neq \beta} R_{\alpha,\beta}$.

9.4.2. We shall now describe a presentation of a group $\mathbf{G}$. This will involve the field k, the root datum Ψ and the structure constants $c_{\gamma,\delta;i,j}$, for roots γ and δ which are contained in some $R_{\alpha,\beta}$. The ingredients of the presentation are the following.

(a) $\mathbf{T} = \mathrm{Hom}(X, k^*)$, the group of homomorphisms of abelian groups (written multiplicatively). There is an isomorphism of groups $\pi : \mathbf{T} \to T$ such that for $\chi \in X$, $t \in \mathbf{T}$ we have $\chi(\pi t) = t(\chi)$ (cf. 3.2.10 (3)). The inverse of π is given by $(\pi^{-1}t)\chi = \chi(t)$ ($t \in T$, $\chi \in X$). For $\chi \in X$ define a homomorphism $\overline{\chi} : \mathbf{T} \to k^*$ by $\overline{\chi}(t) = t(\chi)$. For $\lambda \in X^\vee$ define the homomorphism $\overline{\lambda} : k^* \to \mathbf{T}$ by $\overline{\lambda}(x)(\chi) = x^{\langle \chi, \lambda \rangle}$ ($x \in k$, $\chi \in X$). The Weyl group W acts on $\mathbf{T}$ by $w.t(\chi) = t(w^{-1}.\chi)$ ($w \in W$, $t \in \mathbf{T}$, $\chi \in X$).

(b) For $\gamma \in R_1$, $x \in k$ we have a generator $\mathbf{u}_\gamma(x)$. For $x, y \in k$ we impose the relations

$$\mathbf{u}_\gamma(x + y) = \mathbf{u}_\gamma(x)\mathbf{u}_\gamma(y), \tag{35}$$

$$\mathbf{u}_\gamma(x)\mathbf{u}_\delta(y)\mathbf{u}_\gamma(x)^{-1} = \mathbf{u}_\delta(y) \prod_{i\gamma+j\delta\in R; i,j>0} \mathbf{u}_{i\gamma+j\delta}(c_{\gamma,\delta;i,j}x^i y^j), \tag{36}$$

if $\gamma, \delta \in R_1$ are linearly independent and both lie in some root system $R_{\alpha,\beta}$, where $\alpha, \beta \in D$ and $\alpha \neq \beta$. We also impose the relations

$$\mathbf{t}\mathbf{u}_\gamma(x)\mathbf{t}^{-1} = \mathbf{u}_\gamma(\overline{\gamma}(t)x), \tag{37}$$

if $\mathbf{t} \in \mathbf{T}$, $\gamma \in R_1$, $x \in k$.
(c) For $\gamma \in R_1$ put

$$\mathbf{n}_\gamma = \mathbf{u}_\gamma(1)\mathbf{u}_{-\gamma}(-1)\mathbf{u}_\gamma(1).$$

We require the following relations

$$\mathbf{n}_\gamma \mathbf{u}_\gamma(x)\mathbf{n}_\gamma^{-1} = \mathbf{u}_{-\gamma}(-x) \ (x \in k), \tag{38}$$

$$\mathbf{n}_\gamma^2 = \mathbf{t}_\gamma, \ \text{with } \mathbf{t}_\gamma(\chi) = (-1)^{\langle \chi, \gamma^\vee \rangle} \ (\chi \in X), \tag{39}$$

$$\mathbf{n}_\alpha \mathbf{n}_\beta \mathbf{n}_\alpha \ldots = \mathbf{n}_\beta \mathbf{n}_\alpha \mathbf{n}_\beta \ldots, \tag{40}$$

for $\alpha, \beta \in D$, $\alpha \neq \beta$, where the number of factors in either side is the order $m(\alpha, \beta)$ of $s_\alpha s_\beta$. We also require

$$\mathbf{u}_\gamma(x)\mathbf{u}_{-\gamma}(-x^{-1})\mathbf{u}_\gamma(x) = \overline{\gamma}^\vee(x)\mathbf{n}_\gamma \ (\gamma \in R_1, \ x \in k^*). \tag{41}$$

Using (37) and (39) we see that $\mathbf{n}_\gamma \mathbf{t}\mathbf{n}_\gamma^{-1} \in \mathbf{T}$ with

$$(\mathbf{n}_\gamma \mathbf{t}\mathbf{n}_\gamma^{-1})(\chi) = \mathbf{t}(s_\gamma . \chi) \ (\chi \in X). \tag{42}$$

Let $\mathbf{G}$ be the group generated by $\mathbf{T}$ and the $\mathbf{u}_\gamma(x)$ $(\gamma \in R_1)$, subject to the relations (35),...,(41).

9.4.3. Theorem. *The isomorphism $\pi : \mathbf{T} \to T$ extends to an isomorphism of abstract groups $\pi : \mathbf{G} \to G$ with $\pi(\mathbf{u}_\gamma(x)) = u_\gamma(x)$ $(\gamma \in R_1, \ x \in k)$.*

The relations that we imposed are counterparts of relations holding in G. For (35), (37), (39), (41) see 8.1; (36) is the counterpart of formula (27) in 9.2.1 and (40) of the formula of 9.3.2. The counterpart of (38) for G occurs in the proof of 8.1.4. It follows that π extends to a homomorphism $\pi : \mathbf{G} \to G$ with the last property of the theorem. We must show that π is an isomorphism.

For $\gamma \in R_1$ put $\mathbf{U}_\gamma = \mathrm{Im} \ \mathbf{u}_\gamma$. Then $\pi(\mathbf{U}_\gamma) = U_\gamma$. If γ and δ are two linearly independent roots contained in a root system $R_{\alpha,\beta}$ $(\alpha, \beta \in D)$, denote by $U_{\gamma,\delta}$ (respectively, $\mathbf{U}_{\gamma,\delta}$) the subgroup of G ($\mathbf{G}$) generated by the $U_{i\gamma+j\delta}$ (the $\mathbf{U}_{i\gamma+j\delta}$) with

$i, j \geq 0$, $i + j \geq 1$. Clearly, $\pi(U_{\gamma,\delta}) = U_{\gamma,\delta}$. By 8.2.2 we have a bijective map $\prod U_{i\gamma+j\delta} \to U_{\gamma,\delta}$ (the product is over the pairs (i, j) as above, and the order of the product is prescribed by our given total order on R) and a bijective map $\psi : k^n \to U_{\gamma,\delta}$, where $n = \dim U_{\gamma,\delta}$. Using (36) we obtain an analogous surjective map $\tilde{\psi} : k^n \to \mathbf{U}_{\gamma,\delta}$ with $\psi = \pi \circ \tilde{\psi}$. Since ψ is bijective and the restriction of π to $\mathbf{U}_{\gamma,\delta}$ is surjective it follows that π defines an isomorphism $\mathbf{U}_{\gamma,\delta} \to U_{\gamma,\delta}$.

Next let $\gamma \in D$. If $i\gamma + j\delta \in R$ and $j > 0$, then either $i\gamma + j\delta$, $\delta \in R^+$ or $i\gamma + j\delta$, $\delta \in -R^+$. Let $U'_{\gamma,\delta}$ (respectively $\mathbf{U}'_{\gamma,\delta}$) be the subgroup of G ($\mathbf{G}$) generated by the corresponding groups $U_{i\gamma+j\delta}(\mathbf{U}_{i\gamma+j\delta})$. An argument similar to that of the preceding paragraph shows that π defines an isomorphism $\mathbf{U}'_{\gamma,\delta} \to U'_{\gamma,\delta}$. By (36) both $\mathbf{U}_{\gamma}$ and $\mathbf{U}_{-\gamma}$ normalize $\mathbf{U}'_{\gamma,\delta}$, hence so does $\mathbf{n}_{\gamma}$. Since $\pi(\mathbf{n}_{\gamma} \mathbf{U}_{\delta} \mathbf{n}_{\gamma}^{-1}) = n_{\gamma} U_{\delta} n_{\gamma}^{-1} = U_{s_{\gamma}.\delta}$ we must have $\mathbf{n}_{\gamma} \mathbf{U}_{\delta} \mathbf{n}_{\gamma}^{-1} = \mathbf{U}_{s_{\gamma}.\delta}$ (for γ simple, and γ, δ in a root system $R_{\alpha,\beta}$ with $\alpha, \beta \in D$).

It follows from (40) (see 9.3.3) that there is a map $\phi : W \to \mathbf{G}$ with $\phi(w) = \mathbf{n}_{\alpha_1} \ldots \mathbf{n}_{\alpha_h}$, where $(s_{\alpha_1}, \ldots, s_{\alpha_h})$ is a reduced decomposition of $w \in W$. Consequently, if w lies in the Weyl group $W_{\alpha,\beta}$ of $R_{\alpha,\beta}$ and $\gamma \in R_{\alpha,\beta}$ we have $\phi(w)\mathbf{U}_{\gamma}\phi(w)^{-1} = \mathbf{U}_{w.\gamma}$. If $\gamma \in R$ there exist by 8.2.8 (ii) $w \in W$ and $\alpha \in D$ such that $\gamma = w.\alpha$. Put $\mathbf{U}_{\gamma} = \phi(w)\mathbf{U}_{\alpha}\phi(w)^{-1}$. Using 9.1.4 it follows from what we just established that $\mathbf{U}_{\gamma}$ is well-defined, i.e. is independent of the choice of w and α. Then (37) holds for all $\gamma \in R$. Also, $\mathbf{U}_{w.\gamma} = \phi(w)\mathbf{U}_{\gamma}\phi(w)^{-1}$ for all $w \in W$, $\gamma \in R$.

Next let $\gamma, \delta \in R$ be linearly independent. By 9.1.2 there exist $w \in W$ and $\alpha, \beta \in D$ with $w.\gamma$, $w.\delta \in R_{\alpha,\beta}$. Then, $(\ ,\ )$ denoting a commutator set,

$$(\mathbf{U}_{\gamma}, \mathbf{U}_{\delta}) = \phi(w)^{-1}(\mathbf{U}_{w.\gamma}, \mathbf{U}_{w.\delta})\phi(w),$$

and (36) implies that

$$(\mathbf{U}_{\gamma}, \mathbf{U}_{\delta}) \subset \prod_{i,j \geq 1, i\gamma + j\delta \in R} \mathbf{U}_{i\gamma+j\delta}.$$

Let U (respectively $\mathbf{U}$) be the subgroup of G ($\mathbf{G}$) generated by the U_{γ} ($\mathbf{U}_{\gamma}$) with $\gamma \in R^+$. The argument used twice before shows that π induces a bijection of $\mathbf{U}$ onto U and a bijection of the group $\mathbf{B}$ generated by $\mathbf{T}$ and $\mathbf{U}$ onto the Borel subgroup B of G defined by R^+. Also, if for $w \in W$ we denote by U_w (respectively, $\mathbf{U}_w$) the subgroup of G ($\mathbf{G}$) generated by the U_{γ} ($\mathbf{U}_{\gamma}$) with $\gamma \in R^+$, $w.\gamma \in -R^+$, then π maps $\mathbf{U}_w$ bijectively onto U_w.

Put $\mathbf{C}(w) = \mathbf{B}\phi(w)\mathbf{B}$. Then $\pi(\mathbf{C}(w)) = C(w) = B\dot{w}B$ (as in 8.3). If $\mathbf{s} = (s_1, \ldots, s_h)$ is a reduced decomposition of w with $s_i = s_{\alpha_i}$ ($\alpha_i \in D$), we have as in 8.3.6 (i) a bijection $k^h \times \mathbf{B} \to \mathbf{C}(w)$ sending $(x_1, \ldots, x_h; b)$ to

$$\mathbf{u}_{\alpha_1}(x_1)\phi(s_1)\mathbf{u}_{\alpha_2}(x_2)\phi(s_2)\ldots\mathbf{u}_{\alpha_h}(x_h)\phi(s_h)b$$

and a bijection $\mathbf{U}_{w^{-1}} \times \mathbf{B} \to \mathbf{C}(w)$. It follows that π maps $\mathbf{C}(w)$ bijectively onto $C(w)$.

Also, 8.3.7 carries over to **G**. The fact that $\mathbf{C}(s) \cup \mathbf{C}(e)$ is a group (used in the proof of 8.3.7) follows from (41). As in the proof of 8.3.8 one shows that **G** is the union of the $\mathbf{C}(w)$ ($w \in W$). Using 8.3.8 we conclude that π must be bijective. $\qquad\square$

9.4.4. Exercises. (1) (Notations of 9.4.1.) **T** is as in 9.4.2. Let **G** be the group generated by **T** and symbols $\mathbf{u}_\gamma(x)$, $\mathbf{n}_\alpha$ where $x \in k$, $\alpha \in D$, $\gamma \in R^+$, subject to the relations (35), (36) (for $\gamma, \delta \in R^+$), (37) (for $\gamma \in R^+$), (39) (for $\gamma \in D$), (40), (42) (for $\gamma \in D$). Instead of (41) we impose

$$\mathbf{u}_\gamma(x)\mathbf{n}_\gamma\mathbf{u}_\gamma(x^{-1})\mathbf{n}_\gamma^{-1}\mathbf{u}_\gamma(x) = \overline{\gamma^\vee}(x)\mathbf{n}_\gamma,$$

for $\gamma \in D$, $x \in k^*$. There is an obvious homomorphism $\pi : \mathbf{G} \to G$. Show that it is an isomorphism.

(2) Let F be any field. Let Γ be the group generated by symbols $u(x)$, $v(y)$ (x, $y \in F$), subject to the following relations (A) and (B):

(A) $u(x + y) = u(x)u(y)$, $v(x + y) = v(x)v(y)$ $(x, y \in F)$.
For $x \in F^*$ put $w(x) = u(x)v(-x^{-1})u(x)$, $a(x) = w(x)w(1)^{-1}$.
(B) $w(x)u(y)w(x)^{-1} = v(-x^2 y)$ $(x \in F^*,\ y \in F)$.
    Prove the following facts.

(a) $a(x)u(y)a(x)^{-1} = u(x^2 y)$, $a(x)v(y)a(x)^{-1} = v(x^{-2}y)$ $(x \in F^*,\ y \in F)$.

(b) Let U be the subgroup of Γ generated by the $u(x)$ ($x \in F$) and A the subgroup generated by the $a(x)$ ($x \in F^*$). Show that Γ is the disjoint union of UA and $Ux(1)UA$.

(c) There is a homomorphism $\pi : \Gamma \to \mathbf{SL}_2(F)$ mapping $u(x)$ and $w(1)$ onto the elements $u_1(x)$ and n_1 of the proof of 7.2.4. Show that Ker π is a central subgroup of Γ which is contained in A.

(d) Let Γ' be the group generated by symbols $u'(x)$, $v'(y)$ (x, $y \in F$) subject to relations (A') and (B') as before, and moreover also (with obvious notations)
(C') $a'(xy) = a'(x)a'(y)$ $(x, y \in F^*)$. Show that Γ' is isomorphic to $\mathbf{SL}_2(F)$.

(3) Assume that G is semi-simple. The notations are as in 9.4.1. Let Γ be the group generated by symbols $\mathbf{u}_\gamma(x)$ ($\gamma \in R$, $x \in k$), subject to the relations (35) and (36), where γ and δ are linearly independent roots. Denote by $\pi : \Gamma \to G$ the obvious homomorphism.

(a) Let U be the subgroup of Γ generated by all $\mathbf{u}_\gamma(x)$ with $\gamma \in R^+$. The restriction of π to **U** is bijective.

(b) Put $\mathbf{n}_\gamma(x) = \mathbf{u}_\gamma(x)\mathbf{u}_{-\gamma}(-x^{-1})\mathbf{u}_\gamma(x)$ ($\gamma \in R$, $x \in k^*$). If γ and δ are linearly independent roots then

$$\mathbf{n}_\gamma(x)\mathbf{u}_\delta(y)\mathbf{n}_\gamma(x)^{-1} = \mathbf{u}_{s_\gamma.\delta}(d_{\gamma,\delta}x^{\langle\delta,\gamma^\vee\rangle}y),$$

where $d_{\gamma,\delta}$ is as in 9.2.1. (*Hint*: adapt the proof of 9.2.2.)

(c) Let $\gamma \in R$ and assume that there is $\delta \in R$, linearly independent of γ with $\langle\gamma, \delta^\vee\rangle \neq 0$. Then

$$\mathbf{n}_\gamma(x)\mathbf{u}_\gamma(y)\mathbf{n}_\gamma(x)^{-1} = \mathbf{u}_{-\gamma}(-x^2 y) \ (x \in k^*,\ y \in k).$$

(*Hint*: One may assume that R has rank two and that there is a basis (α, β) with $\gamma \neq \alpha$, β and such that γ is positive, for the corresponding system of positive roots. For $y \neq 0$ there is a relation

$$(\mathbf{u}_\alpha(y_1), \mathbf{u}_\beta(y_2)) = \mathbf{u}_\gamma(y) \prod_{i,j>0; i\alpha+j\beta\neq\alpha} \mathbf{u}_{i\gamma+j\delta}(y_{ij}).$$

Conjugate both sides by $\mathbf{n}_\alpha(x)$ and use (a), (b)).

(d) If R is irreducible of rank > 1 then relation (38) of 9.4.2 is a consequence of the other relations.

The group Γ of this exercise can be defined over an arbitrary field F. For more details see [St1, St4].

9.5. Uniqueness of structure constants

9.5.1. We now recall some facts about root systems, with which we assume the reader to be familiar (see for example [Bou2, Ch. VI]). Let R be a root system in the euclidean vector space V, with a metric as in 7.1.7, and let D be a basis of R. The *Dynkin diagram* $\mathcal{D}$ defined by D is a graph with vertex set D. Two distinct vertices α, β are joined by $\langle \alpha, \beta^\vee \rangle \langle \beta, \alpha^\vee \rangle$ bonds, with an arrow pointing towards the shorter root if α and β have different lengths.

Assume R to be irreducible (8.1.12 (4)). We have the following properties:
(a) $\mathcal{D}$ is connected and is a tree, i.e. contains no circuits.
(b) A multiple bond is either double or triple. If a triple bond occurs then R is of type G_2. At most one double bond can occur and if it does $\mathcal{D}$ is a chain. If multiple bonds occur we have two possible root lengths. We then can speak of *short* and *long* roots.

The connected Dynkin diagrams can be classified. The list of these is as follows.

A_n $(n \geq 1)$ o —— o —— o - ···· o —— o —— o

B_n $(n \geq 2)$ o —— o —— o - ···· o —— o ⇒ o

C_n $(n \geq 3)$ o —— o —— o - ···· o —— o ⇐ o

D_n $(n \geq 4)$ o —— o —— o - ···· o —— o
                                           |
                                           o

E_6 o —— o —— o —— o —— o
                                     |
                                     o

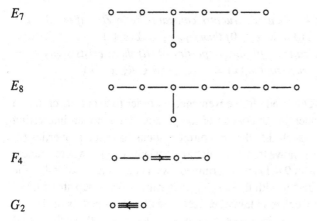

E_7

E_8

F_4

G_2

Let R be arbitrary.

9.5.2. Lemma. *There is a disjoint decomposition $D = D_1 \cup D_2$ such that distinct roots in D_1, respectively D_2, are orthogonal.*

We may assume that R is irreducible. Since $\mathcal{D}$ is a tree it has an endpoint α, which is joined to only one other vertex β. The complement of α in D is a basis of a root system of smaller dimension. By induction we may assume that we have a decomposition $D' = D_1' \cup D_2'$ as in the lemma. If $\beta \in D_1'$ we may take $D_1 = D_1'$, $D_2 = D_2' \cup \{\alpha\}$. $\square$

Let again R be the root system of the reductive group G relative to the maximal torus T, as before. Assume given a decomposition $D = D_1 \cup D_2$ as in 9.5.2. If different root lengths occur in an irreducible component of R, there exists by property (b) of 9.5.1 a unique pair (α, β) of adjacent vertices of the corresponding component of $\mathcal{D}$, such that α is short and β is long. We then assume that $\alpha \in D_1$, $\beta \in D_2$. The assumptions are as in 9.2.1.

9.5.3. Proposition. *Assume that $\mathcal{D}$ contains no triple bonds.*
(i) There exists a realization $(u_\alpha)_{\alpha \in R}$ of R in G such that the corresponding structure constants $c_{\alpha;\beta;i,j}$ have the following properties:
   (a) If $\alpha \in D_1$, $\beta \in D_2$ have the same length and $\alpha + \beta \in R$ then $c_{\alpha,\beta;1,1} = 1$,
   (b) Let $\alpha \in D_1$, $\beta \in D_2$ and $\alpha + \beta \in R$, with α short and β long. Then all structure constants $c_{\alpha,\beta;i,j}$ $(i, j > 0, i\alpha + j\beta \in R)$ equal 1.
The values of the other structure constants are uniquely determined by the requirements (a) and (b) (and the underlying total order of R);

(ii) We may assume that the realization of (i) is such that (ϕ being as in 9.3.3)

$$\phi(w)u_\alpha(x)\phi(w)^{-1} = u_{w.\alpha}(\pm x), \tag{43}$$

for $\alpha \in R$, $w \in W$, $x \in k$. Then all structure constants lie in $\mathbf{Z}.1$. If α, β, $\alpha + \beta$, $\alpha - c\beta \in R$, $\alpha - (c+1)\beta \notin R$ $(c \geq 0)$ then $c_{\alpha,\beta;1,1} = \pm(c+1)$;
(iii) If (u'_α) is another realization with the properties of (ii) there exists signs $\epsilon_\alpha = \pm 1$ $(\alpha \in R)$ with $\epsilon_{-\alpha} = \epsilon_\alpha$, such that $u'_\alpha(x) = u_\alpha(\epsilon_\alpha x)$ $(\alpha \in R$, $x \in k)$.

The assumption about $\mathcal{D}$ is made for convenience, in order to avoid the case of a root system of type G_2 (which in the context of the proposition is not an interesting one). We can normalize u_γ such that the structure constants have the properties (a) and (b) of (i). Then 9.2.5 (ii) shows that the structure constants $c_{\gamma,\delta;i,j}$, where γ and δ lie in a root system $R_{\alpha,\beta}$ (as in 9.4.1) are determined. As a consequence of 9.4.3, the root datum Ψ of (G, T) together with these $c_{\gamma,\delta;i,j}$ determines the group structure of G. From the proof of 9.4.3 we then obtain a realization with the properties of (i).

Assume given the u_γ with γ in a root system $R_{\alpha,\beta}$ $(\alpha, \beta \in D)$, such that (i) holds. For $\gamma \in R_{\alpha,\beta}$ and $w \in W_{\alpha,\beta}$ (the Weyl group of $R_{\alpha,\beta}$) we have (43). This follows by induction on $l(w)$, using 9.3.4 (1) and the fact that the constants $d_{\alpha,\delta}$ and $d_{\beta,\delta}$ $(\delta \in R_{\alpha,\beta})$ are ± 1 (see 9.2.5 (i)). Now let $\gamma \in R$ be arbitrary. There are $w \in W$ and $\alpha \in D$ with $\gamma = w.\alpha$. For $x \in k$ define $u_\gamma(x) = \phi(w)u_\alpha(x)\phi(w)^{-1}$. From 9.3.4 (1) and 9.3.5 we see that u_γ is unique, up to a change of sign as in (iii).

The last assertions of (ii) follow from 9.2.5 (i) if γ and δ lie in a root system $R_{\alpha,\beta}$. The general case then follows by using 9.1.2. We have also established (iii). $\square$

9.5.4. Theorem *(R arbitrary) The structure constants $(c_{\alpha,\beta;i,j})$, for a fixed order on R, are unique up to equivalence.*

This follows from 9.2.5 and 9.5.3.

In the next chapter we shall give an explicit formula for a set of structure constants, in the case that no multiple bonds occur in the Dynkin diagram. See 10.2.

9.6. The isomorphism theorem

9.6.1. Assume that G and G_1 are two connected, reductive, linear algebraic groups over k, with maximal tori T, T_1 and corresponding root data $\Psi = (X, X^\vee, R, R^\vee)$, $\Psi_1 = (X_1, X_1^\vee, R_1, R_1^\vee)$. An *isogeny* $\phi : G \to G_1$ is a surjective homomorphism of algebraic groups with finite kernel. Then Ker ϕ is a central subgroup of G (2.2.2 (4)), which lies in T (7.6.4 (iii)). We have dim G = dim G_1 (5.3.3 (i)).

Assume that $\phi(T) = T_1$. The isogeny ϕ defines homomorphisms of character groups and cocharacter groups $f = f(\phi) : X_1 \to X$, respectively $f^\vee : X^\vee \to X_1^\vee$. We have

$$\langle \chi_1, f^\vee(\lambda) \rangle = \langle f(\chi_1), \lambda \rangle \ (\chi_1 \in X_1, \ \lambda \in X^\vee).$$

Let $\alpha \in R$ and let U_α be as in 8.1.1. Then ϕU_α is a connected, one dimensional, unipotent, subgroup of G_1 which is normalized by T_1. By 8.1.12 (1) there is $\alpha_1 \in R_1$ with $\phi U_\alpha = U_{\alpha_1}$. Conversely, if $\alpha_1 \in R_1$ then $(\phi^{-1}(U_{\alpha_1}))^0$ is a connected, one dimensional, unipotent, subgroup of G normalized by T, hence is a U_α. It follows that there is a bijection $b : R \to R_1$ with $\phi U_\alpha = U_{b\alpha}$ $(\alpha \in R)$.

If ϕ is an isomorphism of algebraic groups then $f(b\alpha) = \alpha$ for all $\alpha \in R$ and f defines an isomorphism of root data $\Psi_1 \to \Psi$, i.e. an isomorphism $X_1 \simeq X$ mapping R_1 onto R and such that its dual maps $R^\vee$ onto $R_1^\vee$.

9.6.2. Theorem. [Isomorphism theorem] *Let f be an isomorphism of Ψ_1 onto Ψ. There exists an isomorphism of algebraic groups $\phi : G \to G_1$ with $\phi T = T_1$ and $f = f(\phi)$. If ϕ' is another isomorphism with these properties there is $t \in T$ such that $\phi'(g) = \phi(tgt^{-1})$ $(g \in G)$.*

Let $(u_\alpha)_{\alpha \in R}$ and $(u_{\alpha_1})_{\alpha_1 \in R_1}$ be realizations of R in G (respectively of R_1 in G_1). Choose total orders on R and R_1 that correspond via f. It follows from 9.5.4 that we may assume that the structure constants $c_{\alpha,\beta;i,j}$ and $c_{f^{-1}\alpha,f^{-1}\beta;i,j}$ are equal. Let $\mathbf{G}$ and $\mathbf{G}_1$ be the abstract groups defined in 9.4.2, for G respectively G_1 (with systems of positive roots which correspond under f). It is clear that f defines an isomorphism $\phi : \mathbf{G} \to \mathbf{G}_1$. If π and π_1 are the isomorphisms for G respectively G_1 of 9.4.3, then $\phi = \pi_1 \circ \phi \circ \pi^{-1}$ is an isomorphism of abstract groups $G \to G_1$. It follows from the uniqueness part 8.3.9 of Bruhat's lemma that the restriction of ϕ to the open set $C(w_0)$ of 8.3.11 is a morphism of $C(w_0)$ to G_1. Also, the restriction of ϕ to a translate $g.C(w_0)$ is a morphism. Since these translates cover G, ϕ is a homomorphism of algebraic groups. Reversing the roles of G and G_1 we see that ϕ^{-1} is also a morphism and that ϕ is as required.

To prove the last part of the theorem it suffices to show that an isomorphism ϕ of G fixing all elements of T and such that $\phi U_\alpha = U_\alpha$ for all $\alpha \in R$, is an inner automorphism defined by an element of T. Let D be a basis of R. By 8.1.4 (iv) there exist $c_\alpha \in k^*$ with $c_\alpha c_{-\alpha} = 1$, such that $((u_\alpha)$ being a realization) $\phi(u_\alpha(x)) = u_\alpha(c_\alpha x)$ $(\alpha \in R, x \in k)$. There exists $t \in T$ with $\alpha(t) = c_\alpha$ for $\alpha \in D$. It follows that the restriction of ψ to T and the $U_{\pm\alpha}$ is conjugation by t. Since these groups generate G (8.2.10), ϕ is conjugation by t. $\qquad\square$

9.6.3. We now consider the case that ϕ is an arbitrary isogeny. Then f is an isomorphism of X_1 onto a subgroup of finite index of X. We have $\phi(u_\alpha(x)) = u_{b\alpha}(h(x))$, where h is a polynomial in one variable, with $h(\alpha(t)x) = (b\alpha(\phi(t)))h(x)$ $(t \in T, x \in k)$. It follows that h is homogeneous. Since also $h(x + y) = h(x) + h(y)$ $(x, y \in k)$, there is a power $q(\alpha)$ of the characteristic exponent p of k such that, up to a constant, $h(T) = T^{q(\alpha)}$. We then have

$$f(b\alpha) = q(\alpha)\alpha, \quad f^\vee(\alpha^\vee) = q(\alpha)(b\alpha)^\vee. \tag{44}$$

If all $q(\alpha)$ equal 1 we say that ϕ is a *central isogeny*. This is the case if char $k = 0$.

We say that a triple $\mu = (f, b, q)$ where f is an isomorphism of X_1 onto a sub-group of finite index of X, b a bijection $R \to R_1$ and q a function $R \to \{p^n\}_{n>0}$ defines a *p-morphism* of Ψ_1 to Ψ if the properties established above hold. If μ comes from an isogeny ϕ we write $\mu = \mu(\phi)$. An example of a p-morphism is a *Frobenius morphism*, with $\Psi_1 = \Psi$, $f = q.\mathrm{id}$, $b = \mathrm{id}$, $q(\alpha) = q$, where q is a power of the prime number p. The notion of a p-morphism makes sense for abstract root data and for an arbitrary natural number p.

Let Ψ and Ψ_1 be root data and let $\mu = (f, b, q)$ be a p-morphism of Ψ_1 to Ψ. Notations are as above. The root system R is assumed to be reduced (7.4.3).

9.6.4. Lemma. *Let* $\alpha, \beta \in R$.
(i) $q(\beta)\langle b\alpha, (b\beta)^{\vee}\rangle = q(\alpha)\langle\alpha, \beta^{\vee}\rangle$;
(ii) $b(s_\alpha.\beta) = s_{b\alpha}(b\beta)$;
(iii) $q(w.\alpha) = q(\alpha)$ *for all* $w \in W$;
(iv) If $\langle\alpha, \beta^{\vee}\rangle > 0$ *and* $q(\alpha) < q(\beta)$ *then* p *equals* 2 *or* 3 *and* $\langle\alpha, \beta^{\vee}\rangle = p$,
$\langle b\alpha, (b\beta)^{\vee}\rangle = 1$, $q(\beta) = pq(\alpha)$.

By (44) we have

$$q(\beta)\langle b\alpha, (b\beta)^{\vee}\rangle = \langle b\alpha, f^{\vee}(\beta^{\vee})\rangle = \langle f(b\alpha), \beta^{\vee}\rangle = q(\alpha)\langle\alpha, \beta^{\vee}\rangle,$$

proving (i). Also, by (44) and (i)

$$f(s_{b\alpha}(b\beta)) = f(b\beta) - \langle b\beta, (b\alpha)^{\vee}\rangle f(b\alpha) =$$

$$= q(\beta)(s_\alpha.\beta) = q(\beta)q(s_\alpha.\beta)^{-1}f(b(s_\alpha.\beta)).$$

Since f is injective, the roots $s_{b\alpha}(b\beta)$ and $b(s_\alpha.\beta)$ differ by the positive factor $q(\beta)q(s_\alpha.\beta)^{-1}$. Since R is reduced, these roots must be equal, whence (ii). It also follows that $q(s_\alpha.\beta) = q(\beta)$, which implies (iii). Finally, (iv) follows from (i) and 7.5.1. $\square$

In the next result G, G_1 and Ψ, Ψ_1 are as in 9.6.1 and p is the characteristic exponent of k.

9.6.5. Theorem. [Isogeny theorem] *Let* $\mu = (f, b, q)$ *be a p-morphism of* Ψ_1 *to* Ψ. *There is an isogeny* $\phi : G \to G_1$ *with* $\phi T = T_1$ *and* $\mu = \mu(\phi)$. *If* ϕ' *is another isogeny with these properties, there is* $t \in T$ *such that* $\phi'(g) = \phi(tgt^{-1})$ $(g \in G)$.

We sketch the proof, without going into all the details. The proof of the uniqueness statement is similar to the proof of the corresponding part of 9.6.2.

There is a unique subgroup X' of the character group X of T containing Im f and such that the finite group X/X' has order prime to p. Let $A \subset T$ be the subgroup of T on which all characters of X' are trivial. Then A is a finite subgroup of T that is contained in the center of G. Put $G' = G/A$ and let Ψ' be the root datum of $(G', T/A)$. The canonical map $\psi : G \to G'$ is an isogeny. We have a factorization $\mu = \mu(\psi) \circ \mu'$, where μ' is a p-morphism of Ψ_1 to Ψ', with which we have to deal. To do this, we may assume that $G = G'$, i.e. that Im f has p-power index in X. If char$(k) = 0$ we have the case of the isomorphism theorem.

Assume now that char $k = p > 0$ (and $G = G'$). Let $\mathbf{G}$ and $\mathbf{G}_1$ be as in the proof of 9.6.2. Then $\mathbf{G}$ is generated by a group $\mathbf{T}$ and symbols $\mathbf{u}_\gamma(x)$ ($\gamma \in R$, $x \in k$), subject to the relations of 9.4.2, and similarly for $\mathbf{G}_1$. If we can prove that there exists a homomorphism $\phi : \mathbf{G} \to \mathbf{G}_1$ such that $(\phi \mathbf{t})(\chi_1) = \mathbf{t}(f \chi_1)$ ($\mathbf{t} \in \mathbf{T}$, $\chi_1 \in X_1$) and $\phi(\mathbf{u}_\gamma(x)) = \mathbf{u}_{b\gamma}(\epsilon_\gamma x^{q(\gamma)})$ ($\gamma \in R$, $x \in k$), with suitable signs ϵ_γ, then we can proceed as in the proof of 9.6.2 to prove the existence of ϕ. First assume that μ is a Frobenius morphism, so that all $q(\gamma)$ are equal. We may assume by 9.5.3 (ii) that all structure constants for G and G_1 lie in the prime field. Then ϕ exists, with $\epsilon_\gamma = 1$ for all $\gamma \in R$.

The crucial part of the proof of the existence of ϕ is to show that the relations (36) can be preserved. But these involve only roots lying in a root system $R_{\alpha,\beta}$ ($\alpha, \beta \in D$). It suffices to deal with the case that R has rank two, which we assume from now on. We may also assume that G is semi-simple and that μ is not a Frobenius morphism. Then we must have one of the cases of 9.6.4 (iv). Since R and R_1 have the same number of roots, they must be isomorphic (by the classification of 9.1.1). We then may assume that $G_1 = G$, $R_1 = R$, and we have by 9.1.1 that R is either of type B_2 and $p = 2$ or of type G_2 and $p = 3$. From 9.6.4 (iv) we can conclude that b interchanges long and short roots.

Now $\mu \circ \mu$ (defined in the obvious way) must be a Frobenius morphism (otherwise we get a contradiction with 9.6.4 (iv)). So μ is a 'square root of a Frobenius morphism'. Let α and β be as in 9.1.1. After composing with an element of the Weyl group we may assume that in type B_2

$$b\alpha = \beta, \ b\beta = \alpha, \ b(\alpha + \beta) = 2\alpha + \beta, \ b(2\alpha + \beta) = \alpha + \beta,$$

and in type G_2

$$b\alpha = \beta, \ b\beta = \alpha, \ b(\alpha + \beta) = 3\alpha + \beta, \ b(2\alpha + \beta) = 3\alpha + 2\beta,$$

$$b(3\alpha + \beta) = \alpha + \beta, \ b(3\alpha + 2\beta) = 2\alpha + \beta.$$

Moreover, $q(\alpha) = p^{n+1}$, $q(\beta) = p^n$, where $n \geq 0$. To prove the existence of ϕ we may assume that $n = 0$ (compose with a Frobenius morphism to deal with the general case). It is now convenient to use the presentation of 9.4.4 (1). With the notations introduced there we have to show that there exists a homomorphism $\phi : \mathbf{G} \to \mathbf{G}$ such that $\phi(\mathbf{u}_\gamma(x)) = \mathbf{u}_{b\gamma}(\epsilon_\gamma x^{q(\gamma)})$, $\phi(\mathbf{n}_\alpha) = \beta^\vee(\epsilon_\alpha)\mathbf{n}_\beta$, $\phi(\mathbf{n}_\beta) = \alpha^\vee(\epsilon_\beta)\mathbf{n}_\alpha$, where $\gamma \in R^+$, $x \in k$, the ϵ_γ being signs which are to be determined. The subgroup of $\mathbf{G}$

generated by the $\mathbf{u}_\gamma(x)$ ($\gamma \in R^+$, $x \in k$) can be explicitly described (see 9.2.6). Using such a description one finishes the proof. For more details about the 'exceptional' isogenies of 9.6.4 (iv) we refer to [St4, §11]]. □

9.6.6. Exercise. Complete the details of the proof of 9.6.5.

Notes

9.1 contains standard material. The treatment in 9.2 of structure constants in the case of a two dimensional root system is inspired by the computations made by Demazure in [SGA3, exp. XXIII, no.3], reproduced in [Hu1, p. 209-215]. We have tried to exploit in these computations the representation theory of $\mathbf{SL}_2$ (see the proof of 9.2.2 (i)).

The useful result 9.3.2 is implicit in the results of [SGA3, exp. XXIII, no.3].

To deal with the structure constants in the general case we use the version 9.4.3 of results of Steinberg (see [St1] or [St4]) on generators and relations for semi-simple groups, generalized to the reductive case. 9.4.3 shows that a reductive group G can be obtained by 'amalgamation' from subgroups with two dimensional root systems, a result due to Curtis [Cu]. The results of Steinberg alluded to are of a wider scope, as they involve an arbitrary ground field.

The main result of this chapter, the isomorphism theorem 9.4.1 is due to Chevalley [Ch4, exp. 24]. See also [Hu1, no. 33]. In these references the group is assumed to be semi-simple (which is the crucial case). In [SGA3, exp. XXIII] the isomorphism theorem is proved for group schemes.

The proof given here involves a study of structure constants, which is a bit technical. Another proof, due to Takeuchi [Tak] and exposed in [Jan1, II.1.14], is more conceptual. It uses the hyperalgebra associated to a reductive group, which englobes higher order infinitesimal invariants of the group at the identity element. We have stuck, however, to the proof given in the previous edition. It involves some points of independent interest, such as the presentation of a reductive group given in 9.4.2.

That structure constants are determined by the underlying root system up to equivalence (9.5.4) was proved by H. Azad [Az] in a different manner.

The isogeny theorem 9.6.5 is also due to Chevalley (see [Ch4, exp. 23-24]).

Chapter 10

The Existence Theorem

10.1. Statement of the theorem, reduction

k is an algebraically closed field. This chapter is devoted to the proof of the following *existence theorem*.

10.1.1. Theorem. *Let* $\Psi = (X, R, X^\vee, R^\vee)$ *be a root datum. There exists a connected reductive linear algebraic group G over k with a maximal torus T, such that the root datum $\Psi(G, T)$ is isomorphic to Ψ.*

By 9.6.2 such a group G is unique up to isomorphism.

10.1.2. The proof is in several steps. (a) A reduction to the case that R spans X and is irreducible. Then the group to be constructed is quasi-simple and adjoint (8.1.11, 8.1.12 (4))).

(b) In that case G will be constructed as a group of automorphisms of its Lie algebra $\mathfrak{g}$ in the case that R is *simply laced*, i.e. that no multiple bonds occur in the Dynkin diagram of R. We first have to construct the Lie algebra. This requires an explicit description of its structure constants.

(c) For arbitrary R the construction of G is reduced to the simply laced case by using automorphisms 'folding'.

We shall now carry out the details. Let Ψ be as before. Assume that X_1 is a subgroup of finite index of X containing R. We can then view $X^\vee$ as a subgroup of finite index of the dual $X_1^\vee$, so $R^\vee$ is a subset of $X_1^\vee$. We thus have a root datum $\Psi_1 = (X_1, R, X_1^\vee, R^\vee)$.

10.1.3. Proposition. *Assume that there exists a connected reductive group G_1 over k, with a maximal torus T_1 such that $\Psi(G_1, T_1) \simeq \Psi_1$. Then there exists a similar pair (G, T) with $\Psi(G, T) \simeq \Psi$.*

An easy argument shows that we may assume there is a prime number l such that $lX \subset X_1$. Then X/X_1 is a vector space over $\mathbf{F}_l$. Choose $\chi_1, \ldots, \chi_m$ in $X - X_1$ such that their images in X/X_1 form a basis of that vector space and let T be a torus with character group X (see 3.2.6). The Weyl group W of R acts on T.

Fix a system of positive roots R^+ in R and let D be the corresponding basis. Let $(u_\alpha)_{\alpha \in R}$ be a realization of R in G_1 (relative to T_1). With the notations of 9.6.3 we have a 1-morphism of Ψ_1 to Ψ. It follows that, if G exists, we may assume that T is a maximal torus of G and that there is a central isogeny (9.6.3) $G \to G_1$ mapping T onto T_1. It also follows that the isogeny defines an isomorphism onto a group U_α ($\alpha \in R$) of the realization of a similar subgroup of G.

There is an obvious homomorphism $\phi : T \to T_1$. Denote by B_1 the Borel subgroup of G_1 defined by R^+ (8.2.4 (i)) and let U be its unipotent radical. Let w_0 be the longest element of the Weyl group W of R (relative to R^+). Put $C(w_0) = B_1 \dot{w}_0 B_1$, as in 8.3.11. This is an open subvariety of G_1. Putting $Y = U\dot{w}_0 U$, the variety $C(w_0)$ is isomorphic to $T_1 \times Y$. It follows that we can view the quotient field $k(G_1)$ (1.8.1) to be $K_1 = k(T_1 \times Y)$. Similarly, the quotient field of the group G to be constructed will be $K = k(T \times Y)$. Then $K = K_1(\chi_1, \dots, \chi_m)$, a finite extension of K_1 of degree l^m. The left translation action of G_1 on $k[G_1]$ defines an action of G_1 as an automorphism group of K_1.

We now define automorphisms of K. The group T acts on $T \times Y$, by translations in the first factor, whence an action of T on K. For $\gamma \in R^+$, $x \in k$ we define an automorphism $\mathbf{u}_\gamma(x)$ of $T \times Y$ by

$$\mathbf{u}_\gamma(x).(t, y) = (t, u_\gamma(\gamma(t)^{-1}x)y)$$

(notice that Y is invariant under left translations by elements of U). Next let $\alpha \in D$. We use the notations of 8.5.9. For $(t, y) \in T \times Y_\alpha$ define

$$\mathbf{n}_\alpha.(t, y) = ((s_\alpha t)\alpha^\vee(-\xi_\alpha(y))^{-1}, \phi_\alpha y).$$

Then $\mathbf{n}_\alpha$ is an automorphism of $T \times Y_\alpha$. It acts on K. Also, the $\mathbf{u}_\gamma(x)$ act as automorphisms on K. Let G be the group of automorphisms of K generated by the automorphisms defined by T, the $\mathbf{u}_\gamma(x)$ ($\gamma \in R^+$, $x \in k$) and the $\mathbf{n}_\alpha$ ($\alpha \in D$). These automorphisms all stabilize K_1 and they induce the automorphisms of K_1 coming from left translations in G_1 by, respectively, the elements of T_1, $u_\gamma(x)$ and n_α. It follows (for example by using 8.2.10) that any $g \in G$ stabilizes K_1 and that the restriction of g to K_1 is induced by left translation by an element of G_1, whence a homomorphism $\phi : G \to G_1$, extending the homomorphism $T \to T_1$. The automorphisms $\mathbf{u}_\gamma(x)$ satisfy the relations (35) and (36) of 9.4 (with γ, $\delta \in R^+$, the structure constants being those of G_1). Moreover we have $t\mathbf{u}_\gamma(x)t^{-1} = \mathbf{u}_\gamma(\gamma(t)x)$ ($t \in T$, $\gamma \in R^+$, $x \in k$) and $\mathbf{n}_\alpha t \mathbf{n}_\alpha^{-1} = s_\alpha(t)$ ($\alpha \in D$, $t \in T$). By a straightforward check one proves the relation (see also 9.4.4 (1))

$$\mathbf{u}_\alpha(x)\mathbf{n}_\alpha\mathbf{u}_\alpha(x^{-1})\mathbf{n}_\alpha^{-1}\mathbf{u}_\alpha(x).(t, y) = \alpha^\vee(x)\mathbf{n}_\alpha.(t, y), \tag{45}$$

where $\alpha \in D$, $x \in k^*$, $t \in T$, for y in a dense open subset of Y (depending on α and x). It follows from 9.4.4 (1) (or from 8.2.10) that ϕ is surjective. If $g \in \mathrm{Ker}\ \phi$ then g is a K_1-automorphism of $K = K_1(\chi_1, \dots, \chi_m)$. Since $\chi_h^l \in K_1$, we then must have $g.\chi_h = \zeta_h \chi_h$, where ζ_h is an lth root of unity. Such g exist, namely $g \in T$ with $\chi_h(g) = \zeta_h$ and $\psi(g) = 1$ for all $\psi \in X_1$ ($1 \le h \le m$). Let p be the characteristic exponent of k. It follows that ϕ is surjective with kernel of order l^m, contained in T, if $l \ne p$. Otherwise ϕ is bijective.

It remains to identify G with an algebraic group. Choose $a \in k[G_1]$ such that $(a\chi_h)^l \in k[G_1]$ for $a \le h \le m$. By 2.3.6 (i) the elements $(g.(a\chi_h))^l = \phi(g).(a\chi_h)^l$

$(g \in G)$ lie in a finite dimensional subspace of $k[G_1]$. Then all translates $g.(a\chi_h)$ lie in a finite dimensional subspace of K. In fact, $a\chi$ lies in $k[T \times Y]$, which is isomorphic to an algebra $R = k[T_1, \dots, T_m, U_1, \dots, U_n, U_1^{-1}, \dots, U_n^{-1}]$. One checks, by looking at degrees, that the set of $r \in R$ such that r^l lies in a fixed finite dimensional subspace of R, spans a finite dimensional subspace.

Let V_1 be a finite dimensional subspace of $k[G_1]$ that generates $k[G_1]$, is stable under left translations and is such that the rational representation of G_1 in V_1 defines an isomorphism of G_1 onto its image (see the proof of 2.3.7). Let V_2 be the finite dimensional subspace of K spanned by all $g.(a\chi_h)$ $(g \in G)$ and put $V = V_1 + V_2$. Then V generates K. Hence G is isomorphic to its restriction to V.

Let $\gamma \in R^+$. Then $x \mapsto \mathbf{u}_\gamma(x)$ defines a rational representation of $\mathbf{G}_a$ in $k[T \times Y]$, whose restriction to V is a finite dimensional rational representation. Also, the restriction to V of the automorphisms defined by the elements of T defines a rational representation of T. For $\alpha \in D$, $x \in k$ define $\mathbf{u}_{-\alpha}(x) = \mathbf{n}_\alpha \mathbf{u}_\alpha(-x)\mathbf{n}_\alpha^{-1}$. It follows from (45) that G is also generated by T and the $\mathbf{u}_\gamma(x)$ $(x \in k)$, where $\gamma \in R^+ \cup (-D)$. We conclude that G (identified with its restriction to V) is generated by a family of connected algebraic groups, hence is itself a closed, connected, subgroup of $GL(V)$ by 2.2.7. The homomorphism ϕ is a restriction to V_1. Since G_1 is reductive, G must also be reductive. Moreover, T must be a maximal torus. We also see that ϕ is a central isogeny.

We claim that the weights of T in V span X. For the weights of T in V_1 span X_1, and the T-translates of $a\chi$ contain a weight in $\chi + X_1$, whence our claim. It follows that $\Psi(G, T)$ is isomorphic to Ψ, which finishes the proof. $\qquad\square$

10.1.4. Exercises. (1) Let G be connected, semi-simple. Show that there is a connected, semi-simple, simply connected group (8.1.11) $\tilde{G}$ together with a central isogeny $\tilde{G} \to G$.
(2) Take $G = \mathbf{SO}_n$ ($n \geq 5$, characteristic $\neq 2$). Show that G is not simply connected. The group $\tilde{G}$ of the preceding exercise is the *spin group* $\mathbf{Spin}_n$.

10.2. Simply laced root systems

10.2.1. Let R be a reduced root system in the real vector space V. We say that R is *simply laced* if for any two linearly independent roots $\alpha, \beta \in R$ we have $\langle \alpha, \beta^\vee \rangle = 0, \pm 1$. The positive definite symmetric bilinear form $(\ .\ )$ on V is as in 7.1.7. Then $\langle \alpha, \beta^\vee \rangle = 2(\beta, \beta)^{-1}(\alpha, \beta)$ $(\alpha, \beta \in R)$.

10.2.2. Lemma. *Assume that R is simply laced. Let $\alpha, \beta \in R$ be linearly independent in R.*
(i) If $\epsilon = \pm 1$ then $\alpha + \epsilon\beta \in R$ if and only if $\langle \alpha, \beta^\vee \rangle = -\epsilon$;
(ii) If R is irreducible there is $w \in W$ with $w.\alpha = \beta$;
(iii) If $\alpha + \beta \in R$ then $(\alpha + \beta)^\vee = \alpha^\vee + \beta^\vee$.

(i) follows from 9.1.3, observing that all root strings must have length ≤ 1. Let D be a basis of R. By 8.2.8 (ii) (see also 8.2.9) we may assume that $\alpha, \beta \in D$. If R is irreducible the connectedness of the Dynkin diagram of R (see 9.5.1) and 8.2.7 imply that there is a chain $\alpha = \alpha_0, \alpha_1, \ldots, \alpha_h = \beta$, with $\alpha_i \in D$, $\langle \alpha_i, \alpha_{i+1}^\vee \rangle = -1$ $(0 \leq i \leq h - 1)$. It follows that we may assume $\langle \alpha, \beta^\vee \rangle = -1$, in which case $(s_\alpha s_\beta).\alpha = \beta$. This proves (ii) and (iii) is a consequence of the last formula of 10.2.1. $\square$

10.2.3. Assume R to be irreducible and simply laced. It follows from 10.2.2 (ii) that we may assume $(\alpha, \alpha) = 2$ for all $\alpha \in R$. Hence $(\alpha, \beta) = 0, \pm 1$ for any two linearly independent roots α, β. Denote by $Q \subset V$ the subgroup of V generated by R. A basis D of R is also a basis of the free abelian group Q. It follows that (x, x) is an even integer for all $x \in Q$. Denote by f a $\mathbf{Z}$-valued bi-additive function on Q satisfying

$$\begin{cases} (x, y) & \equiv \quad f(x, y) + f(y, x) \pmod 2, \\ \frac{1}{2}(x, x) & \equiv \quad f(x, x) \pmod 2, \end{cases}$$

for $x, y \in Q$. Such f exist: take a basis $(e_i)_{1 \leq i \leq n}$ of Q and define f by $f(e_i, e_j) = (e_i, e_j)$ if $1 \leq i < j \leq n$, $f(e_i, e_j) = 0$ if $i > j$, $f(e_i, e_i) = \frac{1}{2}(e_i, e_i)$. Fix a system of positive roots R^+. For $\alpha \in R$ put $\epsilon(\alpha) = \pm 1$ if $\alpha \in \pm R^+$. Let α and β be linearly independent roots. Define

$$\begin{cases} c_{\alpha, \beta} & = \quad 0 \text{ if } \alpha + \beta \notin R, \\ c_{\alpha, \beta} & = \quad \epsilon(\alpha)\epsilon(\beta)\epsilon(\alpha + \beta)(-1)^{f(\alpha, \beta)} \text{ if } \alpha + \beta \in R. \end{cases}$$

If $\alpha \in Q$, $\chi \in Q - R$ put $c_{\alpha, \chi} = c_{\chi, \alpha} = 0$.

10.2.4. Lemma. *(i)* $c_{\alpha, \beta} = -c_{\beta, \alpha}$ *and* $c_{-\alpha, \beta} c_{\alpha, -\alpha + \beta} + c_{\beta, \alpha} c_{-\alpha, \alpha + \beta} = \langle \beta, \alpha^\vee \rangle$; *(ii) If* $\alpha, \beta, \gamma \in R$ *are linearly independent we have* $c_{\alpha, \beta} c_{\alpha + \beta, \gamma} + c_{\beta, \gamma} c_{\beta + \gamma, \alpha}$ $+ c_{\gamma, \alpha} c_{\gamma + \alpha, \beta} = 0$.

If $\alpha, \beta, \alpha + \beta \in R$ we see from 10.2.2 (i) that $f(\alpha, \beta) + f(\beta, \alpha) \equiv (\alpha, \beta) \equiv 1 \pmod 2$. This implies the first relation of (i).
If $\langle \beta, \alpha^\vee \rangle = 0$ we have $c_{-\alpha, \beta} = c_{\beta, \alpha} = 0$. If $\langle \beta, \alpha^\vee \rangle = -1$ then $\alpha + \beta \in R$, $-\alpha + \beta \notin R$ (10.2.2 (i)) and $c_{-\alpha, \beta} = 0$. Thus

$$c_{\beta, \alpha} c_{-\alpha, \alpha + \beta} = -(-1)^{f(\beta, \alpha) + f(-\alpha, \alpha + \beta)} = -1,$$

and the second formula of (i) follows. If $\langle \beta, \alpha^\vee \rangle = 1$ the proof is similar.
To prove (ii) we may assume that $c_{\alpha, \beta} c_{\alpha + \beta, \gamma} \neq 0$. Then by 10.2.2 (i) $(\alpha, \beta) = (\alpha + \beta, \gamma) = -1$, whence either $(\alpha, \gamma) = 0$, $(\beta, \gamma) = -1$ or $(\alpha, \gamma) = -1$, $(\beta, \gamma) = 0$. By symmetry we may assume the first alternative. Then the relation to be proved follows from

$$f(\alpha, \beta) + f(\alpha + \beta, \gamma) + f(\beta, \gamma) + f(\beta + \gamma, \alpha) \equiv$$

$$(\alpha, \beta) + (\alpha, \gamma) + 2f(\beta, \gamma) \equiv 1 \pmod 2.$$

This concludes the proof. □

10.2.5. We now define a Lie algebra $\mathfrak{g}$ over k as follows. Put $\mathfrak{t} = k \otimes Q^{\vee}$ and

$$\mathfrak{g} = \mathfrak{t} \oplus \sum_{\alpha \in R} k e_{\alpha},$$

the e_{α} being linearly independent. The Lie algebra product is defined by

$$[u, u'] = 0, \quad [u, e_{\alpha}] = \langle \alpha, u \rangle e_{\alpha},$$

$$[e_{\alpha}, e_{\beta}] = c_{\alpha, \beta} e_{\alpha + \beta}, \quad [e_{\alpha}, e_{-\alpha}] = 1 \otimes \alpha^{\vee},$$

where $u, u' \in \mathfrak{t}$, $\alpha, \beta \in R$ (with obvious notations). The verification that these formulas define a Lie algebra structure is left to the reader. It uses 10.2.2 (iii) and the preceding lemma (which takes care of the Jacobi identity).

If $a \in \mathfrak{g}$ we define the linear map ad a of $\mathfrak{g}$ by $(\text{ad } a)(b) = [a, b]$. A *derivation* of the Lie algebra $\mathfrak{g}$ is a linear map Δ of $\mathfrak{g}$ such that for $a, b \in \mathfrak{g}$

$$\Delta[a, b] = [\Delta a, b] + [a, \Delta b]. \tag{46}$$

The maps ad a are derivations.

10.2.6. Lemma. *Let Δ be a derivation of $\mathfrak{g}$. There exists a unique $a \in \mathfrak{g}$ with $\Delta = \text{ad } a$.*

Write for $u \in \mathfrak{t}$

$$\Delta u = d(u) + \sum_{\alpha \in R} l_{\alpha}(u) e_{\alpha},$$

where d is an endomorphism of $\mathfrak{t}$ and the l_{α} are linear functions on $\mathfrak{t}$. From $[\Delta u, u'] + [u, \Delta u'] = 0$ we deduce that for all $\alpha \in R$, $u, u' \in \mathfrak{t}$

$$\langle \alpha, u \rangle l_{\alpha}(u') = \langle \alpha, u' \rangle l_{\alpha}(u),$$

from which we conclude that there is $c_{\alpha} \in k$ with $l_{\alpha}(u) = c_{\alpha} \langle \alpha, u \rangle$ $(u \in \mathfrak{t}, \alpha \in R)$. Then $\Delta + \text{ad}(\sum c_{\alpha} e_{\alpha})$ is a derivation mapping $\mathfrak{t}$ into itself. Assume that Δ has this property. Then (46) with $a \in \mathfrak{t}$, $b = e_{\alpha}$ implies that there exist $d_{\alpha} \in k$ with $\Delta e_{\alpha} = d_{\alpha} e_{\alpha}$ and (46) with $a = e_{\alpha}$, $b = e_{\beta}$ gives that $d_{\alpha + \beta} = d_{\alpha} + d_{\beta}$, if $\alpha, \beta, \alpha + \beta \in R$. Also, $d_{-\alpha} = -d_{\alpha}$. These relations imply that there is $u_0 \in \mathfrak{t}$ with $d_{\alpha} = \langle \alpha, u_0 \rangle$. To prove this, choose a system of positive roots R^+ with basis D and take $u_0 \in \mathfrak{t}$ such that the required relations hold for $\alpha \in D$. Then use 8.2.11 (4). We conclude that $\Delta = \text{ad } u_0$, whence the existence part of the lemma.

The uniqueness part follows from the fact that the center of $\mathfrak{g}$ is trivial. The easy proof of this fact is left to the reader. □

10.2.7. Let T be a torus with character group Q. We let T act in $\mathfrak{g}$ by

$$t.u = u, \quad t.e_\alpha = \alpha(t)e_\alpha,$$

where $t \in T$, $u \in \mathfrak{t}$, $\alpha \in R$. This defines an isomorphism of T onto a closed subgroup of $GL(\mathfrak{g})$. For $\alpha \in R$ put $X_\alpha = \operatorname{ad} e_\alpha$. Denote by $X_\alpha^{(2)}$ the linear map of $\mathfrak{g}$ mapping all $u \in \mathfrak{t}$ and all e_β with $\beta \neq -\alpha$ to 0 and $e_{-\alpha}$ to $-e_\alpha$. Then $X_\alpha^2 = 2X_\alpha^{(2)}$ and

$$X_\alpha X_\alpha^{(2)} = X_\alpha^{(2)} X_\alpha = (X_\alpha^{(2)})^2 = 0. \tag{47}$$

Moreover

$$X_\alpha^{(2)}[a, b] = [X_\alpha^{(2)}a, b] + [a, X_\alpha^{(2)}b] + [X_\alpha a, X_\alpha b] \ (a, b \in \mathfrak{g}). \tag{48}$$

We leave it to the reader to check these facts. Define linear maps $u_\alpha(x)$ $(x \in k)$ of $\mathfrak{g}$ by

$$u_\alpha(x) = 1 + xX_\alpha + x^2 X_\alpha^{(2)}.$$

It follows from (47) that $u_\alpha(x + y) = u_\alpha(x)u_\alpha(y)$ $(x, y \in k)$, whence an isomorphism u_α of $\mathbf{G}_a$ onto a closed subgroup U_α of $GL(\mathfrak{g})$. We denote by G the subgroup of $GL(\mathfrak{g})$ generated by T and the U_α $(\alpha \in R)$. By 2.2.7 (i), G is closed and connected.

10.2.8. Proposition. *G is reductive and T is a maximal torus of G. The root system of (G, T) is R.*

It follows from (48) that the elements of U_α are automorphisms of the Lie algebra $\mathfrak{g}$. It is easy to see that the same holds for T. Hence G is a group of automorphisms of $\mathfrak{g}$. Let $q : \mathfrak{g} \otimes \mathfrak{g} \to \mathfrak{g}$ be the map defining the Lie product. Then $q \circ (g \otimes g) = g \circ q$ for all $g \in G$. Using 4.4.14 (ii) one deduces that the elements of the Lie algebra $L(G)$ (a subalgebra of $\mathfrak{gl}(\mathfrak{g})$) are derivations of $\mathfrak{g}$. Using 10.2.6 we conclude that $\dim G \leq \dim \mathfrak{g}$. Now the Lie algebra of U_α is kX_α (by 4.4.15 (1)) so $L(G)$ contains $L(T) \oplus \sum_{\alpha \in R} kX_\alpha$, whence $\dim G \geq \dim T + |R| = \dim \mathfrak{g}$. Hence $\dim G = \dim \mathfrak{g}$ and $L(G)$ is isomorphic to $\mathfrak{g}$.

It follows from 5.4.7 that T is a maximal torus of G. If $\alpha \in R$ the group G_α of 7.1.3 contains U_α and $U_{-\alpha}$. These groups stabilize the subspace spanned by $1 \otimes \alpha^\vee$, e_α, $e_{-\alpha}$ and the group generated by the restrictions of $U_{\pm\alpha}$ to that space is not solvable (in fact this group is isomorphic to $\mathbf{PSL}_2$). Hence α is a root of (G, T) (see 7.4.3). We can now conclude that $\dim R_u(G) = \dim G - \dim T - |R| = 0$, so G is reductive. $\square$

Via our form $(\ ,\ )$ we identify V with its dual. The dual $Q^\vee$ is then identified with a lattice in V containing Q. Then $\alpha^\vee = \alpha$ (see 10.2.1), and $R^\vee = R$. This shows that a root datum $(Q, R, Q^\vee, R^\vee)$ is unique up to isomorphism.

10.2.9. Corollary. *The root datum* $\Psi(G, T)$ *is isomorphic to* $(Q, R, Q^\vee, R^\vee)$.

For $\alpha \in R$ the cocharacter $\alpha^\vee$ is given by $\alpha^\vee(x).u = u$ $(u \in \mathfrak{t})$, $\alpha^\vee(x).e_\beta = x^{\langle \alpha, \beta \rangle} e_\beta$ $(\beta \in R)$, where $x \in k^*$. $\qquad\qquad\square$

10.2.10. Exercises. (1) $(u_\alpha)_{\alpha \in R}$ is a realization of R in G.
(2) The structure constants of the realization of (1), for any order on R (see 9.2.1) are as follows: $c_{\alpha,\beta;i,j} = c_{\alpha,\beta}$ (as in 10.2.3) if $i = j = 1$ and is zero otherwise.

10.3. Automorphisms, end of the proof of 10.1.1

10.3.1. Automorphisms. Assumptions and notations are as in 10.2.3. Let $\mathcal{D}$ be the Dynkin diagram defined by D (9.5.1). It is a graph with vertex set D, two simple roots being the endpoints of an edge if and only if they are not orthogonal (relative to our bilinear form).

Assume that s is a non-trivial automorphism of $\mathcal{D}$ with the following property: two distinct roots in an s-orbit in D are orthogonal. From the classification of irreducible root systems (recalled in 9.5.1) one sees that if R is irreducible s has order 2 or 3. In the first case R is of one of the types A_{2n+1} $(n \geq 1)$, D_n $(n \geq 4)$, E_6 and in the second case it is of type D_4.
The permutation s of D defines a linear map of V, also denoted by s. It stabilizes Q. As a consequence of 8.2.8 (iii) (see also 8.2.9) s stabilizes R and R^+. Since s defines an automorphism of $\mathcal{D}$ (and since R is simply laced) we have $(s.\alpha, s.\beta) = (\alpha, \beta)$ for $\alpha, \beta \in D$. Hence s leaves invariant the bilinear form $(.)$.

We claim that there exist s-invariant bi-additive forms f with the properties of 10.2.3. Such a form is obtained by the procedure of 10.2.3, applied to the basis D of Q, ordered such that elements in an s-orbit are consecutive. (It suffices to see this in the case that R is irreducible, so that s has order 2 or 3. In the first case the verification is straightforward and in the second case one uses that s has only one orbit of length 3.) It follows that s defines an automorphism σ of the Lie algebra $\mathfrak{g}$ of 10.2.5, with $\sigma.(e_\alpha) = e_{s\alpha}$ and $\sigma(1 \otimes \alpha^\vee) = 1 \otimes s.\alpha^\vee$ $(\alpha \in R)$. The automorphism σ of $\mathfrak{g}$ normalizes the group G of 10.2.7. It induces an automorphism of G, also denoted by σ, which stabilizes T. Its restriction to T is the automorphism defined by the automorphism s of the character group Q of T. We have $\sigma(u_\alpha(x)) = u_{s.\alpha}(x)$ $(\alpha \in R,\ x \in k)$.

10.3.2. Lemma. (i) $(1 - s)Q \cap R = \emptyset$;
(ii) If $\alpha, \beta \in R$ and $\alpha - \beta \in (1 - s)Q$ then α and β lie in an s-orbit;
(iii) A root is orthogonal to the other roots in its orbit.

If $x = \sum_{\alpha \in D} x_\alpha \alpha$ is a non-zero element of $(1 - s)Q$, the non-zero coefficients cannot all have the same sign. Hence (i) follows from 8.2.8 (iii).

To prove (ii) we may assume by 8.2.8 (ii) that $\alpha \in D$. Assume that the s-orbit of

α has a elements. In the situation of (ii) we cannot have $\alpha - \beta \in R$ (by (i)), whence $(\alpha, \beta) = 0, -1$ if $\alpha \neq \beta$ (10.2.2 (i)). Similarly for $(\alpha, s.\beta), \ldots, (\alpha, s^{a-1}.\beta)$. We have

$$0 = (\alpha - \beta, (1 + s + \ldots s^{a-1}).\alpha) = 2 - (\alpha, \beta) - \ldots - (\alpha, s^{a-1}.\beta)$$

(recall that the roots in the orbit of α are mutually orthogonal). If α does not lie in the orbit of β, the right-hand side is strictly positive, which gives a contradiction, proving (ii).

To prove (iii) we may assume that R is irreducible and $\alpha \in R^+$. Assume that s has order 2. If $\alpha \in D$, (iii) holds by assumption. If $\alpha \in R^+ - D$, there exist $\beta \in R^+$ and $\gamma \in D$ with $\alpha = \beta + \gamma$ (by 8.2.11 (4)). By an induction we may assume that (iii) holds for β. Using that

$$(\alpha, s.\alpha) = (\beta, s.\beta) + (\gamma, s.\gamma) + 2(\beta, s.\gamma),$$

we conclude that $(\alpha, s.\alpha)$ is even, which can only be if $s.\alpha = \alpha$ or $(\alpha, s.\alpha) = 0$. Assume that s has order 3, that $\alpha \neq s.\alpha$ and that the inner product $(\alpha, s.\alpha)$ is non-zero. It then equals -1. By 10.2.2 (i), $\alpha + s.\alpha \in R$, and

$$(\alpha + s.\alpha, s^2.\alpha) = 2(\alpha, s.\alpha) = -2,$$

which is impossible. We have proved (iii). $\square$

10.3.3. Put $Q^s = Q/(1 - s)Q$. From the fact that s permutes the elements of the basis D of Q, it follows that Q^s has no torsion, hence is a free abelian group. If $\mathcal{O}$ is an s-orbit in R, put $\alpha_{\mathcal{O}} = \alpha + (1 - s)Q$, where $\alpha \in \mathcal{O}$. It follows from 10.3.2 (ii) that this is a correct definition, and $\alpha_{\mathcal{O}} \neq 0$ by 10.3.2 (i). The dual $(Q^s)^\vee$ is the submodule of $Q^\vee$ annihilating $(1 - s)Q$. Define $\alpha_{\mathcal{O}}^\vee = \sum_{\alpha \in \mathcal{O}} \alpha^\vee$. This is an element of $(Q^s)^\vee$. Let R^s and $(R^s)^\vee$ be the set of $\alpha_{\mathcal{O}}$, respectively $\alpha_{\mathcal{O}}^\vee$, $\mathcal{O}$ running through the orbits of s in R.

10.3.4. Lemma. $\Psi^s = (Q^s, R^s, (Q^s)^\vee, (R^s)^\vee)$ *is a root datum.*

If $\mathcal{O}$ and $\mathcal{O}'$ are orbits, we have $\langle \alpha_{\mathcal{O}}, \alpha_{\mathcal{O}'}^\vee \rangle = \sum_{\beta \in \mathcal{O}'} \langle \alpha, \beta^\vee \rangle$, where $\alpha \in \mathcal{O}$. Axiom (RD 1) of 7.4.1 follows from 10.3.2 (iii). We have

$$s_{\alpha_{\mathcal{O}'}}(\alpha_{\mathcal{O}}) = (\prod_{\beta \in \mathcal{O}'} s_\beta).\alpha + (1 - s)Q,$$

proving the first part of axiom (RD 2). We skip the proof of the second part. $\square$

We say that the root datum $(Q^s, R^s, (Q^s)^\vee, (R^s)^\vee)$ (or the root system R^s) is obtained from $(Q, R, Q^\vee, R^\vee)$ (respectively R) by *folding* according to s.

We return to the group G and its automorphism σ, introduced in 10.3.1. We denote fixed point sets for σ by a superscript.

10.3.5. Proposition. $(G^\sigma)^0$ *is a connected, reductive group with maximal torus* T^σ. *The root datum* $\Psi((G^\sigma)^0, T^\sigma)$ *is isomorphic to* Ψ^s.

The character group of the diagonalizable group T^σ is Q^s. This is a free abelian group. Hence T^σ is a torus. By 10.3.2 (i) no root of (G, T) is trivial on T^σ. Using 5.4.7 we see that the centralizer of T^σ in G is T, which implies that T^σ is a maximal torus in $H = (G^\sigma)^0$.
If $\alpha_{\mathcal{O}} \in R^s$ put

$$u_{\alpha_{\mathcal{O}}}(x) = \prod_{\beta \in \mathcal{O}} u_\beta(x) \ (x \in k),$$

the u_β being as in 10.2.7. It follows from 10.3.2 (iii) (using formula (27) of 9.2.1) that $u_{\alpha_{\mathcal{O}}}$ defines an isomorphism of $\mathbf{G}_a$ onto a closed subgroup $U_{\alpha_{\mathcal{O}}}$ of H. An argument similar to one used in the proof of 10.2.8 shows that the $\alpha_{\mathcal{O}}$ are roots of (H, T^σ), whence

$$\dim H \geq |R^s| + \dim T^\sigma. \tag{49}$$

On the other hand, the Lie algebra $L(H)$ is a subalgebra of the fixed point algebra $\mathfrak{g}^\sigma$, whose dimension equals $\dim \mathfrak{t}^\sigma + |R^s|$. Since $\dim T^\sigma$ and $\dim \mathfrak{t}^\sigma$ both equal the number of s-orbits in D, we must have equality in (49). As in the proof of 10.2.8 it follows that H is reductive. Its root system relative to T^σ is R^s. The proposition follows. $\qquad\qquad\square$

10.3.6. Lemma. *Let* $\tilde{R}$ *be an irreducible root system that is not simply laced. There is a simply laced root system* R, *with an automorphism* s *obtained as in 10.3.1, such that* $\tilde{R}$ *is isomorphic to* R^s.

The root system R^s of 10.3.4 has basis $D^s = \{\alpha_{\mathcal{O}} \mid \mathcal{O} \in D\}$ (check this). The corresponding Dynkin diagram can then be read off from D and the action of s on it. We now use the classification (see 9.5.1). The Dynkin diagrams of irreducible root systems that are not simply laced are of the types B_n $(n \geq 2)$ C_n $(n \geq 3)$ F_4, G_2. They can be obtained by folding from the Dynkin diagrams of simply laced root systems, of respective types D_{n+1}, A_{2n+1}, E_6, D_4. Except for the last case, s is unique. In type D_4 there are two possibilities.

10.3.7. We can now prove 10.1.1. Let $Q \subset X$ be the subgroup spanned by R and let X_0 be the subgroup of X orthogonal to $Q^\vee$. Then $Q \oplus X_0$ is a sublattice of X of finite index. By 10.1.3 it suffices to consider the case that $X = Q \oplus X_0$, which is reduced by an easy argument to the cases that either $X = Q$ or $X = X_0$. In the second case we can take G to be a torus. The first case is reduced at once to the situation that R is irreducible. If R is simply laced apply 10.2.8. Otherwise use 10.3.5 and 10.3.6. $\square$

10.3.8. Exercises. G is a connected reductive group over k, with maximal torus T and corresponding root datum $\Psi = (X, R, X^\vee, R^\vee)$.

(1) (a) There exists a connected reductive group $G^\vee$ over k, with a maximal torus $T^\vee$, such that the root datum $\Psi(G^\vee, T^\vee)$ is isomorphic to $\Psi^\vee = (X^\vee, R^\vee, X, R)$.

(b) G is unique up to isomorphism. It is the *dual* of G. We have $(G^\vee)^\vee \simeq G$.

(c) If G is semi-simple and simply connected, respectively adjoint (see 8.1.11) then $G^\vee$ is semi-simple and adjoint, respectively simply connected.

(2) Determine the duals of $\mathbf{GL}_n$, $\mathbf{Sp}_{2n}$, $\mathbf{SO}_n$.

Notes

The most general existence theorem for a reductive group scheme with given root datum is given in [SGA3, exp. XXV]. Chevalley [Ch5] already constructed the adjoint semi-simple group schemes over $\mathbf{Z}$. See also [Bo2, Part A, §4].

A proof of 10.1.1 for the semi-simple case can be found in [St4, §5].

The proof of [SGA3] uses a generalization of a result of Weil, about enlarging an algebraic group germ to an algebraic group. The proofs of [Ch5] and [St4] use representation theory of semi-simple Lie algebras. We need only the adjoint representation, in the case that the root system is simply laced.

The construction in 10.2.5 of the Lie algebra associated to a simply laced root system uses a simple description of Lie algebra structure constants, due to Frenkel and Kac [FK].

For the proof of 10.1.1 we need a result about existence of covering groups (10.1.4). The proof of 10.1.1 given in this chapter can be extended to an existence proof over an arbitrary ground field (see 16.3.3). It could also be adapted to construct a group scheme over $\mathbf{Z}$.

Chapter 11

More Algebraic Geometry

The next chapters will be devoted to rationality questions in the theory of linear algebraic groups, i.e. questions involving ground fields. The present chapter is preparatory. It discusses basic rationality results on algebraic varieties.

11.1. F-structures on vector spaces

11.1.1. In this section, k is a field (not necessarily algebraically closed) and F is a subfield. 'Vector space' will mean vector space over k. An F-*structure* on a vector space V is a subspace V_0 of V of the F-vector space V such that the canonical homomorphism $k \otimes_F V_0 \to V$ is an isomorphism (see 1.3.7). This means that V has a basis whose elements lie in V_0. We shall write $V_0 = V(F)$. A vector space with F-structure will be called an F-*vector space*. A linear map $f : V \to W$ of F-vector spaces is *defined over* F if $f(V(F)) \subset W(F)$. There are some fairly obvious notions and constructions, which we mention briefly. Details can be left to the reader.

Let V be an F-vector space. (a) We have the notion of a *subspace defined over* F or F-*subspace* of V. Such a subspace W has a basis whose elements lie in $V(F)$. The quotient V/W has a canonical F-structure. The canonical map $V \to V/W$ is defined over F.

(b) Let $(V_i)_{i \in I}$ be a family of F-vector spaces. Then $\bigoplus_{i \in I} V_i(F)$ is an F-structure on $\bigoplus_{i \in I} V_i$.

(c) Let W be another F-vector space. The vector space over F spanned by the elements $v \otimes w$ with $v \in V(F)$, $w \in W(F)$ is an F-structure on $V \otimes_k W$. Then $(V \otimes W)(F)$ is isomorphic to $V(F) \otimes_F W(F)$.

(d) The construction of (c) leads to F-structures on exterior powers $\bigwedge^i V$, with $(\bigwedge^i V)(F) = \bigwedge^i(V(F))$.

(e) If $E \subset k$ is an extension of F then the vector space over E generated by $V(F)$ is an E-structure on V. We have $V(E) = E \otimes_F V(F)$.

In the sequel, if we provide quotients, direct sums,... of F-vector spaces with an F-structure, it will always be in the manner described above.

11.1.2. Exercise. V and W are F-vector spaces and $f : V \to W$ is a k-linear map.
(a) If f is defined over F then Ker f and Im f are defined over F.
(b) The converse of (a) is false.
(c) f is defined over F if and only if its graph $\Gamma_f = \{(v, f(v)) \in V \oplus W \mid v \in V\}$ is an F-subspace of $V \oplus W$.

11.1.3. Let V be an F-vector space. Assume that we are given a group of automorphisms Γ of k whose fixed point field is F, i.e.

$$F = \{a \in k \mid \gamma.a = a \text{ for all } \gamma \in \Gamma\}.$$

Then Γ operates on $V = k \otimes_F V(F)$ by $\gamma(a \otimes v) = (\gamma.a) \otimes v$ ($a \in k$, $v \in V(F)$, $\gamma \in \Gamma$). Clearly

$$\gamma.(av) = (\gamma.a)v, \quad \gamma.(v + v') = \gamma.v + \gamma.v', \tag{50}$$

and $V(F)$ is the fixed point set of Γ in V.

11.1.4. Proposition. *A subspace W of V is defined over F if and only if $\Gamma.W = W$.*

Assume that $\Gamma.W = W$ and put $W(F) = \{x \in W \mid \gamma.x = x \text{ for all } \gamma \in \Gamma\}$. This is a subspace of $V(F)$. Let W' be a complementary subspace. We claim that $W(F)$ generates W. If this were not the case there would exist $x \in W - \{0\}$ and $x_1, \ldots, x_n \in W'$, $a_1, \ldots, a_n \in k$ with

$$x = \sum_{1 \le i \le n} a_i x_i.$$

We may assume that $a_1 = 1$ and that n is as small as possible. We have for all $\gamma \in \Gamma$

$$\gamma.x - x = \sum_{2 \le i \le n} (\gamma.a_i - a_i)x_i.$$

The minimality of n implies that $\gamma.x = x$ for all $\gamma \in \Gamma$, i.e. that $x \in W(F)$, which contradicts our assumption. This contradiction establishes our claim and proves one part of the assertion. The other part is obvious. $\qquad\square$

11.1.5. Corollary. *Let V and W be two F-vector spaces and let $f : V \to W$ be a k-linear map. Then f is defined over F if and only if $f \circ \gamma = \gamma \circ f$ for all $\gamma \in \Gamma$.*

Γ operates on V and W in the manner described above. The corollary follows by using 11.1.2 (c). $\qquad\square$

Next assume, moreover, that k is a Galois extension of F, of finite or infinite degree, Γ being the Galois group $\mathrm{Gal}(k/F)$. For the results from Galois theory to be used, we refer to [La2, Ch. VIII] or [Jac5, Ch. 8]. Recall that Γ is a profinite group, i.e., a compact, totally disconnected, topological group. Its topology, the Krull topology, has a basis of open neighborhoods of the identity consisting of subgroups of finite index, viz. the groups $\mathrm{Gal}(k/E)$, where E is a finite extension of F contained in k.

Assume that V is a vector space with continuous Γ-action, V being provided with the discrete topology.

11.1.6. Proposition. *Assume that (50) holds. Then* $V(F) = \{x \in V \mid \gamma.x = x \text{ for all } \gamma \in \Gamma\}$ *is an F-structure on V.*

The continuity assumption implies that V is a union of finite dimensional Γ-stable subspaces. It follows that it suffices to prove the proposition in the case that V is finite dimensional and that k is a finite Galois extension of F. (We leave it to the reader to check these claims.) Let $d = [k : F]$ be the degree. Then Γ has order d. Dedekind's theorem [La2, Ch. VIII, §4] implies that the linear maps of the F-vector space k of the form

$$a \mapsto \sum_{\gamma \in \Gamma} c_\gamma \gamma \qquad (51)$$

with $c_\gamma \in k$ form a space with dimension d^2 over F, hence must be the space of all F-linear maps. The maps of the right-hand side of (51) act in an obvious manner as F-linear maps in V. It follows that we obtain a representation in V of the algebra $End_F(k)$ of all F-linear maps of k, which is isomorphic to the matrix algebra $M_d(F)$. It is known that any finite dimensional representation of $End_F(k)$ is isomorphic to a direct sum $k \oplus \ldots \oplus k$, where $End_F(k)$ acts in each summand in the natural way (see [Jac5, p. 171 and 4.4]). This fact is equivalent to the statement of the proposition. $\square$

11.1.7. There are rather obvious variants of 11.1.4 and 11.1.6, involving supplementary structures. For example, assume that A is a k-algebra with an F-structure. This means (see 1.3.7) that A has a vector space structure $A(F)$, which is an F-subalgebra of A. The Γ-action of 11.1.3 is by automorphisms of the ring A, satisfying (50). Then 11.1.4 implies a criterion for ideals in A to be defined over F. In the case of 11.1.6, if moreover A is a k-algebra and the Γ-action is by ring automorphisms, then $A(F)$ is an F-structure on the k-algebra A.

11.1.8. Exercise. Let k/F be a finite Galois extension with group Γ. Assume given a map $c : \Gamma \to \mathbf{GL}_n(k)$ such that for $\gamma, \delta \in \Gamma$ we have $c(\gamma\delta) = c(\gamma).\gamma(c(\delta))$, Γ acting on $\mathbf{GL}_n(k)$ in the obvious manner. There exists $g \in \mathbf{GL}_n(k)$ with $c(\gamma) = g^{-1}.\gamma(g)$. (*Hint:* using the $c(\gamma)$ define an action of Γ on $V = k^n$ such that 11.1.6 can be applied).

11.1.9. There are counterparts of the previous results involving derivations. Assume now that we are given a Lie algebra L over k of F-derivations (see 4.1.1) of k whose annihilator is F, i.e.

$$F = \{a \in k \mid D.a = 0 \text{ for all } D \in L\}.$$

A *connection* of the Lie algebra L on a vector space V is a k-linear map $c : L \to End_k(V)$ such that

$$c(D).(ax) = (D.a)x + a(c(D).x) \ (a \in k, \ x \in V, \ D \in L). \qquad (52)$$

The connection c is *flat* if c is a homomorphism of Lie algebras, i.e. if

$$c([D, D']) = c(D) \circ c(D') - c(D') \circ c(D) \ (D, D' \in L)$$

(see 4.4.3).
Now assume that V is an F-vector space. There is a unique flat connection $c = c_V$ on V such that

$$c(D).\left(\sum a_i x_i\right) = \sum (D.a_i)x_i \ (D \in L, \ a_i \in k, \ x_i \in V(F)).$$

Then $V(F)$ is the subset of V annihilated by all $c(D)$, $D \in L$. (We leave it to the reader to check these facts.)

11.1.10. Proposition. *A subspace W of V is defined over F if and only if $c(L).W \subset W$.*

11.1.11. Corollary. *Let V and W be two F-vector spaces and let $f : V \to W$ be a linear map. Then f is defined over F if and only if $f \circ c_V(D) = c_W(D) \circ f$ for all $D \in L$.*
The proofs of these results are similar to those of 11.1.4 and 11.1.5 and are left to the reader.   □

11.1.12. Next assume that char $k = p > 0$ and that L is a p-Lie algebra (see 4.4.3). A *p-connection* of the p-Lie algebra L is a connection that moreover satisfies $c(D)^{[p]} = c(D)^p$, where $D \mapsto D^{[p]}$ is the p-operation in L. Now assume that $k \neq F$ is a finite purely inseparable extension of F such that $k^p \subset F$. We denote by $\mathcal{J} = \mathcal{J}_{k/F}$ the p-Lie algebra of F-derivations of k, the Lie product being the commutator and the p-operation ordinary p^{th} power. It is a Lie algebra over F and a vector space over k (but not a Lie algebra over k). Choose $x_1, \ldots, x_d$ in k such that $k = F(x_1, \ldots, x_d)$ and assume that d is as small as possible. Then $x_i^p = a_i \in F \ (1 \leq i \leq d)$.

11.1.13. Lemma. *(i) $(x_1^{n_1} \ldots x_d^{n_d}) \ (1 \leq n_i < p, \ 1 \leq i \leq d)$ is a basis of k over F;*
(ii) There exist $\partial_i \in \mathcal{J}$ with $\partial_i x_j = \delta_{ij} x_i$. We have $\partial_i^{[p]} = \partial_i$, $[\partial_i, \partial_j] = 0 \ (1 \leq i, j \leq d)$;
(iii) $(\partial_i)_{1 \leq i \leq d}$ is a k-basis of $\mathcal{J}$. We have $[k : F] = p^{\dim_k \mathcal{J}}$;
(iv) The annihilator of $\mathcal{J}$ is F.

To prove (i) it suffices to show that the elements in question are linearly independent. If not, we may assume that x_1 is algebraic over $F(x_2, \ldots, x_d)$, of degree $< p$. But since the minimum polynomial of x_1 over F is $T^p - a_1$ this can only be if $x_1 \in F(x_2, \ldots, x_d)$, contradicting the assumption that d is minimal. We have established (i).

It follows from (i) that the F-homomorphism of the polynomial algebra $F[T_1, \ldots, T_d]$ sending T_i to x_i defines an F-isomorphism

$$F[T_1, \ldots, T_d]/(T_1^p - a_1, \ldots, T_d^p - a_d) \simeq k. \tag{53}$$

Let D_i be partial derivation with respect to T_i in $F[T_1, \ldots, T_d]$. Then $T_i D_i$ induces a derivation of the left hand side of (53), which defines a derivation ∂_i of k with the properties of (ii). It follows from (ii) that the ∂_i are linearly independent over k. By (53) we can view k as a module over $A = F[T_1, \ldots, T_d]$. Using that every F-derivation of A annihilates the ideal in the left-hand side of (53), we see that with the notations of 4.1 and 4.2,

$$\mathcal{J} = \operatorname{Der}_F(k, k) \simeq \operatorname{Der}_F(A, k) \simeq \operatorname{Hom}_A(\Omega_{A/F}, k),$$

the last isomorphism coming from 4.2.2 (i). Using 4.2.5 (1) we deduce that $\dim_k \mathcal{J} = d$. Hence (∂_i) is a k-basis of $\mathcal{J}$. Assume that $x \in k$ is annihilated by $\mathcal{J}$. Write x as a linear combination of the basis elements of (i). One deduces from $\partial_i x = 0$ ($1 \leq i \leq d$) that all coefficients of monomials $x_1^{n_1} \ldots x_d^{n_d}$ with at least one non-zero exponent are zero, whence (iv). $\qquad\square$

k being as before, we have the following counterpart of 11.1.6.

11.1.14. Proposition. *Let V be a vector space over k with a flat p-connection c of the p-Lie algebra $\mathcal{J}_{k/F}$. Then*

$$V(F) = \{x \in V \mid c(D).x = 0 \text{ for all } D \in \mathcal{J}_{k/F}\}$$

is an F-structure on V.

We use the notations of the preceding lemma. The $c(\partial_i)$ ($1 \leq i \leq d$) are linear maps of V, viewed as a vector space over the prime field $\mathbf{F}_p$. By the flatness of c and 11.1.13 (ii) these maps commute pairwise. Moreover we have $c(\partial_i)^p = c(\partial_i)$, which implies that the eigenvalues of the $c(\partial_i)$ lie in $\mathbf{F}_p$. For $0 \leq n_i < p$, $1 \leq i \leq d$ put

$$V_{n_1, \ldots, n_d} = \{x \in V \mid c(\partial_i).x = n_i x \text{ for } 1 \leq i \leq d\}.$$

Then V is the direct sum of these $\mathbf{F}_p$-vector spaces and $V(F) = V_{0, \ldots, 0}$. It follows from (52) that

$$V_{n_1, \ldots, n_d} = x_1^{n_1} \ldots x_d^{n_d} V(F),$$

from which we see that $V(F)$ is an F-structure on V, as asserted. $\qquad\square$

We next prove the 'main theorem of Galois theory' for the extension k/F of this section (due to Jacobson, see [Jac5, p. 533-535]).

11.1.15. Proposition. *(i) Let $\mathcal{J}_1$ be a p-subalgebra of $\mathcal{J} = \mathcal{J}_{k/F}$, which is a vector space over k and put*

$$\mathcal{F}(\mathcal{J}_1) = \{x \in k \mid D.x = 0 \text{ for all } D \in \mathcal{J}_1\}.$$

Then $F_1 = \mathcal{F}(\mathcal{J}_1)$ is a subfield of k containing F and $[k : F_1] = p^{\dim \mathcal{J}_1}$;
(ii) $\mathcal{F}$ defines a bijection of the set of p-subalgebras of $\mathcal{J}$ that are vector spaces over k onto the set of subfields of k containing F. The inverse of $\mathcal{F}$ is the map $F_1 \mapsto \mathcal{J}_{k/F_1}$.

For $x \in k$ we have the k-linear function $D \mapsto D.x$ on $\mathcal{J}$. It follows from 11.1.13 that these functions form the dual of the k-vector space $\mathcal{J}$. Let $\mathcal{J}_1$ be as in (i). The preceding remark implies that there is a k-basis $(\partial_i')_{1 \le i \le e}$ of $\mathcal{J}_1$ and a set $(y_i)_{1 \le i \le e}$ of elements of k such that $\partial_i'.y_j = \delta_{ij} y_i$. It follows that, if $D \in \mathcal{J}_1$ annihilates all y_i, we have $D = 0$. Applying this to $D = (\partial_i')^p - \partial_i'$ we conclude that $(\partial_i')^p = \partial_i'$ $(1 \le i \le e)$. Likewise, we see that the ∂_i' commute pairwise. As in the proof of 11.1.14 we deduce that k is the direct sum of simultaneous eigenspaces for the $\mathbf{F}_p$-linear maps ∂_i'. Since $\partial_i'.x = 0$ $(1 \le i \le e)$ is equivalent to $x \in F_1 = \mathcal{F}(\mathcal{J}_1)$, we can also conclude that the elements $y_1^{n_1} \dots y_e^{n_e}$ $(0 \le n_i < p, \ 1 \le i \le e)$ form an F_1 basis of k. This implies (i).

To prove (ii) we have to show that

$$\mathcal{F}(\mathcal{J}_{k/F_1}) = F_1, \ \mathcal{J}_{k/\mathcal{F}(\mathcal{J}_1)} = \mathcal{J}_1, \tag{54}$$

where F_1 is an intermediate field, and where $\mathcal{J}_1$ is as in (i). We have $F_1 \subset F_1' = \mathcal{F}(\mathcal{J}_{k/F_1})$ and by (i) and 11.1.13 (iii)

$$[k : F_1'] = p^{\dim \mathcal{J}_{k/F_1}} = [k : F_1].$$

It follows that $F_1' = F_1$, proving the first formula (54). The proof of the second is similar and is left to the reader.   □

11.1.16. We again have variants of 11.1.10 and 11.1.14. Let A be a k-algebra with an F-structure. Then the maps $c_A(D)$ of 11.1.9 are derivations of A. If A is a k-algebra with a connection c as in 11.1.14, such that all $c(D)$ are derivations, then $A(F)$ is an F-structure of the k-algebra A.

11.1.17. Exercise. Notations are as in 11.1.14. Write $\mathcal{J} = \mathcal{J}_{k/F}$. Let $End_k(V)$ be the space of endomorphisms of V.
(a) For $a \in End_k(V)$, $D \in \mathcal{J}$ put $\gamma(D)(a) = c(D) \circ a - a \circ c(D)$. Then γ is a flat connection of the p-Lie algebra $\mathcal{J}$ on $End_k(V)$.
(b) Fix a basis of V whose elements lie in $V(F)$. The matrix of $\gamma(D)(a)$ relative to this basis is obtained from the matrix of a by applying D to its elements.
    We write $D.a$ for $\gamma(D)(a)$.
(c) Assume given a k-linear map $a : \mathcal{J} \to End_k(V)$ such that

$$a([D, D']) = D.a(D') - D'.a(D),$$

$$(c(D) + a(D))^p = c(D)^p + a(D^p) \ (D, \ D' \in \mathcal{J}).$$

There is an invertible endomorphism g of V such that $a(D) = g^{-1}(D.g)$ for all $D \in \mathcal{J}$.
(*Hint:* apply 11.1.14 to the connection $D \mapsto c(D) + a(D)$).
(d) Using Jacobson's formula (4.4.3) show that the second formula of (c) can be rewritten as a formula expressing $a(D^p)$ as a linear combination of $a(D)^p$ and terms involving the $D^i.a(D)$ ($0 \leq i \leq p - 1$). Consider the particular case that dim $V = 1$.

11.2. F-varieties: density, criteria for ground fields

In this section k denotes again an algebraically closed field and F a subfield. We denote by $\bar{F}$ (F_s) the set of elements in k that are algebraic (respectively, separably algebraic) over F. Then $\bar{F}$ is an algebraic closure of F and F_s a separable closure. Moreover, F_s is a Galois extension of F. Its Galois group is denoted by Γ. The action of Γ extends uniquely to an action of Γ as a group of automorphisms of $\bar{F}$. The fixed point field of Γ on $\bar{F}$ is the field F_i (sometimes denoted by $F^{p^{-\infty}}$) of elements $x \in \bar{F}$ that are purely inseparable over F, i.e such that $x^{p^m} \in F$ for some $m \geq 0$, where p is the characteristic exponent of F. In characteristic 0, we have $F_i = F$. Also, $\bar{F} = (F_i)_s = (F_s)_i$.

11.2.1. We shall establish some basic results about F-varieties. These were introduced in 1.6.14. Let X be an affine variety over k. Recall (1.3.7) that an F-strucure on X is an F-structure on the algebra $k[X]$ in the sense of 11.1.1, which is an F-subalgebra of $k[X]$. As in 1.3.7, we write $F[X]$ for this subalgebra. The next result characterizes the algebras $F[X]$ (see also 1.3.1).

11.2.2. Lemma. *Let A be an F-algebra.*
(i) There is an affine F-variety X (1.4.9) with $A \simeq F[X]$ if and only if the following conditions hold: (a) A is of finite type over F, (b) for any algebraic extension E of F the algebra $E \otimes_F A$ is reduced;
(ii) If (b) holds, the algebra $E \otimes_F A$ is reduced for any extension E of F.

Assume there is an affine F-variety with $A \simeq F[X]$. Since k is algebraically closed, any algebraic extension E of F is F-isomorphic to a subfield of k (see [La2, p. 171]). Since $k[X] \simeq k \otimes_F F[X]$ is reduced, the same is true for $E \otimes_F A$, so (b) holds. An easy argument proves (a).

Now let E be any extension of F and assume that $x = \sum_{i=1}^{n} x_i \otimes a_i \in E \otimes_F A$ is nilpotent. We may assume the a_i to be linearly independent over F. Assume that (b) holds. If h is an F-homomorphism of $F[x_1, \dots, x_n]$ to an algebraic extension E' of F, an argument like that used in the proof of 1.5.2 shows that $h(x_i) = 0$ for all i. It follows from 1.9.6 (2) that all x_i are 0. Hence $E \otimes A$ is reduced and in particular $k \otimes_F A$ is reduced. If (a) holds, this is an affine k-algebra (1.3.1). The lemma follows.

An F-algebra A with the properties (a) and (b) will be called an *affine F-algebra*. $\square$

11.2.3. Exercises. (1) (a) Let A be a reduced F-algebra. Show that for any separable algebraic extension E of F the algebra $E \otimes_F A$ is reduced. (*Hint*: it suffices to deal with the case that $E = F(x)$).

(b) (a) is false without the separability assumption.

(2) There is an obvious category $\mathcal{C}(F, k)$ of k-varieties with an F-structure (see 1.6.14). Let k' be an algebraically closed subfield of k containing F. Show that $\mathcal{C}(F, k)$ and $\mathcal{C}(F, k')$ are equivalent. Notice that $k' = \bar{F}$ is the minimal choice. (*Hint*: formulate the notion of F-variety in terms of F-algebras).

We now come to density results. X is an F-variety.

11.2.4. Lemma. *(i) $X(\bar{F})$ is dense in X;*
(ii) If Y is a closed subvariety of X such that $X(F) \cap Y$ is dense in Y, then Y is defined over F.

We may assume that X is affine. To prove (i) it suffices by 1.3.6 (ii) to show: if $f \in k[X]$ is non-zero there exists $x \in X(\bar{F})$ with $f(x) \neq 0$. Writing $f = \sum_i a_i \otimes f_i$, where $f_i \in \bar{F}[X]$ and the $a_i \in k$ are linearly independent over $\bar{F}$, we see that it suffices to deal with the case that $f \in \bar{F}[X]$, in which case we can apply 1.9.4 (with $A = K = \bar{F}, B = \bar{F}[X]$). This proves (i).

To prove (ii) observe that if the density condition of (ii) holds, the ideal $\mathcal{I}(Y) \subset k[X]$ of functions vanishing on Y must be generated by elements in $\mathcal{I}(Y) \cap F[X]$. $\square$

11.2.5. Lemma. *Assume X to be irreducible. Then $X(F_s)$ is dense in X.*

Since X is a union of open affine F-varieties, we may assume that X is an affine F-variety. We may also assume that $F = F_s$.

By 1.3.6 (ii) it suffices to show that if f is a non-zero element of $k[X]$, there is $x \in X(F)$ with $f(x) \neq 0$. As in the proof of 11.2.4 we see that we may assume $f \in F[X]$. Let E be the quotient field of $F[X]$. It follows from 11.2.2 (ii) that for any algebraic extension K of F, the algebra $K \otimes E$ is reduced. Then 4.2.12 (4) shows that E is separably generated over F. Assume that $E = F(x_1, \ldots, x_m)$, where $x_1, \ldots, x_t$ are algebraically independent over F and x_i is separably algebraic over $F(x_1, \ldots, x_t)$ ($t < i \leq m$). By 11.2.2 (i) there exist affine F-varieties Y and Z with $F[Y] = F[x_1, \ldots, x_m]$, $F[Z] = F[X][x_1, \ldots, x_m]$. We have F-morphisms $Z \to X$, $Z \to Y$, which are birational (5.1.2). It follows as in the proof of 5.1.2 that there exists a non-zero element $f \in F[X]$ with $F[Z]_f \simeq F[X]_f$. An easy argument then shows that $X(F)$ is dense in X if and only if $Z(F)$ is dense in Z, and similarly for Y. It follows that it suffices to prove the theorem for Y.

We then have to show that if g is a non-zero element of $F[Y]$, there is a homomorphism $\phi : F[Y] \to F$ with $\phi(g) \neq 0$. We use induction on $m - t$ (notations as above). If $m = t$ the assertion is easy. Let $m > t$ and let Y' be the F-variety with $F[Y'] = F[x_1, \ldots, x_{m-1}]$. Then x_m is separable over the quotient field $F(Y')$ of $F[Y']$. Let $P \in F[Y'][T]$ be a polynomial with root x_m, of minimal degree, say n. Denote by a and b the leading coefficient and the constant term of P and put $c = a^{\frac{1}{2}n(n-1)} \prod_{i<j} (y_i - y_j)^2$, where the y_i are the roots of P in some extension of E. Then $c \in F[Y']$ (see [Jac4, p. 125]). Finally let d be the product of leading coefficient and constant term of a polynomial in $F[Y'][T]$ of minimal degree with root g. By induction there is a homomorphism $\phi : F[Y'] \to F$ with $\phi(abcd) \neq 0$. By 1.9.3 we can extend ϕ to a homomorphism $F[Y] \to \bar{F}$. But then $\phi(x_m)$ is a root of a polynomial in $F[T]$ without multiple roots, hence must lie in $F_s = F$. $\qquad \square$

11.2.6. Proposition. *The irreducible components of X are defined over F_s.*

Assume that $F = F_s$. By 11.2.4 (i), $X(\bar{F})$ is dense in X. Let $X_1, \ldots, X_s$ be the irreducible components of X and put $U_i = X_i - \bigcup_{j \neq i} X_j$; this is an open subvariety of X whose closure is X_i, and $U_i(\bar{F})$ is dense in X_i $(1 \leq i \leq s)$. It follows from 11.2.4 (ii) that the components are defined over $\bar{F}$, proving the proposition if $p = 0$.

Assume that $p > 0$. There is a finite extension E of $F = F_s$ over which all components X_i are defined. Then E is purely inseparable over F. Let $a \geq 0$ be such that $E^{p^a} \subset F$. Then $E[X]^{p^a} \subset F[X]$. We may assume that X is affine. Let P_i be the prime ideal in $E[X]$ of functions vanishing on the component X_i. Then $\bigcap_j P_j = \{0\}$, $\bigcap_{j \neq i} P_j \neq \{0\}$. Take $f_i \in \bigcap_{j \neq i} P_j - P_i$ and put $g_i = f_i^{p^a}$. Then $g_i \in F[X]$ and the principal open set $D(f_i) = D(g_i)$ (1.3.5) is defined over F. It is contained in U_i and it is dense in X_i. By 11.2.5 and 11.2.4 (ii) we can conclude that X_i is defined over F. $\qquad \square$

11.2.7. Theorem. $X(F_s)$ *is dense in X.*

This is a consequence of 11.2.5 and 11.2.6. $\qquad \square$

$X(F_s)$ and $X(\bar{F})$ are Γ-sets, i.e., sets with a Γ-action. If X is affine, then these are sets of F-homomorphisms of $F[X]$ to F_s, respectively $\bar{F}$, which have obvious Γ-actions. The same is true in general, as one sees by using a covering by open affine F-varieties.

11.2.8. Proposition. *(i) Let Y be a closed subvariety of X. Then Y is defined over F if and only if (a) Y is defined over F_s, (b) there is a subset of $Y(F_s)$ that is dense in Y and is stable under the Γ-action on $X(F_s)$;*
(ii) Let Y be an open subvariety of X. Then Y is defined over F if and only if (a) Y is defined over $\bar{F}$, (b) $Y(\bar{F})$ is a Γ-stable subset of $X(\bar{F})$.

That the conditions of (i) are necessary follows from 11.2.7. Suppose that they are satisfied. We may assume that X is affine. Let $I \subset F_s[X]$ be the ideal of functions vanishing on Y. It follows from (b) that I is Γ-stable. Application of 11.1.6 shows that the set of fixed points of Γ in I is an F-structure on I. This set is also an ideal in $F[X]$. It follows that Y is defined over F, proving (i).

The necessity of the conditions of (ii) is obvious. Assume they are satisfied. We may again assume X to be affine. Let Z be the closure of $X(\bar{F}) - Y(\bar{F})$. First assume that $Z = \emptyset$, i.e. that $Y(\bar{F}) = X(\bar{F})$. Since Y is defined over $\bar{F}$, it is a union of principal open subsets $\bigcup D(f_i)$, the f_i lying in $\bar{F}[X]$. From 1.1.2 (i) we conclude that the ideal in $\bar{F}[X]$ generated by the f_i is all of $\bar{F}[X]$, which implies that $Y = X$, proving (ii).

If $Z \neq \emptyset$ it is a closed subvariety which is defined over $\bar{F}$ by 11.2.4 (ii). Moreover, our assumptions imply that $Z(\bar{F})$ is a Γ-stable subset of $X(\bar{F})$. By 11.2.8 (i) we can conclude that Z is defined over F_i (recall that $\bar{F} = (F_i)_s$). Let Y_1 be the complement of Z in X. It is the union of principal open subsets $D(g_j)$, the g_j lying in $F_i[X]$. There is a p-power q such that all g_j^q lie in $F[X]$. Since $D(g_j) = D(g_j^q)$, Y_1 is a union of principal open subvarieties defined over F, hence is itself defined over F. Y is an open subvariety of the F-variety Y_1, and $Y_1(\bar{F}) = Y(\bar{F})$, by the definition of Y_1. Let Y_2 be an affine open subset of Y_1 defined over F. Then $Y_2(\bar{F}) = (Y_2 \cap Y)(\bar{F})$. We have seen that this implies $Y_2 = Y_2 \cap Y$. It follows that $Y_1 = Y$, proving (ii).   □

11.2.9. Corollary. *Let $\phi : X \to Y$ be a morphism of F-varieties. Then ϕ is defined over F if and only if (a) ϕ is defined over F_s, (b) for $x \in X(F_s)$ and $\gamma \in \Gamma$ we have $\phi(\gamma.x) = \gamma.\phi(x)$.*

The corollary follows from 11.2.8 (i), applied to the graph Z of ϕ (1.6.11), observing that ϕ is defined over F if and only if Z is an F-subvariety of $X \times Y$.   □

11.2.10. Exercises. (1) Let X be an F-variety. The Galois group Γ acts on the set of components of X. A component of X is defined over F if and only if it is fixed under this action.
(2) Define a functor from the category of F-varieties over k to the category of sets with a continuous Γ-action.
(3) Let X be an affine F-variety.
   (a) $F_s[X] = \{f \in k[X] \mid f(X(F_s)) \subset F_s\}$.
   (b) Show that the Γ-set $X(F_s)$ determines $F[X]$

11.2.11. Let X be an F-variety. Assume that E is an extension of F contained in k and that we are given a Lie algebra L over E of derivations of E whose annihilator is F. We use the connection $c = c_{E[X]}$ of L of 11.1.9. For $D \in L$ the map $c(D)$ is an F-derivation of $E[X]$. Let $x \in X(E)$, denote by $M_x \subset k[X]$ the maximal ideal of functions vanishing in x and put $M_x(E) = M_x \cap E[X]$. It follows from 4.1.4 that the E-vector space $T_x X(E)$ of E-rational points of the tangent space $T_x X$

(see 4.1.8) is isomorphic to the dual of the E-vector space $M_x(E)/M_x(E)^2$. The isomorphism is described in the proof of 4.1.4. From the fact that the $c(D)$ $(D \in L)$ are derivations it follows that the map $f \mapsto (c(D).f)(x)$ induces an E-linear map $M_x(E)/M_x(E)^2 \to E$, whence an E-linear map $\mu_x : L \to T_x X(E)$.

We have the following counterpart of 11.2.8 (i).

11.2.12. Proposition. *Let Y be a closed subvariety of X defined over E.*
(i) Let S be a subset of $Y(E)$ with the following properties: (a) S is dense in Y, (b) for all $y \in S$ we have $\mu_y(L) \subset T_y Y$. Then Y is defined over F;
(ii) If Y is defined over F then $\mu_y(L) \subset T_y Y(E)$ for all $y \in Y(E)$.

We may assume that X is affine. Let $I \subset E[X]$ be the ideal of functions vanishing on Y. We have to show that I is defined over F. By 11.1.10, this will follow if we show that $c(L).I \subset I$. This condition is implied by the following one: for $f \in I$, $D \in L$, $y \in S$ we have $(c(D).f)(y) = 0$. Viewing $T_y X(E)$ as the dual of $M_y(E)/M_y(E)^2$, the subspace $T_y Y(E)$ is the annihilator of $I + M_y(E)/M_y(E)^2$, which shows that the last condition is equivalent to: $\mu_y(L) \subset T_y Y(E)$ $(y \in S)$. As this is a consequence of (b) we obtain (i). The proof of (ii) is left to the reader. $\square$

We can now establish tangent space criteria for fields of definition.

11.2.13. Theorem. *Let X be an F-variety and let Y and Z be closed F-subvarieties with a non-empty intersection. Then $Y \cap Z$ is a closed subvariety, which is defined over F if one of the following condition holds: (1) F is perfect, (2) there is a dense open subset U of $Y \cap Z$ such that for $x \in U$ we have $T_x(Y \cap Z) = T_x Y \cap T_x Z$.*

The tangent spaces are viewed as subspaces of $T_x X$. We proceed in several steps.
(a) $Y \cap Z$ is defined over $\bar{F}$. By 11.2.4 (ii) it suffices to show that $X(\bar{F}) \cap (Y \cap Z)$ is dense in $Y \cap Z$. We may assume X to be affine. Let I and J be the ideals in $F[X]$ of functions vanishing on Y respectively Z. Since $Y \cap Z$ is non-empty, $I + J$ is a proper ideal (see 1.1.3). It suffices to show the following: if $f \in k[X]$ does not vanish on $Y \cap Z$, there is $x \in X(\bar{F}) \cap (Y \cap Z)$ with $f(x) \neq 0$. This follows from 1.9.6 (2).
(b) $Y \cap Z$ is defined over F_s. If char $k = p = 0$ or if F is perfect, this follows from (a). Assume that $p > 0$ and that condition (2) holds. By (a), $Y \cap Z$ is defined over a finite extension E of F_s. We have a tower $F_s = E_h \subset E_{h-1} \subset \cdots \subset E_0 = E$ such that $E_i^p \subset E_{i+1}$. We show by induction on i that $Y \cap Z$ is defined over E_i, which will prove (b).

Assume that $Y \cap Z$ is defined over E_i. We apply 11.2.12 (i) with $F = E_{i+1}$, $E = E_i$, $L = \mathcal{J}_{E_i/E_{i+1}}$, as in 11.1.12 (taking into account 11.1.13 (iv)). Since $E_i = (E_i)_s$, we know by 11.2.7 that $(Y \cap Z)(E_i)$ is dense in $Y \cap Z$. We take $S = U \cap (Y \cap Z)(E_i)$ in 11.2.12 (i). If $y \in S$ we have by 11.2.12 (ii) and (2) that $\mu_y L \subset T_y Y \cap T_y Z = T_y(Y \cap Z)$. This shows that condition (b) of 11.2.12 (i) holds and we can conclude

that $Y \cap Z$ is defined over E_{i+1}. It follows that $Y \cap Z$ is defined over F_s.
(c) $Y \cap Z$ is defined over F. We have $(Y \cap Z)(F_s) = Y(F_s) \cap Z(F_s)$ (as subsets of $X(F_s)$). It is immediate that $(Y \cap Z)(F_s)$ is stable under the action on $X(F_s)$ of the Galois group Γ. Now (c) follows from 11.2.8 (i).   □

11.2.14. Corollary. *Let $\phi : X \to Y$ be an F-morphism of irreducible F-varieties. Let $y \in Y(F) \cap \mathrm{Im}\, \phi$.*
(i) If F is perfect then the fiber $\phi^{-1}(y)$ is defined over F;
(ii) Assume that all irreducible components of $\phi^{-1}(y)$ have dimension $\dim X - \dim Y$ and that in each component C of $\phi^{-1}(y)$ there exists a simple point x such that the tangent map $d\phi_x : T_x C \to T_y Y$ is surjective. Then $\phi^{-1}(y)$ is defined over F.

Let $G = \{(x, \phi x) \in X \times Y \mid x \in X\}$ be the graph of ϕ; it is an F-subvariety of $X \times Y$. Then $\phi^{-1}(y)$ is isomorphic to the intersection $G \cap (X \times \{y\})$. Now (i) follows by case (1) of the theorem.

For the proof of (ii) observe first that, by 4.3.6, the set of points x with the property of (ii) is open and dense in C. The assumptions imply that the intersection of the tangent spaces $T_{(x,y)}G$ and $T_{(x,y)}(X \times \{y\})$ has dimension $\dim X - \dim Y = \dim C$, which implies that the intersection is $T_{(x,y)}C$. We are now in case (2) of the theorem. □

11.2.15. Example. Let char $F = p > 0$. Take $X = Y = \mathbf{A}^1$, with the F-structure of 1.4.9. and $\phi x = x^p$. If $y \in F - F^p$ then $\phi^{-1}y = \{y^{\frac{1}{p}}\}$ is not defined over F. In this case $d\phi_x = 0$ for all x.

11.2.16. Exercise. Let X be an F-variety and let α be an automorphism of X defined over F. Assume that the fixed point set $X^\alpha = \{x \in X \mid \alpha(x) = x\}$ is non-empty. If $x \in X^\alpha$ then $d\alpha_x$ is a linear map of $T_x X$; let $(T_x X)^\alpha$ be its set of fixed points. Assume that each component C of X^α contains a dense open subset U such that $(T_x X)^\alpha = T_x C$ for $x \in U$. Then X^α is defined over F.

11.3. Forms

11.3.1. Let X be an F-variety and E a subfield of k containing F. An E-*form* of X is an F-variety Y that is E-isomorphic to X. We denote by $\Phi(E/F, X)$ the set of F-isomorphism classes of E-forms of X. We shall be interested in the case that E is a Galois extension of X. Assume this to be the case and let Γ be the Galois group. Moreover, assume that X is affine. Then Γ acts in $E[X]$ as a group of ring automorphisms (see 11.1.7). The action is continuous, relative to the Krull topology on Γ and the discrete topology on $E[X]$. Denote by $A = \mathrm{Aut}_E(X)$ the group of E-automorphisms of the algebra $E[X]$, which can be identified with the group of E-automorphisms of X, viewed as an E-variety. Then Γ acts continuously on A as a group of automorphisms.

11.3.2. Now let Γ be an arbitrary profinite group (i.e. a compact, totally disconnected topological group) and let A be a group with a continuous action of Γ, relative to the discrete topology of A. So Γ acts as a group of automorphisms of A, and the stabilizer of an element of A is a closed subgroup of Γ of finite index.

A *1-cocycle* of Γ in A is a continuous function $c : \Gamma \to A$ with $c(\gamma\delta) = c(\gamma)(\gamma.c(\delta))$ $(\gamma, \delta \in \Gamma)$. The set of these cocycles is denoted by $Z^1(\Gamma, A)$. Two cocycles c and d are *equivalent* if there is $a \in A$ such that $d(\gamma) = a^{-1}c(\gamma)(\gamma.a)$ for all $\gamma \in \Gamma$. The set of equivalence classes is the *1-cohomology set* $H^1(\Gamma, A)$. It is a *pointed set*, i.e. a set with a special element 1, the class of the constant cocycle $c = 1$. There is a formalism of exact sequences, to which we return in 12.3 (in a special case). We refer to [Se2, Ch. 1, §5] for a more detailed discussion of these matters.

We now assume that Γ and $A = \text{Aut}_E(X)$ are as in 11.3.1, where X is an affine F-variety.

11.3.3. Proposition. *There is a bijection* $\Phi(E/F, X) \to H^1(\Gamma, \text{Aut}_E(X))$ *such that the class of X corresponds to 1.*

We identify $A = \text{Aut}_E(X)$ with the group of isomorphisms of the E-algebra $E[X]$. The group Γ acts on $E[X] = E \otimes_F F[X]$ via the first factor. Then Γ acts on A by $\gamma.\alpha = \gamma \circ \alpha \circ \gamma^{-1}$ $(\gamma \in \Gamma, \alpha \in A)$. Let Y be an E-form of X and let ϕ be an E-isomorphism of $E[Y]$ onto $E[X]$. For $\gamma \in \Gamma$ put $c(\gamma) = \phi \circ \gamma \circ \phi^{-1} \circ \gamma^{-1}$. Then a straightforward check shows that $a \in Z^1(\Gamma, A)$ and that its equivalence class is independent of the choice of the isomorphism ϕ. So we have a map

$$\mu : \Phi(E/F, X) \to H^1(\Gamma, A).$$

It is also straightforward to check that if Y and Z are two E-forms of X whose respective cocycles $a, b \in Z^1(\Gamma, A)$ are equivalent, there is a Γ-equivariant E-isomorphism of $E[Y]$ onto $E[Z]$ which comes from an isomorphism $F[Y] \simeq F[Z]$. It follows that μ is injective. That it maps the class of X to 1 is obvious.

We show that μ is surjective. Let $c \in Z^1(\Gamma, A)$. For $\gamma \in \Gamma$, $f \in E[X]$ define $\gamma \star f = c(\gamma)(\gamma.f)$. The cocycle property implies that $(\gamma, f) \mapsto \gamma \star f$ defines an action of Γ on $E[X]$ to which we can apply 11.1.6. It follows that

$$F[X]_c = \{f \in E[X] \mid \gamma \star f = f \text{ for all } \gamma \in \Gamma\}$$

defines an F-structure on $E[X]$. Moreover, since the elements of A are algebra automorphisms, $F[X]_c$ is an F-algebra. It defines an E-form X_c of X and it is clear that c is the cocycle defined by X_c. It follows that μ is surjective, proving 11.3.3. $\square$

The variety X_c introduced in the proof is said to be obtained from X by *twisting* with the cocycle c.

There are variants of 11.1.3 involving supplementary structures. The case that X is an algebraic group will be discussed in 12.3. The following exercises give some examples.

11.3.4. Exercises. Let E/F be a Galois extension with group Γ.

(1) $H^1(\Gamma, \mathbf{GL_n}(E)) = 1$ (see 11.1.8).

(2) $H^1(\Gamma, \mathbf{Sp_{2n}}(E)) = 1$. (*Hint*: use that two non-degenerate alternating bilinear forms on F^{2n} are isomorphic, see [Jac4, §6.2]).

(3) (char $k \neq 2$) Define a bijection of $H^1(\Gamma, \mathbf{O}_n(E))$ onto the set of isomorphism classes of non-degenerate symmetric bilinear forms on F^n that are E-isomorphic to the standard form $\sum_{i=1}^n x_i y_i$. (In (2) and (3) $\mathbf{Sp_{2n}}(E)$ and $\mathbf{O}_n(E)$ have the obvious meanings.)

(4) Show that 11.3.3 also holds for a projective variety X.

(5) Show that $H^1(\Gamma, E)$ is trivial. (*Hint*: one can assume E/F to be finite, then use Dedekind's theorem.)

11.4. Restriction of the ground field

As before, F is a subfield of the algebraically closed field k. We denote by E a finite extension of F contained in k. In this section we discuss a procedure to associate with an E-variety an F-variety.

11.4.1. We begin with some algebraic constructions. The dual of the F-vector space E is denoted by E' and $\langle \ , \ \rangle$ is the pairing between E and E'. Fix a basis $(x_i)_{1 \leq i \leq n}$ of E and let (x_i') denote the dual basis of E'. Let A be an E-algebra and denote by S the symmetric algebra of the F-vector space $E' \otimes_F A$ (see [Jac5, p. 141–142]). Let I be the ideal in S generated by the elements

$$x' \otimes ab - \sum_{i,j=1}^n \langle x_i x_j, x' \rangle (x_i' \otimes a)(x_j' \otimes b) \text{ and } x' \otimes x - \langle x, x' \rangle.1,$$

where $x \in E, x' \in E', a, b \in A$. (Notice that the sum does not depend on the choice of the basis (x_i)).

If $I \neq S$ we denote by RA or $R_{E/F}A$ the quotient algebra S/I. We then say that RA is the algebra obtained by *restriction of the ground field* to F, that A *admits restriction of the ground field* (to F) or briefly that RA exists. If $I = S$ one could define RA to be the zero ring, but we prefer not to do this. We only admit rings with non-zero identity element.

11.4.2. Proposition. *(i) Assume that RA exists. There is an E-homomorphism $\rho : A \to E \otimes_F RA$ such that the pair (RA, ρ) has the following universal property: for any pair (B, σ) of an F-algebra B and an E-homomorphism $\sigma : A \to E \otimes_F B$ there exists a unique F-homomorphism $\tau : RA \to B$ with $\sigma = (\mathrm{id} \otimes \tau) \circ \rho$.*

(ii) Assume that there exists an F-algebra B together with an E-homomorphism $A \to E \otimes_F B$. Then A admits restriction of the ground field.

Assume that RA exists. For $x' \in E'$, $a \in A$ denote by $u(x', a)$ the image of $x' \otimes a$ in RA. Then u is an F-bilinear function on $E' \times A$ and for $a, b \in A$, $x' \in E$ we have

$$u(x', ab) = \sum_{i,j=1}^{n} \langle x_i x_j, x' \rangle u(x'_i, a) u(x'_j, b), \quad u(x', x) = \langle x, x' \rangle \ (x \in E). \tag{55}$$

The right-hand side of the first formula is independent of the choice of the basis of E. It follows from (55) that

$$\rho a = \sum_{i=1}^{n} x_i \otimes u(x'_i, a)$$

defines an E-homomorphism $A \rightarrow E \otimes_F RA$. Let (B, σ) be as in 11.4.2 (i). There are unique F-linear maps $\sigma_i : A \rightarrow B$ such that

$$\sigma a = \sum_i x_i \otimes \sigma_i a.$$

The $\sigma_i a$ satisfy the relations obtained by replacing in (55) $u(x'_i, a)$ by $\sigma_i a$. Hence there is a homomorphism τ with the required properties, such that $\tau(u(x'_i, a)) = \sigma_i a$. We have proved (i). If there is B as in (ii), a similar argument shows that there is an F-homomorphism $S \rightarrow B$ whose kernel contains I. Hence $I \neq S$, whence (ii). □

11.4.3. Corollary. *RA exists if one of the following conditions holds:*
(a) there exists an E-homomorphism $A \rightarrow E$,
(b) A is an affine E-algebra.

In case (a) apply (ii), with $B = F$. In case (b) take $B = F_s$. By 11.2.7 there is an E-homomorphism $A \rightarrow E_s$. We then use the following lemma.

11.4.4. Lemma. *There exists an E-homomorphism $E_s \rightarrow E \otimes_F F_s$.*

Put $K = E \otimes_F F_s$. If E_1 is a field intermediate between E and F we have

$$K \simeq E \otimes_{E_1} (E_1 \otimes_F F_s),$$

from which we infer by an easy argument that it suffices to prove the lemma in the case that $E = F(x)$. We also may assume that either x is separable over F or that char $F = p > 0$ and x is purely inseparable over F. Let f be the minimum polynomial of x over F. Then $K \simeq F_s[T]/(f)$.

In the first case f is a polynomial with distinct roots in F_s. It follows that K is isomorphic to a direct sum $(F_s)^n$, and the assertion of the lemma is obvious. In the second case K is isomorphic to $F_s(x)$, which is a purely inseparable extension of F_s containing E. If y is separable over E, then a power y^{p^a} is separable over F_s, hence lies in F_s. It follows that y is both separable and purely inseparable over $F_s(x)$, hence lies in that field. Consequently, $E_s = F_s(x)_s$. The lemma follows, and by 11.4.2 (ii) also 11.4.3.   □

11.4.5. Example. We give an example of an algebra A that does not admit restriction of the ground field. Assume that char $F = p > 0$ and that $E = F(x)$ is a purely inseparable extension of degree p, so $x^p \in F$. Let A be an E-algebra containing an element a with $a^p = x$ (we might take, for example, $A = E(x^{\frac{1}{p}})$). Assume that B and σ are as in 11.4.2 (i). Put $\sigma a = \sum x_i \otimes b_i$. Then

$$\sigma(a^p) = x \otimes 1 = 1 \otimes \left(\sum x_i^p b_i^p\right),$$

which is impossible. It follows that RA does not exist.

11.4.6. In the sequel, if we speak of an algebra $R_{E/F}A$, we assume tacitly that A admits restriction of the ground field. A pair (RA, ρ) with the universal property of 11.4.2 (i) is, as usual, unique up to isomorphism. Also, R is part of a partially defined functor of the category of E-algebras to the category of F-algebras. If $\phi : A \to A'$ is a homomorphism of E-algebras then, with the notations of the proof of 11.4.2 the homomorphism $R\phi : RA \to RA'$ is defined by $R\phi(u(x', a)) = u(x', \phi a)$ ($x' \in E'$, $a \in A$). The universal property of 11.4.2 (i) gives a bijection (of sets of algebra homomorphisms)

$$\mathrm{Hom}_E(A, E \otimes_F B) \to \mathrm{Hom}_F(R_{E/F}A, B),$$

which is functorial in A and B. This shows that $R_{E/F}$ is the right adjoint functor (only partially defined) of the tensor product functor $E \otimes_F$.

The next exercises give a number of properties of these functors. The notations are as in 11.4.2 and its proof.

11.4.7. Exercises. (1) If A is of finite type over E then RA is of finite type over F.

(2) Let I be an ideal in A and denote by RI the ideal in RA generated by the elements $u(x', a)$ with $x' \in E'$, $a \in I$. Then $\rho I \subset E \otimes I$ and $R(A/I) \simeq RA/RI$.

(3) Let A' be a second E-algebra. Then $R(A \otimes_E A') \simeq RA \otimes_F RA'$.

(4) Let E_1 be an extension intermediate between E and F. Then $R_{E/F}A \simeq R_{E_1/F}(R_{E/E_1}A)$. More precisely, there is a corresponding isomorphism of (partially defined) functors.

(5) Let $A = E[T]$. Then $RA = \mathrm{Sym}_F(E')$ and ρ with $\rho T = \sum_i x_i \otimes x_i'$ have the universal property of 11.4.2 (i).

(6) If B is an F-algebra we have a multiplicative norm map $N : E \otimes_F B \to B$, defined as follows. If $x \in E \otimes B$ and $x(x_i \otimes 1) = \sum x_j \otimes b_{ji}$ then $Nx = \det(b_{ij})$. Then x is invertible if and only $Nx \neq 0$. Define the multiplicative map $n : A \to RA$ by $n(a) = N(\rho a)$.

(a) Show that for $f \in A$ we have $R(A_f) \simeq (RA)_{n(f)}$. Here $A_f = A[T]/(1 - fT)$, as in 1.4.6. (*Hint*: use the universal property of A_f which says that any homomorphism of A mapping f onto an invertible element comes from a homomorphism of A_f.)

(b) If A is an affine E-algebra then ρ is injective. (*Hint*: if $f \in \mathrm{Ker}\ \rho$ then $n(f) = 0$).

11.4.8. If A is an affine E-algebra $E[X]$, we see from 11.4.3 (b) and 11.4.7 (1) that RA exists, and is an F-algebra of finite type. But RA need not be an affine F-algebra (for an example see 11.4.15 (2)). We shall consider in more detail two special cases: (a) E is separable over F, (b) $p = \mathrm{char}\ F > 0$ and $E = F(x)$ with $x^p \in F$.

Assume E/F to be separable. Denote by Σ the set of F-isomorphisms of E into the algebraic closure $\bar{F}$ of F in k and let $\tilde{E} \subset k$ a field containing all σE ($\sigma \in \Sigma$). (If E/F is a normal extension we can take $\tilde{E} = E$.) For $\sigma \in \Sigma$ define the $\tilde{E}$-algebra B_σ to be the tensor product over E of $\tilde{E}$ and A, where the E-algebra structure of $\tilde{E}$ is given by $(x, y) \mapsto (\sigma x)y$ ($x \in E$, $y \in \tilde{E}$). Then B_1 is the usual tensor product $\tilde{E} \otimes_E A$.

11.4.9. Proposition. *There is an isomorphism* $\alpha : \tilde{E} \otimes_F RA$ *onto the tensor product over* $\tilde{E}$ *of the algebras* B_σ ($\sigma \in \Sigma$) *such that* $\alpha \circ (\mathrm{id} \otimes \rho)$ *is the canonical injection of* B_1 *into the tensor product.*

With the notations of the proof of 11.4.2, we define for $\sigma \in \Sigma$, $a \in A$ elements $u(\sigma, a) \in \bar{F} \otimes_F A$ by

$$u(\sigma, a) = \sum_{i=1}^{n} \sigma x_i \otimes u(x_i', a).$$

It follows from (55) that

$$u(\sigma, ab) = u(\sigma, a)u(\sigma, b), \quad u(\sigma, x) = \sigma x \otimes 1 \ (a, b \in A, \ x \in E). \qquad (56)$$

Now the matrix (σx_i) where $\sigma \in \Sigma$, $1 \le i \le n$ is an invertible square matrix. This follows from Dedekind's theorem, using that $|\Sigma| = [E : F]$. By the proof of 11.4.2 the relations (56) give a presentation of the algebra $\tilde{E} \otimes_F RA$, for the generators $u(\sigma, a)$. The statement of the proposition is an equivalent way of saying this. □

11.4.10. Corollary. *If* E/F *is separable then any* E-algebra admits restriction of the ground field.*

11.4.11. Now assume we are in case (b), i.e that $E = F(x)$ is purely inseparable of degree $p = \mathrm{char}\ k$. By 4.2.8 the E-vector space $\mathrm{Der}_F(E, E)$ of F-derivations of E has dimension one. We fix a non-trivial F-derivation ∂ of E. We shall need the

A-modules of differentials $\Omega_{A/E}$ and $\Omega_{A/F}$ (see 4.2.1). Denote by S_E (S_F) the symmetric algebra $\mathrm{Sym}_A(\Omega_{A/E})$ (respectively, $\mathrm{Sym}_A(\Omega_{A/F})$). Let J be the ideal in S_F generated by the element $\partial x - d_{A/F}x$.

11.4.12. Proposition. *(i) There is a canonical E-isomorphism α of $E \otimes_F RA$ onto $(S_F/I) \otimes_A S_E^{\otimes_A^{p-2}}$ such that $(\alpha \circ \rho)(a) = a.1$ $(a \in A)$;*
(ii) If ∂ can be extended to an F-derivation of A then $E \otimes_F RA$ is (non-canonically) isomorphic to $S_E^{\otimes_A^{p-1}}$.

With the notations of the proof of 11.4.2 define for $0 \le h < p$ maps $u_h : A \to E \otimes_F RA$ by

$$u_h(a) = (h!)^{-1}(\sum_{i=1}^{p} \partial^h x_i \otimes u(x_i', a)).$$

Notice that u_h is independent of the choice of the basis. Using that for $y \in E$

$$y = \sum_i \langle y, x_i' \rangle x_i$$

we obtain from (55) that for $a, b \in A$, $y \in E$

$$u_h(ab) = \sum_{i=0}^{h} u_i(a)u_{h-i}(b), \quad u_h(y) = (h!)^{-1}(\partial^h y). \tag{57}$$

We have $u_0 = \rho$ and u_1 is an F-derivation of A in $E \otimes RA$, viewed as an A-module via ρ. To deal with the set of equations (57) we use a result about derivations.

11.4.13. Lemma. *Let R be an $\mathbf{F}_p$-algebra and let $D_1, \dots D_s$ be derivations of R in an R-algebra S. For $0 \le h < p$, define the map $\Delta_h = \Delta_h(D_1, \dots, D_s)$ of R to S by*

$$\Delta_h(a) = \sum_{i_1 + 2i_2 + \dots + si_s = h} (i_1! \dots i_s!)^{-1} D_1^{i_1}(a) \dots D_s^{i_s}(a).$$

Then

$$\Delta_h(ab) = \sum_{i=0}^{h} \Delta_i(a) \Delta_{h-i}(b) \ (a, b \in R).$$

We make the convention that $\Delta_0(a) = a.1$. For $s = 1$ the lemma is equivalent to Leibniz's rule for higher derivatives of a product. For $s > 1$ we have

$$\Delta_h(D_1, \dots, D_s) = \sum_{si \le h} (i!)^{-1} \Delta_{h-si}(D_1, \dots, D_{s-1}) D_s^i.$$

The asserted formula follows by induction on s, using Leibniz's formula for the D_s^i. $\square$

Continuing with the proof of 11.4.12, we define inductively for $1 \le h < p$ maps d_h of A in the A-algebra $E \otimes RA$ by

$$u_h = d_h + \Delta_h(d_1, \dots, d_{h-1}). \tag{58}$$

We claim that d_h is an F-derivation. This it true for $h = 1$, since $d_1 = u_1$. Assume that $h > 1$ and that $d_1, \dots, d_{h-1}$ are F-derivations. For $i < h$ we have

$$u_i = \Delta_i(d_1, \dots, d_i) = \Delta_i(d_1, \dots, d_{h-1}).$$

Inserting (58) into the first formula (57) and using 11.4.13 we deduce that d_h is also an F-derivation, establishing the claim.

Now we show that for $y \in E$

$$d_1 y = \partial y, \quad d_h y = 0 \ (2 \le h < p). \tag{59}$$

The first formula follows from (57). For $h > 1$ insert (58) into the second formula (57). Using the definition of Δ_h and induction on h one sees that

$$\Delta_h(d_1, \dots, d_{h-1})(y) = (h!)^{-1}(\partial^h y),$$

and the second formula (59) follows.

Next we claim that the matrix $(\partial^h x_i)_{1 \le h, i \le p}$ is non-singular. This is easy to see for the F-derivation ∂_0 of E with $\partial_0(x) = x$, and $x_i = x^{i-1}$. It then also follows for any basis x_i, and ∂_0. An arbitrary derivation ∂ is a multiple $y\partial_0$ with $y \in E$. One checks that the matrix $(\partial^h x_i)$ is the product of a non-singular triangular matrix and $(\partial_0^h x_i)$, whence the claim.

We can now conclude that the A-algebra $E \otimes_F RA$ has the following properties: there exists an F-derivation d_1 of A in $E \otimes RA$ with $d_1 y = \partial y$ for $y \in E$ and E-derivations $d_2, \dots, d_{p-1}$ of A in $E \otimes RA$, such that $E \otimes RA$ is generated by the images of the $d_h(a)$ $(1 \le h < p)$. The fact (obvious from the definitions) that RA is generated by the $u(x', a)$ of 11.4.2, subject to the relations (55), implies that $E \otimes RA$ is universal for the properties just stated, i.e. for any A-algebra B with similar properties there is an A-homomorphism $E \otimes RA \to B$, compatible with the derivations. It follows from the universal properties of modules of differentials (see 4.2.2 (i)), symmetric algebras and tensor products of algebras (see [Jac5, p. 141–142, p. 144]), that $E \otimes RA$ must be isomorphic to the algebra described in 11.4.12 (i). The formula of (i) also follows. The proof shows that α is functorial in A.

Assume that ∂ can be extended to an F-derivation of A, denoted by the same symbol. By 4.2.2 (i) there is $\lambda \in \text{Hom}_A(\Omega_{A/F}, A)$ with $\lambda(d_{A/F}x) = \partial x$. Since ∂x is non-zero, hence invertible in E and A, it follows that $\Omega_{A/F} = A.d_{A/F}x \oplus \text{Ker } \lambda$. Then Ker λ has the universal property characterizing $\Omega_{A/E}$ of 4.2.2, and Ker $\lambda \simeq \Omega_{A/E}$. Since $S_E/J \simeq \text{Sym}_A(\text{Ker } \lambda)$ we obtain (ii). $\square$

11.4.14. Lemma. *(i) The condition of 11.4.12 (ii) is equivalent to: the E-algebra A has an F-structure;*
(ii) If $A = E[T_1, \ldots, T_m]/(f_1, \ldots, f_n)$, such that $m \geq n$ and that the image of $(D_i f_j)_{1 \leq i,j \leq n}$ in the matrix algebra $\mathbf{M}_n(A)$ is invertible, then the condition of 11.4.12 (ii) holds.

In (ii) D_i is partial derivation with respect to T_i. If A has an F-structure then ∂ can be extended (see 11.1.9). The converse follows from 11.1.14. This proves (i).

Assume the situation of (ii).   Let ϕ be the canonical homomorphism $E[T_1, \ldots, T_m] \rightarrow A$. Extend ∂ to a derivation of $E[T_1, \ldots, T_m]$ denoted by the same symbol, with $\partial T_i = 0$ $(1 \leq i \leq m)$. The extendibility of ∂ to a derivation of A means that there exist $a_1, \ldots, a_n$ in A such that

$$\sum_{i=1}^{m} a_i \phi(D_i f_j) + \phi(\partial f_j) = 0 \; (1 \leq j \leq n).$$

The assumption of (ii) implies that this is the case (one can take $a_{n+1} = \cdots = a_m = 0$). $\square$

11.4.15. Exercises. The notations are as in 11.4.12.
(1) A admits restriction of the ground field if and only if $d_{A/F} x$ is not a nilpotent element of S_F.
(2) Assume that F has characteristic 2 and let $A = E[T, U]/(T^2 - U^3)$. Then A admits restriction of the ground field by 11.4.3 (b). Show that RA is not an affine algebra.

We come to the main results of this section. They are geometric consequences of the preceding algebraic results. As before, E/F is a finite field extension.

11.4.16. Theorem. *(i) Let X be an irreducible, smooth, affine E-variety. There exists an irreducible, smooth, affine F-variety ΠX or $\Pi_{E/F} X$, together with a surjective E-morphism $\pi : \Pi X \rightarrow X$ such that the following holds: for any affine F-variety Y together with an E-morphism $\phi : Y \rightarrow X$ there is a unique F-morphism $\psi : Y \rightarrow \Pi X$ with $\phi = \pi \circ \psi$. The pair $(\Pi X, \pi)$ is unique up to isomorphism;*
(ii) If E/F is separable, smoothness and irreducibility may be omitted in the assumptions and conclusions of (i).

Put $A = E[X]$. By 11.4.3 (b) A admits restriction of the ground field to F. The existence of ΠX and π will follow if we show that the F-algebra RA is affine. Using 11.4.7 (4) we see that it suffices to do this in the cases (a) and (b) of 11.4.8. In case (a), when E/F is separable, it follows from 11.4.9, without assuming irreducibility or smoothness, that there is an affine E-variety Z such that $k \otimes_F RA \simeq k[X \times Z]$, π corresponding to the projection morphism $X \times Z \rightarrow Z$. These facts imply (i) for this case, and (ii) also follows.

Now assume case (b). So E/F is purely inseparable of degree $p = \text{char } F$. First assume that A is as in 11.4.14 (ii). Then $\Omega_{A/E}$ is a free A-module, of rank dim X (as follows from 4.2.4).By 11.4.12 (ii) we have that $E \otimes RA \simeq A \otimes_E E[T_1, \ldots, T_d]$, where $d = (p-1) \dim X$, the homomorphism ρ corresponding to the canonical map of A in the tensor product. This description of $E \otimes RA$ implies all assertions of (i), in the case under consideration.

In the general case one sees (using 4.3.2 and 4.3.3) that there is a covering of X by principal open subsets $D(a)$ with a in a subset S of A, such that the algebras $A_a = E[D(a)]$ $(a \in S)$ have the property of 11.4.14 (ii). We have by 11.4.12 that $E \otimes R(A_a) \simeq \text{Sym}_{A_a}(\Omega_{A_a/E})^{\otimes p-1}$. Since $\Omega_{A_a/E} \simeq A_a \otimes_A \Omega_{A/E}$ we can conclude that $E \otimes R(A_a) \simeq (E \otimes RA)_{\rho a}$, from which we see that the last E-algebra is affine. Since the $D(a)$ with $a \in S$ cover X, the ideal generated by S is A. Then the ideal generated by the ρa $(a \in S)$ must be $E \otimes RA$. It follows $E \otimes RA$ is affine. So there is an affine F-variety X with $F[\Pi X] = RA$. Let π be the E-morphism $\Pi X \to X$ defined by ρ. The preceding discussion implies that for $a \in S$ we have $\pi^{-1}(D(a)) \simeq D(a) \otimes \mathbf{A}^d$ and irreducibility and smoothness of ΠX follow. The uniqueness statement of (i) is standard. □

In the following corollaries the notations are as in 11.4.16 (i).

11.4.17. Corollary. dim $\Pi_{E/F} X = [E : F] \dim X$.

Here $[E : F]$ is the degree of the field extension. □

11.4.18. Corollary. *Assume that E/F is purely inseparable. There exists a covering (U_i) by affine open E-subvarieties, the $\pi^{-1}U_i$ being affine open F-subvarieties, together with E-isomorphisms $\phi_i : U_i \times \mathbf{A}^{[E:F](\dim X-1)} \to \pi^{-1}U_i$ such that $\pi \circ \phi_i$ is the first projection.*

This follows from the proof of the theorem. The fact that the $\pi^{-1}U_i$ are defined over F comes from the observation that $(E \otimes RA)_{\rho a} = (E \otimes RA)_{(\rho a)^p}$ and that $(\rho a)^p \in RA$ for all $a \in A$. □

11.4.19. Let X be an E-variety. We say that X *admits restriction of the ground field* to F if there exist an F-variety $\Pi X = \Pi_{E/F} X$ and an E-morphism π, with the properties of 11.4.16 (i). Then ΠX is the variety obtained by restricting the ground field. We also say that ΠX *exists*. The properties of 11.4.16 (i) then hold for any F-variety Y. If $\phi : X \to Y$ is a morphism of E-varieties and if ΠX and ΠY exist, there is an induced morphism $\Pi(\phi) : \Pi X \to \Pi Y$, so we have a (possibly partially defined) functor Π from the category of E-varieties to the category of F-varieties.

11.4.20. Exercises. X is an F-variety, assumed to be affine in the first three exercises and E/F is as before.

(1) In 11.4.16 (i) the assumption and conclusion of irreducibility can be omitted.

(2) $\Pi_{E/F}X$ exists if and only if the ideal of nilpotent elements of $\bar{F} \otimes_F R_{E/F}(E[X])$ is defined over F.

(3) Let E/F be an inseparable quadratic extension.

  (a) If $X = \{(x, y) \in \mathbf{A}^2 \mid xy = 0\}$ then $\Pi_{E/F}X$ exists.

  (b) If $X = \{(x, y) \in \mathbf{A}^2 \mid x^2 - y^3 = 0\}$ then $\Pi_{E/F}X$ does not exist.

(4) (a) If E/F is purely inseparable, then ΠX exists for any irreducible, smooth E-variety X.

  (b) If E/F is separable, then ΠX exists for any projective E-variety. (*Hint*: use a version of 11.4.9 for graded algebras.)

11.4.21. The separable case. Assume that E is a separable extension, contained in F_s. Let $\Gamma = \mathrm{Gal}(F_s/F)$ and $\Delta = \mathrm{Gal}(F_s/E)$ be the respective Galois groups. Then Δ is an open and closed subgroup of Γ, with finite index. We identify the coset space Γ/Δ with the set Σ of F-homomorphisms of E into F_s, introduced in 11.4.8. The group Γ operates on it by left translations. If A is a group we call A-*set* a set on which A acts as a group of permutations. If B is a subgroup of A of finite index and Q is a B-set we define an A-set $P = Ind_B^A Q$ as follows: P is the set of functions $f : A \to Q$ with $f(ab) = b^{-1}.f(a)$ $(a \in A, b \in B)$. The action of A on P is given by $(a.f)(a') = f(a^{-1}a')$ $(a, a' \in A)$. If Q has a supplementary structure that is B-stable, then P has a similar A-stable structure. For example, if Q is a B-module then P is the induced A-module.

Now let X be an affine E-variety. By 11.4.16 (ii) we have an F-variety $Y = \Pi_{E/F}X$.

11.4.22. Proposition. (*i*) *There is an isomorphism* $\rho : X^\Sigma \to Y$ *such that* $\pi \circ \rho$ *is the projection onto the factor defined by* id $\in \Sigma$;
(*ii*) *There is an isomorphism of the* Γ-*set* $Y(F_s)$ *onto* $Ind_\Delta^\Gamma X(F_s)$.

  (i) follows from 11.4.9, with $\tilde{E} = k$.

For $\sigma \in \Sigma$ define B_σ as in 11.4.8, with $\tilde{E} = F_s$, $A = E[X]$. Notice that $E[X]$ is a subring of B_σ, but not necessarily a subalgebra. By 11.4.9, $F_s[Y]$ is the tensor product $\bigotimes_{\sigma \in \Sigma} B_\sigma$. It is generated over F_s by tensor products $\mathbf{f} = \otimes_\sigma f_\sigma$, where f_σ is an element of $E[X] \subset B_\sigma$. It follows from the proof of 11.4.9 that, for $\gamma \in \Gamma$, we have $\gamma.\mathbf{f} = \otimes f_\sigma'$, where $f_\sigma' = f_{\gamma^{-1}.\sigma}$. This describes the Γ-action on $F_s[Y]$.

Let $y \in Y(F_s)$. It is a function on Σ, whose value y_σ is an F_s-homomorphism $B_\sigma \to F_s$. This homomorphism can also be viewed as a ring homomorphism $E[X] \to F_s$ with $y_\sigma(af) = (\sigma.a)y_\sigma(f)$ $(a \in E, f \in E[X])$. Define a function $\phi(y)$: $\Gamma \to X(F_s)$ by $\phi(y)(\gamma)(f) = \gamma^{-1}.(y_{\gamma\Delta}(f))$ for $\gamma \in \Gamma$, $f \in E[X]$. A straightforward check shows that $\phi(y) \in Ind_\Delta^\Gamma X(F_s)$. We have defined a map of Γ-sets $\phi : Y(F_s) \to Ind_\Delta^\Gamma X(F_s)$. For $\sigma \in \Sigma = \Gamma/\Delta$ let $\bar{\sigma} \in \Gamma$ be a representative.

To prove that ϕ is bijective, observe that $\phi(y)$ is uniquely determined by the values $\phi(y)(\bar{\sigma}) \in X(F_s)$ and that these values may be prescribed arbitrarily. $\qquad\square$

Notes

11.1 deals with Galois descent of the ground field. The main results for separable extensions, such as 11.1.6, are standard (see for example [Bo3, AG, §14]). We also establish similar results for purely inseparable extensions of height one, where the role of the Galois group is taken over by the Lie algebra of derivations. We have the analogue 11.1.14 of 11.1.6. The method goes back to Cartier [Car, Ch.2].

The results of 11.1 are applied in 11.2 to obtain criteria for fields of definition for algebraic varieties. The main results are 11.2.8, for Galois extensions (which is a standard result, see [Bo3, AG, 14.4]) and 11.2.12, for purely inseparable extensions (which does not seem to be in the literature in this form). A consequence of these results is the useful criterion 11.2.14 for fields of definition. In 11.2 we also establish the basic density theorem 11.2.7.

The brief section 11.3 is devoted to a proof of the well-known main result 11.3.3 about forms of affine F-varieties.

Restriction of ground fields, the theme of 11.4, is due to Weil. In the context of scheme theory the matter is discussed in [DG, Ch. 1, 1.6]. See also [Oe, App. 2] for a brief review. We give an elementary treatment. It is based on 11.4.2, which gives an explicit presentation of algebras obtained by restriction of the base field, for an arbitrary finite field extension. The main result of 11.4 is 11.4.16.

Chapter 12

F-groups: General Results

In this chapter the results of the preceding one will be applied to algebraic groups. The notations are as in 2.1. 'Linear algebraic group over F' will be abbreviated to 'F-group.' As before, F_s is a separable closure of F.

12.1. Field of definition of subgroups

Let G be an F-group. In 2.2 we already proved that certain subgroups of G are F-subgroups, i.e., are defined over F. For example, if G is connected its commutator subgroup is an F-subgroup (see 2.2.8 (ii)). In the present section we shall discuss other kinds of subgroups.

12.1.1. Proposition. *The identity component G^0 is an F-subgroup.*

By 11.2.6 all irreducible components of G are defined over F_s, and the Galois group Γ of F_s over F permutes them (see 11.2.10 (1)). Since the components are mutually disjoint, the component G^0 containing the point $e \in G(F)$ must be stable under Γ. Applying 11.2.7 and 11.2.8 (i) we see that G^0 is defined over F. $\qquad\square$

Assume that X is a G-space over F (see 2.3.1). Let $x \in X(F)$ and denote by μ the F-morphism $g \mapsto g.x$ of G to X. Its image is the orbit $O = G.x$ and $\mu e = x$. By 1.9.1 (iv) the closure $\bar{O}$ is defined over F and by 2.3.3 (i) O is an open subvariety of $\bar{O}$.

12.1.2. Proposition. *(i) The isotropy group G_x is defined over F if F is perfect or if the tangent map $d\mu_e : T_e G \to T_x O$ is surjective;*
(ii) O is defined over F.

(i) follows from 11.2.14. Notice that in the second case the tangent map $d\mu_g$ is surjective in all points of $g \in G$, so that the condition of 11.2.14 (ii) is satisfied. Also notice that surjectivity of $d\mu_e$ is equivalent to the equality $\dim \operatorname{Ker} d\mu_e = \dim G_x$.

We know that $\bar{O}$ is defined over F. The open subvariety O is defined over $\bar{F}$. This one sees by taking $k = \bar{F}$ (see 11.2.3 (2)). Now (ii) follows from 11.2.8 (ii), with $X = \bar{O}$, $Y = O$, the condition (b) of loc. cit. being clearly satisfied. $\qquad\square$

By 4.3.6 surjectivity of the tangent map $d\mu_e$ is equivalent to separability of μ.

12.1.3. Corollary. *Let $\phi : G \to G'$ be a homomorphism of F-groups. Then $\operatorname{Ker} \phi$ is defined over F if F is perfect or if the tangent map $d\phi_e$ is surjective.*

This follows by applying 12.1.2 (i) to the G-space G', with action $(g, x) \mapsto \phi(g)x$. □

The Lie algebra $\mathfrak{g}$ or $L(G)$ has the F-structure $\mathfrak{g}(F)$ of 4.4.8.
    If $x, y \in G$ we write

$$N_G(x, y) = N(x, y) = \{g \in G \mid gxg^{-1} = y\}$$

this is the *transporter* of x into y. It is a closed subvariety of G. For $x = y$ we obtain the centralizer $Z_G(x) = Z(x)$. If $X, Y \in \mathfrak{g}$ we define, similarly, the transporter of X into Y:

$$N_G(X, Y) = N(X, Y) = \{g \in G \mid \mathrm{Ad}(g)X = Y\}.$$

For $X = Y$ we obtain the centralizer $Z_G(X) = Z(X)$.

12.1.4. Corollary. *Assume G to be connected.*
(i) Let x and y be semi-simple elements of $G(F)$. If $N(x, y)$ is non-empty it is defined over F. In particular, $Z(x)$ is defined over F;
(ii) Let X and Y be semi-simple elements of $\mathfrak{g}(F)$. If $N(X, Y)$ is non-empty it is defined over F. In particular, $Z(X)$ is defined over F.

For the notion of semi-simple element of $\mathfrak{g}$ see 4.4.20.

Apply 12.1.2 to the conjugation action of G on itself. By 5.4.5 (i) the morphism μ is separable. It follows from 12.1.2 (ii) that $Z(x)$ is defined over F. If $Z(x, y) \neq \emptyset$ it is a fiber of μ. Since $d\mu_g$ is surjective for all $g \in G$, 11.2.14 (ii) shows that $Z(x, y)$ is also defined over F. This proves (i). The proof of (ii) is similar. It uses a separability statement similar to that of 5.4.5 (i), which is proved by an argument like the one of the first paragraph of the proof of 5.4.4. □

12.1.5. Proposition. *Let H and K be two F-subgroups of G. Then $H \cap K$ is defined over F if F is perfect or if $L(H \cap K) = L(H) \cap L(K)$.*

This is a consequence of 11.2.13. □

12.1.6. Counterexamples. Over a non-perfect ground field F one has to be careful with fields of definition of group-theoretically defined subsets of G, as is shown by the following examples. To keep them simple we assume that F is a non-perfect field of characteristic two. Similar examples exist in any positive characteristic. Fix $a \in F - F^2$.
(1) Let

$$G = \{(x, y) \in k^2 \mid x^2 - ay^2 \neq 0\},$$

with multiplication

$$(x, y), (x', y') = (xx' + ayy', xy' + x'y).$$

Then G is an F-group, with $F[G] \simeq F[T, U, V]/((T^2 - aU^2)V - 1)$ (notice that $(T^2 - aU^2)V - 1$ is an irreducible polynomial). Define $\phi : G \to \mathbf{G}_m$ by $\phi(x, y) = x^2 - ay^2$. This is an F-homomorphism. The kernel of ϕ of G is the subvariety $\{(x, y) \mid x^2 - ay^2 = 1\}$ of $\mathbf{A}^2$, which is not defined over F. For if it were, 11.2.7 would imply that there existed non-zero $x, y \in F_s$ with $x^2 - ay^2 = 1$, whence $a \in F_s$, which is not the case. This shows that 12.1.3 is not generally true.

(2) In the preceding example, Ker ϕ is isomorphic to $\mathbf{G}_a$, hence is a connected, normal, unipotent subgroup. Since Im $\phi = \mathbf{G}_m$, we see that Ker ϕ is also the unipotent radical of G. So G also provides an example of an F-group whose unipotent radical is not defined over F. This counterexample can be put in a more general context (see 12.4.6 and 12.4.7 (4)).

(3) G and ϕ being as before, let H be the algebraic group with underlying variety $G \times \mathbf{G}_a$, the multiplication being defined by $(g, x)(g', x') = (gg', x + \phi(g)x')$ $(g, g' \in G, x, x' \in k)$. Then the center of H is the subgroup Ker $\phi \times \{0\}$ of H. We have here an example of an F-group whose center is not defined over F.

(4) Let $G = \mathbf{SL}_4$ and take

$$x = \begin{pmatrix} 0 & 0 & 0 & a \\ 0 & 0 & a^{-1} & 0 \\ 0 & 1 & 0 & 0 \\ 1 & 0 & 0 & 0 \end{pmatrix},$$

then $x \in G(F)$. The centralizer of x is the subgroup of G of matrices of the form

$$\begin{pmatrix} x & 0 & 0 & ay \\ 0 & z & t & 0 \\ 0 & at & z & 0 \\ y & 0 & 0 & x \end{pmatrix},$$

with $x, y, z, t \in k$ and $(xz + ayt)^2 - a(xt + yz)^2 = 1$. This group is not defined over F, as is shown by the argument of (1). So centralizers of elements of $G(F)$ need not be defined over F.

(5) Let $G = \mathbf{G}_a^2$. Then $H = \mathbf{G}_a \times \{0\}$ and $K = \{(x, x^2 + ax^4) \mid x \in k\}$ are two F-subgroups of G, whose intersection is not defined over F, giving a counterexample to 12.2.5. In this case $L(H) = L(K)$ is one dimensional and $L(H \cap K) = \{0\}$.

(6) We have already seen in 2.4.11 that semi-simple (and unipotent) parts of elements of $G(F)$ need not lie in $G(F)$.

12.1.7. Perfect ground fields. Now assume that F is perfect. In that case the counterexamples of 12.1.6 disappear.

(a) We have already seen in 12.1.3 and 12.1.5 that kernels of F-homomorphisms and intersections of F-subgroups are defined over F. Also, it follows from 12.1.2 that centralizers of elements of $G(F)$ are defined over F.

(b) The center $C(G) = C$ of G is defined over F. For it follows from 11.2.7 that C is the intersection of the centralizers $Z(x)$, where x runs through $G(F_s)$. By 1.1.5 (ii) C is the intersection of finitely many centralizers. From 12.1.5 we conclude that C is defined over F_s. Since $C(F_s)$ clearly is a subgroup of $G(F_s)$ which is stable under the Galois group Γ, we see from 11.2.8 (i) that C is defined over F.

(c) The semi-simple and unipotent parts of an element of $G(F)$ now also lie in $G(F)$. Viewing G as an F-subgroup of some $\mathbf{GL}_n$ (2.3.7) we see that it suffices to deal with the case of $\mathbf{GL}_n$. It is clear from the theory of Jordan normal forms that, if $x \in \mathbf{GL}_n(F)$, its semi-simple part x_s lies in $G(F_s)$. The uniqueness of the Jordan decomposition implies that x_s is fixed by all elements of the Galois group Γ, hence lies in $\mathbf{GL}_n(F)$. A similar result holds for the Jordan decomposition of elements of $\mathfrak{g}(F)$.

(d) Finally, the unipotent radical $R_u(G)$ is defined over F. It suffices to show that it is defined over F_s, for then 11.2.8 (i) can be applied. First assume that our algebraically closed field k coincides with $\bar{F} = F_s$. Then it is immediate that $R_u(G)$ is defined over F_s. In the general case one sees, by first working over $\bar{F}$, that there exist a Borel subgroup B, respectively a maximal torus T, which are defined over $\bar{F}$, with $T \subset B$. Then B is also a Borel subgroup of G, viewed as an algebraic group over k, as solvability and connectedness are preserved by passing from $\bar{F}$ to k. The unipotent radical of G is the unipotent radical of the intersection I of all Borel subgroups of G (see 6.4.14). I can also be described as the closed subgroup of G whose elements fix all points of the homogeneous space G/B. By 11.2.4 (i) this is also the subgroup whose elements fix all points of $(G/B)(\bar{F})$, i.e. I is the intersection of the Borel subgroups defined over $\bar{F}$. It follows from 12.1.5 that I is defined over $\bar{F}$. Hence its unipotent radical $R_u(G)$ is defined over $F_s = \bar{F}$, which is what we claimed. We can also conclude that the radical $R(G)$ of G is defined over F. For the image of $R(G)$ in the reductive group $H = G/R_u(G)$ is the connected center of H (7.3.1 (i)), which is defined over F by (b) and 12.1.1. $R(G)$ is defined over F as a consequence of 11.2.13.

12.1.8. Exercises. (1) Let A be a finite-dimensional associative algebra over k with an F-structure (see 11.1.7). Assume given an anti-automorphism ι of A which is defined over F. Put

$$A^- = \{x \in A \mid \iota x = -x\}.$$

Let

$$G = \{a \in A \mid a(\iota a) = 1\}$$

this is a linear algebraic group.

(a) Show that the multiplicative group A^* of A is an F-group.

(b) If $\dim_k A^- = \dim G$ then G is defined over F. (*Hint*: consider the action $(x, y) \mapsto xy(\iota x)$ of A^* on itself.)

(2) In (1) take $A = \mathbf{M}_n$. Let $s \in \mathbf{GL}_n(F)$ be a symmetric invertible matrix and put $\iota x = s({}^t x)s^{-1}$. Then the group G of (1) is the orthogonal group defined by s.

(a) If char $F \neq 2$ then G is defined over F. (*Hint*: use 8.1.3 and 7.4.7 to find $\dim G$.)

(b) Show by a counterexample that (a) is not true if char $F = 2$.

(3) Show that the symplectic group $\mathbf{Sp}_{2n}$ is defined over F.

(4) (a) Let $V = k^{2n+1}$ and let Q be the quadratic form on V with

$$Q((\xi_0, \dots, \xi_{2n+1})) = \xi_0^2 + \sum_{i=1}^n \xi_i \xi_{n+i}.$$

Let G be the identity component of isotropy group of Q in $GL(V)$. Show that G is defined over F. (*Hint*: If the characteristic is $\neq 2$ this follows from (2). In characteristic 2 use 7.4.7 (6) and 12.1.2.)

(b) Let $V = k^{2n}$ and let Q be the quadratic form on V with

$$Q((\xi_1, \dots, \xi_{2n})) = \sum_{i=1}^n \xi_i \xi_{n+i}.$$

Let G be as in (a). Show that G is defined over F.

(5) Let G be a connected F-group and σ a semi-simple automorphism of G defined over F. Then its group of fixed points G_σ is defined over F (see 5.4 for semi-simple automorphisms).

12.2. Complements on quotients

Let G be an F-group and H a closed F-subgroup.

12.2.1. Theorem. *A quotient $(G/H, a)$ over F exists and is unique up to a G-isomorphism over F.*

The notion of quotient was defined in 5.5. The theorem is a refinement of 5.5.5. The proof of 5.5.5 can serve for proving 12.2.1, once one knows that there is an F-variety X and a point $x \in X(F)$ with the properties of 5.5.4, such that the G-action on X is defined over F. In the proof of 5.5.4, such a variety X is constructed as a G-orbit. Application of 12.1.2 (ii) shows that it is defined over F. □

12.2.2. Corollary. *If, moreover, H is a normal subgroup of G, then G/H has a structure of F-group.*

This follows from 5.5.10. □

12.2.3. We shall now discuss another quotient construction, in the case that F has characteristic $p > 0$. The Lie algebra $\mathfrak{g}$ of G is a p-Lie algebra (4.4.3), which has an F-structure (4.4.8).

Let $\mathfrak{h}$ be a p-subalgebra of $\mathfrak{g}$ defined over F. A *quotient of G by* $\mathfrak{h}$ is a pair of an affine homogeneous space $G/\mathfrak{h}$ for G over F, together with a point $a \in (G/\mathfrak{h})(F)$ such that
(a) the morphism $\phi : g \mapsto g.a$ of G to $G/\mathfrak{h}$ is bijective and $\mathfrak{h}$ is the kernel of the tangent map $(d\phi)_e : \mathfrak{g} \to T_a(G/\mathfrak{h})$,
(b) for any pair (Y, b) of an affine G-space Y of G over F and a point $b \in Y(F)$ such that the kernel of the differential at e of the morphism $\psi : g \mapsto g.b$ of G to Y contains $\mathfrak{h}$, there exists a unique F-morphism of G-spaces $\chi : G/\mathfrak{h} \to Y$ with $\chi a = b$.

12.2.4. Theorem. *A quotient $G/\mathfrak{h}$ over F exists and is unique up to a G-isomorphism of homogeneous spaces over F.*

The uniqueness part of the theorem is trivial. The construction of the quotient proceeds as follows. Put $A = F[G]$ and denote by $\Delta : A \to A \otimes_F A$ comultiplication (see 2.1.2). By 4.4.8, $\mathfrak{g}(F)$ is the space of F-rational tangent vectors $(T_e G)(F) = \mathrm{Der}_F(A, F_e)$, where $F_e = A/M_e$, M_e being the maximal ideal in A defined by e (see 4.1.3 and 4.1.8). There is a bijection β of $\mathfrak{g}(F)$ onto the space of left invariant F-derivations of A. If $X \in \mathfrak{g}(F)$, $f \in A$ and $\Delta f = \sum f_i \otimes g_i$ then

$$(\beta X)f = -\sum f_i(X g_i), \qquad (60)$$

see 4.4.4 (b). Denote by B the subalgebra of A whose elements are annihilated by all βX, $X \in \mathfrak{h}(F)$. It follows from (60) that $f \in B$ if and only if $\Delta f = \sum f_i \otimes g_i$ with $X g_i = 0$ for all i. This implies that

$$\Delta B \subset A \otimes B. \qquad (61)$$

It is clear that B contains the subalgebra $F[A^p]$ generated by the pth powers of elements of A and that any $f \in A$ is integral over $F[A^p]$. By 5.2.2 A is finite over $F[A^p]$. Since the latter ring is noetherian, we can conclude that B is a finite $F[A^p]$-module, which implies that it is an F-algebra of finite type. Also, B is an affine F-algebra, being a subalgebra of the affine algebra A. By 11.2.2 (i) there is an affine F-variety $G/\mathfrak{h}$ with $F[G/\mathfrak{h}] = B$. The inclusion $B \subset A$ defines a morphism $\phi : G \to G/\mathfrak{h}$. We put $a = \phi e$. By (61) we have a morphism $G \times G/\mathfrak{h} \to G/\mathfrak{h}$. We leave it to the reader to check that it defines a G-action on $G/\mathfrak{h}$ over F and that $\phi(xy) = x.\phi(y)$ $(x, y \in G)$. Since A is integral over B, we have that ϕ is surjective (for example as a consequence of 1.9.3). It follows that $G/\mathfrak{h}$ is a homogeneous space of G.

Let K and L be the quotient fields of A, respectively B. Then $K^p \subset L \subset K$. We conclude from 5.1.5 (i) that ϕ is bijective.

Put $\tilde{\mathfrak{h}} = K \otimes_F \beta \mathfrak{h}$. This is an algebra of derivations of K and from the definition of B we see that

$$L = \{x \in K \mid \tilde{\mathfrak{h}}.x = 0\}.$$

Now let $X \in (\operatorname{Ker}(d\phi)_e)(F)$. Then the derivation X of A in F_e annihilates B. The derivation βX of A also annihilates B and extends to a derivation of K annihilating L. But by 11.1.15 these are the derivations in $\tilde{\mathfrak{h}}$. It follows that $X \in \mathfrak{h}$. We have proved property (a) of 12.2.1.

If Y and b are as in (b), then ψ is defined by an injective algebra homomorphism

$$\psi^* : F[Y] \to A.$$

The assumption that $\mathfrak{h} \subset \operatorname{Ker}(d\psi)_e$ implies that $\operatorname{Im} \psi^* \subset B$. The corresponding inclusion map defines χ. This proves property (b). We leave it to the reader to complete the details. $\square$

12.2.5. Corollary. *In the case of* 12.2.3 *assume, moreover, that $\mathfrak{h}$ is $\operatorname{Ad}(G)$-stable. Then $G/\mathfrak{h}$ has a structure of linear algebraic group over F such that ϕ is a homomorphism. If $\psi : G \to G'$ is a homomorphism of linear algebraic F-groups such that the kernel of the Lie algebra homomorphism $d\psi : \mathfrak{g} \to \mathfrak{g}'$ contains $\mathfrak{h}$, there is a unique F-homomorphism $\chi : G/\mathfrak{h} \to G'$ with $\psi = \chi \circ \phi$.*

Let $\iota : A \to A$ be the antipode (2.1.2). We claim that, under the hypothesis of the corollary, we have $\Delta B \subset B \otimes_F B$, $\iota B \subset B$. It then follows that $G/\mathfrak{h}$ can be given a structure of algebraic group over F such that ϕ is a homomorphism. The last statement of the corollary is then a consequence of property (b) of 12.2.3.

To prove the claim we may and shall assume that $F = k$. If $\mathfrak{h}$ is stable under $\operatorname{Ad}(G)$, we see from 4.4.4 (a) that B is stable under all right translations $\rho(x)$ ($x \in G$). But then $\Delta B \subset B \otimes A$. We have already seen in the proof of 12.2.4 that $\Delta B \subset A \otimes B$, so $\Delta B \subset (B \otimes A) \cap (A \otimes B) = B \otimes B$.

Let $f \in B$ and put $\Delta f = \sum f_i \otimes g_i$, with $f_i, g_i \in B$. We may assume the g_i to be linearly independent over F. Now

$$f(e) = \sum (\iota f_i) g_i.$$

Let $X \in \mathfrak{h}$, and view it as a derivation of A in F_e. Then

$$0 = \sum X(\iota f_i) g_i + \sum (\iota f_i) X(g_i) = \sum X(\iota f_i) g_i,$$

and it follows that all $X(\iota f_i)$ are zero, which means that all ιf_i lie in B. Since (with the previous notations) $\iota f = \sum g_i(e) \iota f_i$, we conclude that $\iota f \in B$. This proves the claim, whence the corollary. $\square$

12.2.6. As in 9.6.1 (where the groups were reductive) we say that a homomorphism of connected algebraic groups $\phi : G \to G'$ is an *isogeny* if it is surjective and has finite kernel. If G and G' are F-groups and ϕ is defined over F, we call ϕ an *F-isogeny*. By 5.3.5 (2) the kernel of ϕ is finite and lies in the center of G.

If the differential $d\phi : \mathfrak{g} \to \mathfrak{g}'$ is bijective, ϕ is separable (4.3.6 (i)). Then ϕ induces an isomorphism of the field $k(G')$ onto a subfield of $k(G)$ over which $k(G)$ is separably algebraic. If ϕ is bijective, then by 5.1.6 (iii) this field extension is purely inseparable. In these two cases, we call the isogenies separable, respectively purely inseparable. The homomorphisms $G \to G/\mathfrak{h}$ of 12.2.5 are examples of purely inseparable isogenies. We call these *elementary*.

12.2.7. Corollary. *Let $\phi : G \to G'$ be an F-isogeny of connected linear F-groups. There exists a factorization of ϕ*

$$G = G_0 \xrightarrow{\phi_0} G_1 \xrightarrow{\phi_1} ...G_s \xrightarrow{\phi_s} G_{s+1} = G',$$

such that
(a) the G_i are connected linear F-groups and the ϕ_i are F-isogenies,
(b) $\phi_0, ..., \phi_{s-1}$ are elementary purely inseparable and ϕ_s is separable.

We view $k(G)$ as a finite extension of $k(G')$. If $d\phi$ is bijective take $s = 0$, $\phi_0 = \phi$. Otherwise, put $G_1 = G/\mathrm{Ker}\ d\phi$ and denote by ϕ_0 the homomorphism of 12.2.5. There is an F-isogeny $G_1 \to G'$ and $k(G') \subset k(G_1) \subset k(G)$. The degree $[k(G_1) : k(G')]$ is less than $[k(G) : k(G')]$. We can now use an induction. $\qquad\square$

12.2.8. Frobenius morphisms. In 12.2.6 we may take $\mathfrak{h} = \mathfrak{g}$. The algebraic group $G/\mathfrak{g}$ has another description, which we now briefly discuss.

If A is an F-algebra, we denote by $\mathbf{F}A$ or $\mathbf{F}_F A$ the F-subalgebra generated by the pth powers of the elements of A. Then $\mathbf{F}_k(k \otimes A) = k \otimes_F \mathbf{F}_F A$ (notice that $k = k^p$). If F is perfect $\mathbf{F}A$ can also be viewed as the algebra whose underlying ring coincides with A, but with scalar multiplication $a.f = a^{\frac{1}{p}} f$ ($a \in F$, $f \in A$). If X is an affine F-variety, there is an affine variety $Fr\ X$, with $F[Fr\ X] = \mathbf{F}_F(F[X])$, and a morphism $Fr : X \to Fr\ X$, the *Frobenius morphism*. There is a corresponding functor on the category of affine F-varieties. We conclude that if G is a linear F-group, so is $Fr\ G$, and $Fr : G \to Fr\ G$ is a homomorphism of F-groups, which is in fact an inseparable isogeny.

12.2.9. Corollary. *Let G be a connected linear F-group. There is an F-isomorphism $\chi : G/\mathfrak{g} \to Fr\ G$ such that $Fr = \chi \circ \phi$, where ϕ is the homomorphism of 12.2.5.*

Put $A = F[G]$, $B = F[G/\mathfrak{g}]$. Then $F[Fr\ G] = F[A^p]$. We have seen in the proof of 12.2.4 that $F[A^p] \subset B$, the inclusion defining the morphism χ. We have to show that these algebras coincide. We may assume that $F = k$. As before, let

K and L be the quotient fields of A, respectively B. Then $K^p \subset L \subset K$. Since F is assumed to be algebraically closed it is easy to see that K is a finite extension of K^p. Also, all derivations of K annihilate K^p, so are K^p-derivations. It follows from 11.1.15, applied to the extension K/K^p, that K^p is, in fact, the subfield of K annihilated by all derivations of K. Using 4.4.4 we conclude that this field coincides with the subfield annihilated by all left invariant derivations. In the proof of 12.2.1 we have seen that this field coincides with L. So $L = K^p$, hence the isogeny χ is birational. It follows from 5.3.4 and 5.2.8 that χ is an isomorphism, proving 12.2.9. $\square$

12.2.10. Exercises. F is a field of characteristic $p > 0$.
(1) Let $p = 2$ and $G = \mathbf{SL}_2$. Denote by T the diagonal torus of G; it is an F-subgroup.

(a) Show that $\mathrm{Ad}(G)$ acts trivially on the Lie algebra $\mathfrak{t}$.

(b) By 12.2.4 there exists an F-group $G/\mathfrak{t}$. Show that it is F-isomorphic to $\mathbf{PSL}_2$. (*Hint*: use 2.1.5 (3).)
(2) (a) Let $A = F[T_1, \ldots, T_n]/I$ be the quotient of a polynomial algebra by an ideal I. Show that the algebra $\mathbf{F}A$ of 12.2.8 is isomorphic to $F[T_1, \ldots, T_n]/J$, where J is the ideal generated by the polynomials obtained from those in I by raising their coefficients to the pth power.

(b) Let X be an affine F-variety. Define the iterated Frobenius morphism

$$Fr^n : X \to Fr^n X.$$

If $F = \mathbf{F}_{p^n}$, the finite field with p^n elements, then Fr^n is F-isomorphic to X.

(c) Let G be as in 12.1.6 (1). Show that $Fr \; G$ is an F-group that is not F-isomorphic to G.
(3) Let $\rho : G \to GL(V)$ be a non-trivial rational representation over F such that $d\rho = 0$. There is a rational representation $\rho_1 : Fr \; G \to GL(V)$ over F such that $\rho = \rho_1 \circ Fr$.

(b) In the case of (a) there is n and a rational representation $\rho_n : Fr^n G \to GL(V)$ over F such that $\rho = \rho_n \circ Fr^n$ and $d\rho_n \neq 0$.
(4) Let T be a torus defined over F. There is an isomorphism $\alpha : Fr \; T \to T$ such that $\alpha \circ Fr$ is the pth power map of T.

12.3. Galois cohomology

As before, G is an F-group. We denote by Γ the Galois group $\mathrm{Gal}(F_s/F)$.

12.3.1. Let $H^1(F, G)$, respectively $Z^1(F, G)$, be the 1-cohomology set $H^1(\Gamma, G(F_s))$, and the set of cocycles $Z^1(\Gamma, G(F_s))$ (defined in 11.3.2). These are pointed sets, with a special element 1. The pointed set $H^1(F, G)$ is the 1-dimensional Galois cohomology set of G. The 0-dimensional Galois cohomology set is defined by $H^0(F, G) = G(F)$, the group of F-rational points of G. Recall (see 2.3.2 (2)) that a *principal homogeneous space* or *torsor* of G over F is a homogeneous space X of G over

F on which G acts simply transitively. An example is $X = G$, with G-action by left translations. A G-variety X over F is a torsor if and only if the F-morphism $(g, x) \mapsto (g.x, x)$ of $G \times X$ to $X \times X$ is an isomorphism. It is F-isomorphic to G if and only if $X(F) \neq \emptyset$.

Let Σ be the set of F-isomorphism classes of torsors of G over F.

12.3.2. Proposition. *There is a bijection of Σ onto $H^1(F, G)$ such that the class of G corresponds to* 1.

This is an application of the results discussed in 11.3. A torsor of G over F has an F_s-rational point (by 11.2.7), hence is F_s-isomorphic to G. The group of the F_s-automorphisms of the torsor G being isomorphic to $G(F_s)$, the proposition follows from 11.3.3 (or rather a version involving supplementary structures). □

12.3.3. If $\phi : H \to G$ is a homomorphism of F-groups, it is immediate from the definitions that there are induced maps $\phi^i : H^i(F, H) \to H^i(F, G)$ $(i = 0, 1)$. Moreover, ϕ^0 is a group homomorphism and ϕ^1 a map of pointed sets. Of course, ϕ^0 is also a map of pointed sets, the special elements being the neutral ones.

Now assume that H is an F-subgroup of G and let i be the inclusion map $H \to G$. We put $H^0(F, G/H) = (G/H)(F)$; this is a set on which the group $G(F)$ acts. It is also a pointed set (with special element H). Let $\pi : G \to G/H$ be the canonical F-morphism. We have an induced map of pointed sets $\pi^0 : H^0(F, G) \to H^0(F, G/H)$. If $x \in (G/H)(F)$ then by 11.2.14 (ii) (using the separability of π) we conclude that $\pi^{-1}(x)$ is defined over F. It follows that there is $g \in G(F_s)$ with $\pi(g) = x$. If $\gamma \in \Gamma$ then $a(\gamma) = g^{-1}(\gamma.g)$ defines a cocycle of Γ in $H(F_s)$, whence a map of pointed sets $\delta^0 : H^0(F, G/H) \to H^1(F, H)$.

12.3.4. Proposition. *(i) The sequence of maps of pointed sets*

$$1 \to H^0(F, H) \xrightarrow{i^0} H^0(F, G) \xrightarrow{\pi^0} H^0(F, G/H) \xrightarrow{\delta^0} H^1(F, H) \xrightarrow{i^1} H^1(F, G)$$

is exact;
(ii) If H is normal, the sequence of pointed sets obtained from the sequence in (i) by adding on the right

$$\xrightarrow{\pi^1} H^1(F, G/H)$$

is exact;
(iii) If H is a subgroup of the center of G, the sequence of pointed sets obtained from the sequence in (i) by adding on the right

$$\xrightarrow{\pi^1} H^1(F, G/H) \xrightarrow{\delta^1} H^2(F, H)$$

is exact.

We define a sequence of pointed sets to be exact if the fiber of a special point of one of the maps is the image of its predecessor. In (iii) $H^2(F, H)$ is defined to be the second cohomology group $H^2(\Gamma, H(F_s))$ of Γ in the abelian group $H(F_s)$, i.e. the quotient of the group of continous functions $a : \Gamma \times \Gamma \to H(F_s)$ satisfying $(\gamma.a(\delta, \epsilon))a(\gamma, \delta\epsilon) = a(\gamma\delta, \epsilon))a(\gamma, \delta)$, modulo those of the form $a(\gamma, \delta) = b(\gamma)(\gamma.b(\delta))b(\gamma\delta)^{-1}$, see [Jac5, p. 356]. The definition of δ^1 is similar to that of δ^0. The proof of the proposition is straightforward, and is left to the reader.   □

12.3.5. Examples. (1) $H^1(F, \mathbf{GL}_n) = 1$, see 11.3.4 (1). In particular, $H^1(F, \mathbf{G}_m) = 1$.

Let $\mathbf{PGL}_n$ be the quotient of $\mathbf{GL}_n$ by the group of scalar multiples of the identity, which is isomorphic to $\mathbf{G}_m$. Using 12.3.4 (iii) we obtain a map

$$\delta^1 : H^1(F, \mathbf{PGL}_n) \to H^2(F, \mathbf{G}_m).$$

Now $\mathbf{PGL}_n(F_s)$ is the automorphism group of the matrix algebra $\mathbf{M}_n(F_s)$ (by the Skolem-Noether theorem, see [Jac5, p. 222]). Using (a version of) 11.3.3 we conclude that $H^1(F, \mathbf{PGL}_n)$ classifies the the isomorphism classes of associative algebras over F that are F_s-isomorphic to the algebra of $n \times n$-matrices. It is well-known that these algebras are the *central simple algebras over* F, i.e. the (finite dimensional) simple associative algebras with center F (for the theory of central simple algebras see [loc. cit., Ch. 4]). The group $H^2(F, \mathbf{G}_m)$ is the *Brauer group* of F. The map δ^1 is also well-known in algebra theory (see [loc. cit., p. 477]).

(2) $H^1(F, \mathbf{G}_a) = 1$, as follows from 11.3.4 (5).

(3) A connected solvable F-group G is called F-*split* if there exists a sequence $\{e\} = G_0 \subset G_1 \subset \cdots \subset G_{n-1} \subset G_n = G$ of closed, connected, normal F-subgroups such that the quotients G_i/G_{i-1} are F-isomorphic to either $\mathbf{G}_a$ or $\mathbf{G}_m$ (the quotients exist by 12.2.34). Using (1), (2) and 12.3.4 (ii) we conclude by induction on the dimension that for such a group G we have $H^1(F, G) = 1$. A particular case is that G is an F-split torus, i.e. a torus over F that is F-isomorphic to a product $(\mathbf{G}_m)^n$.

(4) $H^1(F, \mathbf{Sp}_{2n}) = 1$, see 11.3.4 (2).

(5) (char $F \neq 2$) $H^1(F, \mathbf{O}_n)$ classifies the F-isomorphism classes of n-dimensional non-degenerate symmetric bilinear forms over F, see 11.3.4 (3).

(6) If F is finite then Lang's theorem (4.4.17) implies that $H^1(F, G) = 1$ if G is connected.

12.3.6. Exercises. (1) (char $F \neq 2$) (a) If $A = \{\pm 1\}$, a closed F-subgroup of $\mathbf{G}_m$ then $H^1(F, A) \simeq F^*/(F^*)^2$. (*Hint*: use Galois theory).

(b) As an application of 12.3.4 (i) show that the homomorphism $\mathbf{SL}_2(F) \to \mathbf{PSL}_2(F)$, deduced from the homomorphism of 2.1.5 (3), need not be surjective.

(2) $H^1(F, \mathbf{SL}_n) = 1$. (*Hint*: use the determinant map $\mathbf{GL}_n \to \mathbf{G}_m$).

12.3.7. Forms. If E is an extension of F contained in k, an *E-form of G* is an F-group G' that is E-isomorphic to G, see 11.3.1. By a version of 11.3.3 we have a bijection of the set $\Phi(F, G)$ of F-isomorphism classes of F_s-forms of G onto $H^1(\Gamma, \mathrm{Aut}_{F_s}(G))$, where now $\mathrm{Aut}_{F_s}(G)$ is the group of F_s-automorphisms of the algebraic group G (on which Γ acts). We have a homomorphism $G(F_s) \to \mathrm{Aut}_{F_s}(G)$, sending an element to the inner automorphism that it defines, whence a map Int of $H^1(F, G)$ to $\Phi(F, G)$. An F-form of G whose isomorphism class is in the image of Int is said to be an *inner form*. A form that is not inner is *outer*.

The twisting procedure of the proof of 11.3.3 gives the following explicit description of inner forms. Let $z \in Z^1(F, G)$ be a cocycle in $c \in H^1(F, G)$. There is an F-form G_z of G such that $G_z(F_s) = G(F_s)$, the Γ-action on $G_z(F_s)$ being given by

$$\gamma * g = z(\gamma)(\gamma.g)z(\gamma)^{-1} \ (\gamma \in \Gamma, \ g \in G(F_s)),$$

and the F-isomorphism class of G_z lies in $\mathrm{Int}(c)$.

12.3.8. Examples. (1) $G = \mathbf{G}_a$. Now $\mathrm{Aut}_{F_s}(G)$ is isomorphic to F_s^* (for example, as a consequence of 5.1.8 (1)). It follows from 12.3.5 (1) that all F_s-forms of $\mathbf{G}_a$ are isomorphic. It also follows that, if F is perfect, an F-group which is isomorphic to $\mathbf{G}_a$ is F-isomorphic to $\mathbf{G}_a$. If F is non-perfect this is no longer true. In that case take $a \in F - F^p$ ($p = \mathrm{char}\ F$) and let $G = \{(x, y) \in k^2 \mid x^p = x + ay^p\}$. Then G is an F-subgroup of $\mathbf{G}_a^2$, isomorphic to $\mathbf{G}_a$ over $F(a^{\frac{1}{p}})$. The isomorphisms $\mathbf{G}_a \to G$ are given by $t \mapsto (c^p t^p, a^{-\frac{1}{p}}(c^p t^p - ct))$ with $c \in k^*$. For $c = 1$ this morphism is defined over $F(a^{\frac{1}{p}})$, but for no c is it defined over F.

(2) We have $\mathrm{Aut}_{F_s}(\mathbf{G}_m) = \mathbf{GL}_1(\mathbf{Z}) = \{\pm 1\}$, with trivial Γ-action. It follows that the F_s-forms of $\mathbf{G}_m$ are classified by $H^1(\Gamma, \{\pm 1\})$. For simplicity, assume that char $\neq 2$. Using 12.3.6 (1) we see that there is a bijection of this set onto the set of quadratic extensions of F. If $F(a^{\frac{1}{2}})$ is one, a corresponding F-form of $\mathbf{G}_m$ is

$$G = \left\{ \begin{pmatrix} x & y \\ ay & x \end{pmatrix} \mid x^2 - ay^2 = 1 \right\}.$$

(3) We continue with the example of 12.3.5 (1). The group $\mathbf{PGL}_n$ is the group of inner automorphisms of the algebraic group $\mathbf{GL}_n$. If $z \in Z^1(F, \mathbf{PGL}_n)$ we have an inner F-form $(\mathbf{GL}_n)_z$ of $\mathbf{GL}_n$. Let A be a central simple algebra over F whose isomorphism class corresponds to the cohomology class of z by 12.3.5 (1). There is a division algebra D with center F such that A is isomorphic to a matrix algebra $M_m(D)$ over D (see [Jac5, p. 203]). Then $(\mathbf{GL}_n)_z(F)$ is isomorphic to the group $GL_m(D)$ of invertible $m \times m$-matrices over D. We shall denote an F-group like $(\mathbf{GL}_n)_z$ by $\mathbf{GL}_A$ or $\mathbf{GL}_{m,D}$. We can also define an F-group $(\mathbf{SL}_n)_z = \mathbf{SL}_A = \mathbf{SL}_{m,D}$.

(4) Still let $G = \mathbf{GL}_n$. The automorphism $\sigma : x \mapsto (^t x)^{-1}$ of $\mathbf{GL}_n$ generates a subgroup of order two of $\mathrm{Aut}_{F_s}(\mathbf{GL}_n)$, whence a map of $H^1(\Gamma, \{\pm 1\}) \to \Phi(F, G)$. In more concrete terms we associate in this way to a separable quadratic extension E/F an outer F-form of $\mathbf{GL}_n$, which we denote by $\mathbf{U}_{n,E}$. The notation is explained by the

fact that $\mathbf{U}_{n,E}(F) = U_n(E) = \{x \in \mathbf{GL}_n(F) \mid x(^t\bar{x}) = 1\}$ is a unitary group (the bar denotes the non-trivial automorphism of E/F). Similarly, one defines a group $\mathbf{SU}_{n,E}$.

12.3.9. Exercises. (1) Notations of 12.3.7. For $z \in Z^1(F, G)$ define a bijection $H^1(F, G_z) \to H^1(F, G)$.

(2) In the case of 12.3.4 (i) let $z \in Z^1(F, H)$ have image c in $H^1(F, H)$. There is a bijection of the fiber $(i^1)^{-1}i^1c$ onto the set of orbits of $G_z(F)$ in $(G_z/H_z)(F)$ (where G_z has the obvious meaning).

(3) (a) Prove the facts stated in 12.3.8 (3).

(b) With the notations of 12.3.8 (3) define an F-homomorphism $\nu : \mathbf{GL}_{m,D} \to \mathbf{G}_m$, with kernel $\mathbf{SL}_{m,D}$.

(c) Show that $H^1(\mathbf{SL}_{m,D}) \simeq F^*/\nu(\mathbf{GL}_m(D))$.

(4) (a) Prove the facts stated in 12.3.8 (4).

(b) Let E/F be a separable quadratic extension and let h be a non-singular hermitian $n \times n$-matrix over F relative to the non-trivial automorphism of E/F (see [Jac4, p. 381-384]). Define an outer F-form G of $\mathbf{GL}_n$ such that $G(F)$ is isomorphic to the unitary group defined by E.

12.4. Restriction of the ground field

12.4.1. Let E be a finite extension of F contained in k and let G be a linear algebraic group over E. Since G is smooth, there exists by 11.4.16 (i) and 11.4.20 (1) an affine F-variety $\Pi G = \Pi_{E/F} G$ with the properties of 11.4.16 (i). If $A = E[G]$ then $F[\Pi G]$ is the algebra RA studied in 11.4. Notice that by the formula of 11.4.6 (with $B = F$) there is a bijection $G(E) \to (\Pi G)(F)$.

12.4.2. Proposition. *ΠG is a linear algebraic group over F. There exists a surjective homomorphism of E-groups $\pi : \Pi G \to G$ with the following universal property: if H is an F-group and $\phi : H \to G$ a homomorphism of E-groups, there is a unique homomorphism of F-groups $\psi : H \to \Pi G$ such that $\phi = \pi \circ \psi$.*

Let $\Delta : A \to A \otimes_F A$ and $\iota : A \to A$ be comultiplication and antipode (2.1.2). Since R is a functor (11.4.6) we have morphisms $R(\Delta) : RA \to RA \otimes_F RA$ and $R(\iota) : RA \to RA$. Also, the identity element of G defines an element $e' \in \Pi G(F)$. The properties of 2.1.2 expressing the group axioms, and the fact that the morphism π of 11.4.16 (i) is a group homomorphism now are consequences of functoriality. If H is as stated, there is by 11.4.16 (i) a morphism ψ with the asserted properties. That it is a group homomorphism follows by a formal argument. Details can be left to the reader. $\square$

12.4.3. Corollary. Ker π *contains no non-trivial closed, normal F-subgroups of ΠG.*

Let N be a closed, normal F-subgroup of ΠG contained in Ker π and let ψ : $\Pi G \to (\Pi G)/N$ be the canonical morphism. Denote by π' the E-homomorphism induced by π. The pair $((\Pi G)/N, \pi')$ has the universal property of $(\Pi G, \pi)$. Hence there is an F-homomorphism $\psi' : \Pi G/N \to \Pi G$ such that $\psi' \circ \psi = \mathrm{id}$. This implies that ψ is injective, i.e. that N is trivial.   □

12.4.4. Remarks. (1) There is a somewhat more concrete description of the group ΠG. From 11.4.6 we conclude that there is a bijection of $\Pi G = (\Pi G)(k)$ onto the set of E-homomorphisms $A \to E \otimes_F k$. The latter set is a group, the group multiplication being induced by Δ and the inversion by ι. It is the group $G(E \otimes_F k)$ of $E \otimes_F k$-valued points of G. We have a group isomorphism $\alpha : \Pi G \to G(E \otimes_F k)$. The product homomorphism $E \otimes_F k \to k$ induces a homomorphism $\beta : G(E \otimes_F k) \to G$ and we have $\pi = \beta \circ \alpha$. These homomorphisms are functorial. It follows that if G is a closed E-subgroup of $\mathbf{GL}_n$ the group ΠG is isomorphic to a subgroup of $\mathbf{GL}_n(E \otimes_F k)$.
(2) If G itself is an F-group, the universal property of 12.4.2 (i) shows that there is an F-homomorphism $\sigma : G \to \Pi G$ such that $\pi \circ \sigma = \mathrm{id}$.

12.4.5. If E/F is separable we can describe ΠG as follows, using 11.4.22. Put $\Gamma = \mathrm{Gal}(F_s/F)$, $\Delta = \mathrm{Gal}(F_s/E)$. Then Γ/Δ is the set Σ of F-isomorphisms of E into F_s. It follows from 11.4.22 that (ΠG) is isomorphic to the product G^Σ. Moreover, $\Pi(G)(F_s)$ is isomorphic to $Ind_\Delta^\Gamma G(E_s)$, as a group with Γ-action.

Now assume that E/F is purely inseparable, of characteristic p.

12.4.6. Proposition. *(i) Ker π is a connected unipotent subgroup of ΠG;*
(ii) If G is reductive then Ker π is the unipotent radical of ΠG. It is not defined over F.

It follows from 11.4.18 that Ker π is connected. From 12.4.4 (1) we see that we can view ΠG as a subgroup of $\mathbf{GL}_n(E \otimes_F k)$. Since E/F is purely inseparable, we have $E \otimes k \simeq k \oplus I$, where I is a nilpotent ideal. The homomorphism π is induced by $E \otimes k \to k = (E \otimes k)/I$. It follows that Ker π is isomorphic to a subgroup of $\mathbf{GL}_n(E \otimes k)$ whose matrices are $\equiv 1 \pmod{I}$. Such a group is a p-group. Hence Ker π is a p-group and is unipotent. We have proved (i). The first part of (ii) is now clear and the second part follows from 12.4.3.   □

12.4.7. Exercises. Notations are as in 12.4.2.
(1) If E/F is separable there is a bijection $H^1(E, G) \to H^1(F, \Pi G)$.
(2) (char $F = p > 0$). The Lie algebra of ΠG does not contain any non-trivial p-subalgebra defined over F and stable under $\mathrm{Ad}(\Pi G)$. (*Hint*: use 12.2.6.)
(3) If G is connected (respectively, commutative) then so is ΠG.
(4) Let E/F be a quadratic extension. By 12.4.4 (2) we have an injective homomorphism of F-groups $\sigma : \mathbf{G}_m \to \Pi(\mathbf{G}_m)$. By the previous exercise the latter group is commutative.

(a) Let E/F be separable, with char $F \neq 2$. Show that $(\Pi \mathbf{G}_m)/\mathrm{Im}\ \sigma$ is as in 12.3.8 (2).

(b) If E/F is inseparable, the quotient morphism $\Pi \mathbf{G}_m \to (\Pi \mathbf{G}_m)/\mathrm{Im}\ \sigma$ has no sections over F on $(\Pi \mathbf{G}_m)/\mathrm{Im}\ \sigma$ (see 5.5.7).

(c) If E/F is inseparable and char $F = 2$, then $\Pi \mathbf{G}_m$ is F-isomorphic to a group as in 12.1.6 (1).

Notes

12.1 and 12.2 contain standard results. Most of them are also discussed in [Bo3]. The example 12.1.8 (1) gives a general construction of classical semi-simple groups. The construction goes back to Weil [We2].

The construction in positive characteristics of the quotient of an algebraic group G by a subalgebra of its Lie algebra $\mathfrak{g}$ (12.2.4) is due to Serre [Se1], for arbitrary algebraic groups (not necessarily linear). It can be viewed as the construction of the quotient of G by a certain subgroup scheme.

Thus 12.2.9 can be viewed as saying that $\mathfrak{g}$ is the kernel of Fr, in the sense of group-schemes (see 2.1.6 (a)). More generally, the iterated Frobenius homomorphisms Fr^n of 12.2.10 (2) have kernels that are subgroup schemes of G. More about these higher Frobenius kernels can be found in [Jan1]. 12.2.4 provides a surrogate for the use of group scheme kernels.

The discussion of Galois cohomology in 12.3 has been kept brief. It contains results to be needed later, and some illustrative examples. The reader is referred to [Se2] for more details.

12.4 also deals with 'well-known' results. Restriction of the ground field in the separable case is treated in [We4, Ch. 1]. The inseparable case is discussed briefly in [Oe, App. 3]. The general example 12.2.6 (ii) of a unipotent radical not defined over the ground field of the group that I learned from T. Tamagawa many years ago.

Chapter 13

F-tori

In this chapter we first discuss matters concerning diagonalizable groups and tori involving a ground field. Then we discuss tori in arbitrary F-groups, one of the main results (13.3.6) being the existence of maximal tori in an F-group defined over F. The last section deals with the groups $P(\lambda)$ of 3.2.15. F, F_s, $\bar{F}$ are as in the preceding chapters. The characteristic of F is p. The Galois group $\mathrm{Gal}(F_s/F)$ is denoted by Γ.

13.1. Diagonalizable groups over F

Let T be an F-torus, i.e. an F-group that is a torus. Recall (3.2.12) that T is F-*split* if it is F-isomorphic to a product $(\mathbf{G}_m)^n$.

13.1.1. Proposition. (i) *If E is a purely inseparable finite extension of F such that T is E-split, then T is F-split;*
(ii) *T is F_s-split.*

T is split over $\bar{F}$ (as a consequence of 2.4.2 (ii), for example), which proves the assertions if $p = 0$. So we may assume that $p > 0$. To prove (i) it suffices to deal with the case that $E^p \subset F$. Let X be a character group of T. By 3.2.12 (i) all χ are defined over E. We have to show that they are defined over F.

Let $\mathcal{J}$ be the p-Lie algebra of F-derivations of E. For $D \in \mathcal{J}$ let $c(D) = c_A(D)$ be the connection on $A = E[T]$ that it defines (11.1.9). This is a derivation of A, extending the derivation D of E. Denote by $\Delta : A \to A \otimes_E A$ comultiplication, then $\Delta \circ c_A(D) = c_{A \otimes A}(D) \circ \Delta$. Let $\chi \in X$, then $\Delta\chi = \chi \otimes \chi$ (3.2.5). Putting $f = \chi^{-1}(c(D)\chi)$ we have $\Delta(f) = f \otimes 1 + 1 \otimes f$.
Write $f = \sum_{\psi \in X} c_\psi \psi$. Then

$$\Delta f = \sum c_\psi (\psi \otimes 1 + 1 \otimes \psi) = \sum c_\psi \psi \otimes \psi.$$

This can only be if $c_\psi = 0$ for $\psi \neq 1$. It then readily follows that $f = 0$. Hence $c(D)\chi = 0$ for all $D \in \mathcal{J}$, whence $\chi \in F[T]$ (see 11.1.9). This proves (i), and (ii) is a direct consequence of (i). $\qquad\square$

Let G be a diagonalizable F-group (3.2.1) and denote by $X = X^*(G)$ its character group.

13.1.2. Corollary. *All elements of X are defined over F_s.*

The identity component G^0 is an F-torus (12.1.1). Let C be a coset of G^0 in G. By 11.2.6, C is defined over F_s and by 11.2.5 there is $g \in G(F_s)$ such that $C = gG^0$.

Let n be the order of G/G^0, then n is prime to p if $p > 0$ (3.2.7 (i)). Let $\chi \in X$. By part (ii) of the proposition we have $\chi(g^n) \in F_s$, hence $\chi(g) \in F_s$. It follows that the restriction of χ to C is defined over F_s.   □

13.1.3. A Γ-*module* is an abelian group of finite type M, on which Γ acts continuously, relative to the discrete topology on M. So there is a closed subgroup of finite index of Γ that acts trivially on M.

It follows from 13.1.2 that, if G is a diagonalizable F-group, its character group $X = X^*(G)$ is a Γ-module, the Γ-action being defined by the obvious Γ-action on $F_s[G]$. We have

$$(\gamma . \chi)(g) = \gamma(\chi(\gamma^{-1}g)),$$

for $\gamma \in \Gamma$, $\chi \in X$, $g \in G(F_s)$. Conversely, let M be a Γ-module. If $p > 0$ assume that M has no p-torsion. Define a continuous Γ-action on the group algebra $F_s[M]$ by

$$\gamma . (\sum a(m)e(m)) = \sum \gamma(a(m))e(\gamma.m),$$

notations being as in 3.2.5. By 11.1.6 the fixed point set for this Γ-action is an F-structure $F[M]$ on the algebra $F_s[M]$.

The algebra homomorphisms Δ, ι and e of 3.2.5 are defined over F_s and commute with the Γ-action on $F_s[M]$. Hence they define similar homomorphisms of $F[M]$, denoted by the same symbols. We then have the following result, which generalizes 3.2.6.

13.1.4. Proposition. *(i) $F[M]$ is an affine F-algebra and there is a diagonalizable F-group $\mathcal{G}(M)$ with $F[\mathcal{G}(M)] = F[M]$, such that Δ, ι and e are comultiplication, antipode and identity element of $\mathcal{G}(M)$;*
(ii) There is a canonical isomorphism of Γ-modules $M \simeq X^*(\mathcal{G}(M))$;
(iii) If G is a diagonalizable F-group, there is a canonical isomorphism of F-groups $\mathcal{G}(X^*(G)) \simeq G$.

The proof of 13.1.4 is similar to that of 3.2.6 and is left to the reader.   □

13.1.5. Exercises. (1) Make diagonalizable F-groups and Γ-modules without p-torsion into categories and describe an anti-equivalence between these categories (where now p is the characteristic exponent).

In the next two exercises G is a diagonalizable F-group with character group X. We view X as a left module over the group ring $\mathbf{Z}[\Gamma]$, via the Γ-action of 13.1.3. Similarly, F_s^* is a left $\mathbf{Z}[\Gamma]$-module.
(2) (a) Define an isomorphism $G(F_s) \simeq \mathrm{Hom}(X, F_s^*)$.

   (b) $G(F) \simeq \mathrm{Hom}_{\mathbf{Z}[\Gamma]}(X, F_s^*)$.

   (c) Assume that G is a torus. Make the group $X_*(G)$ of cocharacters (3.2.1) into a right $\mathbf{Z}[\Gamma]$-module. Show that there is an isomorphism $G(F) \simeq X_*(G) \otimes_{\mathbf{Z}[\Gamma]} F_s^*$.

(3) Let F be a finite field with q elements. Then Γ is generated topologically by the Frobenius automorphism ϕ defined by $\phi a = a^q$ $(a \in F_s)$.

(a) $G(F) \simeq \operatorname{Hom}(X/(\phi - q)X, F_s^*)$.

(b) $X/(\phi - q)X$ is isomorphic (non-canonically) to the character group $\operatorname{Hom}(G(F), \mathbf{C}^*)$ of the finite abelian group $G(F)$.

(4) Let $E \subset F_s$ be a finite separable extension of F. Put $\Delta = \operatorname{Gal}(F_s/E)$; this is a closed subgroup of Γ of finite index. Let $M = M_{E/F}$ be the free $\mathbf{Z}$-module with a basis indexed by the cosets $\gamma \Delta$, Γ permuting the basis elements according to left translations. Show that $\mathcal{G}(M)$ is an F-torus, which is isomorphic to the group $\Pi_{E/F}(\mathbf{G}_m)$. (*Hint*: use 12.4.5.)

13.2. *F*-tori

13.2.1. In this section T is an F-torus. We denote by X its character group and by Y its group of cocharacters. We have a pairing $\langle \ , \ \rangle$ between these free modules (3.2.11 (i)). Both X and Y are Γ-modules and the pairing is Γ-stable. The action of Γ on both X and Y factors through a finite quotient A of Γ. Put $V = \mathbf{Q} \otimes X$. This is a vector space over $\mathbf{Q}$, and we have representations of Γ and A in V. Then X is a lattice in V, i.e. a subgroup of V generated by a basis of V.

We say that T is *anisotropic* if it does not contain non-trivial split subtori. Otherwise T is *isotropic*.

13.2.2. Proposition. *(i) T is F-split if and only if Γ acts trivially on X and Y; (ii) T is anisotropic if only if Γ fixes no non-zero element of X and Y; (iii) T is anisotropic if and only any morphism of F-varieties $\mathbf{G}_m \to T$ is constant.*

(i) is a direct consequence of 3.2.12 (i). T is isotropic if and only if there is a non-trivial F-homomorphism $\mathbf{G}_m \to T$, i.e. if Γ fixes some non-zero element of Y. This is the case if and only if the endomorphism $\sum_{\delta \in A} \delta$ of Y is non-zero. An equivalent condition is that the analogous endomorphism of X is non-zero, i.e. that Γ fixes a non-zero element of X. (ii) follows from these facts.

An F-morphism $\mathbf{G}_m \to T$ is given by an F-homomorphism

$$\phi : k[T] \to k[U, U^{-1}]$$

defined over F, (U being an indeterminate). If $\gamma \in X$ let $e(\gamma)$ be the corresponding element of $k[T]$. Then $\phi(e(\gamma))$ is an invertible element of $k[U, U^{-1}]$. These invertible elements are of the form cU^m with $c \in k^*$, $m \in \mathbf{Z}$. We conclude that there is $c_\gamma \in k^*$ and $\lambda \in Y$ with $\phi(e(\gamma)) = c_\gamma U^{\langle \chi, \lambda \rangle}$. Since ϕ is defined over F it follows from 13.1.1 (ii) that the c_γ lie in F_s. Because ϕ commutes with the Γ-actions on $F_s[T]$ and $F_s[U, U^{-1}]$ (11.1.5), we conclude that λ must be fixed by Γ. If T is anisotropic, this can only be if $\lambda = 0$. This proves the 'only if'-part of (iii). The other part of (iii) is obvious. $\qquad \square$

13.2.3. Proposition. *Let T_1 be an F-subtorus of T. There exists an F-subtorus T_2 of T such that $T = T_1.T_2$ and that $T_1 \cap T_2$ is finite.*

Put $X_1 = X^*(T_1)$. We have a surjective homomorphism of Γ-modules $X \to X_1$ (see 3.2.10 (2)); let X_0 be its kernel. The subspace of V generated by X_0 is A-stable and has an A-stable complement V' (since representations of a finite group over $\mathbf{Q}$ are fully reducible by Maschke's theorem, see [Jac5, p. 253]). Let T_2 be the subtorus generated by the images $\lambda(\mathbf{G}_m)$, where λ runs through the elements of Y orthogonal to V'. Then T_2 is a subtorus of T defined over F_s (because of 13.1.1 (ii)). Since V' is Γ-stable, 11.2.8 (i) shows T_2 is defined over F. If $\lambda \in Y$ has image in the intersection $T_1 \cap T_2$ then it is orthogonal to both X_0 and V', hence must be zero. It follows that $T_1 \cap T_2$ is finite. We have dim $T_2 = $ rank $X - $ rank $X_1 = $ dim $T - $ dim T_1. From 5.3.3 (i) we conclude that the product map $T_1 \times T_2 \to T$ is surjective. $\square$

13.2.4. Proposition. *(i) T contains a unique maximal F-split subtorus $T_s = T_{s,F}$ and a unique maximal anisotropic F-subtorus $T_a = T_{a,F}$;*
(ii) $T = T_s.T_a$ and $T_s \cap T_a$ is finite.

If T_1 is an F-split subtorus, then $X_*(T_1)$ must be contained in the fixed point set Y^Γ of Γ in Y. Hence there is a maximal F-split torus T_s, with $X_*(T_s) = Y^\Gamma$. It follows from 13.2.2 (ii) that an F-subtorus T_2 is anisotropic if and only if the sublattice $X_*(T_2)$ of Y is orthogonal to the fixed point set X^Γ. Consequently, there is a unique maximal anisotropic subtorus T_a, such that $X_*(T_a)$ is the annihilator of X^Γ. This proves (i). The assertions of (ii) are proved as the analogous assertions of the previous proposition. $\square$

13.2.5. Exercises.
(1) A subtorus of an F-split torus is defined over F and F-split.
(2) An F-torus is *irreducible* if it does not contain proper non-trivial F-subtori. Show that there exists irreducible F-subtori $T_1, ..., T_h$ of T such that $T = T_1...T_h$ and that all intersections $T_i \cap (T_1...\hat{T}_i...T_h)$ are finite.
(3) (a) For each $n > 0$ there exists, up to $\mathbf{R}$-isomorphism, a unique n-dimensional anisotropic $\mathbf{R}$-torus T_n.
   (b) $T_n(\mathbf{R}) \simeq \mathbf{T}^n$, where $\mathbf{T} = \{z \in \mathbf{C} \mid |z| = 1\}$.
(4) Notations of 13.2.4. If E is a purely inseparable extension of F then $T_{s,E} = T_{s,F}$, $T_{a,E} = T_{a,F}$.

An irreducible F-variety X is *rational (unirational)* over F if its F-quotient field $F(X)$ (see 1.8.1) is a purely transcendental extension of F (respectively, a subfield of a purely transcendental extension of F). A rational variety is unirational, but the converse is not true (this is a subtle matter which we will not go into). For us the importance of unirationality is that it implies the density result of the following lemma.

13.2.6. Lemma. *Let X be an irreducible F-variety.*
(i) X is rational (unirational) if and only if there is an F-isomorphism (respectively, a surjective F-morphism) $U \to U'$, where U is an open F-subvariety of some $\mathbf{A}^n$ and U' an open F-subvariety of X;
(ii) If X is unirational and F is infinite then $X(F)$ is dense in X.

The proof of (i) is like that of 5.1.2 and can be omitted. (ii) follows from (i), observing that if F is infinite $\mathbf{A}^n(F)$ is dense in $\mathbf{A}^n$. □

13.2.7. Proposition. *Let T be an F-torus.*
(i) T is unirational over F;
(ii) If F is infinite then $T(F)$ is dense in T.

(ii) follows from (i) and 13.2.6 (ii). To prove (i) it suffices by 13.2.6 (i) to construct an F-torus T' that is rational over F, together with a surjective homomorphism of F-groups $T' \to T$.

We view the group of cocharacters Y as a right module over the group ring $\mathbf{Z}[A]$ of the finite quotient A of Γ. Then Y is a quotient of a free $\mathbf{Z}[A]$-module Y'. The dual X' of Y' is a free left $\mathbf{Z}[A]$-module and so is a Γ-module, of which X is a Γ-submodule.

We have $A = \Gamma/\Delta$, where Δ is a closed normal subgroup of Γ. Let E be the fixed point set of Δ in F_s; it is a finite Galois extension of F with group A. With the notations of 13.1.5 (4), the Γ-module X' is isomorphic to a product $M_{E/F}^n$. By 13.1.5 (4) we have $T' = \mathcal{G}(X') \simeq \Pi_{E/F}(\mathbf{G}_m)^n$. Since X is a Γ-submodule of X' we have an F-homomorphism $T' \to T$ that is dominant, hence surjective. We show that T' is rational. It suffices to show that $\Pi_{E/F}(\mathbf{G}_m)$ is rational. This amounts to showing that, with the notations of 11.4, the quotient field of $R_{E/F}(E[T, T^{-1}])$ is a purely transcendental extension of F. This is a consequence of 11.4.7 (5), (6). □

13.2.8. Exercises. (1) Let G be an F-group. The subgroup H generated by all F-subtori of G is a connected, unirational F-group. If F is infinite then $H(F)$ is dense in H.
(2) A one dimensional F-torus is rational (see 12.3.8 (2)).
(3) If T is F-split then $H^1(F, T) = 1$ (notations of 12.3.1).

13.3. Tori in *F*-groups

In this section G is a connected F-group. If S and T are subtori of G defined over F, we write $N_G(S, T) = N(S, T) = \{g \in G \mid gSg^{-1} \subset T\}$; this is the *transporter* of S into T. If $S = T$ we obtain the normalizer $N(S)$.

13.3.1. Proposition. *(i) If $N(S, T)$ is non-empty it is defined over F;*
(ii) $N(S)$ and the centralizer $Z(S)$ are defined over F.

To prove (i) it suffices to show that $N(S, T)$ (assumed to be non-empty) is defined over F_s. For then an application of 11.2.8 (i) (and 11.2.7) will show that it is defined over F. First notice that $N(S, T)$ is defined over $\bar{F}$ (work over the algebraically closed field $\bar{F}$). This settles the case $p = 0$, so assume that $p > 0$.

Let C be a component of $N(S, T)$. The rigidity theorem 3.2.8 implies that for fixed $s \in S$ the map $x \mapsto xsx^{-1}$ of C to T is constant, hence C consists of a coset of the centralizer $Z(S)$. Let $U = \{s \in S \mid Z(s) = Z(S)\}$. It follows from the proof of 6.4.3 that this is a non-empty open subset of S. The set U is closed under taking p^{th} powers, as $Z(s) = Z(s^p)$ (it suffices to check this for diagonalizable elements $s \in \mathbf{GL}_n$, which is easy).

By 11.2.7 there exists $s \in U \cap S(F_s)$. Let $c \in C \cap G(\bar{F})$. Then $t = csc^{-1} \in T(\bar{F})$, and C is a component of the set $N(s, t)$ of 12.1.4 (i). There is a p-power q such that $t^q \in T(F_s)$. Replacing s by s^q we see that C is a component of $N(s^q, t^q)$. An application of 12.1.4 (i) now proves what we want. (ii) follows from (i) and 11.2.6, observing that by 6.4.7 (i) $Z(S)$ is the identity component of $N(S)$. $\square$

The main result of the present section is the existence of maximal tori of G defined over F. In the proof we need auxiliary results about the Jordan decomposition in the Lie algebra $\mathfrak{g}$ (4.4.20), to be discussed now. If $X \in \mathfrak{g}$ we denote its centralizer in G by $Z(X)$. The centralizer in $\mathfrak{g}$ is

$$\mathfrak{z}(X) = \{Y \in \mathfrak{g} \mid [X, Y] = 0\}.$$

13.3.2. Lemma. *Let $X \in \mathfrak{g}$.*
(i) If X is semi-simple then $\mathfrak{z}(X)$ is the Lie algebra of $Z(X)$;
(ii) X is semi-simple if and only if it lies in the Lie algebra of a maximal torus;
(iii) $(p > 0)$ X is nilpotent if and only if there is $h \geq 0$ such that $X^{[p^h]} = 0$;
(iv) $(p > 0)$ There is $h > 0$ such that $\mathfrak{g}^{[p^h]}$ is the set of semi-simple elements of $\mathfrak{g}$.

In (iii) and (iv) the map $X \mapsto X^{[p^h]}$ is the hth iterate of the p-power map of $\mathfrak{g}$ (4.4.3).

The proof of (i) is along the lines of the proof of 5.4.4 (ii) and is left to the reader. If X is semi-simple then by (i) X lies in the Lie algebra of $Z(X)^0$. By 4.4.21 (1) the latter group cannot be unipotent, so it contains a non-trivial maximal torus T. The centralizer C of T in $Z(X)^0$ is a Cartan subgroup and by 5.4.7 the Lie algebra $\mathfrak{c}$ contains X. It follows from 6.4.2 (i), 6.3.2 (ii) and 4.4.21 (1) that the semi-simple elements of $\mathfrak{c}$ are those of $\mathfrak{t}$, so $X \in \mathfrak{t}$. The torus T lies in a maximal torus of G, which must centralize X, hence lies in $Z(X)^0$. Then T is a maximal torus of G, and (ii) follows.

Using 2.3.7 (i) and 4.4.9, the proof of (iii) is reduced to the case that $G = \mathbf{GL}_n$, in which case the result follows from 4.4.10 (3). Similarly, one sees that there is h such that $\mathfrak{g}^{[p^h]}$ is contained in the set of semi-simple elements of $\mathfrak{g}$. That it coincides with that set follows from (ii), observing that, if G is a torus, we have $\mathfrak{g}^{[p]} = \mathfrak{g}$ (use 4.4.10 (2)).   □

Let $X \in \mathfrak{g}$ be semi-simple and let T be a maximal torus with $X \in \mathfrak{t}$. Denote by P the set of non-zero weights of T in $\mathfrak{g}$, acting via the adjoint action. We say that X is *central* if $Z(X) = G$.

13.3.3. Lemma. *(i) X is central if and if only if it lies in the center of $\mathfrak{g}$ or equivalently, if and only if it is annihilated by all differentials $d\alpha$, $\alpha \in P$;*
(ii) $(p = 0)$ X is central if and only if it lies in the Lie algebra of the center of G.

The first assertion of (i) follows from 13.3.2 (i). Let $\alpha \in P$. The differential $d\alpha$ is a linear function on $\mathfrak{t}$. If $Y \in \mathfrak{g}$ is a weight vector for α, then $\mathrm{Ad}(t)Y = \alpha(t)Y$ $(t \in T)$, which implies that we have $[X, Y] = d\alpha(X)Y$ (use 4.4.15 (1), (3)). Since $\mathfrak{g}$ is the direct sum of the weight spaces of T, for the weights in P and the zero weight, the remaining assertion of (i) follows.

The intersection C of the kernels of all α in P is a subgroup of T lying in the center of G (by 5.4.7). Consider the homomorphism of tori $T \to (\mathbf{G}_m)^P$ sending t to $(\alpha(t))_{\alpha \in P}$. If $p = 0$ this is a separable morphism, and we can conclude that the Lie algebra of its kernel C is the intersection of the kernels of the $d\alpha$, $\alpha \in P$. This implies (ii).   □

13.3.4. Lemma. *(i) Assume that all semi-simple elements of $\mathfrak{g}$ are central. If G is not nilpotent then $p > 0$, all maximal tori have the same Lie algebra and all weights of a maximal torus T of G in $\mathfrak{g}$ are divisible by p in the character group $X^*(T)$;*
(ii) If $\mathfrak{g}$ contains non-central semi-simple elements, there is a non-empty open subset U of $\mathfrak{g}$ such that the semi-simple part of an element of U is non-central;
(iii) If $\mathfrak{g}$ contains non-central semi-simple elements, and if F is infinite then $\mathfrak{g}(F)$ contains such elements. The centralizers in $\mathfrak{g}$ of these elements of $\mathfrak{g}(F)$ span $\mathfrak{g}$.

Let T be a maximal torus and let P be as before. Then $P \neq \emptyset$ if and only if G is not nilpotent (by 5.4.7 and 6.3.2). If in that case all semi-simple elements of $\mathfrak{g}$ are central, we have by the preceding lemma that $d\alpha = 0$ for all $\alpha \in P$. This means that $p > 0$ and that $P \subset pX^*(T)$. From the conjugacy of maximal tori it follows that all maximal tori have Lie algebra $\mathfrak{t}$. We have proved (i).

For $X \in \mathfrak{g}$ let $P(T, X) \in k[\mathfrak{g}]$ be the characteristic polynomial of the linear map

$$\mathrm{ad}X : Y \mapsto [X.Y]$$

of $\mathfrak{g}$. Then

$$P(T, X) = T^d + \sum_{i \geq 1} f_i(X) T^{d-i},$$

where $d = \dim G$ and $f_i \in k[\mathfrak{g}]$ $(i \geq 1)$. From the proof of 2.4.4 and 4.4.20 we see that $P(T, X_s) = P(T, X)$. It follows from (i) that $X \in \mathfrak{g}$ has central semi-simple part if and only if it is annihilated by all f_i. Hence, if $\mathfrak{g}$ contains non-central semi-simple elements, not all f_i are zero. We can then take for the open set U of (ii) the union of the principal open sets $D(f_i)$. If F is infinite, $\mathfrak{g}(F)$ is dense in $\mathfrak{g}$, which implies the first statement of (iii). The last one follows from the observation that the union of the centralizers in question contains the dense subset $U \cap \mathfrak{g}(F)$ of $\mathfrak{g}$.   □

We shall say in this section that G is *bad* if G is not nilpotent and if all semi-simple elements of $\mathfrak{g}$ are central. Otherwise G is *good*.

13.3.5. Lemma. *Assume that G is a bad F-group. There exists a good F-group G' and a purely inseparable F-isogeny $\pi : G \to G'$ such that for any F-torus T' of G' the group $\pi^{-1}T'$ is an F-torus of G.*

Assume that G is bad. By 13.3.4 (i) we know that $p > 0$ and that the set $\mathfrak{t}$ of semi-simple elements of $\mathfrak{g}$ is a p-subalgebra that is $\mathrm{Ad}(G)$-stable. It is defined over F, being the image of $\mathfrak{g}$ under an F-morphism (by 13.3.2 (iv)). By 12.2.5 there exists a purely inseparable isogeny ϕ of G onto an F-group G_1, such that $\mathfrak{t}$ is the kernel of the Lie algebra homomorphism $d\phi : \mathfrak{g} \to \mathfrak{g}_1$.

Let T be a maximal torus of G (over k). By 13.3.4 (i) its Lie algebra is $\mathfrak{t}$. The restriction ϕ_T of ϕ to T is a purely inseparable isogeny of T onto a maximal torus T_1 of G_1. From 12.2.5 and 12.2.9, applied to T, we see that ϕ_T factors through the torus $Fr\ T$.

Let α be a non-zero weight of T in $\mathfrak{g}$. Since $\mathrm{Ker}\ d\phi = \mathfrak{t}$, the corresponding weight space $\mathfrak{g}_\alpha$ is mapped by $d\phi$ onto a weight space $(\mathfrak{g}_1)_{\alpha_1}$ for T_1 in $\mathfrak{g}_1$. By 13.3.4 (i) the non-zero weights of T are divisible by p in $X^*(T)$. It follows from 12.2.10 (4) that if p^a is the highest power of p dividing α, then α_1 is divisible in $X^*(T_1)$ at most by p^{a-1}. If G_1 is bad we continue in the same manner. It is clear after a finite number of steps that we obtain a purely inseparable isogeny π of G onto a good F-group G'.

To prove the last property it suffices to show that if G_1 is as before and if T_1 is an F-torus in G_1 then $\phi^{-1}T_1$ is defined over F. Let $\Gamma_\phi = \{(g, \phi g) \subset G \times G_1 \mid g \in G\}$ be the graph of ϕ. Then $\phi^{-1}T_1$ is isomorphic to $\Gamma_\phi \cap (G \times T_1)$. We show that this intersection of subgroups of $G \times G_1$ is defined over F. By 12.1.5 it suffices to show that the intersection of the corresponding Lie algebras is isomorphic to the Lie algebra of $\phi^{-1}T_1$. By 13.3.4 (i) this amounts to showing that

$$\Gamma_{d\phi} \cap (\mathfrak{g} \times \mathfrak{t}_1) = \mathfrak{t} \times \{0\}, \tag{62}$$

where $\Gamma_{d\phi}$ is the graph of $d\phi$. Now the semi-simple part of any element of $\mathfrak{g}$ lies in $\mathfrak{t}$. It then follows from 4.4.20 (ii) that $\mathrm{Im}\ d\phi$ consists of nilpotent elements, which implies (62).   □

13.3.6. Theorem. *Assume that F is infinite.*
(i) G contains maximal tori defined over F;
(ii) G is generated by the Cartan subgroups (i.e centralizers) of the maximal tori of (i).

First assume that G is nilpotent. There is a unique maximal torus T, viz. the set of semi-simple elements of G (6.3.2 (i)). That T is defined over F follows from 11.2.8 (i) if $p = 0$ and from the fact that there is $h \geq 0$ with $T = G^{p^h}$ if $p > 0$. We skip the details.

Assume that G is not nilpotent and good. Then G is generated by the groups $Z(X)^0$, where X runs through the non-central semi-simple elements in $\mathfrak{g}(F)$. For by 13.3.2 (i) and 13.3.4 (iii) the Lie algebra of the subgroup generated by these centralizers is $\mathfrak{g}$. Let H be such a centralizer, it is a proper F-subgroup containing a maximal torus of G (13.3.2 (ii)). By induction on $\dim G$ we may assume the theorem to be true for H. But then 13.3.6 also follows.

If G is bad we use an isogeny $G \to G'$ onto a good group G' as in 13.3.5. The theorem holds for G' and 13.3.5 implies that it also holds for G. Observe that by 6.4.7 (i) and 13.3.1 the Cartan subgroups of (ii) are connected F-subgroups. $\square$

13.3.7. Remark. 13.3.6 also holds in the case that F is finite. Then (i) is an easy consequence of Lang's theorem 4.4.17, as the reader may check. For a proof of (ii) in that case see [BoS, 2.9].

By the theorem the maximal F-subtori of G are the maximal tori defined over F.

13.3.8. Corollary. *(i) A semi-simple element of $G(F)$ lies in a maximal F-torus;*
(ii) A semi-simple element of $\mathfrak{g}(F)$ lies in the Lie algebra of a maximal F-torus.

This follows from part (i) of the theorem (and 13.3.7), applied to the centralizer of the element in question, using 12.1.4. $\square$

13.3.9. Corollary. *Let F be infinite.*
(i) If the Cartan subgroups of maximal F-tori are unirational varieties over F then G is unirational over F and $G(F)$ is dense in G;
(ii) If G is a reductive F-group, then G is unirational over F and $G(F)$ is dense in G.

From 13.3.6 (ii) and 2.2.7 we see that there is a surjective F-morphism $C_1 \times \cdots \times C_n \to G$, where the C_i are Cartan subgroups of maximal F-tori. This implies the unirationality of G. For the density statement see 13.2.6 (ii). If G is reductive, then Cartan subgroups are maximal tori (7.6.4 (ii)) and we can apply 13.2.7 (i). If F is finite then G is still unirational over F, see [loc. cit., 7.12]]. $\square$

In the next corollary we anticipate results from the next chapter (whose proof does not depend on the results of this chapter).

13.3.10. Corollary. $G(F)$ *is dense in* G *when* F *is perfect and infinite.*
It follows from 14.3.10 and 14.3.8 that in this case the Cartan subgroups of 13.3.6 (ii) are unirational. So 13.3.9 (i) can be applied.   □

For a non-perfect field the density result is not generally true, see 13.3.13 (2). It is obviously false if F is a finite field and $G \neq 1$.

Let G_t be the subgroup of G generated by its maximal tori.

13.3.11. Proposition. *(i)* G_t *is the smallest closed, connected, normal, subgroup of* G *whose factor group is unipotent;*
(ii) Assume that F *is infinite.* G_t *is defined over* F *and is generated by the maximal F-tori of* G.

Let H be the subgroup of G_t generated by the maximal tori of G defined over $\bar{F}$. Then H is closed and connected, and is defined over $\bar{F}$ (2.2.7). Since $G(\bar{F})$ is dense in G, it follows from 6.4.1 that H is a normal subgroup and by 13.3.8 (i) any semi-simple element of $G(\bar{F})$ lies in H. We conclude from 2.4.8 (ii) (with $k = \bar{F}$) that $(G/H)(\bar{F})$ does not contain semi-simple elements $\neq 1$. It follows that G/H is unipotent. This implies that any maximal torus of G is contained in H, whence $H = G_t$. Now (i) follows.

Assume that F is infinite. Let L be the subgroup of G generated by the maximal F-tori of G. It is a closed, connected, F-subgroup of G_t. We shall prove that $L = G_t$, from which (ii) will follow. First we dispose of the case $p = 0$. Then G_t is defined over F_s. Clearly, $G_t(F_s)$ is a Γ-stable subgroup of $G(F_s)$. By 11.2.8 (i), G_t is defined over F. Using induction on $\dim G$ we see that we may assume $G_t = G$. It follows from 13.3.10 that L is a normal subgroup of G. We have a surjective homomorphism $\mathfrak{g} \to L(G/L) = \mathfrak{g}/\mathfrak{l}$, which induces a surjective F-homomorphism $\mathfrak{g}(F) \to (\mathfrak{g}/\mathfrak{l})(F)$. A semi-simple element of $(\mathfrak{g}/\mathfrak{l})(F)$ is the image of a semi-simple element of $\mathfrak{g}(F)$ (by 4.4.20 (ii) and 12.1.7 (c)). Since by 13.3.8 (ii) all semi-simple elements of $\mathfrak{g}(F)$ lie in $\mathfrak{l}$ we conclude that the elements of $(\mathfrak{g}/\mathfrak{l})(F)$ are nilpotent. From 13.3.6 (i) and 13.3.8 (ii) we conclude that G/L is unipotent. Since $G = G_t$ this implies $L = G$.

Now assume that $p > 0$. We do not yet know that G_t is defined over F. But we do know, as before, that all semi-simple elements of $\mathfrak{g}(F)$ lie in $\mathfrak{l}(F)$. This implies that there is a p-power q such that $(\mathfrak{g}(F))^{[q]} \subset \mathfrak{l}(F)$. Since $\mathfrak{g}(F)$ is dense in $\mathfrak{g}$, we can conclude that all semi-simple elements of $\mathfrak{g}$ lie in $\mathfrak{l}$. Let T be a maximal torus of G. Its Lie algebra $\mathfrak{t}$ lies in $\mathfrak{l}$. Then there is a maximal torus T_1 of L, with Lie algebra $\mathfrak{t}$ (as a consequence of 13.3.8 (ii)). After conjugation of T by an element of L we may assume that T_1 is defined over F. Then $\mathfrak{t}$ is defined over F. If $\mathfrak{t}(F)$ contains a non-central element X, we can pass to its centralizer, which contains both T and T_1. By induction on dim G we can then conclude that $T \subset L$.

Now assume that all elements of $\mathfrak{t}(F)$ are central. We may also assume that G is not nilpotent. Let $\pi : G \to G'$ be as in 13.3.5. We know that (with obvious notation) $L' = (G')_t$. From 13.3.5 we see that $L = \pi^{-1}L'$, and it is clear that $G_t = \pi^{-1}(G')_t$. The asserted equality $L = G_t$ follows. $\square$

13.3.12. Corollary. *(F infinite) (i)* G_t *is unirational over* F *and* $G_t(F)$ *is dense in* G_t;
(ii) G is generated by G_t *and one Cartan F-subgroup;*
(iii) If some Cartan F-subgroup of G is unirational over F then so is G. In that case $G(F)$ *is dense in* G.

(i) is proved as 13.3.9 (i), taking into account 13.2.7 (i). From the conjugacy of maximal tori in G_t it follows that two Cartan F-subgroups of G are conjugate by an element of G_t. Now (ii) follows from 13.3.6 (ii) and (iii) is proved as (i). $\square$

13.3.13. Exercises. (1) (a) If T and T' are maximal F-tori of G then the F-varieties $G/N_G(T)$ and $G/N_G(T')$ are canonically isomorphic.
(b) Define an F-variety of maximal tori $\mathcal{T}_G$ such that for any extension E of F there is a bijection of $\mathcal{T}_G(E)$ onto the set of maximal tori of G defined over E, with good functorial properties.
(2) Let $p > 0$ and assume that F is a field of rational functions $F_0(T)$. Define

$$G = \{(x, y) \in \mathbf{G}_a^2 \mid x^p = x + Ty^p\};$$

this is a connected one dimensional F-subgroup of $\mathbf{G}_a^2$. Show that $G(F)$ is finite (hence is not dense in G).
(3) Let $G = \mathbf{GL}_2$, $F = \mathbf{Q}$. Show that the number of $G(F)$-conjugacy classes of maximal F-tori of G is infinite.

13.4. The groups $P(\lambda)$

13.4.1. G is a connected F-group. $X_*(G)$ is the set of cocharacters of G (homomorphisms $\mathbf{G}_m \to G$). Let $X_*(G)(F)$ be the subset of those defined over F. Recall that $-\lambda$ is defined by $(-\lambda)(a) = \lambda(a)^{-1}$ $(a \in k^*)$ (3.2.1). If λ is a cocharacter, we denote by $P(\lambda)$ or $P_G(\lambda)$ the closed subgroup formed by the $x \in G$ such that $\lim_{a \to 0} \lambda(a)x\lambda(a)^{-1}$ exists. It was introduced in 3.2.15. If G is reductive these subgroups are the parabolic subgroups, as was shown in 8.4.5.

We denote by $U(\lambda)$ or $U_G(\lambda)$ the normal subgroup of $P(\lambda)$ formed by the $x \in P(\lambda)$ for which the limit of the preceding paragraph equals e. The centralizer of Im λ is denoted by $Z(\lambda)$; it is a closed subgroup of $P(\lambda)$. The group $\mathbf{G}_m$ acts in $\mathfrak{g}$ via Ad $\circ \lambda$. The weights are integers. We denote by $\mathfrak{g}_0(\lambda)$ the zero weight space, which is the Lie algebra of $Z(\lambda)$ (5.4.7), and by $\mathfrak{g}_+(\lambda)$ ($\mathfrak{g}_-(\lambda)$) the sum of the weight spaces with strictly positive (respectively: negative) weight. Then $\mathfrak{g}$ is the direct sum of the subspaces $\mathfrak{g}_0$, $\mathfrak{g}_+(\lambda)$, $\mathfrak{g}_-(\lambda)$. If λ is defined over F these spaces are F-subspaces of $\mathfrak{g}$.

13.4.2. Theorem. *Let $\lambda \in X_*(G)(F)$.*
(i) $P(\lambda)$, $Z(\lambda)$, $U(\lambda)$ are connected F-subgroups and $U(\lambda)$ is a unipotent normal subgroup of $P(\lambda)$. Moreover, the product morphism $Z(\lambda) \times U(\lambda) \to P(\lambda)$ is an F-isomorphism of varieties;
(ii) $L(P(\lambda)) = \mathfrak{g}_0(\lambda) \oplus \mathfrak{g}_+(\lambda)$ and $L(U(\pm\lambda)) = \mathfrak{g}_\pm(\lambda)$;
(iii) The product morphism $U(-\lambda) \times P(\lambda) \to G$ is a bijection onto an open subset of G.

First let $G = \mathbf{GL}_n$. We may assume that $\lambda(a) = \mathrm{diag}(a^{m_1}, \ldots, a^{m_n})$, with $m_1 \geq \cdots \geq m_n$. It is then straightforward to prove the first statements of (i). The last point of (i) follows from 8.4.3, using 8.4.5. The proof of (ii) is also straightforward. (iii) follows from (ii). By considering the tangent map at (e, e), using that $\mathfrak{g} = L(P(\lambda)) \oplus L(U(-\lambda))$, one sees that the image of the product map is dominant (4.3.6 (i)). As it is an orbit for an action of $U(-\lambda) \times P(\lambda)$, it must be open in G by 2.3.3 (i). The bijectivity follows from $U(-\lambda) \cap P(\lambda) = \{e\}$.

Now let G be arbitrary. Assume that it is a closed F-subgroup of $H = \mathbf{GL}_n$. Then λ is also a cocharacter of H. We first prove (ii). We have

$$L(P_G(\lambda)) = L(P_H(\lambda) \cap G) \subset L(P_H(\lambda)) \cap \mathfrak{g} = \mathfrak{g}_0(\lambda) + \mathfrak{g}_+(\lambda), \qquad (63)$$

since we already know (ii) for H. Likewise,

$$L(U_G(\pm\lambda)) \subset \mathfrak{g}_\pm(\lambda). \qquad (64)$$

Next observe that (by (iii) for H) $U = U_H(-\lambda)P_H(\lambda) \cap G$ is an open subset of G, containing e. By 5.5.3 there exists a rational representation $\rho : H \to GL(V)$ such that G is the stabilizer of a line kv. Let $g \in U$. By (i) and (iii) for H we can write $g = xyz$, with $x \in U_H(-\lambda)$, $y \in U_H(\lambda)$, $z \in Z_H(\lambda)$. There is $c \in k^*$ with $\rho(g)v = cv$ and v is a weight vector of $\mathbf{G}_m$ for the representation $\rho \circ \lambda$, since $\mathrm{Im}\, \lambda \in G$. It follows that for $a \in k^*$

$$c\rho(\lambda(a)x^{-1}\lambda(a)^{-1})v = \rho(\lambda(a)y\lambda(a)^{-1}z)v. \qquad (65)$$

By an easy computation in $H = \mathbf{GL}_n$, with λ as in the beginning of the proof, we see that the right-hand side of (65) is a polynomial function of a and that the left-hand side is a polynomial function of a^{-1}. These polynomial functions must be constant. It follows that the left-hand side equals cv and the right-hand side $\rho(z)v$, whence $z \in G$. Also, $\rho(x)v = \rho(y)v = v$, and x, $y \in G$. We have shown than $U_G(-\lambda)P_G(\lambda)$ contains an open subset of G, whence

$$\dim U_G(-\lambda) + \dim P_G(\lambda) = \dim G = \dim \mathfrak{g}_-(\lambda) + (\dim \mathfrak{g}_0(\lambda) + \dim \mathfrak{g}_+(\lambda)).$$

From (63) and (64) we conclude that (ii) holds.

(ii) implies (iii), as in the case of $\mathbf{GL}_n$. It follows from (iii) that $P_G(\lambda)$ and $U_G(\lambda)$ are irreducible. Since $U_H(\lambda)$ is a unipotent normal subgroup of $P_H(\lambda)$, $U_G(\lambda)$ has similar properties. The proof of the last point of (i) is similar to that of (iii).

It remains to be proved that $P_G(\lambda)$ and $U_G(\lambda)$ are defined over F. Since $P_G(\lambda) = G \cap P_H(\lambda)$ it suffices by 12.1.5 to show that

$$L(P_G(\lambda)) = L(P_H(\lambda)) \cap \mathfrak{g}.$$

But this is a consequence of (ii). It also follows that $U_G(-\lambda)$ is defined over F. $\square$

Let $\phi : G \to G'$ be a separable, surjective homomorphism of F-groups. Then $\lambda' = \phi \circ \lambda$ is a cocharacter of G' defined over F.

13.4.3. Corollary. $\phi(P_G(\lambda)) = P_{G'}(\lambda')$, $\phi(U_G(\lambda)) = U_{G'}(\lambda')$.

Consider the first equality. Clearly, the left-hand side is contained in the right-hand side. Since ϕ is separable the differential $d\phi$ is surjective. Using 13.4.2 (ii) for G and G' we conclude that $\phi(P_G(\lambda))$ and $P_{G'}(\lambda')$ have the same Lie algebra. It follows that these connected groups must coincide. The same argument proves the other equality. $\square$

13.4.4. Corollary. *If G is solvable then the product morphism*

$$U(-\lambda) \times P(\lambda) \to G$$

is an F-isomorphism of varieties.

It suffices to prove that $G = U(-\lambda)P(\lambda)$ (check this). We may assume that $F = k$. Fix a maximal torus T of G that contains Im λ (6.3.6 (i)). We can assume that $G \neq T$. Let N be a closed, normal, subgroup of G contained in the center of G_u and isomorphic to $\mathbf{G}_a$ (6.3.4). By induction on dim G we may assume that the assertion holds for $G' = G/N$ and the cocharacter λ' of G' induced by λ. Then 13.4.3 implies that $G = U_G(-\lambda)P_G(\lambda)N$. Applying 13.4.2 (ii) for the group $T.N$, and its cocharacter defined by λ, we see that N is a subgroup of one of the groups $Z(\lambda)$, $U_+(\lambda)$, $U_-(\lambda)$. The corollary follows. $\square$

13.4.5. If H is a connected closed subgroup of G that is normalized by Im λ, we define the groups $Z_H(\lambda)$, $U_H(\pm\lambda)$ to be the intersections with H of the analogous subgroups of G. Then 13.4.2 and 13.4.4 hold for such a group H. This follows, for example, by applying 13.4.2 to the group H_1 whose underlying variety is $H \times \mathbf{G}_m$, with multiplication $(x, a)(y, b) = (x\lambda(a)y\lambda(a)^{-1}, ab)$ $(x, y \in H,\ a, b \in k^*)$, a semi-direct product of H and $\mathbf{G}_m$. We skip the details.

As a particular case, let H be the unipotent radical R of G. We then have the closed subgroups $U_R(\pm\lambda)$.

13.4.6. Corollary. *The product map* $U_R(-\lambda) \times P_R(\lambda) \to R$ *is bijective.*

This follows from 13.4.4. $\square$

13.4.7. Exercises. λ is a cocharacter of G such that $\mathrm{Im}\lambda$ is non-central.
(1) Conjugation in G induces an action of G on $X_*(G)$. We have $P(g.\lambda) = g P(\lambda)g^{-1}$, $U(g.\lambda) = gU(\lambda)g^{-1}$.
(2) Notations of 3.2.1. For $n > 0$ we have $P(n\lambda) = P(\lambda)$, $U(n\lambda) = U(\lambda)$.
(3) (a) We have $G = U(\lambda)U(-\lambda)P(\lambda)$.
    (b) Show that the normalizer $N_G(P(\lambda))$ coincides with $P(\lambda)$.
(*Hint* : Consider $\{g \in U(-\lambda) \mid g(\mathrm{Im}\ \lambda)g^{-1} \subset P(\lambda)\}$).
(4) If $G = U(-\lambda).P(\lambda)$ then G is solvable. (*Hint*: reformulate in terms of reductive groups.)
(5) $U(\lambda)$ is contained in the group G_t of 13.3.11.

Notes

The material of 13.1 and 13.2 is familiar. It can also be found in [Bo3, Ch. III]. We have given a proof of 13.1.1 (ii) using the inseparable Galois descent of 11.1. For another, more direct, proof see [loc.cit.,8.10].

The main result 13.3.6 is due to Chevalley in characteristic 0 (see [Ch2]) and to Grothendieck in general [SGA3, exp.XIV]. The proof given here is a version of the one given in [BoS] (also exposed in [Bo3, §18]). It avoids the use of regular elements of a Lie algebra. A more extensive discussion of the variety of maximal tori of 13.3.9 (1) can be found in [loc.cit.]. There one also finds a proof that of the fact that this variety is rational over the ground field (another result of Grothendieck [SGA3, exp.XIV]).

The results of 13.4.2 are due to Borel and Tits. See the note [BoT3] (which does not contain proofs).

Chapter 14

Solvable F-groups

This chapter is about solvable groups over a ground field F. The emphasis is on F-split solvable groups and their properties. In 14.4.3 we prove the conjugacy over F of two maximal F-tori of a solvable F-group. k, F, ... are as before. G is a connected, solvable, linear algebraic group over F. Its unipotent radical -which by 6.3.3 (ii) coincides with the set of its unipotent elements- is denoted by G_u.

14.1. Generalities

14.1.1. Lemma. *There exists a sequence* $\{e\} = G_0 \subset G_1 \subset \dots \subset G_{n-1} \subset G_n = G$ *of closed, connected, normal F-subgroups of G such that each quotient group* G_i/G_{i-1} $(1 \le i \le n)$ *is either a torus or an elementary unipotent group.*

By 12.2.3 such quotients are F-groups. Recall that in 3.4.1 a connected unipotent group was defined to be elementary if it is abelian and if its elements have order dividing the characteristic p, when $p > 0$. Using 2.2.8 the proof of 14.1.1 is reduced to the case that G is abelian, which we now assume. If $p > 0$ the groups $G^{(p^m)}$ of 2.2.9 (3) give a descending sequence of connected F-subgroups. For large m they coincide with the unique maximal torus of G (cf. the beginning of the proof of 13.3.6). The assertion now readily follows. If $p = 0$ the unipotent radical G_u is defined over F (12.1.7 (d)). Then take $n = 2$, $G_1 = G_u$. $\qquad \square$

We say that G is *trigonalizable over F* or *F-trigonalizable* if there is an F-isomorphism of G onto a closed F-subgroup of a group $\mathbf{T}_n$ of upper triangular $n \times n$-matrices, with its obvious F-structure. The group G is trigonalizable over k, by the Lie-Kolchin theorem 6.3.1.

14.1.2. Proposition. *G is F-trigonalizable if and only if G_u is defined over F and G/G_u is an F-split torus.*

Let G be F-trigonalizable. We may assume it to be an F-subgroup of the group $\mathbf{T}_n$. Let $\mathbf{D}_n \subset \mathbf{T}_n$ be the group of diagonal matrices; it is an F-split torus. Denote by $\phi : G \to \mathbf{D}_n$ the F-homomorphism that maps an element of G onto its diagonal part. Its image is a torus and its kernel is G_u. Also, Ker $d\phi$ is the set of nilpotent elements of $\mathfrak{g}$, which contains the Lie algebra $L(G_u)$ (see 4.4.21 (1)). Moreover $\mathfrak{g}/L(G_u)$ consists of semi-simple elements. It follows that Ker $d\phi = L(G_u)$. Application of 11.2.14 (ii) shows that $G_u = $ Ker ϕ is defined over F. The F-torus $T = G/G_u$ is F-isomorphic to a subtorus of the F-split torus $\mathbf{T}_n$, hence is F-split (see 13.2.5 (1)).

Now assume that G_u is defined over F and that G/G_u is F-split. We assume that G is an F-subgroup of $\mathbf{GL}_n$, so that it acts linearly on the F-vector space $V = k^n$. We claim that there is $v \in V(F) - \{0\}$ that is a simultaneous eigenvector for the elements of G. Then trigonalizability will follow (check this). Let W be the fixed point set of G_u in V. It is a non-zero subspace, as a consequence of 2.4.12. By 11.2.7, W is the fixed point set of $G_u(F_s)$, which set is defined over F_s. Since it is obviously Γ-stable, it is defined over F by 11.1.4. The F-split torus G/G_u acts linearly in W. The claim now follows from 3.2.12 (ii). $\square$

14.1.3. Corollary. *Assume G to be trigonalizable. If $\phi : G \to \mathbf{GL}_n$ is an F-homomorphism then* Im ϕ *is trigonalizable.*

By 13.3.6 (i) (and 13.3.7) G contains a maximal torus S defined over F. We see from the proposition that S is F-split, being F-isomorphic to G/G_u. Then ϕS is a maximal torus of Im ϕ defined over F and F-split. Moreover $\phi(G_u)$ is the unipotent radical of Im ϕ and is defined over F. By the proposition, Im ϕ is F-trigonalizable. $\square$

G is *F-split* if there is a sequence (G_i) as 14.1.1 such that all successive quotients are F-isomorphic to either $\mathbf{G}_a$ or $\mathbf{G}_m$ (see 12.3.5 (3)).

14.1.4. Proposition. *Assume G to be F-split. Then G is F-trigonalizable.*

Assume that the F-split group G is an F-subgroup of $\mathbf{GL}_n$. It suffices to show that there is $v \in F^n - \{0\}$ which is a common eigenvector for the elements of G. By an induction we may assume $((G_i)$ being as above) that the proposition is true for G_{n-1}. Then $(G_{n-1})_u$ is defined over F by 14.1.2, Let $V \subset k^n$ be the fixed point set of $(G_{n-1})_u$. As in the proof of 14.1.2 one sees that it is a non-zero subspace which is defined over F. We shall show that $V(F)$ contains a common eigenvector v. We are reduced to proving the theorem in the case, that $(G_{n-1})_u$ is trivial, i.e. that G_{n-1} is a torus. Then either G is a torus, in which case the assertion follows from 3.2.12 (ii), or G/G_{n-1} is F-isomorphic to $\mathbf{G}_a$. In the latter case the weight spaces for the split F-torus G_{n-1} are G-stable and are defined over F. Replacing V by such a weight space, we are reduced to proving the proposition in the case of $\mathbf{G}_a$. This case is covered by 14.1.2. The proposition is proved.

14.1.5. Remark. We shall see below (14.3.10) that if F is perfect a unipotent connected F-group is F-split. It follows from 14.1.2 that in that case the notions of F-trigonalizable and F-split solvable groups coincide. Over a non-perfect field this is not true, as the example of 12.3.8 (1) shows.

14.1.6. Exercise. If G is F-split then so is G_u. Moreover, G/G_u is an F-split torus.

For F-split groups we have the following version over F of Borel's fixed point theorem 6.2.6.

14.1.7. Theorem. *Let G be a connected, solvable F-group that is F-split and let X be a projective F-variety with a G-action over F. If $X(F) \neq \emptyset$ there is a point of $X(F)$ that is fixed by all elements of G.*

To prove this we shall construct inductively for each i a point $x_i \in X(F)$ fixed by the elements of G_i (the groups G_i are as before). Take for x_0 a point of $X(F)$. Assume x_{i-1} has been found. If it is fixed by G_i take $x_i = x_{i-1}$. Otherwise, the map $g \mapsto g.x_{i-1}$ defines an F-morphism $G_i/G_{i-1} \to X$. This quotient is F-isomorphic to an F-open subset of $\mathbf{P}^1$. By 6.1.4 this morphism extends to an F-morphism of $\mathbf{P}^1$ to X. We can then take for x_i the image of the point ∞ of $\mathbf{P}^1$. $\qquad\square$

14.2. Action of $\mathbf{G}_a$ on an affine variety, applications

14.2.1. Let $G = \mathbf{G}_a$ and let X be an irreducible affine G-variety over F. We put $A = F[X]$. The action $m : G \times X \to X$ is defined by an algebra homomorphism $m^* : A \to A[T]$ (we have identified $F[\mathbf{G}_a] \otimes_F A$ with the polynomial algebra $A[T]$). For $f \in A$ write

$$m^* f = \sum_{h \geq 0} (D_h f) T^h.$$

The D_h are F-linear maps $A \to A$. For given f almost all $D_h f$ are zero. The fact that m is a $\mathbf{G}_a$-action is equivalent to the following properties of the D_h:

$$\begin{cases} D_0 & = \ \text{id}, \\ D_h(fg) & = \ \sum_{i+j=h}(D_i f)(D_j g) \ (f, g \in A), \\ D_h D_i & = \ (h+i, i) D_{h+i}, \end{cases} \qquad (66)$$

where $(h+i, i) = (h+i)!(h!i!)^{-1}$ is a binomial coefficient.

We see from (66) that D_1 is an F-derivation of A. If the characteristic p is zero, the last formula (66) shows that $D_h = (h!)^{-1} D_1^h$. So D_1 is a locally nilpotent linear map.

Now let $p > 0$. Write $h = \sum_{i \geq 0} h_i p^i$ with $0 \leq n_i < p$. Using 3.4.2 we deduce from the last formula (66) that

$$D_h = \prod_{i \geq 0} (h_i!)^{-1} D_{p^i}^{h_i},$$

and that $D_h^p = 0$. For $a \in k$, $f \in k[X] = k \otimes_F A$ we write $a.f = s(a)f$ where s is as in 2.3.5. Then

$$m^*(a.f) = \sum (D_h f)(T - a)^h = \sum D_h(a.f) T^h, \qquad (67)$$

whence

$$D_h(a.f) = \sum_{i \geq h}(i, h)(-a)^{i-h}D_i f. \tag{68}$$

14.2.2. Proposition. *Assume that G acts non-trivially on X. There exists an affine F-variety Y with the following properties:*
(a) there is an F-isomorphism ϕ of $G \times Y$ onto an open F-subvariety of X,
(b) there is an F-morphism $\psi : G \times Y \to G$ such that for $a, b \in k,\ y \in Y$

$$\psi(a + b, y) = \psi(a, y) + \psi(b, y), \quad a.\phi(b, y) = \phi(\psi(a, y) + b, y).$$

Put

$$k[X]_0 = \{f \in k[X] \mid a.f = f \text{ for all } a \in k\}.$$

This is the subalgebra of elements of $k[X]$ annihilated by all D_h, $h > 0$. It is defined over F, with F-structure $A_0 = A \cap k[X]_0$. The non-triviality of the action of G implies that $A_0 \neq A$. Let $u \in A$ be such that $m^*u \in A[T]$ has minimal strictly positive degree n. From (67) we see that for $a \in k$ the degree of the polynomial $m^*(a.u) - m^*u \in k[X][T]$ is strictly smaller than n, hence this polynomial must be constant. This means that $D_h(a.u) = D_h u$ for $h \geq 1$. From (68) we conclude that

$$(i, h)D_i u = 0 \ (0 < h < i). \tag{69}$$

First assume that $p > 0$ (the more complicated case). Using 3.4.2 (ii) we see from (69) that $D_h u = 0$ unless h is a power of p. In particular, n is a p-power p^r. By (66) we have for $h > 0$, $s \geq 0$

$$D_h(D_{p^s}u) = (h + p^s, h)D_{h+p^s}u = 0,$$

because either $h + p^s$ is not a p-power or $(h + p^s, h) = 0$. It follows that the coefficients of $m^*u - u$ lie in A_0. Let $c = D_{p^r}u$ be the leading coefficient. The principal open subset $D(c) = \{x \in X \mid c(x) \neq 0\}$ is an affine open F-subvariety of X that is G-stable. Replacing X by U we may assume that $c = 1$.

We claim that $A = A_0[u]$. Let $f \in A$ and assume that $m^* f$ has degree $N > 0$. Then $D_h f = 0$ for $h > N$. Let p^s be the highest power of p dividing N. Using (66) we see that

$$m^*(D_{N-p^s}f) = \sum_{0 \leq h \leq p^s}(N - p^s + h, h)(D_{N-p^s+h}f)T^h.$$

It follows from 3.4.2 (ii) that $(N, p^s) \not\equiv 0 \pmod{p}$. Hence the right-hand side is a polynomial of degree p^s, consequently $p^s \geq p^r$. The last formula (66) shows that the leading coefficient d of m^*f lies in A_0. It follows that the degree of $m^*(f - du^{p^{-r}N})$ is strictly smaller than the degree of m^*f. The claim follows by induction.

The A_0-homomorphism $F[T] \otimes_F A_0 \to A$ sending T to u is surjective. It is also injective: a relation $\sum a_i u^i = 0$ with coefficients in A_0 can only be the trivial one (apply m^*). It follows that A_0, being a quotient of A and a subalgebra of A, is an affine algebra over F. Let Y be the affine F- variety with $F[Y] = A_0$ (11.2.2 (i)). Then $F[G \times Y] \simeq A$, whence an isomorphism $\phi : G \times Y \to X$. Define $\psi : G \times Y \to G$ by $\psi(a, y) = \sum_{h \geq 0}(D_{p^h}u)(y)a^{p^h}$. Then ϕ and ψ are as required, as a straightforward check shows.

In the case that $p = 0$ the proof is similar, but easier. Now the degree of m^*u equals one. We skip the details. $\square$

14.2.3. Proposition. *Let H be a connected, solvable, F-group and let G be a closed, normal, F-subgroup that is F-isomorphic to $\mathbf{G}_a$. Assume that X is an affine F-variety that is a homogeneous space for H over F. If G acts non-trivially on X, there exists a homogeneous space Y of H over F, together with an F-isomorphism $\phi : G \times Y \to X$ and an F-homomorphism $\chi : G \to G$, such that for $g, g' \in G$, $y \in Y$*

$$g.\phi(g', y) = \phi(\chi(g)g', y)$$

and the composite morphism

$$\tau : X \xrightarrow{\phi^{-1}} G \times Y \xrightarrow{pr_2} Y$$

is H-equivariant.

We write the group operation in G multiplicatively. We proceed as in the proof of 14.2.2. The notations are as in that proof. Let again n be the degree of m^*u. Consider the set of all $u' \in A$ such that m^*u' has degree n and let I be the ideal generated in A_0 by their leading coefficients. We claim that $I = A_0$. It suffices to prove this if $F = k$. The group H operates on I via translations, and it follows from the Lie-Kolchin theorem (6.3.1) that there is $c \in I$ that is an eigenvector for all elements of H. As H acts transitively on X, this can only be if c has no zeros on X, hence is invertible in A_0, proving the claim. We may then choose u such that m^*u has leading coefficient 1. The transitivity of the H-action implies that the coefficients of $m^*u - u$ are constants. If ψ is as in 14.2.2, we have now $\psi(g, y) = \chi(g)$, where χ is an F-homomorphism $G \to G$.

It remains to be shown that Y is a homogeneous space and that the morphism τ is H-equivariant. The last point follows from the fact that $k \otimes A_0 = k[Y]$ is stable under translations by elements of H. Since H acts transitively on X and τ is surjective, H must also act transitively on Y. $\square$

14.2.4. Exercises. (1) Let X be a homogeneous space for $\mathbf{G}_a$ over F that is not reduced to a point. There exists an F-isomorphism $\phi : \mathbf{G}_a \to X$ and a surjective F-homomorphism $\chi : \mathbf{G}_a \to \mathbf{G}_a$ such that $a.\phi(b) = \phi(\chi(a) + b)$ $(a, b \in k)$.

(2) (a) Prove an analogue of 14.2.2 for $G = \mathbf{G}_m$. (*Hint:* see 3.2.16 (3)).

(b) Prove an analogue of the result of the preceding exercise for $\mathbf{G}_m$.

(3) (a) Let G be a connected nilpotent F-group and let T be its unique maximal torus, defined over F (see the proof of 13.3.6). Assume that there exists a homogeneous space X for G over F together with an F-isomorphism $\phi : T \times X \to G$ such that for $t, t' \in T$, $x \in X$ we have $t.\phi(t', x) = \phi(tt', x)$. Show that the unipotent radical G_u is defined over F.

(b) Using the example of 12.1.6 (2) show that 14.2.3 is not true for $G = \mathbf{G}_m$.

For the next theorem we need an auxiliary result, in which G may be an arbitrary linear algebraic group over F and H a closed F-subgroup. Let $\pi : G \to G/H$ be the canonical F-morphism.

14.2.5. Lemma. *If G/H is affine, then $F[G/H] = \{f \in F[G] \mid f(gh) = f(g)$ for all $g \in G$, $h \in H\}$.*

This is true if $F = k$ (see the proof of 5.5.5). To prove the lemma it suffices to show: if G/H is affine and if $f \in F[G/H]$, then the function $\bar{f} \in k[G]$ with $\bar{f}(g) = f(\pi g)$ lies in $F[G]$. It is enough to show that $\bar{f} \in F_s[G]$ (as $\bar{f}$ is invariant under the Galois group Γ). It follows from 11.2.14 (ii) that $\pi(G(F_s)) = (G/H)(F_s)$, whence

$$\bar{f}(G(F_s)) = f(\pi(G(F_s))) \subset F_s.$$

This implies that $\bar{f}$ is defined over F_s (11.2.10 (3)). $\square$

A *section* of π on G/H over F (briefly, an F-*section*) is an F-morphism $\sigma : G/H \to G$ such that $\pi \circ \sigma = \mathrm{id}$ (see 5.5.7). From now on G is again a connected, solvable F-group.

14.2.6. Theorem. *Let H be a connected, unipotent F-subgroup of G that is F-split.*

(i) There exists an F-section for $\pi : G \to G/H$;

(ii) G is F-isomorphic (as a variety) to $G/H \times H$.

If $\sigma : G/H \to G$ is an F-section, then $\phi(x, h) = \sigma(x)h$ defines an F-isomorphism $G/H \times H \to G$, which shows that (ii) is a consequence of (i) (see also 5.5.7). In fact, (i) and (ii) are equivalent, as is easily seen.

We prove (i) by induction on $\dim H$, starting with $H = \{e\}$. Assume that $\dim H > 0$ and let H_1 be a closed, normal, F-subgroup of H such that H/H_1 is F-isomorphic to $\mathbf{G}_a$. We may assume that there exists an F-section

$$\sigma_1 : G/H_1 \to G.$$

It defines an F-isomorphism $\phi_1 : G/H_1 \times H_1 \to G$, as above.

Now G/H_1 is an F-variety, on which G acts on the left and H/H_1 on the right, whence an action of the group $G \times (H/H_1)$. An application of 14.2.3 shows that there exists a homogeneous space Y of $G \times (H/H_1)$ over F and an F-isomorphism $\phi : (H/H_1) \times Y \to G/H_1$. Consider the F-morphism $\theta : G \to Y$ that is the composition of the canonical map $G \to G/H_1$ and the morphism $\tau : G/H_1 \to Y$ of 14.2.3. Then θ a separable F-morphism that is constant on the cosets of H and G-equivariant. It factorizes through G/H, whence an F-morphism $\rho : G/H \to Y$ that is separable, bijective and G-equivariant, hence is an isomorphism (5.3.2 (iii)). It follows that there exists an F-isomorphism $H/H_1 \times G/H \to G/H_1$. The F-isomorphism $\phi_1 : G/H_1 \times H_1 \to G$ induces an isomorphism $H/H_1 \times H_1 \to H$. Putting these facts together we obtain an F-isomorphism $\psi : G/H \times H \to G$ such that $(\mathrm{pr}_1 \circ \psi^{-1})(g) = gH$ ($g \in G$), whence an F-section. $\qquad \square$

We write $\mathbf{A}^{m,n} = \mathbf{G}_m^m \times \mathbf{G}_a^n$ and view this as an open F-subvariety of $\mathbf{A}^{m+n}$.

14.2.7. Corollary. *Assume G to be F-split. As an F-variety, G is isomorphic to $\mathbf{A}^{m,n}$, with $m = \dim G/G_u$, $n = \dim G_u$. In particular, if G is unipotent it is F-isomorphic to $\mathbf{A}^n$.*

G_u is F-split by 14.1.6. Applying the theorem to G (or applying 6.3.5 (iv)) we see that G is F-isomorphic to $G/G_u \times G_u$. Now G/G_u is an F-split torus (14.1.6), isomorphic to $(\mathbf{G}_m)^m$. Repeated application of the theorem shows that G_u is F-isomorphic to $\mathbf{G}_a^n$. $\qquad \square$

14.2.8. Exercise. $(F = k)$ Let H be a connected, closed subgroup of G. The variety G/H is isomorphic to some $\mathbf{A}^{m,n}$.

14.3. F-split solvable groups

14.3.1. Elementary unipotent F-groups. Let G be a connected elementary unipotent F-group (see 3.4.1). As in 3.3.1 we denote by $\mathcal{A}(G)$ the set of additive functions on G (homomorphisms of algebraic groups $G \to \mathbf{G}_a$) and by $\mathcal{A}(G)(F)$ the subset of those defined over F. Assume that $p > 0$. Then $\mathcal{A}(G)(F)$ is a left module over the non-commutative ring $R = R(F)$ of 3.3.1. In 3.4 we discussed the case $F = k$. We now deal with an arbitrary ground field F.

14.3.2. Lemma. *G is F-isomorphic to $\mathbf{G}_a^n$ if and only if $\mathcal{A}(G)(F)$ is a free R-module of rank n.*

By 3.3.5, $\mathcal{A}(\mathbf{G}_a^n)$ is a free R-module. It remains to prove the 'if'-part. Assume that $\mathcal{A}(F)$ is a free R-module, with basis $(f_1, \dots, f_n)$. By 3.4.8 (a) we have $F[G] = F[f_1, \dots, f_n]$ and by 3.3.6 (ii) the f_i are algebraically independent over F. It follows that $g \mapsto (f_1(g), \dots, f_n(g))$ defines an isomorphism of G onto $\mathbf{G}_a^n$. □

14.3.3. Lemma. *A submodule of a free R-module with countable basis is free.*

Let M be a free R-module with basis $(e_i)_{i \geq 1}$. For $m \in M$ denote by $e_i'(m)$ its ith component relative to this basis. Denote by M_i the submodule with basis $(e_j)_{j \geq i}$. Let N be a non-zero submodule of F. Then $e_i'(N \cap M_i)$ is a left ideal I_i in R, which by 3.3.3 (i) is principal, say $I_i = Rr_i$. Denote by f_i an element of N with $e_i'(f_i) = r_i$. We leave it to the reader to check that the f_i such that $r_i \neq 0$ form a basis of N. □

14.3.4. Examples of free R-modules. (a) If F is perfect then any finitely generated torsion free R-module is free. See 3.3.3 (iii).
(b) Let $M = F[T_1, \dots, T_n, T_1^{-1}, \dots, T_n^{-1}]$, with the R-action given by $T.f = f^p$. Then M is a free R-module, a basis being the set of monomials $T_1^{h_1} \dots T_n^{h_n}$, where either all h_i are 0 or the greatest common divisor of the integers h_i $(1 \leq i \leq n)$ is not divisible by p.
(c) Let E/F be a separable finite extension of F. Then $R(E)$ is a free module over $R = R(F)$. For let (x_i) be a basis of E over F. Then $(x_i^{p^n})$ is also a basis of E over F, for any $n \geq 0$ (see the definition of separability given before 4.2.10). It follows that (x_i) (viewed as a subset of $R(E)$) is a basis of $R(E)$ over R.

14.3.5. If $p = 0$ we have $R = F$. Then 14.3.2 also holds; the proof is the same. 14.3.3 and the examples of 14.3.4 are trivial.

14.3.6. There is an analogue of 13.1.4 for elementary abelian F-groups, which we briefly indicate. Assume that $p > 0$. Let M be a left R-module of finite type. Denote by S the symmetric algebra of M over F and by I the ideal in S generated by the elements $T.m - m^p$ $(m \in M)$. Put $F[M] = S/I$. Then $F[M]$ is an affine algebra. The homomorphisms $S \to S \otimes_F S$ and $S \to S$ with $m \mapsto m \otimes 1 + 1 \otimes m$, respectively $m \mapsto -m$ and the augmentation homomorphism $S \to F$, induce homomorphisms $\Delta : F[M] \to F[M] \otimes F[M], \iota : F[M] \to F[M]$ and $e : F[M] \to F$, which are the ingredients of an F-group $\mathcal{G}(M)$ (see 2.1.2). This group is elementary abelian, and $\mathcal{A}(\mathcal{G}(M))(F) \simeq M$. In fact, we have an anti-equivalence between the categories of left R-modules and of elementary abelian F-groups. One has a similar (trivial) result for $p = 0$.

14.3.7. Exercises. (1) Work out the details of 14.3.6.
(2) Assume that F is a non-perfect field. Recall that the additive group of the ring $R = R(F)$ of 3.3.1 is $F[T]$. View R as a left vector space over F.
    (a) RT^m is a two-sided ideal in R. The quotient R/RT^m has dimension m over F.

It is a local ring, i.e. it has a unique maximal ideal. Put $F_n = \{x \in k \mid x^{p^n} \in F\}$, this is a subfield of k. Let M be a left R-module such that the $R(F_n)$-module $F_n \otimes_F M$ is free of rank one.

(b) $T^n M$ is a free R-module of rank one.

(c) $M/T^n M$ is a free R/RT^n-module of rank one. (*Hint*: Show that the image in $M/T^n M$ of an element of $M - TM$ is a basis element).

(d) Let e be a basis element of $T^n M$ and let $f + T^n M$ be a basis element of $M/T^n M$. There exist elements $a_0, a_1, \ldots, a_r \in F$ with $a_0 \neq 0$ such that $T^n f = (a_0 + a_1 T + \cdots + a_r T^r)e$.

(e) M is isomorphic to a quotient $(R \oplus R)/S$, where S is the free submodule generated by $(T^n, a_0 + \cdots + a_r T^r)$ (notations of (d)).

(f) Let G be an F-group isomorphic to $\mathbf{G}_a$. Then G is F-isomorphic to an F-subgroup of $\mathbf{G}_a^2$ of the form

$$\{(x, y) \in \mathbf{G}_a^2 \mid y^{p^n} = a_0 x + a_1 x^p + \cdots + a_r x^{p^r}\},$$

notations being as before. (Russell's theorem, see [Rus] or[Kam]).

(3) Let F be an infinite field and let S be an F-torus that is F-split.

(a) Let ϕ be a homomorphism $S(F) \to \mathbf{GL}_n(R)$ such that

$$\phi(s) = \sum \chi(s) A_\chi,$$

where the sum is over a finite set of characters of S, the A_χ being matrices with entries in R. Show that Im ϕ lies in the subgroup $\mathbf{GL}_n(F)$.

(b) Let V be an F-vector space with a locally finite action of S over F. Assume that $V(F)$ has a structure of left R-module, such that $a.(s.v) = s.(a.v)$ for $a \in R$, $s \in S(F)$, $v \in V(F)$. Show that if $V(F)$ is a free R-module of finite rank it has a basis consisting of weight vectors for S.

(c) Let F be any field and let G be a connected solvable F-group with the following properties: (1) it contains a maximal torus S defined over F and F-split, (2) the unipotent radical G_u is defined over F and is F-isomorphic to a product $\mathbf{G}_a^n$. For each character χ of S there exists a unique F-subgroup G_χ of G_u, isomorphic to a product $\mathbf{G}_a^{n_\chi}$, which is normalized by S and is such that the Lie algebra $L(G_\chi)$ is the χ-weight space for S in $L(G)$. Moreover, G_u is the direct product of the non-trivial G_χ.

We come now to the main results on F-split solvable groups.

14.3.8. Theorem. (*i*) *G is F-split if and only if there exists a dominant F-morphism $\mathbf{A}^{m,n} \to G$, for some m and n;*
(*ii*) *G is unipotent and F-split if and only if there exists a dominant F-morphism $\mathbf{A}^n \to G$, for some n;*
(*iii*) *A unipotent group G is F-split if and only if it is F_s-split.*

That the conditions of (i) and (ii) are necessary follows from 14.2.7. For the condition of (iii) this is obvious. We prove the sufficiency. First consider (i). From 2.2.6

it follows that if G satisfies the condition of (i), the same holds for the commutator subgroup (G, G), which is a connected F-subgroup by 2.2.8. If $p > 0$ the same is true for the subgroup $G^{(p)}$ generated by the pth powers of elements of G. An easy argument shows that there is a sequence $\{e\} = G_0 \subset G_1 \subset \cdots \subset G_{n-1} \subset G_n = G$ as in 14.1.1 such that all G_i satisfy the condition of (i). Moreover, it follows from 14.2.5 that the successive quotients G_i/G_{i-1} $(1 \le i \le n)$ also satisfy this condition. Such a quotient H is either an F-torus or an elementary abelian group over F. We claim that H is split.

If H is a torus, H is F_s-split (13.1.1 (ii)). The condition of (i) implies that $F[H]$ is a subalgebra of an algebra $A = F[T_1, ..., T_a, T_1^{-1}, ..., T_a^{-1}]$. A character of H is an invertible element of $F_s[H]$, hence of $F_s \otimes_F A$. But the invertible elements of the last algebra are of the form $cT_1^{h_1}...T_a^{h_a}$, with $c \in F_s^*$. A character χ of H lies in $F_s[H]$. It follows that for $\gamma \in \Gamma = \mathrm{Gal}(F_s/F)$ we have that the character $\gamma.\chi$ is a scalar multiple of χ. Since characters have the value 1 at e, χ is fixed by the elements of Γ and is defined over F (11.1.4). By 3.2.12 (i) H is F-split.

If H is elementary abelian, then the R-module $\mathcal{A}(H)(F)$ of additive functions on H is a submodule of the R-module A, which is free by 14.3.4 (b). Then H is F-split by 14.3.3 and 14.3.2. We have proved the claim made above. It shows that G is a successive extension of F-split groups. We still have to prove that this implies that G is split.

As in the proof of 14.1.4 one shows that such a successive extension is F-trigonalizable. Hence the unipotent radical G_u is defined over F (14.1.2). The algebra $F[G/G_u]$ being a subalgebra of $F[G]$ by 14.2.5, the torus G/G_u must be F-split by what we already know. Let T be a maximal F-torus of G (13.3.6 (i) or 13.3.7). Then G is F-isomorphic as a variety to $T \times G_u$ (by 6.3.5 (iv)), from which we see that G_u also satisfies the condition of (i). As in the first paragraph of the proof, one sees that there exists a connected, elementary F-subgroup N of the center of G_u which is normal in G and satisfies the condition of (i). N is F-split by what we already proved and is normalized by T. Also, G/N satisfies the condition of (i). Induction on dim G reduces the proof to the case that G_u is an elementary F-group. In that case we obtain, as in the proof of 6.3.4, an additive function $f \in \mathcal{A}(G_u)(F)$ that is a simultaneous eigenvector for the elements of T. We may also assume that $df \ne 0$ (check this). Then Ker f is defined over F by 11.2.14 (ii), and so is (Ker $f)^0$ (12.1.1). We can now proceed by induction. We have proved (i).

If the condition of (ii) holds then G is F-split by (i). Now $F[G]$ does not contain non-constant invertible elements. Then the maximal torus T of the proof of (i) must be trivial and G is unipotent.

By the arguments of the first paragraph, we are reduced to proving (iii) in the case that G is elementary abelian, which we now assume. The R-module $\mathcal{A}(G)(F)$ is a submodule of a free $R(F_s)$-module. By 14.3.4 (c) the latter is a free R-module. Application of 14.3.2 shows that G is split, proving (iii). $\square$

14.3.9. Corollary. *Assume that G is F-split and is not a torus. There exists a closed, normal F-subgroup of G that is F-isomorphic to* $\mathbf{G}_a$ *and lies in the center of G.*

This refinement of 6.3.4 is a corollary of the proof of 14.3.8 (i).   □

14.3.10. Corollary. *Assume that F is perfect. Then* G_u *is F-split.*

The proof is reduced to the case of an elementary abelian group, in which case one uses 14.3.4 (a).   □

The next proposition is a variant of 14.2.7.

14.3.11. Proposition. *Let G be F-split and let T be a maximal F-torus of G.*
(i) There exists an F-isomorphism of varieties $\phi : G_u \to \mathbf{A}^n$ *with* $\phi(e) = 0$ *and a rational representation* ρ *of T in* k^n *over F such that* $\phi(\mathrm{Int}(t).g) = \rho(t).\phi(g)$ *($g \in G$, $t \in T$);*
(ii) For $x, y \in \mathbf{G}_a^n$ *we have* $\phi(\phi^{-1}(x)\phi^{-1}(y)) = x + y + \sum_{i \geq 2} F_i(x, y)$, *where* $F_i : \mathbf{A}^n \times \mathbf{A}^n \to \mathbf{A}^n$ *is a polynomial map of degree i;*
(iii) The weights of T for ρ *are the weights of T in* $L(G)$.

Int(t) is the inner automorphism defined by $t \in T$. Let N be as in 14.3.9. We fix an isomorphism of F-groups $\psi : N \to \mathbf{G}_a$. There is a character χ of T such that $\psi(tnt^{-1}) = \chi(t)\psi(n)$ ($n \in N$, $t \in T$).

The action Int of T on G induces an action I of T on G/N. Let σ be a an F-section for $G \to G/N$ (14.2.6). From the definition of sections it follows that there is an F-morphism $z : T \times G/N \to \mathbf{A}^1$ such that for $x \in G/N$, $t \in T$

$$\mathrm{Int}(t)^{-1}.(\sigma(I(t).x)) = \sigma(x)\psi^{-1}(z(t, x)).$$

It follows that

$$z(tt', x) = z(t, x) + \chi(t)^{-1}z(t', I(t).x). \tag{70}$$

Let X be the character group of T. Since T is F-split (14.1.6), all characters are defined over F (3.2.12 (i)). Write

$$z(t, x) = \sum_{\theta \in X} \theta(t)z_\theta(x),$$

with $z_\theta \in F[G/N]$. Insert this into (70). By Dedekind's theorem on linear independence of characters the terms on both sides not involving t' are equal. We obtain that

$$z(t, x) = z_1(x) - \chi(t)^{-1}z_1(I(t).x).$$

Put $\tau(x) = \sigma(x)\psi^{-1}(z_1(x))$. Then τ is again a section. It defines an F-isomorphism $G \rightarrow G/N \times \mathbf{A}^1$. This isomorphism is T-equivariant, T operating on G by Int, on G/N by I and on $\mathbf{A}^1$ by the character χ. We may assume that the assertion of the proposition holds for G/N. Using the T-equivariant isomorphism $G \rightarrow G/N \times \mathbf{A}^1$ it is now straightforward to establish the proposition for G. $\square$

14.3.12. Exercises. G is a connected solvable F-group.
(1) G is F-split if and only if G_u is defined over F and F-split and G/G_u is an F-split torus.
(2) Assume G to be F-split. Let $\phi : G \rightarrow G'$ be a surjective homomorphism of F-groups. Then G' is connected, solvable and F-split.
(3) Let H and K be two connected F-subgroups of G that are F-split.
    (a) The connected F-subgroup (H, K) of 2.2.8 is F-split.
    (b) The same is true for the connected F-subgroups $H^{(n)}$ of 2.2.9 (3).

14.3.13. Theorem. *Assume that G is F-split. If X is an affine F-variety that is a homogeneous space for G, then $X(F) \neq \emptyset$.*

First let G be a torus and let X be its character group. G acts on $A = F[X]$ by translations and as in 3.2.13 we have a decomposition into weight spaces $A = \bigoplus_{\chi \in X} A_\chi$. If $x \in X$ the map $g \mapsto g.x$ defines a surjective morphism $G \rightarrow X$, compatible with translations. It follows that $k[X]$ is isomorphic to a translation invariant subspace of $k[G]$, from which we conclude that the spaces A_χ have dimension ≤ 1. Let $X' = \{\chi \in X \mid \dim A_\chi = 1\}$. Then X' is closed under addition (see 3.2.13). Also, if $\chi \in X'$ then the ideal $A_\chi k[X]$ cannot be proper, as X does not contain proper G-stable subvarieties. It follows that X' is a subgroup of X. Let $(\chi_i)_{1 \leq i \leq d}$ be a basis of X' and let a_i be a non-zero element of A_χ. There is an isomorphism of A onto the algebra of Laurent polynomials $F[T_1, \ldots, T_d, T_1^{-1}, \ldots, T_d^{-1}]$ sending a_i to T_i. It is now clear that $X(F) \neq \emptyset$. The argument gives, in fact, that the homogeneous space X is F-isomorphic to the F-split torus $\mathcal{G}(X')$ (see 13.1.4), which is a homomorphic image of G. The action on $\mathcal{G}(X')$ comes from the translations in G.

If G is not a torus, let N be as in 14.3.9. Applying 14.2.3 we obtain a homogeneous space Y of the F-split group G/N over F such that X is F-isomorphic to $N \times Y$. The assertion now follows by induction on $\dim G$, using 14.2.4 (1). $\square$

14.3.13 generalizes the result of 12.3.5 (3), which is equivalent to it in the case of torsors (principal homogeneous spaces).

14.4. Structural properties of solvable groups

Recall that G is a connected solvable F-group.

14.4.1. Theorem. *(i) There is a unique maximal closed, connected, F-split, F-subgroup $G_{us,F}$ of G_u. It is a normal subgroup of G. Moreover, $G_{us,F} = G_{us,F_s}$;*
(ii) $(G/G_{us,F})_{us,F} = \{e\}$ and $G/G_{us,F}$ is nilpotent;
(iii) If $\phi : \mathbf{A}^1 \to G$ is an F-morphism then Im $\phi \subset \phi(0)G_{us,F}$;
(iv) If G is generated by its F-tori then G_u is defined over F and F-split.

According to 2.2.7 (ii) the closed, connected F_s-split unipotent F_s-subgroups of G generate a closed, connected, unipotent F_s-subgroup G_1, which is F_s-split by 14.3.8 (ii). Identify G, G_u, G_1 with their groups of $\bar{F}$-rational points. The Galois group $\Gamma = \mathrm{Gal}(F_s/F)$ operates on G. If $\gamma \in \Gamma$ we have that $\gamma.G_u$ is a connected, normal, unipotent, closed subgroup of G, hence must coincide with G_u. So Γ stabilizes G_u. A similar argument shows that Γ stabilizes G_1 and also $G_1(F_s)$. Now 11.2.7 and 11.2.8 (i) imply that G_1 is defined over F. It is F-split by 14.3.8 (iii). We can also conclude that G_1 is normalized by $G(F_s)$, and 11.2.7 implies that G_1 is normal. It follows that $G_{us,F} = G_1$ has the properties of (i).

Let ϕ be as in (iii). To prove (iii) we may assume that $\phi(0) = e$. Using 2.2.6 and 14.3.8 (ii) we conclude that Im ϕ generates a connected, unipotent F-split F-subgroup of G, which must be contained in $G_{us,F}$, whence (iii).

Let ψ be an F-morphism $\mathbf{A}^1 \to G/G_{us,F}$. Application of 14.2.6 (i) shows that we can lift ψ to an F-morphism $\mathbf{A}^1 \to G$. Then (iii) shows that ψ is constant. In particular, $G/G_{us,F}$ cannot contain any F-split unipotent F-subgroup, hence $(G/G_{us,F})_{us,F} = \{e\}$, proving the first part of (ii).

If G is generated by its F-tori it follows from 2.2.7 and 13.1.1 (ii) that there exists a dominant F_s-morphism $\mathbf{G}_m^n \to G$. From 14.3.8 (i) we conclude that G is F_s-split, so G_u is defined over F_s and F_s-split (14.1.6). G_u is defined over F by 11.2.8 (i) and is F-split by 14.3.8 (iii), proving (iv).

It remains to prove the last point of (ii). It follows from what we already proved that the subgroup of $G/G_{us,F}$ generated by its F-tori is a torus, which is the only maximal F-torus. If it is a proper subtorus then F must be infinite (use 14.3.10) and 13.3.6 (ii) then shows that it is a central torus, which implies that $G/G_{us,F}$ is nilpotent, completing the proof of (ii). $\square$

14.4.2. Corollary. *Let G be a connected unipotent F-group and let T be an F-torus that acts on G as a group of automorphisms over F. Assume that e is the only element of G fixed by all elements of T. Then G is F-split.*

We denote by H the semi-direct product of T and G. So $H = T \ltimes G$, with multiplication $(t, g)(t', g') = (tt', (t')^{-1}gt'g')$ $(g, g' \in G, t, t' \in T)$. Then T is a maximal torus of H, which coincides with its centralizer. We may assume that F is infinite (if F is finite 14.3.10 applies). From 13.3.6 (ii) we see that H is generated by its maximal F-tori. By 14.4.1 (iv) we conclude that the unipotent radical G of H is F-split. $\square$

14.4.3. Theorem. *Two maximal F-tori S and T of G are conjugate by an element of* $G(F)$.

We may assume that G is generated by its F-tori. Then G_u is defined over F and F-split by 14.4.1 (iv). We may also assume that G_u is non-trivial. Let N be a subgroup of G_u with the properties of 14.3.9. It follows from 14.2.6 (i) that the canonical map $G(F) \to (G/N)(F)$ is surjective. By induction we may assume that the assertion is true for G/N. As in the proof of 6.3.5, we are reduced to proving the assertion in the case that $G_u = N$. Then either N centralizes S, in which case $S = T$, or there is a unique $g \in G$ with $gSg^{-1} = T$. From 13.3.1 (i) we infer that $g \in G(F_s)$. Its uniqueness implies that it is invariant under Γ, hence lies in $G(F)$.   □

14.4.4. Proposition. *Let C be a Cartan subgroup of G defined over F. Then G is F-split if and only if C is F-split.*

We may assume that F is infinite. If C is F-split then it follows from 14.4.3 that all Cartan subgroups defined over F are F-split. From 2.2.7, 13.3.6 (ii) and 14.2.7, we deduce that G satisfies the condition of 14.3.8 (i), hence is F-split.

Conversely, assume that G is F-split. Let T be a maximal F-torus of G; it is F-split. There is a cocharacter $\lambda \in X_*(T)(F)$ such that, with the notations of 13.4.1, we have $C = Z(\lambda)$ (see 7.1.2). By 13.4.4 we have an isomorphism of F-varieties $U(-\lambda) \times Z(\lambda) \times U(\lambda) \to G$, from which we see that there is a surjective F-morphism $G \to C$. Applying 14.3.8 (i) we conclude that C is F-split.   □

Let H be an arbitrary connected F-group.

14.4.5. Proposition. *(i) H contains a maximal, connected, solvable (respectively: unipotent), normal F-subgroup $R_F H$ (respectively: $R_{u,F} H$);*
(ii) The maximal F-split unipotent subgroup $R_{us,F} H$ of $R_F H$ is a normal F-subgroup of H;
(iii) If $R_{us,F} H = \{e\}$ then $R_{u,F} H$ centralizes all F-subtori of H;
(iv) $R_{u,F_s} H = R_{u,F} H$, $R_{us,F_s} H = R_{us,F} H$;
(v) If F is perfect then $R_{u,F} H = R_{us,F} H = R_u H$, the unipotent radical of H.

It follows from 2.2.7 that the connected, normal F-subgroups of H contained in the radical RH (respectively, in $R_u H$) generate a similar subgroup, whence (i). As in the proof of 14.4.1 (i) one shows that the F_s-subgroup $R_{u,F_s} H$ is defined over F and is normal in H. It contains $R_{u,F} H$. Since the second group is contained in the first one, we have the first assertion of (iv). A similar argument, using 14.4.1 (i), proves the second assertion. We have proved (ii) and (iv).

Let $R_{us,F} H = \{e\}$ and let S be an F-torus in H. Let $K = S.R_{u,F} H$; this is a connected solvable F-group with maximal torus S. By 14.4.1 (ii) it must be nilpotent, whence (iii).

Finally, (v) follows from 14.3.10. $\qquad\square$

The groups R_F, $R_{u,F}H$, $R_{us,F}H$ are, respectively, the *F-radical, unipotent F-radical, split unipotent F-radical* of H. The *F*-group H is called *F-reductive* if $R_{u,F}H = \{e\}$.

14.4.6. Lemma. *H is F-reductive if and only if $R_u H$ does not contain any non-trivial connected F-subgroup.*

Assume that $R_u H$ contains a non-trivial connected *F*-group K. Then the conjugates of K by elements of $H(F_s)$ generate a connected normal *F*-subgroup contained in $R_u H$, which shows that $R_{u,F}H$ is non-trivial. $\qquad\square$

Notes

Most of the results of this Chapter are due to Rosenlicht [Ros1, Ros2]. See also [Bo3, §15].

14.2.2 is a particular case of [Ros1, Theorem 1]. We have given an elementary proof.

The main results on connected, solvable *F*-groups, discussed in 14.3 and 14.4, come from [Ros2]. The proofs of the results on *F*-split groups in 14.3 given here make use of the algebra R.

14.4.1 (iii) is due to Tits [Ti4]. The 'only if' part of the characterization 14.4.4 of *F*-split groups is proved in [BoT2, 1.6].

The group $R_{us,F}H$ of 14.4.5 was introduced in [BoT3].

Chapter 15
F-reductive Groups

In this chapter we discuss general F-groups. An important general result is the conjugacy over F of maximal F-split tori (15.2.6). In 15.3 we introduce the root datum of an F-reductive group. In the case of reductive F-groups the proofs of several results (such as 15.1.3 and 15.3.4) are easier. The notations are as in the previous chapters. G is a connected F-group.

15.1. Pseudo-parabolic F-subgroups

15.1.1. A *pseudo-parabolic F-subgroup* P of G is a subgroup of the form $P = P(\lambda)$, where λ is a cocharacter of G defined over F, $P(\lambda)$ is as in 13.4 and $R_{u,F}G$ is the unipotent F-radical (14.4.5). It follows from 13.4.2 (ii) that P is a connected F-subgroup of G. Observe that, as a consequence of 13.4.3 and 14.4.5 (iii), we have $R_{u,F}G \subset Z(\lambda).R_{us,F}G$, from which we see that $P = P(\lambda).R_{us,F}G$. The pseudo-parabolic subgroups will play a role in the theory of F-groups similar to that of the parabolic subgroups in the theory of k-groups.

15.1.2. Lemma. *(i) G contains proper pseudo-parabolic F-subgroups if and only if $G/R_{u,F}G$ contains non-central F-subtori that are F-split;*
(ii) If G is reductive or if F is perfect the pseudo-parabolic subgroups are the parabolic F-subgroups.

To prove (i) we may and shall assume that $R_{u,F}G$ is trivial. Assume that λ is a cocharacter of G defined over F such that $P(\lambda) = G$. By 13.4.2 (i), $U(\lambda)$ is a unipotent normal F-subgroup of G which is trivial by 14.4.6. From 13.4.2 (i) we see that $Z(\lambda) = G$, so that Imλ lies in the center of G. It follows that, if proper pseudo-parabolic F-subgroups do not exist, then all F-split F-subtori, of G lie in the center of G. Conversely, if there exist non-central F-split F-subtori, there exist λ with $Z(\lambda) \neq G$ (check this). Using 13.4.2 (ii) we conclude that $P(\lambda) \neq G$. We have proved (i).

Let G be reductive. It follows from 8.4.5 that a pseudo-parabolic F-subgroup is parabolic. If $F = k$ we already proved the converse in 8.4.5. Now let P be a parabolic F-subgroup of G and let T be a maximal torus in P defined over F. Denote by X the character group of T and by Y the group of its cocharacters. The elements of X and Y are defined over F_s and the Galois group $\Gamma = \text{Gal}(F_s/F)$ acts on these groups via a finite quotient A (see 13.2.1). Let $R \subset X$ be the root system of (G, T). The non-zero weights of T in the Lie algebra $L(P)$ form a subset R' of R, which is stable under A. Let Λ be the set of $\lambda \in Y$ such that

$$R' = \{\alpha \in R \mid \langle \alpha, \lambda \rangle \geq 0\}.$$

It follows from the proof of 8.4.5 that $\Lambda \neq \emptyset$ and that for $\lambda \in \Lambda$ we have $P = P(\lambda)$. The group A stabilizes Λ.

If $\lambda \in \Lambda$ the Γ-stable element $\mu = \sum_{\delta \in \Delta} \delta.\lambda$ also lies in Λ. The cocharacter μ is defined over F and $P = P(\mu)$, showing that P is a pseudo-parabolic F-subgroup. If F is perfect then $R_{u,F}G = R_u G$ and $G/R_{u,F}G$ is reductive by 14.4.5 (v). It follows from what we already proved that the assertion of (ii) holds (using 6.2.7 (i)) $\quad\square$

Let $P = P(\lambda)$ be as before and denote by $\pi : G \to G/P$ the canonical morphism.

15.1.3. Theorem. *Let F be infinite. Then π has local sections over F.*

This means that G/P is covered by open F-subvarieties on each of which there is a section defined over F, see 5.5.7 and 14.2.6.

We claim that we may assume that $G(F)$ is dense in G. If G is reductive, this is the case by 13.3.9 (ii). In the general case, let G_t be the subgroup of G generated by the F-subtori of G. Then $G_t(F)$ is dense in G_t (13.3.12 (i)).

Put $S = \text{Im } \lambda$. We can assume that this is a one dimensional F-subtorus of G. It is a maximal torus of the solvable F-subgroups $S.U(\pm\lambda)$ (notations of 13.4) and coincides with its centralizer in these groups. We conclude from 13.3.6 (ii) that $U(\pm\lambda) \subset G_t$. Also note that as a consequence of 14.4.2 the groups $U(\pm\lambda)$ are F-split. 13.3.12 (ii) implies that $G = G_t.Z(\lambda)$. Denote by $G^{\sharp}$ the identity component of the closure of $G(F)$. By 2.2.4 (i), 11.2.4 (ii) and 12.1.1 this is a connected F-subgroup of G that contains G_t, whence $G = G^{\sharp}.Z(\lambda)$. Using 14.2.7 we see that $G^{\sharp}$ also contains $R_{us,F}G$. It follows that $G^{\sharp} \cap P$ is a pseudo-parabolic F-subgroup of $G^{\sharp}$ and that the inclusion $G^{\sharp} \to G$ defines a bijective, $G^{\sharp}$-equivariant F-morphism $\phi : G^{\sharp}/G^{\sharp} \cap P \to G/P$. Using 13.4.2 (ii) we see that the tangent map $(d\phi)_x$ at the image x of e is bijective. By 5.3.2 (iii) ϕ is an isomorphism. Since $G^{\sharp}(F)$ is dense in $G^{\sharp}$ the claim follows.

Assuming $G(F)$ to be dense, it suffices to produce one non-empty open F-subvariety U of G/P and a section $\sigma : U \to G$ of π over F. For then the translates $g.U$ ($g \in G(F)$) cover G/P, and on each translate there is an obvious section.

Using 5.5.11 (i) and 13.4.3 we see that we may assume G to be F-reductive. By 13.4.2 (iii) the product map defines an F-isomorphism of $U(-\lambda) \times P$ onto an open F-subvariety of G. Then $U = \pi(U(-\lambda))$ is an open F-subvariety of G/P and π induces an isomorphism $U(-\lambda) \simeq U$. The inverse of this isomorphism is a section over F. $\quad\square$

The theorem is also true if F is finite (see [BoT1, 3.25]). We shall not use this result.

15.1.4. Corollary. *(F arbitrary) The canonical map $G(F) \to (G/P)(F)$ is surjective.*

If F is infinite this is an immediate consequence of the theorem. If F is finite it is a consequence of Lang's theorem 4.4.17, applied to P, as the reader may check.   □

We also record the following result, which we encountered in the course of the proof of 15.1.2.

15.1.5. Corollary. *The connected unipotent F-groups $U(\pm\lambda)$ are F-split.*

15.2. A fixed point theorem

15.2.1. Let P be a pseudo-parabolic F-subgroup of G. We shall prove in this section a theorem for G/P similar to the fixed point theorem 14.1.7. Recall that it asserts that a connected, solvable, F-split, F-group H that acts on a projective F-variety X over F has a fixed point in $X(F)$ if $X(F) \neq \emptyset$. The proof uses the property 6.1.4 of projective F-varieties. However, the proof works already if X has the following weaker property:

(P_F) *Any F-morphism $\mathbf{A}^1 - \{0\} \to X$ extends to an F-morphism $\mathbf{A}^1 \to X$.*

We shall prove in this section that the F-variety G/P has this property. If G is not reductive, this provides an example of a non-complete F-variety with property (P_F). A trivial example of such a variety is the non-split form G of $\mathbf{G}_a$ of 12.3.8 (1). It follows from 14.3.8 (i) that any F-morphism $\mathbf{A}^1 - \{0\} \to G$ is constant, hence is extendible to $\mathbf{A}^1$. The reader who is only interested in reductive groups should continue with 15.2.5.

We require some results about property (P_F).

15.2.2. Proposition. *Let X be an F-variety.*
(i) If X has property (P_E) for some extension E of F, then it has property (P_F);
(ii) If X is a closed F-subvariety of an F-variety that has property (P_F), then so has X;
(iii) A projective F-variety has property (P_F);
(iv) Let $\phi : X \to Y$ be a surjective morphism of irreducible F-varieties with finite fibres. If Y has property (P_F) then so has X;
(v) Let E be a finite purely inseparable extension of F and assume that $X = \Pi_{E/F}Y$, where Y is an irreducible, smooth E-variety that has property (P_E). Then X has property (P_F).

In (v) $\Pi_{E/F}$ is as in 11.4.19 (see also 11.4.20 (i)). (i) and (ii) are immediate and (iii) follows from 6.1.4. In the situation of (iv), let

$$\psi : \mathbf{A}^1 - \{0\} \to X$$

be an F-morphism. If Y has property (P_F) then $\phi \circ \psi$ can be extended to a morphism $\mathbf{A}^1 \to Y$. We have to prove that ϕ can be extended to $\mathbf{A}^1$. To do this we may assume that $F = k$. Replacing X and Y by open subsets, we may also assume by 5.2.6 that X and Y are affine and that X is finite over Y. Let α and β be the homomorphisms $k[X] \to k[T, T^{-1}]$ and $k[Y] \to k[T]$ defined by ψ and the extension of $\phi \circ \psi$. Since X is finite over Y, there exists for any $f \in k[X]$ an equation

$$f^n + g_1 f^{n-1} + \cdots + g_n = 0,$$

with $g_i \in k[Y]$. Hence

$$(\alpha f)^n + (\beta g_1)(\alpha f)^{n-1} + \cdots + (\beta g_n) = 0.$$

The normality of $k[T]$ (5.2.9 (2)) implies that $\alpha f \in k[T]$, for all $f \in k[X]$. Hence ψ is extendible, and (iv) follows.

If Z is an irreducible, smooth affine E-variety and W an affine F-variety, we have by 11.4.16 a bijection of the set of E-morphisms $W \to Z$ onto the set of F-morphisms $W \to \Pi_{E/F} Z$, which is functorial in W. This remains true for an arbitrary irreducible, smooth F-variety Z, which can be obtained by glueing together affine ones (see 11.4.20 (4)). It follows that in the case of (v) there is a bijection of the set of F-morphisms $\mathbf{A}^1 - \{0\} \to X$ onto the set of E-morphisms $\mathbf{A}^1 - \{0\} \to Y$ carrying extendible morphisms to extendible morphisms. This implies (v). $\qquad\square$

15.2.3. Theorem. *The F-variety G/P has property (P_F).*

If G is reductive or if F is perfect, then by 15.1.2 (ii) P is a parabolic subgroup and G/P is a projective F-variety. The theorem holds by 15.2.2 (iii). So we may assume that F is non-perfect, hence infinite. By 15.2.2 (i) it suffices to deal with the case that $F = F_s$, so that $G(F)$ is dense in G. Finally, we may also assume that G is F-reductive.

Put $X = G/P.R_u G$. There is a finite purely inseparable extension E of F such that $R_u G$ is defined over E and E-split. Then $X = G/P.R_u G$ is a projective E-variety and we have the canonical E-morphism $\phi : G/P \to X$. By 11.4.20 (4), $Y = \Pi_{E/F} X$ exists. So we have a surjective E-morphism $\pi : Y \to X$ and an E-morphism $\psi : G/P \to Y$ such that $\phi = \pi \circ \psi$, as in 11.4.19.

15.2.4. Lemma. (i) Im ψ *is a closed F-subvariety Z of Y;*
(ii) For any $y \in Y(F)$ the set $\psi^{-1} y \cap (G/P)(F)$ is finite;
(iii) $\psi : G/P \to Z$ is separable.

It suffices to prove this with G/P replaced by an affine open F-subvariety U, occurring in a covering of G/P by such subvarieties, X being replaced by ϕU and Y by $\Pi_{E/F}(\phi U)$. We use the covering of the proof of 15.1.3, whose sets are $G(F)$-translates of the set $\pi U(-\lambda)$ of that proof. Notice that we need only a finite number of these translates. We conclude that it suffices to prove assertions similar to (i) and (ii) with G/P replaced by an F-split, connected, unipotent, F-subgroup V of G, X

by the image $\bar{V}$ of V in $G/R_u G$, which is an E-variety, π being the canonical E-homomorphism. Y is to be replaced by $\Pi_{E/F} \bar{V}$. By 12.4.2 the morphism ψ is now a group homomorphism, hence has closed image (2.2.5 (ii)). This proves (i).

If a set as in (ii) were infinite, we could conclude that the kernel K of ϕ contained infinitely many elements of $V(F)$. But this kernel lies in $R_u G$. It would follow from 11.2.4 (ii) and 12.1.1 that $R_u G$ contained a non-trivial connected F-subgroup, contradicting 14.4.6. We have proved (ii).

To prove (iii) observe that by 14.2.6 (ii) we have an E-isomorphism $V \simeq \bar{V} \times K$. Now ψ has a factorization

$$V \xrightarrow{\sigma} \Pi V \simeq Y \times \Pi K \xrightarrow{\mathrm{pr}_1} Y,$$

the first morphism being as in 12.4.4 (2). We have $\mathrm{Ker}\, d\sigma = 0$ and $\dim \mathrm{Im}\, \psi = \dim G/P$. The separability of ψ follows.   □

We can now prove 15.2.3. It follows from 15.2.4, 11.2.7 and 11.2.14 (ii) that for $y \in Y(F)$ the fibers $\psi^{-1} y$ are finite. Using that $Y(F)$ is dense in Y, we conclude from 5.2.6 that all fibres of ψ are finite. By 15.2.2, Y has property (P_F). Now apply 15.2.2 (iv).   □

15.2.5. Theorem. *Let H be a connected, solvable, F-split F-subgroup of G. There exists $g \in G(F)$ such that $g H g^{-1} \subset P$.*

As was pointed out in 15.2.1 it follows from 15.2.3 that H has a fixed point in $(G/P)(F)$. Then use 15.1.3. If G is reductive use 14.1.7 instead of 15.1.3.   □

15.2.6. Theorem. *Two maximal F-split F-subtori are conjugate by an element of $G(F)$.*

Let S and S' be two maximal F-split F-subtori. If G contains a proper pseudo-parabolic F-subgroup, then S and S' are both $G(F)$-conjugate to a subtorus of P, by the preceding theorem. The assertion follows by induction. If no such P exists it follows from 15.1.2 (i) that S and S' are maximal F-split F tori of the solvable F-group $R_F G$. Now the assertion follows from 14.4.3 and 13.2.4 (i).   □

15.2.7. Exercise. In 15.2.2 and 15.2.3 one can replace property (P_F) by the following stronger property of an F-variety X:

$(P_F)'$ *Any F-morphism of an F-open subset of $\mathbf{P}^1$ can be extended to an F-morphism $\mathbf{P}^1 \to X$.*

15.3. The root datum of an F-reductive group

15.3.1. In this section G is a connected F-reductive group. We fix a maximal F-split F-subtorus S of G. Its dimension is the F-rank of G. We denote by $_F X$ and

$_FX^\vee$ the group of characters, respectively cocharacters, of S and by $\langle\ ,\ \rangle$ the pairing between them. By 13.2.2 (i) the characters and cocharacters of S are defined over F. We assume that S is non-trivial, so $_FX \neq \{0\}$.

If H is a subgroup of G normalized by S, we denote by $R(H, S)$ the set of non-zero weights of S in $L(H)$. The characters in $R(G, S)$ are the *roots* of G relative to S, or the *F-roots* of G. They form a subset $_FR$ or $_FR(G)$, $_FR(G, S)$ of $_FX$. The *Weyl group of G relative to F* is the finite group $_FW = N_G(S)/Z_G(S)$ (see 3.2.9). We view it as a group of automorphisms of $_FX$. It will follow from 15.3.8 that $_FR$ (if non-empty) is a root system (the *relative root system*) with Weyl group $_FW$ (the *relative Weyl group*). By 5.4.7, $_FR$ is empty if and only if S is non-central.

Let T be a maximal F-torus in G containing S. Such a torus exists, as a consequence of 13.3.6 (i) (or 13.3.7), applied to $Z_G(S)$. We denote by X the character group of T, by $R \subset X$ the root system of G relative to T (7.4.3) and by $W = N_G(T)/Z_G(T)$ its Weyl group. If H is an arbitrary connected F-group we define the root system $_FR(H)$ to be that of the F-reductive group $H/R_{u,F}H$.

15.3.2. Lemma. *(i) Let T' be any F-subtorus of G. The centralizer $Z_G(T')$ is a connected, closed, F-reductive F-subgroup of G;*
(ii) If T is a maximal torus of G containing S, the root system of $(Z_G(S), T)$ consists of the roots of (G, T) whose restriction to S is trivial.

It follows from 6.4.7 (i) and 13.3.1 (ii) that $Z_G(T')$ is a connected, closed F-subgroup. As a consequence of 7.6.4 (ii), $R_{u,F}(Z_G(T'))$ lies in R_uG. It then follows from 14.4.6 that $Z_G(T')$ is F-reductive. This proves (i). By (i), $Z_G(S)$ is connected and F-reductive. Then (ii) is a consequence of 5.4.7. □

15.3.3. Lemma. *Let $a \in\ _FR$.*
(i) The torus $S_a = (\mathrm{Ker}\ a)^0$ is defined over F;
(ii) $G_a = Z_G(S_a)$ is a connected, closed, F-reductive, F-subgroup with maximal split F-subtorus S and $_FR(G_a, S) \subset\ _FX$ consists of rational multiples of a.

There is a maximal integer $n > 0$ such that there exists $\chi \in\ _FX$ with $a = n\chi$. Then $S_a = (\mathrm{Ker}\ \chi)^0$. The subgroup $\mathbf{Z}\chi$ is a direct summand of $_FX$, which implies that $\chi : S \to \mathbf{G}_m$ is separable. By 12.1.3 and 12.1.1, S_a is defined over F, whence (i). (ii) follows from the preceding lemma and 5.4.7. □

By 13.1.1 (ii), T is a maximal F_s-split torus of G, so $_{F_s}X = X$. As usual, p is the characteristic of F.

15.3.4. Theorem. *(i) $R \subset\ _{F_s}R \subset \frac{1}{2}R$;*
(ii) $_{F_s}R = R$ if $p \neq 2$ or if G is reductive.

If G is reductive, which is the case if $p = 0$, then (ii) (and (i)) hold. So we may assume that $p > 0$. Using 14.4.5 (iv) we see that we may also assume that $F = F_s$. We take $S = T$. If $\alpha \in R$ is a weight of T, acting in $L(G/R_uG)$, then α is also a weight of T in $L(G)$, whence $R \subset {}_FR$. Assume that $a \in {}_FR$ is not a rational multiple of a root in R and let λ be a cocharacter of T with $\langle \alpha, \lambda \rangle > 0$. The group $U = U_{G_a}(\lambda)$ is a connected, unipotent F-subgroup normalized by T, and $R(U, T) \cap R = \emptyset$. Then U must lie in R_uG and by 14.4.6 we have $U \subset R_{u,F}G$, a contradiction. So an element of ${}_FG$ is a rational multiple of a root of R.

We may now assume that there is $\alpha \in R$ such that $G = G_\alpha$ (notation of 7.1.3). There is $\chi \in X$ such that the positive rational multiples of α lying in X are positive integral multiples of χ. From property (RD 1) of the root datum of G/R_uG (7.4.1), we see that either $\alpha = \chi$ or $\alpha = 2\chi$.

Let $V = U_{G_a}(\lambda)$, where λ is as before. Then V is F-split (15.1.5). The characters in $R(V, T)$ are strictly positive integral multiples of χ. By 14.3.11, applied to the F-group $T.V$, there is an F-isomorphism $\phi : V \to \mathbf{A}^n$ and a rational representation of T in k^n over F with the properties (ii) and (iii) of 14.3.11. By property (ii) we have for $x, y \in \mathbf{A}^n$

$$\phi(\phi^{-1}(x)\phi^{-1}(y)\phi^{-1}(x)^{-1}\phi^{-1}(y)^{-1}) = \Phi(x, y),$$

where $\Phi : \mathbf{A}^n \times \mathbf{A}^n \to \mathbf{A}^n$ is a polynomial map without constant or linear term. It follows from 14.3.11 (iii) that the weights of T on the Lie algebra of the commutator subgroup (V, V) (which is defined over F by 2.2.8 (ii)) are sums of at least two weights of T in $L(V)$. Hence the weights in $L((V, V))$ are multiples $m\chi$ with $m \geq 2$.

Assume that $\alpha = \chi$. We can then conclude that (V, V) is an F-subgroup contained in R_uG, which by 14.4.6 must be trivial. A similar argument shows that the subgroup $V^{(p)}$ generated by the p-th powers of elements of V is trivial. Hence V is an elementary unipotent F-group. By 14.3.7 (3) it is a product of F_s-subgroups V_ψ such that T acts on $L(V_\psi)$ via the character ψ. If $\psi \neq \alpha$ we must have $V_\psi \subset R_uG$. It follows from 14.4.6 that $V = V_\alpha$, which implies that ${}_FR = R = \{\pm\alpha\}$.

Now let $\alpha = 2\chi$ and $p \neq 2$. Take $t \in T$ such that $\chi(t) = -1$. Then t acts on V by conjugation and the set $\Sigma = \{tvt^{-1}v^{-1} \mid v \in V\}$ lies in R_uG (check this). By 14.4.6 this can only be if V centralizes t. But this means that $R(V, T)$ consists of multiples of α. One can now argue as before to establish that ${}_FR = R = \{\pm\alpha\}$.

There remains the case that $\alpha = 2\chi$ and $p = 2$. In that case the same argument gives that (V, V) is an elementary unipotent F-group, such that T acts on its Lie algebra via α, if (V, V) is non-trivial. Also, $(V, (V, V))$ must be trivial. So (V, V) lies in the center of V. The group $V^{(2)}$ has the same property. Let V_1 be the subgroup of V generated by (V, V) and $V^{(2)}$. Then V_1 is a closed, connected, central F-subgroup of V and, if it is non-trivial, then T acts on $L(V_1)$ via α. If V_1 is trivial one can argue as before. So assume this is not the case. Then V/V_1 is an elementary unipotent F-group. It is a product of groups $(V/V_1)_\psi$, as in 14.3.7 (3). If $\psi \neq \chi, 2\chi$ then by 14.3.11 (ii) the elements $\phi^{-1}(x)$, with x in the ψ-weight space of

T, would generate an F-subgroup of $R_u G$, which is impossible by 14.4.6. It follows that $R(V, T) \subset \{\frac{1}{2}\alpha, \alpha\}$. The theorem follows. $\qquad\square$

It has been shown by Tits that the exceptional case $R \neq {}_{F_s}R$ can occur, see [Ti6, 1991-1992, p. 128].

Let $a \in {}_F R$. The F-groups G_a and S_a are as in 15.3.3.

15.3.5. Lemma. *(i) There exists $n \in (N_{G_a}S)(F)$ that induces in S/S_a a non-trivial automorphism;*
(ii) If n is as in (i) then $(N_G T)(F_s) \cap Z_{G_a}(S)n \neq \emptyset$.

Let $G' = G_a/S_a$, an F-group with one dimensional maximal F-split torus $S' = S/S_a$. Let λ be a cocharacter of G' over F with Im $\lambda = S'$. We then have the F-subgroups $P(\pm\lambda)$ and $U(\pm\lambda)$ of G'. It follows from 13.4.2 (ii) that they are non-trivial. By 15.1.5 the solvable F-group $H = S.U(-\lambda)$ is F-split. By 15.2.5 there is $g \in G'(F)$ such that $gHg^{-1} \subset P(\lambda)$ and it follows from 15.2.6 (applied to G') that we may assume g normalizes S'. If g centralized S' we would have $U(-\lambda) = gU(-\lambda)g^{-1} \subset P(\lambda)$, which is not the case. It follows from 12.3.4 (i) and 13.2.8 (3) that the canonical homomorphism $G_a(F) \to (G_a/S_a)(F)$ is surjective. So there is $n \in G_a(F)$ with image g. This element has the property of (i). T and nTn^{-1} are maximal F-tori of the F-group $Z_{G_a}(S)$. They are F_s-split. By 15.2.6 there exists $h \in Z_{G_a}(S)(F_s)$ with $hnT(hn)^{-1} = T$. Then $hn \in (N_G T)(F_s) \cap Z_{G_a}(S)n$, whence (ii). $\qquad\square$

15.3.6. The Weyl group W of (G, T) acts in X and in the real vector space $V = \mathbf{R} \otimes_{\mathbf{Z}} X$. Notice that by 13.3.1 (ii), 12.1.1 and 11.2.7 the elements of W can be represented by elements of $G(F_s)$.

As in 7.1..7, introduce a positive definite symmetric bilinear form $(\ ,\ )$ on V that is W-invariant. We identify V with its dual space via this form. Then the dual $X^\vee$ of X (the group of cocharacters of T) can be identified with the lattice of $\lambda \in V$ with $(\lambda, X) \subset \mathbf{Z}$. The group $_F X^\vee$ of cocharacters of S is a subgroup of $X^\vee$. If $_F X^\perp$ is the annihilator of $_F X^\vee$ in X (viewed as the dual of $X^\vee$), restriction of characters induces an isomorphism $X/_F X^\perp \simeq {}_F X$ (see 3.2.10 (4)). Then $_F V = \mathbf{R} \otimes_F X$ is the orthogonal complement in V of $\mathbf{R} \otimes_F X^\vee$. The map $\pi : V \to {}_F V$, induced by restriction of characters of T to S, is, orthogonal projection onto the subspace $_F V$ and $_F R = \pi R - \{0\}$. Also, $_F X$ is a subgroup of $_F V$ and $_F X^\vee = \{\lambda \in {}_F V \mid (\lambda, {}_F X) \subset \mathbf{Z}\}$.

By the formula of the proof of 7.1.8 we have for $\alpha \in R$

$$\alpha^\vee = 2(\alpha, \alpha)^{-1}\alpha.$$

For $a \in {}_F X$ we put $a^\vee = 2(a, a)^{-1}a$. Let $_F R^\vee$ be the set of $a^\vee$, $a \in {}_F R$.

The element n of 15.3.5 defines an element $s_a \in {}_F W$ of order two.

15.3.7. Lemma. *(i) s_a is the orthogonal reflection of ${}_F V$ defined by a;*
(ii) ${}_F W$ is generated by the s_a, $a \in {}_F R$.

(i) follows from 15.3.5. The proof of (ii) is similar to that of 7.1.9. Details can be left to the reader.

15.3.8. Theorem. $({}_F X, {}_F R, {}_F X^\vee, {}_F R^\vee)$ *is a root datum.*

We denote the quadruple of the proposition by ${}_F \Psi$ or ${}_F \Psi(G, S)$. It follows from 15.2.6 that it is unique up to isomorphism. To prove the theorem we have to verify the axioms (RD 1) and (RD 2) of 7.4.1. (RD 1) is obvious. By 15.3.7 (i) we have for $v \in_F V$

$$s_a v = v - 2(a, a)^{-1}(v, a)a,$$

whence for $x \in {}_F X$, $y \in {}_F X^\vee$

$$s_a x = x - \langle x, a^\vee \rangle a, \quad s_{a^\vee} y = y - \langle a, y \rangle a^\vee,$$

as required in 7.4.1.

It remains to prove (RD 2). Let $a \in {}_F R$. Using 15.3.5 (ii) we see that there is $w \in W$ such that for $x \in {}_F V$ we have $s_a x = wx$. Let $b \in {}_F R$ and take $\beta \in R$ with $\pi \beta = b$. Then

$$s_a.b = s_a.(\pi \beta) = \pi(s_a.\beta) = \pi((s_a w)(w^{-1}.\beta)).$$

$\gamma = w^{-1}.\beta$ lies in R put $c = \pi \gamma$. Since $s_a w$ fixes all elements of ${}_F V$ and stabilizes the orthogonal complement of ${}_F V$ in V, we have $\pi(s_a w(\gamma)) = c$, whence $s_a.b = c \in {}_F R$. It follows that $s_a({}_F R) = {}_F R$. Similarly, one shows that $s_{a^\vee}({}_F R^\vee) = {}_F R^\vee$. We have proved the theorem.

15.3.9. Non-reduced root systems. The root system ${}_F R$ can be non-reduced, i.e. it may happen that $a, 2a \in {}_F R$ (for an example see 15.3.10 (2)). We review some facts about non-reduced root systems (see also [Bou2, Ch. VI, 1.4]). Let R be a root system in a real vector space V, as in 7.4.1 (we write in V instead of V'). The form $(\,,\,)$ is as usual. Then for $\alpha \in R$ we have $\alpha^\vee = 2(\alpha, \alpha)^{-1}\alpha$.

(a) *Let $\alpha \in R$. If an integral multiple $m\alpha$ lies in R then $m \in \{\pm 1, \pm 2\}$.*
This follows from $2 = \langle m\alpha, (m\alpha)^\vee \rangle \in m\mathbf{Z}$.

(b) *If $\alpha, 2\alpha \in R$ then $\langle \beta, \alpha^\vee \rangle$ is even for all $\beta \in R$.*
We have $(2\alpha)^\vee = \frac{1}{2}\alpha$. Now use that $\langle \beta, (2\alpha^\vee) \rangle \in \mathbf{Z}$.

(c) Let R_i be the set of $\alpha \in R$ such that $\frac{1}{2}\alpha \notin R$. This the the set of *indivisible* roots of R . It is a reduced root system. A *basis* D *of* R is one of R_i. It follows from 8.2.8

that any root of R is an integral linear combination $\sum_{\alpha \in D} n_\alpha \alpha$ with all $n_\alpha \geq 0$ or ≤ 0. It is clear how to define systems of simple roots in R.

(d) Assume that R is irreducible and non-reduced, let D be a basis of R and let $\mathcal{D}$ be the corresponding Dynkin diagram (9.5.1). It follows from 8.2.8 (ii) that there is $\alpha \in D$ with $2\alpha \in R$. (b) shows that α is a long root and that the corresponding vertex of $\mathcal{D}$ is joined to other vertices only by double bonds. The classification of irreducible root systems shows that if R_i is irreducible it must be of type B_n (see 7.4.7). The corresponding non-reduced root system is denoted by BC_n $(n \geq 1)$. With the notations of 7.4.7, the root system of type BC_n is $\{\pm\epsilon_i, \pm 2\epsilon_i, \pm\epsilon_i \pm \epsilon_j \mid 1 \leq i, j \leq n, \ i \neq j\}$.

15.3.10. Examples. (1) Assume that char $F \neq 2$. Let $s \in \mathbf{GL}_n(F)$ be symmetric and define the F-automorphism θ of $\mathbf{GL}_n$ by $\theta x = s({}^t x)^{-1} s^{-1}$. By 12.1.8 (2) the fixed point group $\{x \in \mathbf{GL}_n \mid \theta x = x\}$ is defined over F. This is the orthogonal group of s, a form of $\mathbf{O}_n$. Its identity component G is a form of $\mathbf{SO}_n$.

Let $V = k^n$. The symmetric matrix s defines a quadratic form Q on the F-vector space $V(F) = F^n$, by $Q(v) = vs({}^t v)$ (where $v \in V(F)$ is viewed as a $1 \times n$-matrix). We may assume, by passing to another basis of V, that with $v = (a_1, ..., a_n)$, $v_0 = (a_{r+1}, \dots, a_{n-r})$ we have

$$Q(v) = a_1 a_n + \cdots + a_r a_{n-r+1} + Q_0(v_0),$$

where Q_0 is an anisotropic quadratic form defined by a symmetric $(n-2r) \times (n-2r)$-matrix s_0. This means that $Q_0(v_0) = 0$ only if $v_0 = 0$. The integer r is the *Witt index* of Q or s (see [Jac4, p. 351]). The group G is the special orthogonal group defined by Q. We denote by G_0 the corresponding group defined by s_0.

Let A be a split F-subtorus of G and let $v \in V(F)$ be a weight vector of A in $V(F)$, with weight χ. For $a \in A$ we have $Q(v) = Q(a.v) = \chi(a)^2 Q(v)$, which implies that $Q(v) = 0$ if $\chi \neq 0$. It follows that if Q is anisotropic the group G does not contain non-trivial split subtori.

r being as before, let $S \subset \mathbf{GL}_n$ be the subgroup of diagonal matrices of the form

$$\mathrm{diag}(x_1, \dots, x_r, 1, \dots, 1, x_r^{-1}, \dots, x_1^{-1}),$$

with $x_i \in k^*$. It is an F-split subtorus of G. The centralizer $Z_G(S)$ is F-isomorphic to $S \times G_0$. It follows that S is a maximal F-split subtorus of G. The Lie algebra $L(G)$ is the space of $n \times n$-matrices x such that xs is skew-symmetric. One checks that the root system $_F R$ is of type B_r unless $n = 2r$, in which case it is of type D_r. This example generalizes 7.4.7 (3), (4).

(2) In the case of the preceding example let $E = F(\sqrt{\alpha})$ be a quadratic extension of F. Put

$$s^\sharp = \begin{pmatrix} s & 0 \\ 0 & -\alpha s \end{pmatrix}.$$

Let $G^\sharp \subset \mathbf{GL}_{2n}$ be the orthogonal group of $s^\sharp$ and let H be the centralizer in $G^\sharp$ of

$$\begin{pmatrix} 0 & 1_n \\ \alpha 1_n & 0 \end{pmatrix}.$$

Then H is defined over F. The elements of H lie in $G^\sharp$ and have the form

$$\begin{pmatrix} x & y \\ \alpha y & x \end{pmatrix}. \tag{71}$$

$H(F)$ is the unitary group of the Hermitian form H on $V(E)$ defined by $H(v) = vs^t\bar{v}$, the bar denoting the non-trivial automorphism of E/F. If the Hermitian form H_0 defined by s_0 is anisotropic then r is the Witt index of H (see [Dieu, Ch. I, §11]). The elements of the form (71), with x in the torus S and $y = 0$, form a maximal F-split torus S' in H. H is an F-form of $\mathbf{GL}_n$ (compare with 12.3.8 (4)). If $g \in \mathbf{GL}_n$ denote by ϕg the matrix of the form (71) with $2x = s(^tg)^{-1}s^{-1} + g$, $2\sqrt{\alpha}y = -s(^tg)^{-1}s^{-1} + g$. Then ϕ is an E-isomorphism $\mathbf{GL}_n \to H$. Using this fact one can describe the action of S' on $L(H)$. If $2r < n$ the root system of (H, S') is non-reduced of type BC_r, otherwise it is of type C_r.

15.3.11. Exercises. (1) Complete the details in 15.3.10.

(2) Assume that $F = F_s$ and that E/F is a finite extension. Let G be a connected reductive F-group and let T be a maximal F-torus of G. Denote by G' the F-group $\Pi_{E/F}G$ of 12.4.2. By 12.4.4 (2) there is an injective F-homomorphism $\sigma : G \to G'$.

   (a) G' is F-reductive and $T' = \sigma T$ is a maximal F-torus of G'.

   (b) Deduce from 15.3.4 that if char $F \neq 2$ the root data of (G, T) and (G', T') are isomorphic.

15.4. The groups $U_{(a)}$

15.4.1. The assumptions and notations are as in the preceding section. So S is a maximal F-split F-subtorus of the connected F-reductive group G and $_FR$ is the relative root system. Let $a \in {}_FR$. As before, write $S_a = (\mathrm{Ker}\, a)^0$, $G_a = Z_G(S_a)$; these are F-subgroups of G. Define $U'_{(a)} = U_{G_a}(a^\vee)$ (notation of 13.4.1).

15.4.2. Lemma. *(i) $U'_{(a)}$ is a connected, unipotent, F-subgroup of G that is F-split. It is normalized by S;*

(ii) The Lie algebra $L(U'_{(a)})$ is the sum of the weight spaces of S in $L(G)$ corresponding to the roots which are positive rational multiples of a;

(iii) If λ is a cocharacter of S with $\langle a, \lambda \rangle > 0$ then $U'_{(a)} \subset U_G(\lambda)$.

(i) follows from 13.4.2 (i) and 15.1.5, and (ii) from 5.4.7 and 13.4.2 (ii). Let λ be as in (iii). Define $\mu \in X_*(S)$ by

$$2\lambda = \langle a, \lambda \rangle a^\vee + \mu.$$

Then $\langle a, \mu \rangle = 0$, which means that Im μ lies in the center of G_a. Using 13.4.7 (1) we see that

$$U_{G_a}(\lambda) = U_{G_a}(2\lambda) = U_{G_a}(\langle a, \lambda \rangle a^\vee) = U_{G_a}(a^\vee) = U'_{(a)}.$$

It follows that $U'_{(a)} \subset U_G(\lambda)$, as asserted. □

Since $_F R$ is a -possibly non-reduced- root system, the set of roots that are positive rational multiples of a is either $\{a\}$ or $\{a, 2a\}$ or $\{\frac{1}{2}a, a\}$, as follows from the facts reviewed in 15.3.9. In the first two cases we put $U_{(a)} = U'_{(a)}$.

15.4.3. Lemma. *Assume that* $\frac{1}{2}a$, $a \in {}_F R$. *There exists a unique connected, F-split, F-subgroup* $U_{(a)}$ *of the center of* $U_{(\frac{1}{2}a)}$ *whose Lie algebra is the weight space in* $L(G)$ *of the root a.*

Put $V = U_{(\frac{1}{2}a)}$ and let V_1 be the commutator subgroup (V, V) if $p = 0$ or the group generated by (V, V) and $V^{(p)}$ if $p > 0$. Then V_1 is a connected F-subgroup normalized by S. An argument like the one in the proof of 15.3.4, using 14.3.11, shows that a is the only weight of S in the Lie algebra $L(V_1)$. Moreover, V_1 lies in the center of V, and $\tilde{V} = V/V_1$ is an elementary unipotent F-group. By 14.3.7 (3), $\tilde{V}$ is a direct product $\tilde{V}_{\frac{1}{2}a} \times \tilde{V}_a$, such that $L(\tilde{V}_\chi)$ is the χ-weight space of S. Let $U_{(a)}$ be the inverse image of $\tilde{V}_a$ for the map $V \to \tilde{V}$. This is an F-subgroup of V (by 11.2.14 (ii)), whose Lie algebra is the a-weight space in $L(V)$. As before, one shows that it is a central subgroup of V. It is F-split as a consequence of 14.4.2. The uniqueness of $U_{(a)}$ follows from the uniqueness of the groups G_χ of 14.3.7 (3), observing that the image of $U_{(a)}$ in $\tilde{V}$ must be $\tilde{V}_a$. □

We have now defined for any root $a \in {}_F R$ an F-split unipotent group $U_{(a)}$, whose Lie algebra is the sum of the weight spaces in $L(G)$ for the positive integral multiples of a.

Next let $\lambda \in X_*(S)$ be an arbitrary cocharacter. We have the F-groups $P(\lambda)$, $Z(\lambda)$ $U(\lambda)$ of 13.4.1. Then $P(\lambda) = Z(\lambda).U(\lambda)$.

15.4.4. Lemma. *(i)* $U(\lambda)$ *is generated by the* $U_{(a)}$ *with* $\langle a, \lambda \rangle > 0$;
(ii) $Z(\lambda)$ *is F-reductive. It is generated by* $Z(S)$ *and the* $U_{(a)}$ *with* $a \in {}_F R$, $\langle a, \lambda \rangle = 0$.

By 15.4.2 (i) the groups $U_{(a)}$ of (i) are contained in $U(\lambda)$. The group U' generated by them is a closed, connected F-subgroup of $U(\lambda)$ (2.2.7 (ii)). It follows from 13.4.2 (ii) that the Lie algebras of U' and $U(\lambda)$ coincide. Hence the groups coincide, proving (i).

From 7.6.4 (i) we see that $R_u Z(\lambda) \subset R_u G$, from which the first point of (ii) follows. It is obvious that $Z(S) \subset Z(\lambda)$. If a is as in (ii), then Im λ lies in the center of the group G_a, whence $U_{(a)} \subset Z(\lambda)$. Now (ii) follows by comparing Lie algebras, as in the proof of (i) (using 5.4.7). □

Fix a system of positive roots $_F R^+$ in $_F R$ and let D be the corresponding basis of $_F R$. Let I be a subset of D. It is a basis of a subsystem $_F R_I$. Denote by L_I the subgroup of G generated by $Z(S)$ and the G_a with $a \in {}_F R_I$. Denote by U_I the subgroup generated by the $U_{(a)}$ with $a \in {}_F R^+ - {}_F R_I$ and let P_I be the subgroup generated by L_I and U_I.

15.4.5. Lemma. (i) P_I is a pseudo-parabolic F-subgroup of G;
(ii) U_I is a closed, connected, normal, unipotent, F-subgroup, which is F-split;
(iii) L_I is a closed, connected F-reductive F-subgroup of P_I and the product morphism $L_I \times U_I \to P_I$ is an F-isomorphism of varieties.

Choose a cocharacter λ of S such that $\langle a, \lambda \rangle = 0$ for $a \in I$ and $\langle a, \lambda \rangle > 0$ for $a \in D - I$. It follows from the preceding lemma that $U_I = U(\lambda)$, $L_I = Z(\lambda)$. Now most of the statements follow from 13.4.2. L_I is F-reductive by 15.4.4 (ii) and U_I is F-split as a consequence of 15.1.5 and 14.3.8 (ii). □

The groups P_I are the *standard pseudo-parabolic subgroups* (relative to S and D).

15.4.6. Theorem. (i) Any pseudo-parabolic F-subgroup of G is G(F)-conjugate to a standard one P_I, and I is unique;
(ii) Two minimal pseudo-parabolic F-subgroups are G(F)-conjugate.

(i) generalizes 8.4.3 (iv). Let λ be a cocharacter of G over F and put $P = P(\lambda)$. By 15.2.6 we may assume that Im $\lambda \subset S$. The set of $a \in {}_F R$ with $\langle a, \lambda \rangle > 0$ is contained in a system of positive roots $_F \tilde{R}^+$ of $_F R$. In fact, with the notations of 15.3.6 we may take

$$_F \tilde{R}^+ = \{a \in {}_F R \mid (a, x) > 0\},$$

where $x \in {}_F V$ is close to λ and not orthogonal to any root of $_F R$. Since $_F W$ acts transitively on the systems of positive roots (see 8.2.8) there is $w \in {}_F W$ with $w._F \tilde{R}^+ = {}_F R^+$. By 15.3.5 (i) and 15.3.7 (ii), w can be represented by an element of $G(F)$. It follows that we may assume $_F \tilde{R}^+ = {}_F R^+$. Take $I = \{a \in D \mid \langle a, \lambda \rangle > 0\}$. Then $P = P_I$. Except for the uniqueness statement, we have proved (i).

It is clear from the definitions that if $I, J \subset D$ and $I \subset J$, we have $P_I \subset P_J$. It follows that $P_\emptyset$ is the minimal standard pseudo-parabolic F-subgroup. Then (ii) is a consequence of what we already proved.

Now let P_I and P_J be two standard pseudo-parabolic subgroups and suppose that there is $g \in G(F)$ with $P_I = g P_J g^{-1}$. Then $P_\emptyset$ and $g P_\emptyset g^{-1}$ are two minimal pseudo-parabolic F-subgroups of G, contained in P_I. By (ii), applied to the F-reductive group L_I, they are conjugate by an element of P_I. Hence we may assume that g normalizes $P_\emptyset$. By 13.4.7 (3) we have $g \in P_\emptyset$, whence $P_I = P_J$. Then $I = J$, as one sees by considering the root systems of these groups relative to S. □

15.4.7. We obtain from the preceding results the following explicit description of minimal pseudo-parabolic F-subgroups. Let S be a maximal F-split subtorus of G, with corresponding root system ${}_F R$. Fix a system of positive roots ${}_F R^+$. Then the subgroup generated P by $Z(S)$ and the $U_{(a)}$ with $a \in {}_F R^+$ is a minimal pseudo-parabolic F-subgroup and any such subgroup is obtained in this manner.

Now assume, moreover, that S is a maximal torus of G, i.e. that, with previous notations, $T = S$. (This is the case if $F = F_s$.) Then the image of P in $G/R_u G$ is a Borel group. It follows that P is solvable. By 15.4.5 (ii) it is F-split. By the next exercise it is a maximal connected, solvable, F-subgroup, which is F-split (a pseudo-Borel subgroup).

15.4.8. Exercise. Let G be an arbitrary F-group, containing a maximal torus defined over F and F-split. Two maximal connected, solvable, F-split, F-subgroups of G are conjugate by an element of $G(F)$.

15.5. The index

We keep the assumptions of the preceding sections. $\pi : V \to {}_F V$ is as in 15.3.6. Recall that by 15.3.4 the root system ${}_{F_s} R$ coincides with the root system R of (G, T) if $p \neq 2$ or if G is reductive. These two root systems have the same Weyl group W. Let ${}_F R^+$ be a system of positive roots in ${}_F R$, defining the basis ${}_F D$.

15.5.1. Lemma. (i) *Let* $\alpha \in {}_{F_s} R$ *and assume* $a = \pi\alpha \neq 0$. *Then* $a \in {}_F R$;
(ii) *There exists a system of positive roots* R^+ *in* ${}_{F_s} R$ *with the following property:* $\alpha \in R^+$ *if and only if* $a = \pi\alpha \in {}_F R^+$.

(i) is clear from the definition of π. There is $x \in {}_F V$ such that

$${}_F R^+ = \{ a \in {}_F R \mid (a, x) > 0 \}$$

(see 7.4.5). We choose $y \in V$ orthogonal to ${}_F V$ such that $x + y$ is not orthogonal to any root of R and that if $\alpha \in {}_{F_s} R$, $\pi\alpha = a \in {}_F R$ then $(\alpha, x + y)$ and (a, x) have the same sign (the last condition is fulfilled if y is sufficiently close to 0). Then

$$R^+ = \{ \alpha \in {}_{F_s} R \mid (\alpha, x + y) > 0 \}$$

is a system of positive roots R^+ in R with the property of (ii). □

15.5.2. Let R^+ be as in 15.5.1 (ii). Denote by D the basis defined by R^+ and by $\mathcal{D}$ or $\mathcal{D}_G$ the corresponding Dynkin diagram (9.5.1). The Galois group $\Gamma = \mathrm{Gal}(F_s/F)$

acts linearly in V via a finite quotient A and it follows from 13.2.2 (i) that $_F V$ is the subspace of V whose elements are fixed by all $s \in A$. Also, Γ acts as an automorphism group on $_{F_s} R$ and on the Weyl group W. We may assume that our bilinear form $(\ ,\ )$ is A-invariant (by an argument like that used in 7.1.7 to obtain a W-invariant form), hence it is Γ-invariant. If $\gamma \in \Gamma$ then $\gamma.R^+$ is also a system of positive roots of $_{F_s} R$, and there is a unique $w_\gamma \in W$ with $w_\gamma(\gamma.R^+) = R^+$ (see 8.2.8). Then $w_\gamma(\gamma.D) = D$. For $\gamma \in \Gamma$, $\alpha \in D$ put

$$\tau(\gamma)(\alpha) = w_\gamma(\gamma.\alpha).$$

It is immediate that τ defines a continuous homomorphism of Γ to the permutation group of D and the automorphism group $\mathrm{Aut}(\mathcal{D})$ of the Dynkin diagram $\mathcal{D}$.

We have the map $\pi : {}_{F_s} R \to {}_F R \cup \{0\}$. Put $R_0 = \{\alpha \in R \mid \pi\alpha = 0\}$. It is a subsystem of R with basis $D_0 = D \cap R_0$. Let W_0 be its Weyl group and denote by $\mathcal{D}_0$ be the full subgraph of $\mathcal{D}$ with vertex set D_0.

15.5.3. Proposition. *(i)* $\tau(\Gamma)$ *stabilizes* D_0 *and* $D - D_0$;
(ii) $\mathcal{D}_0$ *is the Dynkin diagram* $\mathcal{D}_{Z(S)}$ *(relative to the system of positive roots* $R(Z(S), T) \cap R^+$*);*
(iii) We have $\pi(D - D_0) = {}_F D$. *If* $\alpha, \beta \in D - D_0$ *then* $\pi\alpha = \pi\beta$ *if and only if* α *and* β *lie in the same* $\tau(\Gamma)$*-orbit.*

Let $\alpha = \sum_{\delta \in D} n_\delta \delta \in {}_{F_s} R$. Then $\pi\alpha = \sum n_\delta(\pi\delta)$. It follows that $\pi\alpha = 0$ if and only if $\alpha \in R_0$. If $\alpha \notin R_0$ we have $\pi\alpha \in {}_F R$ and as a consequence of 15.5.1 (ii) a positive root of $_F R$ is a positive integral linear combination of roots in $\pi(D - D_0)$, which implies the first point of (iii).

Now let $\gamma \in \Gamma$. Then $\pi(\gamma.\alpha) = \pi\alpha$. We conclude from the preceding observations and 15.5.1 (i) that $\gamma.R_0 = R_0$ and $\gamma(R^+ - R_0) = R^+ - R_0$. There is $w_\gamma \in W_0$ with $w_\gamma(\gamma.D_0) = D_0$. Since W_0 stabilizes $R^+ - R_0$ we have that $w_\gamma(\gamma.R^+) = R^+$. It follows that w_γ is the element of 15.5.2. We also see that $\tau(\Gamma)$ stabilizes D_0 and $D - D_0$. proving (i).

The root system $_{F_s} R(Z(S), T)$ consists of the roots of $_{F_s} R$ whose restriction to S is trivial, i.e. of the $\alpha \in R$ with $\pi\alpha = 0$. We have seen that these are the roots of R_0, whence (ii).

With the previous notations, the orthogonal projection $\pi : V \to {}_F V$ is given by

$$\pi = |A|^{-1} \sum_{s \in A} s.$$

Since the elements w_γ lie in W_0, we see that if α and β are as in (iii) we have

$$\sum \tau(\gamma)(\alpha) = \sum \tau(\gamma)(\beta).$$

the summation being over a set of representatives in Γ of the elements of A. This observation implies the last part of (iii). $\square$

15.5.4. Corollary. *If $\gamma \in \Gamma$, $\alpha \in D - D_0$ then $\gamma.\alpha = \tau(\gamma).\alpha + \sum_{\delta \in D_0} n_\delta \delta$, where the n_δ are positive integers.*

This follows from the proof of (i). □

15.5.5. $_F R^+$ being fixed, we had chosen R^+ with the property of 15.5.1 (ii). Let $\tilde{R}^+$ be another system of positive roots on R with that property. There is $w \in W$ with $\tilde{R}^+ = w.R^+$. Then $w.R^+ - R_0 = R^+ - R_0$, which implies that if $\alpha \in R^+$ and $w^{-1}.\alpha \in -R^+$ we have $\alpha \in R_0$. From 8.3.2 (i) we conclude that $w \in W_0$. Let $\tilde{D}$ and $\tilde{D}_0$ be the basis defined by $\tilde{R}^+$ and the corresponding basis of R_0. It follows from the preceding remarks and 15.5.4 that there is a unique bijection of D onto $\tilde{D}$, mapping D_0 onto $\tilde{D}_0$ and commuting with the respective Γ-actions.

Another system of positive roots $_F R^+$ in $_F R$ is of the form $w._F R^+$, with $w \in {}_F W$. From 15.3.5 (ii) and 15.3.7 (ii) we conclude that there is $w_1 \in W$ that stabilizes $_F V$ and whose restriction to $_F V$ is w. It follows that $w_1.R^+$ is a system of positive roots having the property of 15.5.1 (ii) relative to $w._F R^+$. We again have a bijection of D onto the basis defined by $w_1.R^+$, with similar properties. The triple (D, D_0, τ) is the *index* of G, relative to F, S, T. In fact, the index does not depend on S and T.

Let T' be another maximal F-torus containing S, with root system R'. By 13.3.1 (i) and 11.2.7, T' is conjugate to T by an element of $Z(S)(F_s)$. We obtain an isomorphism $\phi : R \to R'$. Then $(R')^+ = \phi R^+$ is a system of positive roots in R', with basis $D' = \phi D$. Put $D'_0 = \phi D_0$, $\tau' = \phi \circ \tau \circ \phi^{-1}$. Then the index of G relative to F, S, T' is isomorphic to (D', D'_0, τ') (we skip the straightforward check). One sees similarly, as a consequence of 15.2.6, that the index is independent of the choice of the maximal F-split torus S. So the index depends only on F, up to isomorphism.

In chapter 17 we shall describe the possible indices of reductive F-groups. The following lemmas will be useful.

15.5.6. Lemma. *If $_{F_s} R$ is irreducible then so is $_F R$.*

For the notion of irreducibility see 8.1.12 (4). Let R^+ be a system of positive roots in $_{F_s} R$, as in 15.5.1 (ii), with basis D. So $\pi R^+ = {}_F R^+$ is a system of positive roots in $_F R$, with basis $\pi D = {}_F D$.

Assume that $_{F_s} R$ is irreducible. Then R^+ contains a unique highest root $\tilde{\alpha}$, characterized by the property that $\langle \tilde{\alpha}, \delta^\vee \rangle \geq 0$ for all $\delta \in D$ (see [Bou2, p. 165]). If $\alpha \in R^+$ is another positive root, there is $\delta \in D$ with $\langle \alpha, \delta \rangle < 0$. It follows that $s_\delta.\alpha = \alpha + m\delta$, where $m > 0$. We conclude that for any $\alpha \in R^+$ there is a chain of positive roots $\alpha = \alpha_1, ..., \alpha_s, \alpha_{s+1} = \tilde{\alpha}$, such that $\alpha_{i+1} - \alpha_i$ is a positive multiple of a simple root ($1 \leq i \leq s$). Define $\tilde{a} \in {}_F R$ by $\tilde{a} = \pi \tilde{\alpha}$. It follows from 15.5.1 that if $a \in {}_F R^+$ there is a similar chain $a = a_1, ..., a_s, a_{s+1} = \tilde{a}$. This can only be if $_F R$ is irreducible.

□

15.5.6 has the following complement.

15.5.7. Lemma. *In the case of 15.5.6, the highest root of $_FR$ is $\tilde{a} = \pi\tilde{\alpha}$, where $\tilde{\alpha}$ is the highest root of $_{F_s}R$.*

Put $\tilde{\alpha} = \sum_{\delta\in D} n_\delta\delta$. Then if $\alpha = \sum_{\delta\in D} m_\delta\delta$ is any other positive root, we have $n_\delta \geq m_\delta$ for all $\delta \in D$ (see loc. cit.). This implies that the highest root $\tilde{\alpha}$ is characterized by the chain property of the proof of 15.5.6. Applying this to $_FR$ we obtain 15.5.7.   □

If R^+ is as in 15.5.1 (i) there is a unique element $w_0 \in W$ with $w_0.R^+ = -R^+$ (by 8.2.4 (ii)). Then $w_0D = -D$, and $\iota.\alpha = -w_0.\alpha$ defines a permutation of D, the *opposition involution*. We have $\iota^2 = \mathrm{id}$.

15.5.8. Lemma. *ι stabilizes the index (D, D_0, τ).*

This means that ι stabilizes D_0 and commutes with all $\tau(\gamma)$ ($\gamma \in \Gamma$). The argument of the second paragraph of 15.5.5 shows that there is $w_1 \in W$ that stabilizes $_FV$ such that $w_1._FR^+ = -_FR^+$. By the first paragraph of 15.5.5, with $\tilde{R}^+ = -w_1^{-1}.R^+$ we conclude that there is $w_2 \in W_0$ with $w_2.R^+ = -w_1^{-1}.R^+$, whence $w_0 = w_1w_2$. Since w_1 commutes with π it stabilizes R_0 and the same is obviously true for w_2. Hence w_0 stabilizes R_0 and ι stabilizes $R_0 \cap D = D_0$. w_γ being as in 15.5.2 ($\gamma \in \Gamma$) it follows from its definition that the permutations w_0 and $w_\gamma\gamma$ of R commute. Hence ι commutes with all $\tau(\gamma)$. The lemma is proved.   □

15.5.9. Exercise. Describe the index of the groups in the examples of 15.3.10 (1), (2).

Notes

In the case of a reductive F-group G the results of this chapter are contained in the fundamental paper [BoT1] of Borel and Tits. For general groups most of the results also are due to them. See [BoT3], see also [Ti6, 1991-1992, 1992-1993].

An F-variety with the property (P_F) of 15.2.1 could be viewed as being 'complete relative to F', in a weak sense. In [BoT3] a better notion of completeness is used, in which in the definition of P_F the variety $\mathbf{A}^1 - \{0\}$ is replaced by Spec $F((T))$ and $\mathbf{A}^1$ by Spec $F[[T]]$. A proof of the corresponding version and a discussion of its consequences is contained in [Ti6, 1992-1993, p. 117-119]. One also finds in this reference, and in [Ti6, 1991-1992] a discussion of the problem of classification of non-reductive F-reductive groups, which we have not gone into. The classification of reductive F-groups will be discussed in Chapter 17.

Chapter 16

Reductive F-groups

This chapter is a continuation of the preceding one. We now consider the case of reductive groups. Some basic results about parabolic subgroups are established. The chapter is mainly devoted to a discussion of versions over F of the isomorphism and existence theorems. G is a connected, reductive F-group.

16.1. Parabolic subgroups

Since G is reductive, the pseudo-parabolic subgroups of G are its parabolic F-subgroups (15.1.2 (ii)). Let P be such a group. Recall (see 8.4.4) that a *Levi subgroup* of P is a closed subgroup L such that the product map $L \times R_u P \to P$ is bijective.

16.1.1. Proposition. *(i) The unipotent radical $R_u P$ is defined over F and F-split;*
(ii) There exist Levi subgroups of P defined over F. Two such subgroups are conjugate by a unique element of $R_u P(F)$;
(iii) P contains a maximal F-split F-subtorus of G.

In the proof of 15.1.2 (ii) it was shown that there is a cocharacter μ of G defined over F, such that $P = P(\mu)$. By 13.4.2 (i) the group $Z(\mu)$ is a connected F-group which is reductive by 7.6.4 (i). Also, $U(\mu)$ is a connected, unipotent F-group, which is F-split by 15.1.5. Since it is normal in P it must coincide with $R_u P$, which thus has the properties asserted in (i).

It follows from 13.4.2 (i) that $Z(\mu)$ is a Levi subgroup of P which is defined over F, proving the first part of (ii).

Let L be any Levi subgroup of P defined over F. Then L contains maximal tori of G and by 13.3.6 (i) (respectively, 13.3.7) it contains a maximal F-torus T. By 8.4.4, L is uniquely determined by T. The argument used in the proof of 15.1.2 then gives that there is a cocharacter ν of G defined over F such that $L = Z(\nu)$. By 13.1.1 (ii) the torus T is split over F_s. The identity component C_L of the center of L is a subtorus of T, which is also defined over F_s (since all cocharacters of T are defined over F_s, see 13.2.2 (i)). Also, $C_L(F_s)$ is stable under the Galois group Γ. It follows from 11.2.8 (i) that C_L is defined over F. Since $L = Z(\nu)$ we have a fortiori $L = Z_G(C_L)$. Using 8.4.4 and the conjugacy of maximal tori of P we see that there is $x \in R_u P$ with $L = x Z(\mu) x^{-1}$ (where μ is as before). If $u \in R_u P = U(\mu)$ normalizes $Z(\mu)$ it centralizes $Z(\mu)$, as a consequence of 13.4.2 (i). It follows that the element x is unique. So it is the only element of the transporter $N(C_{Z(\mu)}, C_L)$. By 13.3.1 (i) it must lie in $G(F)$. This proves the second part of (ii).

P contains a minimal parabolic F-subgroup. The description of these groups given in 15.4.7 shows that (iii) holds. $\quad\square$

16.1.2. Lemma. *Let Q be another parabolic F-subgroup of G. The intersection $P \cap Q$ is an F-subgroup containing a maximal F-split F-subtorus of G.*

To prove that $P \cap Q$ is defined over F it suffices by 12.1.5 to show that $L(P \cap Q) = L(P) \cap L(Q)$. Since obviously $L(P) \cap L(Q) \subset L(P) \cap L(Q)$, it also suffices to show that

$$\dim L(P \cap Q) = \dim(L(P) \cap L(Q)). \tag{72}$$

It follows from 8.3.10 that $P \cap Q$ contains a maximal torus T of G. Now (72) follows from 8.1.12 (1). So $P \cap Q$ is defined over F.

By 8.4.6 (2), $R = (P \cap Q)R_u P$ is a parabolic subgroup, with normal subgroup $R_u P$. Because $P \cap Q$ and $R_u P$ are defined over F, the same holds for R. By 16.1.1 (iii), R contains a maximal F-split F-subtorus of G. Then the same must be true for $P \cap Q$, as one sees by passing to the quotient $R/R_u P$. $\square$

Let P be a minimal parabolic F-subgroup of G. Fix a maximal F-split F-subtorus S of G contained in P, and let $_F W = N_G(S)/Z_G(S)$ be the relative Weyl group (15.3.1). By 15.3.5 (i) and 15.3.7 (ii) the elements of $_F W$ can be represented by elements of $N_G(S)(F)$. For $w \in {}_F W$ let $\dot{w} \in N_G(S)(F)$ be a representative. Put $C(w) = P(F)\dot{w}P(F)$.

16.1.3. Theorem. [Bruhat's lemma for $G(F)$) $G(F)$] *is the disjoint union of the sets $C(w)$, $w \in {}_F W$.*

Let $g \in G(F)$. By 16.1.2 the intersection $P \cap gPg^{-1}$ contains a maximal F-split F-subtorus, which by 15.2.6 can be written as $x^{-1}Sx$ with $x \in P(F)$. This means that $(xg)^{-1}S(xg) \subset P$. Another application of 15.2.6 gives $y \in P(F)$ such that xgy normalizes S. It follows that there is $w \in {}_F W$ with $g \in C(w)$.

It remains to prove the uniqueness of w. Let T be a maximal F-torus of P containing S, let $L = Z_G(S)$ (a Levi group of P) and denote by W and W_L the Weyl groups of (G, T), respectively (L, T). By 15.3.7 (ii) we can represent the elements of $_F W$ by elements of $N_G(T)$. It follows that $_F W$ can be viewed as a subgroup of W which normalizes W_L. Now if $w, z \in {}_F W$ and $C(w) = C(z)$ we have $P\dot{w}P = P\dot{z}P$. It follows from 8.4.6 (3) that $z \in W_L w W_L = w W_L$. Hence $z^{-1}w \in W_L \cap {}_F W$. Since S is central in L we must have $z = w$, finishing the proof of 16.1.3. $\square$

Remark. There is a more precise result: $(P(F), N_G(S)(F))$ is a Tits system in $G(F)$, see [BoT1, p. 101] (for Tits systems see [Bou2, Ch.IV, §2]).

16.1.4. Exercise. The notations are as in 16.1.1. Let T be a maximal F-torus of P. Show that the unique Levi subgroup of P containing T (see 8.4.4) is defined over F.

16.2. Indexed root data

16.2.1. Let $\Psi = (X, R, X^\vee, R^\vee)$ be a root datum (7.4.1). Let D be a basis of R, with a corresponding system of positive roots R^+. Then $(R^+)^\vee$ is a system of positive roots in $R^\vee$ let $D^\vee$ be the corresponding basis. We call the quadruple $\Psi_0 = (X, D, X^\vee, D^\vee)$ a *based root datum*.

An *indexed root datum* over F is a sextuple

$$(X, D, X^\vee, D^\vee, D_0, \tau),$$

where $(X, D, X^\vee, D^\vee)$ is a based root datum, D_0 is a subset of D, and τ is a continuous action of the Galois group $\Gamma = \mathrm{Gal}(F_s/F)$ on the group X, stabilizing the sets D and D_0.

Let again S and T be a maximal F-split F-torus in G and a maximal F-torus containing S. We have the root datum $\Psi(G, T) = (X, R, X^\vee, R^\vee)$ of 7.4.3. (By 15.3.4 (ii) it coincides with the root datum $_{F_s}\Psi$ of 15.3.8). Fix a basis D of R. Let (D, D_0, τ) be the index of G, defined in 15.5.5. Then $_i\Psi = {}_i\Psi(G) = (X, D, X^\vee, D^\vee, D_0, \tau)$ is the *indexed root datum* of G over F. The observations of 15.5.5 show that is independent of the choice of S and T, up to isomorphism.

We say that our reductive F-group G is *split* or F-split if $S = T$, i.e. if G contains a maximal F-torus which is F-split. In that case $D_0 = \emptyset$ and τ is trivial. G is *quasi-split* over F if $D_0 = \emptyset$ and *anisotropic* if $D_0 = D$. Clearly, if G is split it is quasi-split.

16.2.2. Proposition. *(i) G is quasi-split over F if and only if there exists a Borel subgroup of G that is defined over F;*
(ii) G is anisotropic if and only if none of its proper parabolic subgroups is defined over F.

G is quasi-split if and only if the restriction map $X^*(T) \to X^*(S)$ is injective. Using 5.4.7 we conclude that G is quasi-split if and only if $Z(S) = T$. If this is so, choose a cocharacter λ of S with $Z(\lambda) = Z(T) = S$ (notation of 13.4.1). It follows from 13.4.2 (i) that $P(\lambda)$ is a solvable F-group, which is parabolic by 15.1.2 (ii). From 6.2.7 (i) we conclude that $P(\lambda)$ is a Borel subgroup defined over F. Conversely, by 15.1.2 (ii) a subgroup of G with these properties is of the form $P(\lambda)$, for some F-cocharacter λ of G. The torus $\mathrm{Im}\,\lambda$ is F-split. An application of 15.2.6 shows that we may assume $\mathrm{Im}\,\lambda \subset S$, i.e. that λ is a cocharacter of S. Since $P(\lambda)$ is solvable, the connected, reductive group $Z(\lambda)$ must be a torus T. It follows that $Z(S) = T$. Hence G is quasi-split. This proves (i). (ii) is a consequence of 15.1.2. $\qquad\square$

16.2.3. Central isogenies. The notations are as in 16.2.1. Let G_1 be another connected, reductive F-group and let $\phi : G \to G_1$ be a central F-isogeny (9.6.3). We put $S_1 = \phi S$, this is a maximal F-split F-subtorus of G_1. Then $T_1 = \phi T$ is a maximal

F-torus of G_1 containing S_1. Let $(X_1, R_1, X_1^\vee, R_1^\vee)$ be the root datum of (G_1, T_1). The isogeny ϕ defines an isomorphism f of X_1 onto a subgroup of finite index of X. Its transpose $f^\vee$ is an isomorphism of $X^\vee$ onto a subgroup of finite index of $X_1^\vee$. Moreover (see 9.6.3) there is a bijection $b : R \to R_1$ such that $f(b\alpha) = \alpha$ ($\alpha \in R$). To simplify notations we view X_1 as a subgroup of X and f as the inclusion map. Then $R_1 = R$ and $b = \mathrm{id}$. The indexed root datum of G_1 is $(X_1, D, X_1^\vee, D^\vee, D_0, \tau)$.

The Galois group $\Gamma = \mathrm{Gal}(F_s/F)$ acts on X, and X_1 is a Γ-stable subgroup of X containing R.

16.2.4. Lemma. *Let X_1 be a subgroup of finite index of X that is Γ-stable and contains R. There exists a central isogeny $\phi : G \to G_1$, giving rise to the subgroup X_1 in the manner described in 16.2.3.*

There exists an F-torus T_1 with character group X_1, together with an F-isogeny $\psi : T \to T_1$, such that ϕ induces the inclusion map $X_1 \to X$ (see 13.1.5 (1)). By an easy induction one is reduced to proving the lemma in the case that X/X_1 is an elementary abelian l-group, for some prime l. When l is not the characteristic p of F, then $\mathrm{Ker}\, \psi$ is a finite subgroup A of $T(F_s)$ isomorphic to X/X_1. Moreover A is Γ-stable, hence defined over F (11.2.8 (i)), and central in G. Then $G_1 = G/A$, together with the canonical homomorphism $G \to G_1$, is as required.

Now let $l = p$. Then $\mathfrak{a} = \mathrm{Ker}\, d\psi$ is a p-subalgebra of the Lie algebra $L(T)$. Moreover, $\mathfrak{a}$ is defined over F and is centralized by all elements of $\mathrm{Ad}(G)$. We have a quotient $G_1 = G/\mathfrak{a}$ (see 12.2.4). Then G_1 and the homomorphism $\phi : G \to G_1$ of 12.2.4 are as required.   □

16.2.5. As an application of the lemma, we show how to describe the based root datum of G in terms of the based root data of a semi-simple group and a torus.

The commutator subgroup $G_1 = (G, G)$ is a connected, semi-simple F-subgroup. Its root datum is described in 8.1.9. We use the notations introduced there. The maximal torus T_1 of G_1 introduced in 8.1.8 (iii) is defined over F. Its maximal F-split subtorus S_1 (see 13.2.4 (i)) lies in S, and must be a maximal F-split subtorus of G_1 (as one sees by passing to the semi-simple F-group $G/R(G)$). The indexed root datum of G_1, relative to T_1, S_1 and the Borel subgroup B_1 defined by D is

$$_i\Psi_1 = (X/(Q^\vee)^\perp, D, \tilde{Q}^\vee, D^\vee, D_0, \tau).$$

The centralizer $K = Z_{G_1}(S)$ is generated by T_1 and the subgroups U_α (see 8.1.1), where α runs through the root system R_{D_0} with basis D_0. We call K the *kernel* of G. It is an anisotropic F-group. Let $C = R(G)$ be the connected center of G, an F-subtorus of G. By 8.1.8 (ii) its character group is $X/\tilde{Q}$, so is determined by the indexed root datum. (But its structure as a Γ-module is not determined by $_i\Psi$.) The triple $(_i\Psi, K, C)$ will be used to classify reductive F-groups.

The product map $G_1 \times C \to G$ is a central isogeny. In 8.1 we have described the root datum of the first group, let X' be the corresponding character group. 8.1.10

describes the character group X as a subgroup of X'. It follows from 16.2.4 that the F-group G is determined by the F-groups G_1 and C.

This discussion shows that the problem of describing all based root data of connected reductive F-groups G reduces to two particular cases: G is semi-simple and G is a torus. The second case was taken care of in 13.1. In the sequel we shall concentrate on the first case.

16.2.6. Restriction of the ground field. Let E be a finite extension of F contained in F_s. Let G_1 be a connected, reductive E-group. We then have for G_1 the objects of 16.2.1 (maximal E-split torus, indexed root datum...). They will be denoted as in 16.2.1, with a suffix 1. Let $G = \Pi_{E/F} G_1$ be the F-group obtained from G_1 by restriction of the ground field (12.4.2). Put $\Delta = \mathrm{Gal}(F_s/E)$ and let Σ be the set of F-isomorphisms of E into F_s. We identify it with Γ/Δ. As we saw in 12.4.5, G is isomorphic to G_1^Σ from which it follows that G is a connected, reductive F-group.

$\Pi_{E/F} S_1$ is an F-subtorus of G. Denote by S its maximal F-split subtorus (13.2.4 (i)). Write $T = \Pi_{E/F} T_1$. Then T is a maximal torus of G containing S.

16.2.7. Lemma. *S is a maximal F-split F-subtorus of G. We have* $\dim S = \dim S_1$.

Clearly, S is an F-split F-subtorus. The equality $\dim S = \dim S_1$ is equivalent to the assertion that the maximal split F-subtorus of $\Pi_{E/F}(\mathbf{G}_m)^n$ has dimension n. It follows by using 11.4.7 (3) that it suffices to prove this in the case $n = 1$, in which case the asserted equality is a consequence of 13.1.5 (4) and 13.2.2 (i). Similarly, one sees that, π being as in 12.4.2, we have $\pi S = S_1$. Now let S' be an F-split F-subtorus of G containing S. Then $\pi S'$ is an E-split E-subtorus of G_1 containing S_1, hence equal to it. If $S' \neq S$ then $\mathrm{Ker}\,\pi$ would contain a non-trivial subtorus of S', which would be defined over F (by 13.2.2 (i)), contradicting 12.4.3. Hence S is maximal F-split. $\square$

The indexed root datum ${}_i\Psi$ of G is described as follows. We use the notations of 11.4.22.

16.2.8. Lemma. *(i) We have $X = X_1^\Sigma$ and similarly for $X^\vee$, D, $X^\vee$, $D^\vee$, D_0;*
(ii) τ is the permutation representation of Γ on D^Σ induced by the permutation representation τ_1 of Δ on D.

This follows from 11.4.22. $\square$

16.2.9. Exercises. (1) The notations are as in 16.2.1 and G is semi-simple. Assume that the root system of G is a direct sum of isomorphic irreducible ones, which are permuted transitively by $\mathrm{Im}\,\tau$. Then there is a finite separable extension E/F and a connected, semi-simple E-group G_1 with irreducible root system such that $G = \Pi_{E/F} G_1$.

(2) Assume that F is finite. Then G is quasi-split. (*Hint*: use Lang's theorem 4.4.17).

(3) G is split over F_s.

(4) Notations as in 16.2.5. Assume that T_1 is F-split, i.e. that $S_1 = T_1$. We have an injective homomorphism of Γ-modules $X \to X'$. Show that the induced Γ-action on the finite group X'/X is trivial.

16.3. F-split groups

16.3.1. The notations are as in 16.2.1. Assume that G is F-split. In that case the groups $U_{(a)}$ of 15.4 are the groups U_α ($\alpha \in R$) of 8.1.1. By 15.4.2 (i) these groups are defined over F. It follows that we may choose a realization $(u_\alpha)_{\alpha \in R}$ of R in G (see 8.1) such that all u_α are defined over F. Fix a system of positive roots R^+ in R, with basis D, and let B be the Borel subgroup of G generated by T and the U_α with $\alpha \in R^+$ (8.2.4 (i)). Then B is defined over F. Any Borel subgroup containing T can be obtained in this manner (by 8.1.3 (i)). The indexed root datum $_i\Psi$ of G is of the form $(X, D, X^\vee, D^\vee, \emptyset, \mathrm{id})$, so is determined by the based root datum $\Psi = (X, D, X^\vee, D^\vee)$. It is determined by B and T. We shall also denote this based root datum by $\Psi(G, B, T, F)$.

Let G_1 be another F-split group, with data T_1, B_1... as before, determining a based root datum Ψ_1. Assume that $\phi : G \to G_1$ is an F-isomorphism that maps B onto B_1 and T onto T_1. As in the situation of 9.6.1 one sees that ϕ defines an isomorphism $f = f(\phi)$ of Ψ_1 onto Ψ (the notion of isomorphism of based root data is clear).

The following theorem is a version over F of the isomorphism theorem 9.6.2.

16.3.2. Theorem. *Let f be an isomorphism of Ψ_1 onto Ψ. There exists an F-isomorphism $\phi : G \to G_1$ with $\phi T = T_1$, $\phi B = B_1$ and $f = f(\phi)$. If ϕ' is another F-isomorphism with these properties, there is $t \in T$ with $\alpha(t) \in F$ for $\alpha \in D$, such that $\phi'(g) = \phi(tgt^{-1})$ ($g \in G$).*

Choose realizations $(u_\alpha)_{\alpha \in R}$ of R in G and $(u_{\alpha_1})_{\alpha_1 \in R_1}$ of the root system R_1 of (G_1, T_1) such that all u_α and u_{α_1} are defined over F. Proceeding as in the proof of 9.6.2 we see that there exists an isomorphism $\phi : G \to G_1$ such that $\phi \circ u_\alpha = u_{f^{-1}\alpha}$ ($\alpha \in R$). It will be shown that ϕ is defined over F.

The elements n_α ($\alpha \in D$) of 8.1.4 (i) lie in $G(F)$, since u_α and $u_{-\alpha}$ are defined over F. It follows that the elements of the Weyl group W of (G, T) can be represented by elements of $N(T) \cap G(F)$. Let $w \in W$ have reduced decomposition $(s_{\alpha_1}, ..., s_{\alpha_h})$. Put $\dot{w} = n_{\alpha_1}...n_{\alpha_h}$. This element of $G(F)$ is uniquely determined by w (9.3.3). Moreover, $\phi(\dot{w}) \in G_1(F)$. The open set $C(w_0)$ of 8.3.11 is defined over F and so are its translates $\dot{w}.C(w_0)$ ($w \in W$). By 8.5.10 (1) they cover G. The restriction of ϕ to $C(w_0)$ is defined over F, and also its restriction to the sets $\dot{w}.C(w_0)$. It follows that ϕ is defined over F, proving the first part of the theorem.

To prove the last part, observe that by 9.6.2 there exists $t \in T$ with $\phi'(g) = \phi(tgt^{-1})$ $(g \in G)$. The automorphism $\text{Int}(t)$ of G is defined over F. By considering its action on the F-groups U_α one sees that $\alpha(t) \in F$ for $\alpha \in R$ or (which amounts to the same) for $\alpha \in D$. Conversely, if this the case then the restriction of $\text{Int}(t)$ to the open F-subvarieties $\dot{w}.C(w_0)$ is defined over F. As we saw above these cover G, and it follows that $\text{Int}(t)$ is defined over F, finishing the proof of 16.3.2. $\square$

The next theorem is a version over F of the existence theorem 10.1.1.
16.3.3. Theorem. *Let* $\Psi = (X, D, X^\vee, D^\vee)$ *be a based root datum. There exists a connected, reductive F-group G, which is F-split, containing a Borel subgroup B is defined over F and a maximal F-torus $T \subset B$, which is F-split, such that* $\Psi = \Psi(G, B, T, F)$.

The proof of 10.1.1 carries over. First, there is the following refinement of 10.1.3. Let X_1 be a subgroup of X of finite index containing D. There is an F-torus T_1 with character group X_1. We then have a based root datum $\Psi_1 = (X_1, D_1, X_1^\vee, D_1^\vee)$, where $D_1 = D$.

16.3.4. Proposition. *Assume that there exist (G_1, B_1, T_1) with $\Psi_1 = \Psi(G_1, B_1, T_1, F)$. Then there exists a similar triple (G, B, T) with $\Psi = \Psi(G, B, T, F)$.*

In the proof of 10.1.3 we constructed a k-group G. The construction can be carried out over F. In particular, the vector spaces V_1 and V_2 introduced at the end of the proof of 10.1.3, can be taken such that they are defined over F and that the restriction of G to $V = V_1 + V_2$ is an F-subgroup of $GL(V)$. We leave the details to the reader. $\square$

Using 16.3.4 one reduces the proof of 16.3.3 to the case that D spans X (as in 10.3.7). In the case that the root system R with basis D is simply laced, the existence is proved as in 10.2. In fact, the Lie algebra $\mathfrak{g}$ introduced in 10.2.5 is obviously defined over the prime field contained in F, and the same is true for the subgroups U_α of 10.2.7. It follows that the group G of 10.2.8 is defined over F, settling the case that R is simply laced. If R is not simply laced, we proceed as in 10.3. The groups T^σ and U_{α_O} introduced in the proof of 10.3.5 can be taken to be defined over F. It follows that the group generated by them, which is the group whose existence we have to prove, is also defined over F. $\square$

16.3.5. Let G, B, T be as in 16.3.1. So G is F-split. Let $Q \subset X$ be the subgroup of X spanned by D. The dual $Q^\vee$ is $X^\vee/Q^\perp$, where $Q^\perp$ is the annihilator of Q. It follows from 8.1.8 (iii) that we may identify $D^\vee$ with its image in $Q^\vee$. By 16.3.3 there is a connected, semi-simple F-split group G^{ad} with based root datum $(Q, D, Q^\vee, D^\vee)$. The group G^{ad} is the *adjoint group* of G. Let C be the identity component of the center of G. The F-group G/C is semi-simple. Its character group is the rational closure of $\tilde{Q}$ in X. We have an F-isogeny $G/C \to G^{\mathrm{ad}}$ and an F-homomorphism $\pi : G \to G/C \to G^{\mathrm{ad}}$.

Put $\bar{T} = \pi T$, this is a maximal F-torus in G^{ad}. Fix a realization $(u_\alpha)_{\alpha \in R}$ of R in G, such that all u_α are defined over F. Put $\bar{u}_\alpha = \pi \circ u_\alpha$. Then $(\bar{u}_\alpha)$ is a similar realization of R in G^{ad} and the restriction of π to $U_\alpha = \mathrm{Im}\, u_\alpha$ defines an isomorphism of U_α onto its image $\bar{U}_\alpha$.

Put $T(F)^\sharp = \{t \in T(\bar{F}) \mid \alpha(t) \in F \text{ for } \alpha \in D\}$.

16.3.6. Lemma. *(i) $T(F)^\sharp$ normalizes $G(F)$;*
(ii) $\bar{T}(F) = \pi T(F)^\sharp$;
(iii) $G^{\mathrm{ad}}(F)$ is generated by $\pi G(F)$ and $\bar{T}(F)$.

It follows from 16.1.3 that $G(F)$ is generated by the groups $T(F)$ and $U_\alpha(F)$ ($\alpha \in R$). Since these groups are normalized by $T(F)^\sharp$, we have (i). To prove (ii) observe that π defines a surjective homomorphism $T \to \bar{T}$ and that Q is the character group of $\bar{T}$. Then (iii) follows from 16.1.3, applied to G^{ad}. $\square$

Let $(\mathrm{Int}\, G)(F)$ be the group of inner automorphisms of G defined over F.

16.3.7. Lemma. *There is an isomorphism $\psi = \psi(F) : (\mathrm{Int}\, G)(F) \to G^{\mathrm{ad}}(F)$ that is functorial in F.*

Fix a Borel group B of G defined over F and contains T. Let $a \in (\mathrm{Int}\, G)(F)$. By 15.2.5 there exists $g \in G(F)$ such that $aB = gBg^{-1}$ and by 14.4.3 we may assume that, moreover, $aT = gTg^{-1}$. By 6.4.9 and 6.3.6 (ii) there is $t \in T$ such that a is the inner automorphism $\mathrm{Int}(g.t)$. Since a is defined over F and $g \in G(F)$, the inner automorphism $\mathrm{Int}(t)$ is defined over F. By considering its action on the groups U_α one sees that $t \in T(F)^\sharp$. Moreover, the pair (g, t) is determined by a up to a change $(g, t) \mapsto (gt_1^{-1}, t_1 t)$, where $t_1 \in T(F)$. Using 16.3.6 (ii) we see that $\psi a = \pi(gt)$ is well-defined and lies in $G^{\mathrm{ad}}(F)$. If $a' = \mathrm{Int}(g'.t')$ is a similar decomposition of another inner F-automorphism, we have for the product aa' the decomposition $aa' = \mathrm{Int}(g(tg't^{-1}).tt')$. Notice that $tg't^{-1} \in G(F)$ by 16.3.6 (i). It follows that ψ is a homomorphism $(\mathrm{Int}\, G)(F) \to G^{\mathrm{ad}}(F)$.

If $\psi a = e$ we must have $g \in T$, by the arguments of the beginning of the proof, applied to G^{ad}. We may then take $g = e$, and we have $\alpha(t) = 1$ ($\alpha \in R$). This implies that t lies in the center of G (by 8.1.12 (3)), so that $a = \mathrm{id}$. Hence ψ is injective. That it is surjective follows from 16.3.6 (iii). The functoriality in F follows from the definitions. $\square$

Let $\mathcal{D}$ be the Dynkin diagram defined by D (see 9.5.1) and denote by A the finite group of its automorphisms. An automorphism of G stabilizing B and T and fixing the elements of the connected center C induces an automorphism of $\mathcal{D}$. Let A_0 be the subgroup of A of automorphisms of $\mathcal{D}$ obtained in this manner (we may have $A_0 \neq A$, see 16.3.9 (2)). If G is adjoint then $A = A_0$, as in that case D is a basis of $X = Q$.

16.3.8. Lemma. *Assume that G is semi-simple.*
(i) Let $\sigma \in A_0$. There exists a unique F-automorphism $\tilde{\sigma}$ of G such that for $\alpha \in D$ we have $\tilde{\sigma}(u_\alpha(x)) = u_{\sigma.\alpha}(x)$ $(x \in k)$;
(ii) For $\sigma, \tau \in A_0$ we have $(\sigma\tau)\tilde{} = \tilde{\sigma}\tilde{\tau}$;
(iii) If a is an automorphism of G defined over F and fixing the elements of C, there exist unique $a' \in (\text{Int } G)(F)$ and $\sigma \in A_0$ such that $a = a' \circ \tilde{\sigma}$.

It follows from 16.3.2 that if $\sigma \in A_0$ there are $c_\alpha \in F^*$ such that there is an F-automorphism of G sending $u_\alpha(x)$ to $u_{\sigma.\alpha}(c_\alpha x)$ $(\alpha \in D, \ x \in k)$. Conjugating by a suitable element of $T(F)^\sharp$ we obtain $\tilde{\sigma}$, for which all c_α equal 1. Then $\tilde{\sigma} \circ u_{-\alpha}$ is uniquely determined (by 8.1.4 (iii)). The uniqueness of $\tilde{\sigma}$ follows from 8.1.5 (i). This proves (i) and (ii) is a direct consequence of (i).

Let B be as in the proof of 16.3.7 and let D be the basis of R determined by B according to 7.4.6. Proceeding as in the proof of 16.3.7, we see that there is $g \in G(G)$ such that $a' = \text{Int}(g) \circ a$ fixes both B and T. Then there exist $t \in T(F)^\sharp$ and $\sigma \in A$ with $a' = \text{Int}(t) \circ \tilde{\sigma}$. Moreover, σ is unique. These facts imply (iii). $\qquad\square$

The preceding lemma, applied to G^{ad}, shows that A acts on G^{ad} as a group of F-automorphisms. Assume G to be semi-simple and let **Aut** G be the semi-direct product of G^{ad} and A_0, i.e. $G^{\text{ad}} \times A_0$, with multiplication $(x, \sigma)(y, \tau) = (x(\tilde{\sigma}.y), \sigma\tau)$. It is clear that **Aut** G is an F-group, with identity component G^{ad}. We also write **Inn** $G = G^{\text{ad}}$. It follows from 16.3.8 (iii) that $(\textbf{Aut } G)(F)$ is the group of F-automorphisms of G. We call **Aut** G (**Inn** G) the algebraic group of automorphisms (respectively, inner automorphisms) of G. For general reductive groups an algebraic group of automorphisms does not exist, see 16.3.9 (4).

16.3.9. Exercises.
(1) The notations are as in 16.3.5. Assume G to be semi-simple. Then X/Q is finite, and is isomorphic to a product of cyclic groups $\prod_{i=1}^{s} \mathbf{Z}/n_i\mathbf{Z}$.
    (a) $\pi G(F)$ is a normal subgroup of $G^{\text{ad}}(F)$.
    (b) The quotient $G^{\text{ad}}(F)/\pi G(F)$ is isomorphic to $\prod_{i=1}^{s} F^*/(F^*)^{n_i}$.
    (c) Consider the case $G = \textbf{PSL}_2$.
    (d) $\pi G(F_s) = G^{\text{ad}}(F_s)$ if and only if the characteristic p does not divide the order of X/Q.
(2) The notations are as in 16.3.8.
    (a) Let G be semi-simple and simply connected. Show that $A_0 = A$.
    (b) Assume that G is semi-simple of type D_{2n} $(n \geq 2)$. Show that X can be chosen such that $A_0 \neq A$.
(3) Let T be a torus over k of dimension > 1. Show that the group of automorphisms

of the algebraic group T cannot have a structure of algebraic group.

(4) The notations are as in 16.2.1. Assume that G is semi-simple and F-split. Let I be a subset of D, let $P_I \supset B$ be the parabolic subgroup defined by I (8.4.3) and let L_I be its Levi subgroup containing T. These are F-subgroups of G. Denote by $\mathbf{Aut}_I G$ ($\mathbf{Inn}_I G$) the subgroup of $\mathbf{Aut}\, G$ formed by the automorphisms (respectively: inner automorphisms) of G that stabilize P_I and L_I.

(a) $\mathbf{Inn}_I G$ is the image in $\mathbf{Inn}\, G$ of L_I and $\mathbf{Aut}_I G$ is the semi-direct product of $\mathbf{Inn}\, G$ and the stabilizer of I in the group A_0 of 16.3.8.

(b) $\mathbf{Aut}_I G$ and $\mathbf{Inn}_I G$ are closed F-subgroups of $\mathbf{Aut}\, G$.

16.4. The isomorphism theorem

16.4.1. Notations are as in 16.2.5. We have the indexed root datum $_i\Psi$ of G, relative to S and T, and two F-groups, viz. the kernel $K = Z_{G_1}(S)$ and the connected center C. We have $T = T_1.C$. Let X (X_1, Y) be the character group of T (respectively, T_1, C). These are Γ-modules, and X is a Γ-stable subgroup of finite index of the Γ-module $X_1 \oplus Y$. The indexed root datum of K is $_i\Psi' = (X_1, D_0, X_1^\vee, D_0, D_0^\vee, \tau)$. By 8.1.8 (iii) we have $X_1 = X/(Q^\vee)^\perp$. A triple $(_i\Psi, K, C)$ consisting of an indexed root datum $_i\Psi = (X, D, X^\vee, D^\vee, D_0, \tau)$, a connected reductive F-group K with indexed datum $_i\Psi' = (X_1, D_0, X_1^\vee, D_0^\vee, D_0, \tau)$ and an F-torus C with character group Y is *admissible* if:

(a) X is a subgroup of finite index of $X_1 \oplus Y$, such that the projections on the two summands induce isomorphisms $X + Y/Y \simeq X_1$, $X + X_1/X_1 \simeq Y$.

(b) X is a Γ-stable subgroup of $X_1 \oplus Y$.

A triple $(_i\Psi, K, C)$ coming from an F-group G is admissible. That condition (b) is fulfilled is clear and condition (a) is a consequence of 8.1.10.

Let G and G_1 be two connected reductive F-groups, determining admissible triples $(_i\Psi, K, C)$, $(_i\Psi_1, K_1, C_1)$ (relative to data S, T, B and S_1, T_1, B_1). An F-isomorphism $\phi : G \to G_1$ with $\phi S = S_1$, $\phi T = T_1$, $\phi B = B_1$ defines an isomorphism $f(\phi) : (_i\Psi_1, K_1, C_1) \to (_i\Psi, K, C)$. The notion of isomorphism of admissible triples is the obvious one.

The following theorem generalizes 9.6.2.

16.4.2. Theorem. *(Isomorphism theorem) Let* $f : (_i\Psi_1, K_1, C_1) \to (_i\Psi, K, C)$ *be an isomorphism. There is an isomorphism* $\phi : G \to G_1$ *with* $\phi S = S_1$, $\phi T = T_1$, $\phi B = B_1$ *such that* $f = f(\phi)$.

Consider the central isogeny $(G, G) \times C \to G$. By 8.1.8 and 8.1.10 we can describe the based root datum of G in terms of the based root data of (G, G) and C, and similarly for G_1. An application of 16.2.4 shows that it suffices to consider the case that G is either a semi-simple group or a torus. The second case being easy, we may assume that G and G_1 are semi-simple. An application of 16.3.2 shows that there is an F_s-isomorphism $\phi : G \to G_1$ with $\phi S = S_1$, $\phi T = T_1$, $\phi B = B_1$. We can assume that the restriction of ϕ to K defines an F-isomorphism $K \to K'$, similarly for the restriction of ϕ to T. We shall show that there exists $t \in T$ such that $g \mapsto \phi(tgt^{-1})$ is defined over F.

Let $(u_\alpha)_{\alpha \in R}$ and $(v_{\alpha_1})_{\alpha_1 \in R_1}$ be realizations of R in G, respectively, of R_1 in G_1. Denote the corresponding subgroups of G and G_1 by U_α, respectively V_{α_1}. Since G and G_1 are split over F_s (16.2.9 (3)) these groups are defined over F_s (see 16.3.1). We have $\phi(U_{f(\alpha_1)}) = V_{\alpha_1}$ ($\alpha_1 \in R_1$). With the notations of 15.5.2, let $w_\gamma \in W_0$ be the element with $w_\gamma(\gamma.D_0) = D_0$. That $w_\gamma \in W_0$ was established in the proof of 15.5.3. It was also established that for $\alpha \in D - D_0$ we have

$$\gamma.\alpha = w_\gamma^{-1}.(\tau(\gamma)(\alpha)).$$

The same holds for $\alpha \in D_0$ and hence, since D is a basis of R, for all $\alpha \in R$. A similar result holds in R_1. Under the isomorphism of the Weyl groups W_0 of (K, T) and $(W_0)_1$ of (K_1, T_1) defined by f, the element $w_\gamma \in W_0$ corresponds to its counterpart in $(W_0)_1$, for all $\gamma \in \Gamma$.

For $\gamma \in \Gamma$ put $c(\gamma) = \phi^{-1} \circ \gamma \circ \phi \circ \gamma^{-1}$ (compare with the proof of 11.3.3). This is an F_s-isomorphism $G \to G_1$. It fixes the elements of T and stabilizes all U_α ($\alpha \in R$). It follows from 16.3.2 and 16.3.7 that there is $t_\gamma \in \bar{T}(F_s)$ (the maximal torus of $G^{ad} = \mathbf{Int}\ G$ of 16.3.5) such that $c(\gamma)$ is the inner automorphism $\text{Int}(t_\gamma)$ defined by t_γ.

$\bar{T}$ is defined over F. Its character group is the root lattice Q and the Γ-action on Q is the one defined by the F-structure of T. Since ϕ induces an F-isomorphism $K \to K_1$, its action on the U_α with $\alpha \in R_0$ commutes with the Γ-action (11.2.9). This implies (by 8.1.1) that $\alpha(t_\gamma) = 1$ for $\alpha \in D_0$, i.e that the t_γ lie in the intersection of the kernels of the $\alpha \in D_0$. This is a torus T' whose character group is isomorphic to the sublattice Q' of Q with basis $(\alpha)_{\alpha \in D - D_0}$ and the $c(\gamma)$ define a cocycle $c \in Z^1(F, T')$ (see the proof of 11.3.3). Using 15.5.4 we conclude that the action of $\gamma \in \Gamma$ on Q' is via $\tau(\gamma)$. This shows that T' is direct a product of F-tori indexed by the orbits of $\tau(\Gamma)$ on $D - D_0$: for each such orbit the character group of the corresponding torus has as a basis the roots of that orbit. It follows from 13.1.5 (4) that T' is a product of tori of the form $\Pi_{E/F}\mathbf{G}_m$, where E/F is a finite separable extension. We conclude from 12.4.7 (1) and 12.3.5 (1) that $H^1(F, T') = 1$. Hence there is $t \in T'(F_s)$ with $t_\gamma = t^{-1}(\gamma.t)$. Then $\phi \circ \text{Int}(t^{-1})$ is an isomorphism $G \to G_1$ defined over F. $\quad\square$

16.4.3. We now shall relate the indexed root datum of a semi-simple group and the twisting procedure of 11.3.3.

Let H be a connected, semi-simple F-group with maximal F-torus T. Let $\Psi = (X, R, X^\vee, R^\vee)$ be the root datum of (H, T). By 16.3.3 there exists an F-split, connected, semi-simple F-group G with root datum Ψ. From 13.1.1 (ii) we see that H is F_s-split, hence is F_s-isomorphic to G by 16.3.2. With the terminology of 12.3.7, H is an F_s-form of G. By loc.cit. there is a bijection of the set of F-isomorphism classes of connected, reductive F-groups with root datum Ψ onto the Galois cohomology set $H^1(\Gamma, \mathrm{Aut}_{F_s}(G)) = H^1(F, \mathbf{Aut}\ G)$.

We recall the twisting procedure used to set up the bijection. The elements of $H^1(F, \mathbf{Aut}\ G)$ are represented by cocycles $z \in Z^1(F, \mathbf{Aut}\ G) = Z^1(\Gamma, (\mathbf{Aut}\ G)(F_s))$. There is an F-form G_z of G, such that $G_z(F_s) = G(F_s)$, the Γ-action on $G_z(F_s)$ being given by

$$\gamma \star g = z(\gamma)(\gamma.g) \ (\gamma \in \Gamma, \ g \in G(F_s)). \tag{73}$$

Any F-form H of G is F-isomorphic to such a G_z. Moreover, the class of z in $H^1(F, \mathbf{Aut}\ G)$ is uniquely determined by H.

Let $\Phi(F, \Psi)$ be the set of isomorphism classes of connected, semi-simple F-groups with root datum Ψ.

16.4.4. Lemma. *There is a bijection* $\Phi(F, \Psi) \to H^1(F, \mathbf{Aut}\ G)$.

This follows from 11.3.3 and 16.3.8. $\square$

16.4.5. We denote by $(\mathbf{Aut}\ G)_z$ the form of $\mathbf{Aut}\ G$ obtained by twisting $\mathbf{Aut}\ G$ with the cocycle $z \in Z^1(F, \mathbf{Aut}\ G)$, the group acting on itself by inner automorphisms. So $(\mathbf{Aut}\ G)_z(F_s) = (\mathbf{Aut}\ G)(F_s)$, with Γ-action given by

$$\gamma \star a = z(\gamma)(\gamma.a)z(\gamma)^{-1} \ (\gamma \in \Gamma, \ a \in (\mathbf{Aut}\ G)(F_s)).$$

One defines similarly the F-group $(\mathbf{Inn}\ G)_z$ of inner F-automorphisms of G_z.

16.4.6. Lemma. *The group of F-automorphisms (respectively, inner F-automorphisms) of G_z is isomorphic to* $(\mathbf{Aut}\ G)_z(F)$ $((\mathbf{Inn}\ G)_z(F))$.

This follows from the definitions by a straightforward check. $\square$

16.4.7. We assume, as we may, that $H = G_z$, as in 16.4.3. Let S be a maximal F-split torus in H and $T \supset S$ a maximal F-torus containing S. Choose a set of positive roots R^+ in the root system R of (G, T) with the property of 15.5.1 (ii), let D be the basis determined by R^+ and let $B \supset T$ be the corresponding Borel group. We have an indexed root datum $_i\Psi = (X, D, X^\vee, D^\vee, D_0, \tau)$ as in 16.2.1. From 15.4.7 we see that we may assume that the standard parabolic subgroup $P = P_{D_0} \supset B$ is a minimal parabolic F-subgroup of H.

Let T_0 be a maximal F-torus of G that is F-split and let $B_0 \supset T_0$, a Borel group of G defined over F. There is $g \in G(F_s)$ with $B_0 = gBg^{-1}$, $T_0 = gTg^{-1}$. Then $P_0 = gP_{D_0}g^{-1}$ is defined over F. After replacing z by the equivalent cocycle $\gamma \mapsto g^{-1}z(\gamma)(\gamma.g)$ we have the case that $T = T_0$, $P_0 = P$. So these groups are subgroups both of H and G, but with different F-structures.

Let L be the Levi subgroup of P containing T. Then L is defined over F (16.1.4), for both F-structures. The connected center C_L of L is the identity component of the intersection of the kernels Ker α, $\alpha \in D_0$. We view it as an F-subgroup of G.

We have the ingredient τ of $_i\Psi$, a homomorphism $\Gamma \to A$. It follows from the definition of τ (15.5.5) that its image lies in the group A_0 of 16.3.8. Since Im τ stabilizes D_0 (15.5.3 (i)), and since all elements of A_0 are defined over F, we see from 16.3.8 (i) that we can view τ as a homomorphism of Γ to the group of F-automorphisms of G, which we view as cocycles of Γ with values in the group $\mathbf{Aut}_{F_s} G$ ($\mathbf{Aut}_{F_s} L$) of F_s-automorphisms of G. So we have a twisted F-group G_τ, as in 11.3.3. Similarly, we have a twisted group L_τ. We also have a group $(\mathbf{Inn}_{D_0} G)_\tau$ and a map $z \mapsto \bar{z}$ of $Z^1(F, (\mathbf{Inn}_{D_0} G)_\tau)$ to $Z^1(F, (\mathbf{Inn}\ L)_\tau)$ (the notation is as in 16.3.9 (4)).

16.4.8. Lemma. *(i) G_τ and L_τ are quasi-split F-groups;*
(ii) There is a cocycle $z \in Z^1(F, (\mathbf{Inn}_{D_0} G)_\tau)$ such that $H = (G_\tau)_z$;
(iii) The kernel of H is F-isomorphic to $(L_\tau)_{\bar{z}}$.

It is clear that B is a Borel subgroup of G_τ defined over F. This implies that G_τ is quasi-split by 16.2.2 (i), and similarly for L_τ. This proves (i).

The Γ-actions on D for H and G_τ are the same, which implies that H is an inner F-form of G_τ, i.e. that $H = G_\tau$ with $z \in Z^1(F, \mathbf{Inn}\ G)_\tau)$ (use 16.3.8). So $z(\gamma) = \mathrm{Int}(g_\gamma)$ ($\gamma \in \Gamma$). Since S is an F-split subtorus of H, it follows from (61) that the g_γ centralize $S(F_s)$. Because $L = Z(S)$ we have $g_\gamma \in L$. The proof of (iii) is straightforward.   □

Now let G be an arbitrary connected, reductive F-group. The elements of $H^1(F, \mathbf{Inn}\ G)$ define by 11.3.3 isomorphism classes of F-forms of G. These forms are the *inner forms* of G (12.3.7).

16.4.9. Proposition. *G is an inner F-form of a quasi-split F-group.*

An easy argument, which is left to the reader, reduces the proof to the case that G is semi-simple, in which case the assertion follows from 16.4.8.   □

16.5. Existence

16.5.1. By 16.4.2 a connected, reductive F-group G is determined by an admissible triple $(_i\Psi, K, C)$. We turn to the question of which admissible triples exist. We say that an admissible triple is *representable* if it comes from an F-group. The argument of the first paragraph of the proof of 16.4.2 reduces the existence problem to the case that G is semi-simple. This we assume from now on. Then C can be omitted, and we have a pair $(_i\Psi, K)$ of a based root datum $_i\Psi$ and an anisotropic F-group. Such a pair is *representable* if it comes from an F-group. We shall give conditions for representability.

16.5.2. Let G be a connected, semi-simple F-group that is F-split. The notations are as in 16.2.1. Denote by $\Psi_0 = (X, D, X^\vee, D^\vee)$ the based root datum of G. Let $_i\Psi$ be an indexed root datum with underlying based root datum Ψ_0, and other ingredients D_0 and τ. Let P_{D_0} be the standard parabolic subgroup of G defined by D_0 and let L be its Levi subgroup containing T, it is a connected, reductive F-subgroup that is F-split. Assume that K is an anisotropic F-form of L. The analysis made in 16.4 leads to a criterion for representability of the pair $(_i\Psi, K)$. The quasi-split forms G_τ and L_τ of G and L are as in 16.4.8.

We have the twisted group of inner automorphisms $(\mathbf{Inn}\, L)_\tau$ of 16.4.5. There is an obvious homomorphism $(\mathbf{Inn}_{D_0} G)_\tau \to (\mathbf{Inn}\, L)_\tau$ inducing a map

$$\phi : H^1(F, (\mathbf{Inn}_{D_0} G)_\tau) \to H^1(F, (\mathbf{Inn}\, L)_\tau).$$

We may assume that there is $z \in Z^1(F, (\mathbf{Inn}\, L)_\tau)$ such that $K = (L_\tau)_z$. The class of z in $H^1(F, (\mathbf{Inn}\, L)_\tau)$ is denoted by c.

16.5.3. Lemma. $(_i\Psi, K)$ *is representable if and only if $c \in \mathrm{Im}\,\phi$.*

The condition is necessary by 16.4.8 (iii). Conversely, suppose it is satisfied. Take $z \in Z^1(\Gamma, (\mathbf{Inn}_{D_0} G)_\tau(F_s))$ whose cohomology class has image c under ϕ. The twisted group $G_1 = (G_\tau)_z$ is an F-group that represents $(_i\Psi, K)$. In fact, G_1 contains an F-split F-torus S whose centralizer L_1 is an F-group isomorphic to K. Moreover, S is the maximal F-split torus in the connected center of L_1. Then S must be a maximal F-split F-subtorus of G_1, and L_1 is the kernel of G_1. The indexed root datum is as required.   □

16.5.4. Let L be as before and let M be the commutator group (L, L); it is a semi-simple F-group. We have the canonical F-homomorphism $\pi : G \to \mathbf{Inn}\, G$, and $\mathbf{Inn}_{D_0} G = \pi L$. The inclusion $M \to L$ induces a map

$$\psi : H^1(F, (\pi M)_\tau) \to H^1(F, (\mathbf{Inn}\, L)_\tau)$$

(the τ-twists being as before). For any semi-simple F-group M we say that an element of $H^1(F, M)$ is *anisotropic* if its image in $H^1(F, \mathbf{Inn}\, M)$ represents an anisotropic F-form of M.

We have the following criterion for representability, to be used in the next chapter.

16.5.5. Proposition. $(_i\Psi, K)$ *is representable if and only if $c \in \mathrm{Im}\,\psi$. If this is so then $H^1(F, (\pi M)_\tau)$ must contain anisotropic elements.*
There is an exact sequence of F-groups

$$1 \to (\pi M)_\tau \to (\pi L_1)_\tau \to T' \to 1,$$

where T' is an F-torus (the τ-twists being as before). In fact, the torus T' is isomorphic to the torus denoted by T' in the last part of the proof of 16.4.2 (check this).

By what we proved there we have $H^1(F, T') = 1$. We then obtain from the exact sequence of 12.3.4 (ii) a surjective map

$$H^1(F, (\pi M)_\tau) \to H^1(F, (\mathbf{Inn}_{D_0} G)_\tau).$$

The first point of the proposition now follows from 16.5.3. The second point is clear.

□

In the case of 16.5.5 assume that G is simply connected (8.1.11). In applying 16.5.5 one has to identify πM. The following lemma is useful. With the notations 8.1.11, the cocenter $C^*(G)$ is the fundamental group P/Q. By 8.4.6 (6), M is also simply connected.

16.5.6. Lemma. *(i) If there is a factorization* $M = M_1 \times M_2$ *over* F *such that the orders of the cocenters* $C^*(G)$ *and* $C^*(M_1)$ *are relatively prime, then* π *induces an* F-*isomorphism* $M_1 \to \pi M_1$;
(ii) If $C^*(G)$ *is trivial then* $M \simeq \pi M$.

The group M_1 of (i) is semi-simple and simply connected. The restriction $\pi|_{M_1}$ is bijective, as a consequence of 8.1.12 (8). If the characteristic p does not divide the order of C_{M_1} then $\pi|_{M_1}$ is separable and is an isomorphism by 5.3.3. If p divides that order then it does not divide the order of C_G, so that π is separable. Then $d\pi$ is an isomorphism of Lie algebras, and so is $d\pi|_{M_1}$. Again, $\pi|_{M_1}$ is separable, and must be an isomorphism. This proves (i). The proof of (ii) is similar. □

In 16.5.5 we assumed the index (D, D_0, τ) to be given. We say that this index is *representable* if it comes from an F-group. 15.5.7 and 15.5.8 give necessary conditions for representability of an index. Another such condition is given in the next lemma.

Let D_1 be an (Im τ)-stable subset of D containing D_0.

16.5.7. Lemma. *If the index* (D, D_0, τ) *is representable then the same is true for the index* $(D_1, D_0, \tau|_{D_1})$.

Let T_1 be the identity component of the intersection

$$\bigcap_{\alpha \in D - D_1} (\text{Ker } \alpha)$$

and denote by G_1 the centralizer $Z_G(T_1)$; it is a connected reductive F-group whose index is $(D_1, D_0, \tau|_{D_1})$. We skip the details.

16.5.8. By 16.5.1 the question of representability of admissible triples can be reduced to the question of representability of pairs $(_i\Psi, K)$ in the semi-simple case, i.e. in the case that $X = \tilde{Q}$ (notations of 8.1).

We may assume that the based root datum is one of a simply connected group, i.e. that $X = P$. In fact, let G be a semi-simple F-group representing $(_i\Psi, K)$ and assume that it is a twist H_z, where H is split and $z \in Z^1(F, \textbf{Aut } H)$. It follows from 16.3.4 that there is a simply connected F-group H_1 which is F-split, together with an F-isogeny $H_1 \to H$. Also, the definition of **Aut** H in 16.3 shows that there is an injection **Aut** $H \to$ **Aut** H_1, whence an injection $Z^1(F, \textbf{Aut } H) \to Z^1(F, \textbf{Aut } H_1)$. Viewing z as an element of the second set, we have a twisted group $G_1 = (H_1)_z$, which is simply connected, and an isogeny $G_1 \to G$. The based root datum $(_i\Psi)_1$ of G_1 can be described in terms of $_i\Psi$, and using the more refined twisting procedure of 16.4, the kernel K_1 of G_1 can be described in terms of K. It follows that representability of $(_i\Psi, K)$ implies representability of $((_i\Psi)_1, K_1)$. The converse is also true by 16.2.4.

16.5.9. Assume that we are in the simply connected case. If $(_i\Psi, K)$ is represented by an F-group G, then applying 16.2.9 (1) we see that G is a direct product of F-groups each of which is of the form $\Pi_{E_h/F} G_h$, where E_h is a finite separable extension of F, and G_h is a quasi-simple E_h-group (i.e. has an irreducible root system). Moreover, G_h is simply connected. The core of the existence problem is thus the description of the admissible pairs $(_i\Psi, K)$ in the quasi-simple case. In the next chapter we shall describe, for each connected Dynkin diagram, the corresponding indices that are representable over some field F.

Notes

The results of 16.1 are due due to Borel and Tits [BoT1].

The idea of classifying F-groups by indexed root data is due to Tits ([Ti2], see also [Ti7] and [Sat]). In particular, the isomorphism theorem 16.4.2 (in the semi-simple case) and the representability condition 16.5.5 are contained in [Ti2]. In [Ti2, p. 50] another criterion for representability is given, involving representation theory.

[Ti2] contains tables of the possible indices, without proofs. For more details see [Selb], [Ti7] and the next chapter.

Chapter 17

Classification

The notations are as in the previous chapter. We assume that G is quasi-simple. We do not assume that G is simply connected. Sometimes it is more convenient to work with another isogeny type.

As before, the indexed root datum of G is $(X, D, X^\vee, D^\vee, D_0, \tau)$ (relative to S and T). As G is quasi-simple, X is the rational closure $\tilde{Q}$ of the subgroup Q spanned by D and the root system R is irreducible. We assume known the classification of irreducible root systems. We shall use the detailed information given in [Bou2, p.250-276]. The numbering of the vertices of the irreducible Dynkin diagrams is as in [loc. cit.].

In the present chapter we address the question of giving, for each type of irreducible root system R or Dynkin diagram $\mathcal{D}$, a concrete description of the corresponding F-groups G. There is a satisfactory answer to the question in the case of a 'classical' root system (of one of the types A_n, B_n, C_n, D_n), in terms of classical (orthogonal, unitary,...) groups. But for some of the five exceptional types the answer is not (yet) known. In particular, a concrete description of all anisotropic groups over a given field F is not known. But we shall describe in all cases which indices (D, D_0, τ) are possible, for some F.

We denote by H a split F-group with the same root datum as G. With notations as in 16.5, we assume that G is a twisted group H_z, where $z \in Z^1(F, \mathbf{Aut}\ H)$. If τ is trivial or, equivalently, if $z \in Z^1(F, \mathbf{Inn}\ H)$ then G is an inner form of H, or is of *inner type*. Otherwise G is of *outer type*.

We shall deal separately with each type of irreducible root system R. We denote by $A(R)$ or A the automorphism group of the Dynkin diagram $\mathcal{D}$ and by C^* or $C^*(R)$ the fundamental group P/Q. This is the cocenter of G if G is simply connected (8.1.11).

The table at the end of this chapter gives, for each of the irreducible root systems, the subsets D_0 of the vertex set of the Dynkin diagram which can occur, for some field F.

17.1. Type A_{n-1}

17.1.1. Inner types. For the root system of type A_{n-1} see [Bou2, p.250-251]. The Dynkin diagram is

$$\underset{\circ}{\overset{1}{\circ}} \!\!-\!\! \underset{\circ}{\overset{2}{\circ}} \!\!-\cdots-\!\! \underset{\circ}{\overset{n-2}{\circ}} \!\!-\!\! \underset{\circ}{\overset{n-1}{\circ}}$$

The automorphism group A is trivial if $n = 2$ and has order 2 if $n > 2$. The group C^* is cyclic of order n.

Assume that G is adjoint and of inner type. We may assume that $H = \mathbf{Inn}\ H$ is the quotient $\mathbf{PGL}_n$ of $\mathbf{GL}_n$ by its center (see 7.4.7 (2)). From 12.3.5 (1) it follows that $H^1(F, \mathbf{PGL}_n)$ classifies the associative k-algebras with an F-structure which are isomorphic to $\mathbf{M}_n$ over F_s. If $\mathcal{A}$ is such an algebra then $\mathcal{A}(F)$ is a central simple algebra over F, i.e. a simple associative algebra with center F. It is known that there is a division algebra K with center F such that $\mathcal{A}(F)$ is isomorphic to a matrix algebra $M_m(K)$ (see [Jac5, p. 203]). Using 12.3.8 (3) we see that we may assume that G is the quotient of $\mathbf{GL}_{m,K}$ by the subgroup Z of scalar matrices. Let d be the degree of K. Then $\dim_F K = d^2$ and $n = dm$.

It is convenient to describe the indexed root datum of $G_1 = \mathbf{GL}_{m,K}$. Recall that $G_1 = (\mathbf{GL}_n)_z$, where $z \in Z^1(F, \mathbf{PGL}_n)$ defines our simple associative algebra. Assume it to be such that $G_1(F)$ is the group $GL_m(K)$ of invertible $m \times m$-matrices with entries in K. Let T be the maximal F_s-torus of G_1 such that $T(F_s)$ is the diagonal torus of $\mathbf{GL}_n(F_s) = G_1(F_s)$. We use the notations of 7.4.7 (1). Put $\alpha_i = \epsilon_i - \epsilon_{i+1}$ ($1 \le i \le n-1$). Then $D = (\alpha_1, \dots, \alpha_{n-1})$ is a basis of the root system R (see 8.2.11 (1)). The corresponding Borel group of G_1 is the subgroup of upper triangular matrices. Let $S \subset T$ be the intersection of the kernels of the α_i with d not dividing i, i.e. S is the subgroup of G_1 of the elements of the form $\mathrm{diag}(x_1, \dots, x_n)$ with $x_1 = \cdots = x_d$, $x_{d+1} = \cdots = x_{2d}, \dots$. The centralizer $Z(S)$ is isomorphic to $(\mathbf{GL}_d)^m$.

17.1.2. Lemma. *(i) The cocycle z can be taken to lie in the image of $Z(S)$ in H;*
(ii) S is a maximal F-split F-subtorus of G;
(iii) $Z_{G_1}(S)$ is F-isomorphic to $(\mathbf{GL}_{1,K})^m$.

The proof is straightforward. □

We obtain the following classification result.

17.1.3. Proposition. *(i) For inner type A_{n-1} the possible subsets $D - D_0$ of D are of the form $(\alpha_d, \alpha_{2d}, \dots)$, where d divides n;*
(ii) For D_0 as in (i) the index (D, D_0, triv) can be realized over the field F if and only if there exists a division algebra of degree d dividing n, with center F;
(iii) If the condition of (ii) is satisfied the root system ${}_F R$ is irreducible of type $A_{d^{-1}n-1}$.

17.1.4. Outer types. Assume that G is adjoint and of outer type. We take H as in 17.1.1. Assume now that the homomorphism τ of 15.5.2 is non-trivial. Then $n \ge 3$. Recall that τ is a homomorphism of the Galois group $\Gamma = \mathrm{Gal}(F_s/F)$ to the group A of automorphisms of the Dynkin diagram, which now has order ≤ 2. It is generated by the permutation sending α_i to α_{n-i} ($1 \le i \le n-1$) (the notations are as in 17.1.1). It comes from the outer automorphism $x \mapsto ({}^t x)^{-1}$ of $\mathbf{GL}_n$.
$\Delta = \mathrm{Ker}\ \tau$ is a closed subgroup of Γ of index 2. Let $E = E_\tau \subset F_s$ be the set

of elements fixed by all of Δ; it is a separable quadratic extension of F. We have a cocycle $z \in Z^1(F, \textbf{Aut } H)$ such that $G = H_z$. By the definition of E, the image u of z under the canonical map

$$Z^1(F, \textbf{Aut } G) \rightarrow Z^1(E_\tau, \textbf{Aut } H)$$

lies in the subset $Z^1(E, \textbf{Inn } H) = Z^1(E, \textbf{PGL}_n)$. By what we established in 17.1.1, there is a central simple algebra $M = M_m(K)$ over E (where $n = dm$), such that G is E-isomorphic to the quotient of $\textbf{GL}_{m,K}$ by the subgroup of diagonal matrices.

There is an associative algebra $\mathcal{A}$ defined over E such that $M = \mathcal{A}(E)$. We have $\mathcal{A}(F_s) = M_n(F_s)$ and the action of the Galois group Δ is the twisted action given by

$$s.x = u_s(sx) \quad (s \in \Delta, \ x \in M_n(F_s)),$$

where $u_s \in \textbf{PGL}_n(F_s)$ and $s \in \Delta$ act on matrices in the obvious manner. Now $\textbf{Aut } H(F_s)$ is generated by $(\textbf{Inn } H)(F_s)$ and the automorphism induced by the automorphism $x \mapsto ({}^tx)^{-1}$ of $\textbf{GL}_n$. It follows that, for any $\alpha \in \textbf{Aut } H(F_s)$, there exists a unique involution (or anti-automorphism) ι_α of $M_n(F_s)$ such that $\iota_\alpha(x) = (\alpha x)^{-1}$ if x is invertible.

Let σ be the non-trivial automorphism of E_τ/F, and let $s \in \Gamma$ represent it. Then $m \mapsto z_s(sm)$ defines an automorphism a of the group of invertible elements of M, which is independent of the choice of s, as follows from the cocycle property of z. Moreover, there is an involution ι of M inducing on F the automorphism σ such that $am = (\iota m)^{-1}$ if $m \in M$ is invertible. An involution of a central simple algebra $M_m(K)$, which induces on the center a non-trivial automorphism, is an *involution of the second kind*. In that case there is an involution of the second kind κ of K and an invertible element $h \in M$ such that

$$\iota m = h.{}^t(\kappa m).h^{-1} \quad (m \in M), \tag{74}$$

see [Scha, Ch. 8, §7]. Then $h.{}^t(\kappa h)^{-1}$ lies in the center of K. Using 12.3.5 (1) we see that we may assume h to be κ-hermitian, i.e. ${}^th = \kappa h$. Let $U(h)$ be the unitary group of h, i.e.

$$U(h) = \{g \in M \mid g.h.{}^t(\kappa g) = h\}.$$

The group of F-rational points $G(F)$ is isomorphic to $U(h)$.

17.1.5. We now have associated with our group G a division algebra K with center F with an involution κ of K of the second kind and a non-singular κ-hermitian matrix $h \in M = M_m(K)$.

Conversely, if we have such a triple (K, κ, h), then (74) defines an involution of the second kind of M. We obtain a finite dimensional algebra $\mathcal{A}$ with an F-structure and an involution ι of $\mathcal{A}$ defined over F. Forgetting the F-structure, we may take

$\mathcal{A} = \mathbf{M}_n \oplus \mathbf{M}_n$, with $\iota(a, b) = ({}^t b, {}^t a)$ (check this). An application of 12.1.8 (1) shows that

$$G = \{a \in \mathcal{A} \mid a(\iota a) = 1\}$$

is an F-group. It is isomorphic to $\mathbf{GL}_n$.

Put $V = K^m$. Then h defines a non-degenerate κ-hermitian form Φ on V by $\Phi(v, w) = vh(\kappa^t w))$, where $v, w \in V$ are viewed as $1 \times m$-matrices, the action of κ on V being componentwise. The values $\Phi(v, v)$ $(v \in V)$ lie in F. The Witt index r of Φ is defined (see [Dieu, I, §11]). We may assume that the following holds: if $v = (a_1, \ldots, a_m)$, $v_0 = (a_{r+1}, \ldots, a_{m-r})$ and $w = (b_1, \ldots, b_m)$, $w_0 = (b_{r+1}, \ldots, b_{m-r})$ then

$$\Phi(v, w) = a_1(\kappa(b_m)) + \cdots + a_r(\kappa(b_{m+1-r})) + \Phi_0(v_0, w_0),$$

where Φ_0 is an anisotropic hermitian form defined by a hermitian $(m-2r) \times (m-2r)$-matrix (anisotropy means that $\Phi_0(v_0, v_0) = 0$ implies $v_0 = 0$).
There exists an F-form G_1 of $\mathbf{GL}_n$ obtained by twisting with the cocycle z. Then G is a quotient of G_1 by a central torus. Let E be the center of K; it is a separable quadratic extension of F. By 17.1.1 we know that over E our group G_1 is isomorphic to $\mathbf{GL}_{m,K}$. Let S_1 be the E-torus denoted by S in 17.1.1 and denote by S the subtorus of S_1 formed by the elements $\mathrm{diag}(x_1, \ldots, x_n) \in S_1$ such that

$$x_1 = \ldots = x_d, \ x_{d+1} = \ldots = x_{2d}, \ldots, x_{(r-1)d+1} = \ldots = x_{rd},$$

$$x_i = 1 \text{ for } rd + 1 \le i \le (m-r)d,$$

$$x_i = x_{n+1-i}^{-1} \text{ for } i > (m-r)d.$$

Then S is a maximal F-split F-subtorus of G_1. We find the following result. D and E_τ are as before.

17.1.6. Proposition. *(i) For outer type A_{n-1} the possible subsets $D - D_0$ are of the form $(\alpha_d, \ldots, \alpha_{rd}, \alpha_{n-rd}, \ldots, \alpha_{n-d})$, where $d \mid n$ and $2rd \le n$ (α_{n-rd} should be omitted if $2rd = n$);*
(ii) For given τ and D_0 as in (i) the index (D, D_0, τ) can be realized over F if and only if there exists a division algebra K with center E_τ, an involution κ of K of the second kind over F, together with a non-degenerate κ-hermitian form over K of dimension $d^{-1}n$ and of index r;
(iii) If the conditions of (ii) are satisfied, the root system $_F R$ is irreducible of type BC_r if $2rd < n$ and of type C_r otherwise.

17.2. Types B_n and C_n

17.2.1. Type B_n.
The root system of type B_n $(n \geq 2)$ is described in [Bou2, p. 252-253]. The Dynkin diagram is

The automorphism group A is trivial and C^* has order 2.

First assume that char $F \neq 2$. Let $H = \mathbf{SO}_{2n+1}$. Then $H = \mathbf{Inn}\ H = \mathbf{Aut}\ H$. By 7.4.7 (3) it is a connected, semi-simple group of type B_n. It follows from 12.1.8 (2) that H is defined over F and F-split. With the notations of 7.4.7 (3), define $\alpha_i = \epsilon_i - \epsilon_{i+1}$ $(1 \leq i \leq n-1)$ and put $\alpha_n = \epsilon_n$. Then $D = (\alpha_1, \ldots, \alpha_n)$ is a basis of the root system R (see 8.2.11 (1)). By 12.3.5 (5) we have that $H^1(F, \mathbf{O}_{2n+1})$ classifies the isomorphism classes of $(2n+1)$-dimensional non-degenerate symmetric bilinear forms over F. Similarly, $H^1(F, \mathbf{SO}_{2n+1})$ classifies the isomorphism classes of quadratic forms Q defined as in 15.3.10 (1) by a symmetric matrix $s \in \mathbf{GL}_{2n+1}(F)$ whose determinant is a square in F (check this). It follows that we may take $G = H_z$ to be the special orthogonal group defined by a non-degenerate quadratic form Q over F. Such a form comes from a symmetric matrix s, as before. Multiplying s by a suitable scalar we can satisfy the determinant condition. The associated orthogonal group remains the same. If r is the Witt index of Q (see [Jac4, p. 351]) then the F-rank of G equals r, see 15.3.10 (1).

17.2.2.
Assume that char $F = 2$ and define H to be the group of 7.4.7 (6). Again, $H = \mathbf{Inn}\ H = \mathbf{Aut}\ H$. It is a connected, semi-simple group of type B_n, which is defined over F by 12.1.8 (4); moreover H is F-split. From 7.4.7 (6) we see that there is an inseparable isogeny $H \rightarrow \mathbf{Sp}_{2n}$ defined over F. Since $\mathbf{Sp}_{2n}$ is connected (see 2.2.9 (1)) the same holds for H.

Recall that the *defect* of an odd dimensional quadratic form q on an F-vector space A in characteristic 2 is the codimension of the radical R of the alternating form $(x, y) = q(x+y) + q(x) + q(y)$ $(x, y \in A)$ (see [Dieu, Ch. I, §16]). q is *non-degenerate* if the restriction of q to R takes the value 0 only in 0. An application of 11.3.3 shows, as in 15.3.10 (1), that $H^1(F, H)$ classifies the isomorphism classes of non-degenerate quadratic forms over F of defect one, which take the value 1 on the radical R.

We can proceed as in 15.3.10 (1). We have a quadratic form Q on $V(F)$. Its Witt index r is the dimension of a maximal subspace of $V(F)$ on which Q is identically zero. Using [loc.cit.] we see that the results are as in the case of characteristic $\neq 2$.

We obtain the following.

17.2.3. Proposition.
(i) For type B_n the possible subsets $D - D_0$ of D are of the form $(\alpha_1, \ldots, \alpha_r)$, with $r \leq n$;

(ii) For D_0 as in (i) the index (D, D_0, triv) can be realized over F if and only there exists a $(2n + 1)$-dimensional non-degenerate quadratic form over F that has Witt index r;

(iii) If the conditions of (ii) are satisfied the root system $_FR$ is irreducible, of type B_r.

For the last point see 15.3.10 (1). Also notice that in the present case τ is always trivial.

17.2.4. Type C_n. The root system of type C_n $n \geq 2$) is described in [Bou2, p. 254-255]. The Dynkin diagram is

$$\overset{1}{\underset{\circ}{}} \!\!-\!\! \overset{2}{\underset{\circ}{}} - \cdots - \overset{n\text{-}2}{\underset{\circ}{}} \!\!-\!\! \overset{n\text{-}1}{\underset{\circ}{}} \!\!\Longleftarrow\!\! \overset{n}{\underset{\circ}{}}$$

The group A is trivial, and C^* has order 2.

Take $H = \mathbf{Sp}_{2n}$. This is a connected, semi-simple group of type C_n (see 7.4.7 (5)), defined over F (see 12.1.8 (3)) and F-split. Moreover, it is simply connected. The imbedding $H \to \mathbf{GL}_{2n}$ induces an F-isomorphism of **Inn** H onto a closed F-subgroup of $\mathbf{PGL}_{2n}$ (check this), whence an injection $\phi : Z^1(F, \mathbf{Inn}\ H) \to Z^1(F, \mathbf{PGL}_{2n})$. We twist the algebra $\mathbf{M}_{2n}$ with the cocycle ϕz, obtaining an associative algebra $\mathcal{A}$ over k with an F-structure. Recall that $\mathcal{A}(F_s) = \mathbf{M}_{2n}(F_s)$ and that the action of the Galois group $\Gamma = \mathrm{Gal}(F_s/F)$ on $\mathcal{A}(F_s)$ is obtained by twisting with the cocycle ϕz. Then $\mathcal{A}(F)$ is a central simple algebra over F.

Define $j \in \mathbf{M}_{2n}$ by

$$j = \begin{pmatrix} 0 & 1_n \\ -1_n & 0 \end{pmatrix}$$

(where 1_n is the $n \times n$ identity matrix) and put $\iota x = j({}^tx)j^{-1}$ $(x \in \mathbf{M}_{2n})$. Then ι is an involution of $\mathbf{M}_{2n}$ defined over F, and

$$H = \{x \in \mathbf{GL}_{2n} \mid \iota x = x^{-1}\}.$$

It follows that ι commutes with the Γ-action on $\mathcal{A}(F_s)$ and determines an involution of $\mathcal{A}$ defined over F.

As in 17.1.1 we have that $\mathcal{A}(F)$ is a matrix algebra $M = M_m(K)$ over a division algebra K with center F. The involution ι induces an F-linear involution or *involution of the first kind* of $\mathbf{M}_m(K)$. There exists an involution of the first kind κ of K and an invertible element $h \in M$ such that (74) holds (as in 17.1.4). Then $\epsilon = h.{}^t(\kappa h)^{-1}$ lies in the center of M, i.e. in F, and it is easily seen that $\epsilon = \pm 1$. So $h \in M$ is an ϵ-hermitian (i.e. hermitian or anti-hermitian) matrix with respect to the involution κ of K.

We write $G = U(K, \kappa, h)$. The group of its F-rational points is the unitary group defined by h, i.e.

$$G(F) = \{g \in M \mid gh(\iota g) = h\}.$$

17.2.5. We now make a digression about involutions. Let $M = M_m(K)$ be a central simple algebra, as above, and let ι be an involution of the first kind of M. As before, we have an involution κ of K. We may assume that $\mathcal{A} = k \otimes_F M = M_N$. There is $h \in \mathbf{GL}_N$ with ${}^t h = \epsilon h$, where $\epsilon = \pm 1$, such that

$$\iota a = h({}^t a)h^{-1} \quad (a \in \mathcal{A}).$$

First assume that char $F \neq 2$. There exists an invertible element $x \in \mathbf{GL}_N$ such that either $h = x({}^t x)$ or $h = xj({}^t x)$ (where j is as before, with $N = 2n$), see [Jac4, p. 340, p. 334]. This implies that, up to isomorphism, we may assume that $h = 1_N$ or $h = j$. We say that ι is *orthogonal* in the first case and *symplectic* in the second case.

If $\epsilon = \pm 1$ we write $M_{\iota, \epsilon}$ for the ϵ-eigenspace of ι. The proofs of the next two lemmas are straightforward. Assume that κ is non-trivial.

17.2.6. Lemma. *(i) The involution ι is orthogonal (symplectic) if and only if $\dim M_{\iota, 1} = \frac{1}{2}N(N + 1)$ (respectively, $\frac{1}{2}N(N - 1)$);*
(ii) If ι is symplectic (orthogonal) then either the involution κ of K is symplectic and h is hermitian (respectively, anti-hermitian) or κ is orthogonal and h is anti-hermitian (respectively, hermitian).

Let K and κ be as in 17.2.4. Since κ is non-trivial, $K_{\kappa, -1} \neq \{0\}$. Let $a \in K_{\kappa, -1} - \{0\}$ and put $\kappa_1 = \mathrm{Int}(a) \circ \kappa$. This is an involution of K.

17.2.7. Lemma. *(i) $K_{\kappa, \epsilon} = K_{\kappa_1, -\epsilon}$;*
(ii) If h as in (74) is ϵ-hermitian relative to κ then ha^{-1} is $-\epsilon$-hermitian relative to κ_1 and $U(K, \kappa, h) = U(K, \kappa_1, ha^{-1})$.

It follows that in characteristic $\neq 2$ we can view the group $G(F)$ of 17.2.4 as a unitary group over a division algebra K with an involution of the first kind, either of a hermitian matrix relative to a symplectic involution, or of an anti-hermitian matrix relative to an orthogonal involution. In the classification of groups of type C_n, we prefer to work with symplectic involutions.

Now assume that char $F = 2$. In that case the argument of the proof of [Jac4, Th. 6.5, p. 338] gives that, if $h \in \mathbf{GL}_N(F_s)$ is symmetric, there is $x \in \mathbf{GL}_N(F_s)$ and an integer s such that $N - s$ is even and

$$h = x(d \oplus j_{N-s})({}^t x),$$

where d is an $s \times s$ diagonal matrix (j_{N-s} having the obvious meaning). If $s = 0$ the involution ι of 17.2.5 is said to be *symplectic*. The next lemma characterizes symplectic involutions. M and ι are as before.

17.2.8. Lemma. *(char $F = 2$) (i) ι is symplectic if and only if there is $x \in M$ with $x + \iota x = 1$;*

(ii) Let $M = M_m(K)$ and let ι be given by (74). Then ι is symplectic if and only if there is $y \in M$ with $h = y + {}^t(\kappa y)$. If $\kappa \neq$ id this is equivalent to: κ is a symplectic involution of K.

We assume that ι is as in (74). First let $F = F_s$. We may then take $M = \mathbf{M}_N$, and $\iota a = h({}^t a)h^{-1}$, where $h = d \oplus j_{N-s}$. If $s \neq 0$ a matrix $x + \iota x \in \mathbf{M}_N$ has a diagonal element 0. It follows that ι is symplectic if the condition of (i) is satisfied. Conversely, if ι is symplectic, a computation in $\mathbf{M}_N$ (which we leave to the reader) shows that there is $x_0 \in F_s \otimes M$ with $x_0 + \iota x_0 = 1$. For $\gamma \in \Gamma$ the element $x_\gamma = \gamma.x_0 + x_0$ lies in the space $\mathrm{Ker}(1 + \iota)$, which is defined over F. Application of 11.3.4 (5) shows that there is $z \in \mathrm{Ker}(1 + \iota)$ with $x_\gamma = \gamma.z + z$. Then $x = x_0 + z$ is as required.

If x is as in (i) then $h = xh + {}^t(\kappa(xh))$. The first point of (ii) readily follows. Now assume that $\kappa \neq$ id, i.e. that K is non-commutative, and let y be as in (ii). Put $V = K^m$. Then h defines a non-degenerate hermitian form Φ on V by $\Phi(v, w) = vh(\kappa({}^t w))$, where $v, w \in V$ are viewed as $1 \times m$-matrices, the action of κ on V being componentwise. For $v \in V$ we have $\Phi(v, v) = a + \kappa a$, where $a = vy(\kappa({}^t v))$. If $\Phi(v, v) \neq 0$ for some $v \in V$, κ is symplectic by what we already proved. We are left with the case that $\Phi(v, v) = 0$ for all v. Then $\Phi(v, w) = \Phi(w, v)$ for all $v, w \in V$. Take them such that $\Phi(v, w) = 1$, let $a \in K$ and replace in the equality v by av. We conclude that $\kappa a = a$, i.e. that $\kappa =$ id. $\square$

17.2.9. In all characteristics we can take $G = U(K, \kappa, h)$, where K is a division algebra K with center F with a symplectic involution κ, where h is a non-singular hermitian matrix relative to κ. The degree d of K is a power 2^s (see [Jac5, Ex. 1, p. 493]). Conversely, if we have such a triple (K, κ, h), then (74) defines a symplectic involution of M. We obtain a finite dimensional algebra $\mathcal{A}$ with an F-structure and an involution ι of $\mathcal{A}$ defined over F. Using 12.1.8 (1) we see that

$$G = \{a \in \mathcal{A} \mid a(\iota a) = 1\}$$

is an F-group isomorphic to $\mathbf{Sp}_{2n}$.

Define a non-degenerate Hermitian form Φ on $V = K^m$ as in the proof of 17.2.8. The values $\Phi(v, v)$ $(v \in V)$ lie in $\mathrm{Im}(1 + \kappa)$. This was established in the proof of 17.2.8 if char $F = 2$, and it is obvious if char $F \neq 2$. This means that Φ is a trace-valued hermitian form (forme tracique) in the sense of [Dieu, Ch. I, §10]. Then its Witt index is defined (see [loc.cit., Ch. I, §11]). We may assume that the following holds. If $v = (a_1, \dots, a_m)$, $v_0 = (a_{r+1}, \dots, a_{m-r})$ and $w = (b_1, \dots, b_m)$, $w_0 = (b_{r+1}, \dots, b_{m-r})$ then

$$\Phi(v, w) = a_1(\kappa(b_m)) + \cdots + a_r(\kappa(b_{m+1-r1})) + \Phi_0(v_0, w_0),$$

where Φ_0 is an anisotropic hermitian form defined by a hermitian $(m-2r) \times (m-2r)$-matrix (anisotropy means, as before, that $\Phi_0(v_0, v_0) = 0$ implies $v_0 = 0$).

We have an isomorphism $\phi : k \otimes_F K \simeq M_d$. We can then identify $M = M_m(K)$ with a subspace of $M_m(M_d) = M_{2n}$ (recall that $2n = dm$). Let $S \subset \mathbf{GL}_{2n}$ be the subgroup of diagonal matrices $\text{diag}(x_1, \ldots, x_{2n})$ with

$$x_1 = \cdots = x_d, \ x_{d+1} = \cdots = x_{2d}, \ldots, x_{(r-1)d+1} = \cdots = x_{rd},$$

$$x_i = 1 \text{ for } rd + 1 \leq i \leq (m-r)d,$$

$$x_i = x_{2n+1-i}^{-1} \text{ for } i > (m-r)d.$$

Then S is an F-split subtorus of G. By arguments similar to those used in 15.3.10 (1) one shows that S is a maximal F-split subtorus of G.

The isomorphism ϕ can be taken to be such that the group of diagonal matrices $\text{diag}(x_1, \ldots, x_{2n})$ with $x_i = x_{2n+1-i}^{-1}$ is a maximal torus T of H. The basis D of the root system is $(\alpha_1, \ldots, \alpha_n)$, where $\alpha_1, \ldots, \alpha_{n-1}$ are as before and $\alpha_n = 2\epsilon_n$.

The preceding analysis leads to the following result. We make the convention that the identity map of F is also a symplectic involution κ and that the κ-hermitian forms over F are the alternating ones (the Witt index of a non-degenerate form of dimension $2r$ being r).

17.2.10. Proposition. *(i) For type C_n the possible subsets $D - D_0$ of D are of the form $(\alpha_d, \ldots, \alpha_{rd})$, where d is a power of 2 dividing $2n$;*
(ii) For D_0 as in (i) the index (D, D_0, triv) can be realized over the field F if and only if there exists a division algebra K of degree d with center F with a symplectic involution κ of the first kind, together with a non-degenerate κ-hermitian form over K of dimension $2d^{-1}n$ and of index r;
(iii) If the conditions of (ii) are satisfied the root system $_F R$ is irreducible, of type BC_r if $r < d^{-1}n$ and of type C_r if $r = d^{-1}n$.

17.3. Type D_n

17.3.1. The root system of type D_n ($n \geq 4$) is described in [Bou2, p. 256-257]. The Dynkin diagram is

The group A has order 2 if $n > 4$ and is isomorphic to the symmetric group S_3 if $n = 4$. The cocenter C^* is the product of two cyclic groups of order 2 if n is even and is cyclic of order 4 if n is odd.

First assume that char $F \neq 2$. Take $H = \mathbf{SO}_{2n}$, see 7.4.7 (4). By 12.1.8 (2), H is defined over F. The maximal torus T being as in 7.4.7 (4), the basis D of the root system of (G, T) is $(\alpha_1, \ldots, \alpha_n)$, where $\alpha_1, \ldots, \alpha_{n-1}$ are as before and $\alpha_n = \epsilon_{n-1} + \epsilon_n$. If $n > 4$ the non-trivial element a of A permutes α_{n-1} and α_n and fixes the other simple roots. If $n = 4$ the elements of A fix α_2 and permute the other three roots arbitrarily. In this section we will consider only F-groups of type D_4 whose index (D, D_0, τ) is such that Im τ has order ≤ 2, as in the case $n > 4$. We may then assume that for $n = 4$ we have Im $\tau \subset \{1, a\}$. The outer forms of groups of type D_4 where Im τ has order ≥ 3 are called *trialitarian*. They pertain to the exceptional groups, discussed below.

17.3.2. Still assume that char $F \neq 2$. The group $H = \mathbf{SO}_{2n}$ is a subgroup of index two of $H_1 = \mathbf{O}_{2n}$. The automorphism a of the Dynkin diagram is induced by conjugation by an element of H_1. It follows that for $n > 4$

$$\mathbf{Aut}\, H = \mathbf{PO}_{2n} = \mathbf{O}_n / \{\pm 1\}.$$

We may also assume this to be the case if $n = 4$ in the non-trialitarian case. We can assume that $G = H_z$ where $z \in Z^1(F, \mathbf{PO}_{2n})$. The inclusion $\mathbf{O}_{2n} \subset \mathbf{GL}_{2n}$ induces an injective map

$$Z^1(F, \mathbf{Aut}\, H) \to Z^1(F, \mathbf{PGL}_{2n}).$$

As in 17.2.4 we obtain an algebra $\mathcal{A}$ over k with an F-structure, and an involution ι of $\mathcal{A}$ defined over F. We have to replace j by

$$s = \begin{pmatrix} 0 & 1_n \\ 1_n & 0 \end{pmatrix}$$

Now $\iota x = s({}^t x)s$; it is an orthogonal involution. Assume that $\mathcal{A}(F)$ is a matrix algebra $M = M_m(K)$ over a division algebra K with center F. Then ι induces an involution of the first kind of M. As before, K has an involution of the first kind κ and there is h such that ι satisfies (74). Since ι is orthogonal, h must be hermitian (by 17.2.6 (ii)) Again, the degree d of K is a power of 2 and $2n = dm$.

Proceeding as in 17.2.9 we obtain a hermitian form Φ on K^m. We have a maximal F-split torus S and a maximal torus T of H containing S, defined as in loc.cit. D being as in 17.3.1 we have $D_0 = \{\alpha_d, \ldots, \alpha_{rd}\}$.

17.3.3. To describe completely the index of G we have to identify the homomorphism $\tau : \Gamma \to \{1, a\}$. If τ is non-trivial it defines a quadratic extension E_τ, as in 17.1.4. If τ is trivial we put $E_\tau = F$. Let $E_\tau = F(\sqrt{\xi})$. Then the element $\delta(\tau) = \xi(F^*)^2 \in F^*/(F^*)^2$ is uniquely determined (and determines τ).

Denote by Nr : $M \to F$ the reduced norm function on M (the restriction to M of the determinant function on $\mathbf{M}_{2n}$).

17.3.4. Proposition. *(i) $M_{\iota,-1}$ contains invertible elements of M;*
(ii) If $x \in M_{\iota,-1}$ is invertible then $(-1)^n \mathrm{Nr}(x) \in \delta(\tau)$.

Let $\mathbf{Alt}_{2n} \subset \mathbf{M}_{2n}$ be the space of alternating matrices. Then $\mathcal{A}_{\iota,-1} = s\mathbf{Alt}_{2n}$. There is a non-zero polynomial function Pf on $\mathbf{Alt}_{2n}$ (the Pfaffian) such that $\det(y) = \mathrm{Pf}(y)^2$ ($y \in \mathbf{Alt}_{2n}$). For $a \in A$ we have

$$\mathrm{Pf}(ay(^t a)) = \det(a)\mathrm{Pf}(y). \tag{75}$$

(see [Jac4, p. 334-335]). Then (i) is immediate if F is infinite. If F is finite then $K = F$, $M = M_{2n}(F)$, and h is a symmetric matrix. We leave the check of (i) in that case to the reader.

The Γ-action on $\mathcal{A}(F_s)$ is of the form

$$\gamma \star a = g_\gamma(\gamma a)g_\gamma^{-1}, \tag{76}$$

where $g_\gamma \in \mathbf{O}_{2n}(F_s)$ has image $z(\gamma)$ in $\mathbf{Aut}(H)(F_s)$. It follows from (75) that for $a \in M_{\iota,-1}$

$$\mathrm{Pf}(\gamma \star a) = \epsilon(\gamma)\mathrm{Pf}(a),$$

where $\epsilon(\gamma) = 1$ if $\tau(\gamma) = 1$ and $\epsilon(\gamma) = -1$ otherwise. This implies that $\mathrm{Pf}(a) \in \sqrt{\xi}\, F$, where ξ is as in 17.3.3. This implies (ii). $\qquad \square$

$\delta(\tau)$ is the *discriminant* $\delta(M)$ of the simple algebra with involution of the first kind M, as defined in [Kn, §6].

17.3.5. Exercise. Assume that $M = M_{2n}(F)$ with $\iota m = h(^t m)h^{-1}$, where h is symmetric. Show that $\delta(\tau) = (-1)^n \det(h)(F^*)^2$.

17.3.6. Now assume that char $F = 2$. Let $V = k^{2n}$ and let Q_0 be the quadratic form on V with

$$Q_0((\xi_1, ..., \xi_{2n})) = \sum_{i=1}^{n} \xi_i \xi_{n+i}.$$

Let H be the identity component of the isotropy group of Q_0 in $GL(V)$. Then H is a semi-simple group of type D_n by 7.4.7 (7) and by 12.1.8 (4) it is defined over F. The inclusion $H \subset H' = \mathbf{Sp}_{2n}$ induces an injective F-homomorphism $\mathbf{Inn}\, H \to \mathbf{Inn}(H')$. Let $z' \in Z^1(F, H')$ be the image of z under the obvious map. As in 17.2.4 we obtain an algebra $\mathcal{A}$, with a symplectic involution of the first kind ι over F, such that $\mathcal{A}(F)$ is a matrix algebra $M = M_m(K)$.

17.3.7. To deal with the F-forms of H we have to establish some facts about quadratic forms. For the moment, the characteristic is arbitrary.

If $x \in \mathbf{M}_{2n}$ then $Q_x(v) = vxs({}^t v)$ defines a quadratic form on V (identified with the space of $1 \times (2n)$-matrices), which is 0 if and only if $x \in \mathrm{Im}(1 - \iota)$, where $\iota x = s({}^t x)s)$. We obtain an injective linear map of the quotient $\mathcal{Q} = \mathbf{M}_{2n}/\mathrm{Im}(1 - \iota)$ to the space of quadratic forms on V, which is easily seen to be surjective. We identify the two vector spaces. They have obvious F-structures.

We have a linear map $\sigma : \mathcal{Q} \to \mathrm{Im}(1 + \iota)$ induced by $y \mapsto y + \iota y$. If Q_0 is as in 17.3.6 we have $\sigma Q_0 = 1$.

Now let $M = M_m(K)$ be a central simple algebra over F of dimension d^2, with an involution of the first kind ι as in 17.3.2. Put $\mathcal{Q}_m(K) = M/(1 - \iota)M$. We call this F-vector space the *space of m-dimensional quadratic forms* over the division algebra K with involution κ (of symplectic type if char $F = 2$ and of orthogonal type with dm even otherwise), see [Ti5]. We have, as before, an F-linear map $\sigma : \mathcal{Q}_m(K) \to \mathrm{Im}(1 + \iota)$. We say that $Q \in \mathcal{Q}_m(K)$ is *non-degenerate* if σQ is invertible. Then $h = \sigma Q$ is the *associated element* of Q in $\mathrm{Im}(1 + \iota)$. In that situation let $l \in M$ represent Q. Since $l + \iota l = h = \sigma Q$ is invertible, we have an involution $\iota' = \mathrm{Int}(h) \circ \iota$. Putting $l' = lh^{-1}$ we have

$$l' + \iota'(l') = 1.$$

Then l' defines a non-degenerate quadratic form Q' relative to ι', whose associated element is 1.

Let l be as before. For $v \in K^m$ (viewed as an $1 \times m$-matrix) define a function with values in $K/\mathrm{Im}(1 - \kappa)$ by

$$Q(v) = vl(\kappa({}^t v)) + (1 - \kappa)K.$$

This is a *quadratic form* Q on K^m. The orthogonal group $O(Q)$ is defined in the obvious way.

If char $F \neq 2$ then σ is bijective and we can identify $\mathcal{Q}_m(K)$ with the space $M_{\iota,1}$ of hermitian elements of M and $K/\mathrm{Im}(1 - \kappa)$ with the fixed point set of κ in K. There is a hermitian form Φ on K^m as in 17.3.2 such that $Q(v) = \Phi(v, v)$ and the orthogonal group of Q coincides with the unitary group of Φ.

Now assume that char $F = 2$. Let t be the reduced trace function on $\mathcal{A}$. It is the coefficient of T^{n-1} in $\mathrm{Nr}(T.1 + a)$. Denote by $q(a)$ the coefficient of T^{n-2}. If $\xi_1, \ldots, \xi_{2n}$ are the eigenvalues of a (viewed as elements of $\mathbf{M}_{2n}$) then

$$t(a) = \sum_i \xi_i, \quad q(a) = \sum_{i<j} \xi_i \xi_j.$$

$t(q)$ is a linear (respectively, quadratic) form on $\mathcal{A}$ defined over F.

17.3.8. Lemma. *Let $x, y \in \mathcal{A}$.*
(i) $q(x + y) = q(x) + q(y) + t(x)t(y) - t(xy)$;

(ii) $q(x + \iota x) = t(x)^2$;

(iii) *If* $l + \iota l = 1$ *then* $q(l + x + \iota x) = q(l) + t(x) + t(x)^2$.

We may assume that $\mathcal{A} = \mathbf{M}_{2n}$. Over a field of characteristic 0 we have $q(x) = \frac{1}{2}(t(x)^2 - t(x^2))$, from which the formula of (i) follows immediately. In that case the formula can be viewed as a polynomial identity in two sets of matrix variables, with integral coefficients. Reducing modulo 2 we see that the formula holds in characteristic 2, whence (i).

We may assume that $\iota x = j({}^t x)j$, as in 17.2.4. To prove (ii) write $x \in \mathbf{M}_{2n}$ in the form

$$x = \begin{pmatrix} a & b \\ c & d \end{pmatrix},$$

where $a, b, c, d \in \mathbf{M}_n$. Then

$$\iota x = \begin{pmatrix} {}^t d & {}^t b \\ {}^t c & {}^t a \end{pmatrix},$$

and

$$x + \iota x = \begin{pmatrix} u & v \\ w & {}^t u \end{pmatrix},$$

where v and w are alternating and $t(u) = t(x)$. A reduction argument, as in the proof of (i), shows that $q(x + \iota x) = t(u^2) + t(vw) = t(u)^2 + t(vw)$. Since v and w are alternating we have $t(vw) = 0$. (ii) follows.

If l is as in (iii) it follows from (i) and (ii) that

$$q(l + x + \iota x) = q(l) + t(x)^2 + t(l(x + \iota x)) + t(l)t(x + \iota x).$$

Then (iii) follows by observing that $t(x + \iota x) = 0$ and $t(l(x + \iota x)) = t((l + \iota l)x) = t(x)$. □

For $a \in k$ put $\wp a = a + a^2$. Let Q be a non-degenerate quadratic form with associated element h. Choose l with $l + \iota l = h$. The *discriminant* $\delta(Q) \in F/\wp F$ is the coset $n + q(lh^{-1}) + \wp F$. It follows from 17.3.8 (iii) that this definition is independent of the choice of the representative l.

17.3.9. Exercise. Let $K = F$, $V = F^{2n}$. Let Q be the quadratic form on V with

$$Q((\xi_1, ..., \xi_{2n})) = \sum_{i=1}^{n}(a_i \xi_i^2 + b_i \xi_{i+n}^2 + \xi_i \xi_{i+n}).$$

Show that $\delta(Q) = \sum_{i=1}^{n} a_i b_i + \wp F$ (this is the *Arf invariant* of Q, see [Scha, Ch. 9, §4]).

17.3.10. Let $Q \in \mathbf{Q}_m(K)$ be the quadratic form defined by l, as before. The orthogonal group $O(Q)$ is the group of all invertible elements $x \in M$ such that $\iota x = x^{-1}$ and $xlx^{-1} + l = y + \iota y$, with $y \in M$. By 17.3.8 (iii) we have

$$q(l) = q(xlx^{-1}) = q(l) + t(y) + t(y)^2.$$

It follows that $t(y) \in [0, 1]$. The *Dickson invariant* $\Delta(x)$ (see [loc.cit.]) is $t(y)$.

17.3.11. Lemma. Δ *is a surjective F-homomorphism* $O(Q) \to \{0, 1\}$, *whose kernel is the identity component* $O(Q)^0$.

Recall that $O(Q)$ is defined over F. That Δ is a homomorphism follows from its definition. To prove the other statements we may work over k, in which case we identify $O(Q)$ with the group H of 17.3.6. We write the elements of $\mathbf{M}_{2n}$ as 2×2-matrices, as in the proof of 17.3.8. We may take

$$l = \begin{pmatrix} 1 & 0 \\ 0 & 0 \end{pmatrix}.$$

Then $O(Q)$ is the set of

$$x = \begin{pmatrix} a & b \\ c & d \end{pmatrix}$$

such that $a.^t d + b.^t c = 1$ and that $a.^t b$ and $c.^t d$ are alternating. Moreover, $\Delta(x) = \mathrm{tr}(b.^t c)$. It is not hard to exhibit x with $\Delta(x) = 1$: it suffices to do this for $n = 1$, in which case one takes $a = d = 0$, $b = c = 1$.

It remains to prove that Ker $\Delta = H$. It is clear that $H \subset$ Ker Δ. We have seen that H is semi-simple of type D_n. Let $x \in O(Q)$ normalize H. There is $y \in H$ such that xy normalizes a maximal torus T of H. Now $O(Q)$ is a subgroup of $\mathbf{Sp}_{2n}$ and T is also a maximal torus of $\mathbf{Sp}_{2n}$, by 7.4.7 (4), (7). The Weyl group of (H, T) has index 2 in the Weyl group of $(\mathbf{Sp}_{2n}, T)$ (see [Bou2, p. 255]). It follows that H has index ≤ 2 in $O(Q)$, which implies that Ker $\Delta = H$. $\square$

Let Q be as in 17.3.10. A subspace S of K^m is *singular*, if the restriction of Q to S is zero. Then W is isotropic for the hermitian form Φ of 17.3.7. The dimension of a maximal singular subspace of K^m is the *Witt index* of Q. If it is zero then Q is *anisotropic*. Proceeding as in 17.2.9 we obtain that $D - D_0 = (\alpha_d, \ldots, \alpha_{rd})$. As before, we have to identify the homomorphism $\tau : \Gamma \to \{\pm 1\}$, or the extension E_τ. Now there is $\xi \in F$ with $E_\tau = F(\{\wp^{-1}\xi\})$. The element $\delta(\tau) = \xi + \wp F \in F/\wp F$ is well-defined.

17.3.12. Proposition. $\delta(\tau) = \delta(Q)$.

Let $\mathbf{M} = \mathbf{M}_{2n}$, with involution ι_0 defined by $\iota_0(a) = j({}^t a)j$. The quadratic from Q_0 of 17.3.6 is the one defined by

$$l_0 = \begin{pmatrix} 1_n & 0 \\ 0 & 0 \end{pmatrix}.$$

An arbitrary quadratic form Q over a division algebra is obtained from Q_0 by twisting with a cocycle $z \in Z^1(F, O(Q_0))$. With the notations of 17.3.2 we have $\mathcal{A}(F_s) = M_{2n}(F_s)$, with Γ-action given by (76). For $\gamma \in \Gamma$

$$g_\gamma l_0 g_\gamma^{-1} = l_0 + a_\gamma,$$

where $a_\gamma \in \mathrm{Im}(1 + \iota_0)$. By 17.3.11 we have $D(a_\gamma) \in \{0, 1\}$. Also, for $\gamma, \delta \in \Gamma$

$$a_{\gamma\delta} = a_\gamma + \gamma \star a_\delta.$$

It follows from 11.1.6 and 11.3.4 (5) that there exists $a \in \mathrm{Im}(1 + \iota_0)$ such that $a_\gamma = \gamma \star a + a$. Then $l = l_0 + a$ lies in $\mathbf{M} = \mathcal{A}(F)$ and defines the quadratic form Q. By 17.3.8 (iii)

$$q(l) = q(l_0) + t(a) + t(a)^2.$$

Now $q(l_0) = n$ and $t(a_\gamma) = \gamma(t(a)) + t(a)$. It follows that $\gamma(t(a)) = a$ if $\tau(\gamma) = 1$ and $\gamma(t(a)) = a + 1$ otherwise. We conclude that $\delta(\tau) = t(a) + t(a)^2 + \wp F$. $\square$

What we have established implies the following classification result for type D_n (non-trialitarian if $n = 4$).

17.3.13. Proposition. *(i) The possible subsets $D - D_0$ are of the form $(a_d, ..., a_{rd})$, where d is a power of 2 dividing $2n$;*
(ii) For D_0 as in (i) the index (D, D_0, τ) can be realized over F if and only if there exists a quadratic form Q over a division algebra with center F with an involution of the first kind, of degree d such that Q has dimension $2d^{-1}n$ and index r and that $\delta(Q) = \delta(\tau)$;
(iii) If the conditions of (ii) are satisfied, the root system $_F R$ is of type B_r if $r < 2d^{-1}n$ and of type D_r if $r = 2d^{-1}n$.

The involution ι of (ii) is of the first kind and orthogonal if char $F \neq 2$ and symplectic otherwise.

17.3.14. Exercise. If in 17.3.13 (i) we have $rd = n - 1$, then K is a quaternion algebra and τ is non-trivial.

17.4. Exceptional groups, type G_2

We now assume that our root system R is of one of the exceptional types E_6, E_7, E_8, F_4, G_2. In several cases there is an explicit description of G as an automorphism group of an algebraic structure. An exhaustive discussion of what is known, with complete proofs, is beyond the scope of this book. But I shall try to review the existing results, with references to the literature. We start with type G_2, where the picture is clear.

17.4.1. For the root system of type G_2 see [Bou2, p. 274-275]. The Dynkin diagram is

$$\overset{1}{\circ} \Lleftarrow \overset{2}{\circ}$$

The groups A and C^* are trivial. Recall that H is a split F-group of type G_2. Then **Aut** $H = H$, hence $H^1(F, H)$ classifies the F-forms G of H.

17.4.2. Proposition. *G is either split or anisotropic.*

The numbering of the vertices of $\mathcal{D}$ identifies D with $\{1, 2\}$. Describe the roots by their coordinates relative to the simple roots. Then the highest root is $(3, 2)$. Since A is trivial an index has the form (D, D_0, triv). If G is not split or anisotropic then $D_0 = \{1\}$ or $\{2\}$ and $_F R$ is of rank 1, i.e. is of type A_1 or BC_1. In that case the highest root coefficient is 2. From 15.5.7 we see that the only possible case is $D_0 = \{2\}$. Then we apply the criterion of 16.5.5. The group H is simply connected. It follows from 8.4.6 (6) that the same holds for the group M of 16.5.5. Hence M is F-isomorphic to $\mathbf{SL}_2$. By 16.5.6 we have $\pi M \simeq M$. Since $H^1(F, \mathbf{SL}_2)$ is trivial (see 12.3.6 (2)) it does not contain anisotropic elements. We conclude from 16.5.5 that this case is impossible. The proposition follows. $\qquad\square$

We next discuss the explicit description.

17.4.3. Octonion algebras. A *composition algebra* C over F is a (not necessarily associative) finite dimensional k-algebra with identity element, having an F-structure, that is provided with a non-degenerate quadratic form N (the *norm form*) such that

$$N(xy) = N(x)N(y) \ (x, \ y \in C),$$

and that N is defined over F. If char $F = 2$, assume that the bilinear form associated to N is non-degenerate (which means that N is F_s-isomorphic to the form Q_0 of 17.3.6). For the basic facts on composition algebras see [Jac4, 7.6]. More details can be found in [Kn, Ch. VIII] and [SpV]. One shows that dim $C = 1, 2, 4, 8$. We assume from now on that dim $C = 8$. In this case C is an *octonion algebra* or *Cayley algebra*.

We introduce a special octonion algebra $\mathbf{O}$. As a vector space

$$\mathbf{O} = \mathbf{M}_2 \oplus \mathbf{M}_2.$$

If

$$x = \begin{pmatrix} a & b \\ c & d \end{pmatrix} \in \mathbf{M}_2$$

put

$$\bar{x} = \begin{pmatrix} d & -b \\ -c & a \end{pmatrix}.$$

Then $x\bar{x} = \det(x)$. The multiplication in $\mathbf{C}$ is defined by

$$(x, y)(u, v) = (xu + \bar{v}y, vx + y\bar{u})$$

and

$$N((x, y)) = \det(x) - \det(y).$$

A straightforward check shows this defines indeed an octonion algebra. There is an obvious F-structure on C such that the algebra product and the quadratic form N are defined over F. Let G be the automorphism group of $\mathbf{O}$; clearly this is a linear algebraic group over k, which is a closed subgroup of the F-group $GL(\mathbf{O}) \simeq \mathbf{GL}_8$.

17.4.4. Theorem. *(i) G is a connected simple group of type G_2;*
(ii) G is defined over F and F-split.

We sketch a proof. The orthogonal group of N is an F-group (see 17.3.2 and 17.3.6) and so is its identity component $SO(N)$, which is quasi-simple of type D_4. Consider the set S of triples (t, t_1, t_2) of elements of $SO(N)$ such that for all $x, y \in C$

$$t(xy) = t_1(x)t_2(y).$$

It defines a closed subgroup of $SO(N)^3$. One shows that S is an F-group, which is F-isomorphic to the split simply connected F-group $\mathbf{Spin}_8$ of type D_4. See [SpV] or [Kn, §35] (where it is assumed that char $F \neq 2$). Moreover $(t, t_1, t_2) \mapsto (t_2, t_1, t)$ defines an outer automorphism θ of S of order 3, which is defined over F. The automorphism group G is a closed subgroup of S, viz. the fixed point group S^θ of θ. By [SpV] G is an F-group. Using the results of 10.3 one shows G is a split F-group of type G_2, hence is isomorphic to H. $\qquad\square$

17.4.5. Any octonion algebra over F is F_s-isomorphic to $\mathbf{O}$ (by [SpV, 1.8]). The principle of 11.3.3 shows $H^1(F, G)$ classifies the isomorphism classes of octonion algebras over F. Hence there is a bijection between isomorphism classes of F-forms of H and isomorphism classes of octonion algebras over F. By [loc.cit., 1.6.2] two octonion algebras over F are F-isomorphic if and only if their norm forms are F-isomorphic.

Assume that char $F \neq 2$ and let ϕ_α ($\alpha \in F^*$) be the norm form of the two-dimensional algebra $F[T]/(T^2 - \alpha)$. The quadratic forms that can occur as norm forms of octonion algebras are isomorphic to tensor products

$$f_{\alpha,\beta,\gamma} = \phi_\alpha \otimes \phi_\beta \otimes \phi_\gamma,$$

with $\alpha, \beta, \gamma \in F^*$. Such a form (*threefold Pfister form*) is either anisotropic or of maximal Witt index. In the first case the octonion algebra is a division algebra, in the second case it is F-isomorphic to $\mathbf{O}$. The corresponding automorphism groups are anisotropic, respectively split. We conclude that, if char $F \neq 2$, the F-forms of H are classified by threefold Pfister forms. The next result uses that fact.

We have $H^1(F, \mathbf{Z}/2\mathbf{Z}) = F^*/(F^*)^2$. For $a \in F^*$ let $[a]$ be its image in $H^1(F, \mathbf{Z}/2\mathbf{Z})$. Associate to the form $f_{\alpha,\beta,\gamma}$ the cup product $[\alpha] \cup [\beta] \cup [\gamma] \in H^3(F, \mathbf{Z}/2\mathbf{Z})$. We obtain a map $\phi : H^1(F, H) \to H^3(F, \mathbf{Z}/2\mathbf{Z})$.

17.4.6. Proposition. ϕ *is injective.*

See [Kn, 34.40] or [Se3, p. 16]. This implies that the F-forms of a group of type G_2 are classified by a cohomological invariant, lying in $H^3(F, \mathbf{Z}/2\mathbf{Z})$.

17.5. Indices for the types F_4 and E_8

17.5.1. We next discuss the types F_4 and E_8. As in the case of type G_2, the groups A and C^* are trivial (because of this there are fewer technicalities than in types E_6 and E_7). Assume that H is a split F-group of type F_4 or E_8. In [Bou2, p. 268-273] the root systems and their properties are described. The Dynkin diagrams are

$$\overset{1}{\circ} - \overset{2}{\circ} \Rightarrow \overset{3}{\circ} - \overset{4}{\circ}$$

and

$$\overset{1}{\circ} - \overset{3}{\circ} - \overset{4}{\circ} - \overset{5}{\circ} - \overset{6}{\circ} - \overset{7}{\circ} - \overset{8}{\circ}$$
$$\underset{2}{\circ}$$

$H^1(F, H)$ classifies the F-forms of H, since the groups A and C^* are trivial. We describe the possible indices (D, D_0, triv). This amounts to identifying the possible subsets D_0 of D. We proceed as in the proof of of 17.4.2. Identify D with $\{1, 2, 3, 4\}$ and $\{1, 2, 3, 4, 5, 6, 7, 8\}$. We describe the roots by their coordinates relative to the simple roots.

17.5.2. Proposition. *(i) In type F_4 the possibilities for $D - D_0$ are:* $\emptyset$, $\{4\}$, D; *(ii) In the second case the root system $_F R$ is of type BC_1.*

The first and the last possibility correspond to an anisotropic group, respectively to the split group. The highest root $\tilde{\alpha}$ of the root system of type F_4 is $(2, 3, 4, 2)$. By 15.5.6 and 15.5.7 the set D_0 must be such that the coordinates of $\tilde{\alpha}$ with respect to the simple roots in $D - D_0$ are the highest root coefficients of an irreducible root system of rank $|D - D_0|$ (the root system $_F R$). The tables of [Bou2] show that, in an irreducible root system of rank ≤ 3, the highest root coefficients are ≤ 3 and that a coefficient 3 only occurs in type G_2. It follows that, for a group G that is not split or anisotropic, $D - D_0$ can only be one of the following sets

$$\{1\}, \ \{4\}, \ \{1, 2\}, \ \{1, 4\}, \ \{2, 4\}.$$

We have to show that, except for the second one, these sets are impossible.

Assume we had an index with $D - D_0 = \{1\}$. We apply the criterion of 16.5.5. The group H being simply connected, 8.4.6 (6) implies that the same holds for M. By 16.5.6 we have $\pi M \simeq M$. The group M is now quasi-simple and simply connected of type C_3, hence isomorphic to $\mathbf{Sp_6}$. By 12.3.5 (4) we have $H^1(F, M) = 1$, which shows that $H^1(F, M)$ cannot contain anisotropic elements. By 16.5.5 this case cannot exist. A similar argument rules out the other cases where $D - D_0 \neq \{4\}$ ($\mathbf{Sp_6}$ being replaced by $\mathbf{SL_3}$, $\mathbf{Sp_4}$, $\mathbf{SL_2} \times \mathbf{SL_2}$, respectively). This proves (i).

If $D - D_0 = \{4\}$ the relative root system $_F R$ is of rank 1, hence of type A_1 or BC_1. Since the highest root coefficient is 2 (by 15.5.7) we are in the second case. This proves (ii). $\qquad\square$

17.5.3. A straightforward application of 16.5.5 shows that an index with $D_0 = \{1, 2, 3\}$ exists over F if and only if $H^1(F, \mathbf{Spin_7})$ contains an anisotropic element, where $\mathbf{Spin_7}$ is the simply connected, quasi-simple F-group of type B_3. This is equivalent to the existence of 7-dimensional anisotropic quadratic forms over F with trivial Hasse invariant. We shall meet similar conditions below, in the discussion of indices for type E_8. A further analysis of such conditions would require results from the theory of quadratic forms. We shall not go into these matters. More details can be found in [Selb].

In 17.6 we shall discuss the realization of F_4 as an automorphism group of an algebraic structure. This will lead to an explicit description of the anisotropic forms of groups of type F_4.

17.5.4. Proposition. *(i) In type E_8 the possibilities for $D - D_0$ are:*

$$\emptyset, \ \{1\}, \ \{8\}, \ \{1, 8\}, \ \{7, 8\}, \ \{1, 6, 7, 8\}, \ D;$$

(ii) The respective root systems $_F R$ in the cases that $D - D_0 \neq \emptyset$, D are irreducible of types BC_1, BC_1, BC_2, G_2, F_4.

We proceed as in the case of F_4. Now the highest root is $(2, 3, 4, 6, 5, 4, 3, 2)$. A consideration of highest root coefficients of $_F R$, as before, gives the following non-trivial possibilities for $D - D_0$:

$$\{1\}, \ \{8\}, \ \{1, 2\}, \ \{1, 3\} \ \{1, 7\}, \ \{1, 8\}, \ \{2, 8\}, \ \{7, 8\},$$

$$\{1, 2, 3, 8\}, \ \{1, 2, 6, 8\}, \ \{1, 3, 7, 8\}, \ \{1, 6, 7, 8\}.$$

The argument used to discard the cases not listed in the proposiition is similar to the one used for type F_4 as before: application of 16.5.5 and (in this case) use of the triviality of $H^1(F, \mathbf{SL}_n)$, for various n. We skip the details.

To prove (ii) observe that by 15.5.7 we know the highest root of the (by 15.5.6) irreducible root system $_F R$. In the present cases the highest root determines the root system uniquely. $\square$

17.5.5. Denote by G_{D-D_0} an F-form of H with index (D, D_0, id). Then 16.5.5 gives conditions on F for the existence of G_{D-D_0}, which we now review. Of course, the split group G_D exists over any field F.

(a) $G_{\{1\}}$ ($G_{\{1,8\}}$) exists if and only if $H^1(F, \mathbf{Spin}_{14})$ (respectively, $H^1(F, \mathbf{Spin}_{12})$) contains anisotropic elements. This is equivalent to the existence over F of anisotropic quadratic forms in the appropriate dimension with trivial invariant δ (see 17.3) and trivial Hasse invariant. The existence question is discussed in [Selb, V,5.2].

(b) $G_{\{1\}}$ exists if and only if $H^1(F, K)$ contains anisotropic elements, where K is the simply connected, split group of type E_7. The existence question is discussed in [Ti7].

(c) $G_{\{7,8\}}$ exists if and only if $H^1(F, K)$ contains anisotropic elements, where now K is simply connected, split, of type E_6. A criterion for existence is contained in 17.6.5.

(d) $G_{\{1,6,7,8\}}$ exists if and only if $H^1(F, \mathbf{Spin}_8)$ contains anisotropic elements.

(e) A general description of anisotropic F-forms is not known. Nor does one know necessary and sufficient conditions on F for each F-form of G to be isotropic.

(f) The classification of F-forms of groups of type E_8 is tantamount to the classification of F-forms of the corresponding Lie algebra, by the next proposition. There are explicit constructions of Lie algebras of type E_8, which realize some of the possible indices (see [Fre, II], [Ti3]). Let H be F-split of type E_8, with Lie algebra $\mathfrak{h}$. This is a Lie algebra with an F-structure.

17.5.6. Proposition. $H^1(F, H)$ *classifies the F-forms of* $\mathfrak{h}$.

By the principle of 11.3.3 this will follow if we show that H is the automorphism group of $\mathfrak{h}$, and is defined over F.

Let S be the space of k-homomorphisms $\mathrm{Hom}(\mathfrak{h} \otimes_k \mathfrak{h}, \mathfrak{h})$. The group $M = GL(\mathfrak{h})$ acts on it. S and M have obvious F-structures, and the action is defined over F. The Lie algebra structure on $\mathfrak{h}$ is an element $s \in S(F)$, and the automorphism group of $\mathfrak{h}$ is the isotropy group M_s. Application of 12.1.2 shows that M_s is defined over F if the fixed point set $L(M)_s$ of s in the Lie algebra $L(M) = \text{End}(\mathfrak{h})$ coincides with $L(M_s)$. Now $L(M)_s$ is the space of derivations of $\mathfrak{h}$, i.e. of linear maps D of $\mathfrak{h}$ such that

$$D[u, v] = [Du, v] + [u, Dv] \ (u, v, \in \mathfrak{h})$$

(to see this observe that $S \simeq \mathfrak{h}^* \otimes \mathfrak{h}^* \otimes \mathfrak{h}$, where $\mathfrak{h}^*$ is the vector space dual of $\mathfrak{h}$ and use 4.4.14 (ii)).

The construction of semi-simple groups with a given simply laced root system in 10.2 shows that H is a subgroup of M_s (proof of 10.2.8) and that $\dim L(M_s) = \dim \mathfrak{h}$ (10.2.6). Hence

$$\dim H \leq \dim M_s = \dim L(M_s) = \dim \mathfrak{h} = \dim H,$$

and $\dim H = \dim M_s$. It follows that H is the identity component of M_s, so that M_s normalizes H. But since H has no outer automorphisms, the cosets of M_s/H can be represented by linear maps centralizing $\text{Ad}(H)$ and also $\text{ad}(\mathfrak{h})$. Using the description of $\mathfrak{h}$ given in 10.2 one sees that such a map must be a scalar multiplication. Being an automorphism of $\mathfrak{h}$ it must be the identity.

We have proved that $H \simeq M_s$. Also, $\dim L(M)_s = \dim H$. The proposition follows from 12.1.2. $\qquad\qquad\qquad\qquad\qquad\qquad\qquad\qquad\qquad\qquad\qquad\qquad\qquad\qquad\Box$

17.6. Descriptions for type F_4

17.6.1. As in the case of type G_2, the classification of groups of type F_4 is tantamount to the classification of certain algebraic structures, which we call *Albert algebras*. In characteristic $\neq 2$ these are closely related to the exceptional Jordan algebras. We review the relevant results.

Let $V = (\mathbf{M}_3)^3$. This is a 27-dimensional vector space, with an obvious F-structure. Let d and t be the determinant and trace functions on $\mathbf{M}_3$. Denote by N the cubic form on V defined by

$$N(x) = d(x_0) + d(x_1) + d(x_2) - t(x_0 x_1 x_2).$$

There is a unique quadratic map $\nu : \mathbf{M}_3 \to \mathbf{M}_3$ such that for $a \in \mathbf{GL}_3$ we have

$$a^{-1} = d(a)^{-1} \nu(a).$$

We denote by n the quadratic map $V \to V$ with

$$n(x) = (\nu(x_0) - x_1 x_2, \nu(x_2) - x_0 x_1, \nu(x_1) - x_2 x_0).$$

Then n is defined over F. For $x, y \in V$ put $x \times y = n(x + y) - n(x) - n(y)$. This defines a symmetric bilinear map $V \times V \to V$, which is defined over F. Finally, define the non-degenerate symmetric bilinear form σ on V by

$$\sigma(x, y) = t(x_0 y_0 + x_1 y_1 + x_2 y_2).$$

The ingredients N, n, σ are defined over F.

We define

$$N(x, y) = (dN)_x(y).$$

This is a polynomial function on $V \times V$, homogeneous quadratic in x and linear in y, and

$$N(x + y) = N(x) + N(x, y) + N(y, x) + N(y).$$

Then

$$N(x, y, z) = N(x + y, z) - N(x, z) - N(y, z)$$

defines a symmetric trilinear form on V with $N(x, x, x) = 6N(x)$. Let $e \in V(F)$ be the point $(1, 0, 0)$. We have the following properties for $x, y \in V$:

(i) $N(e) = 1$, $n(e) = e$;

(ii) $\sigma(x, y) = N(e, x)N(e, y) - N(e, x, y)$;

(iii) $N(x, y) = \sigma(nx, y)$;

(iv) $n(nx) = N(x)x$.

(i) is clear. For proofs of the other formulas see [Spr1, p. 61-63].

We call, more generally, *Albert algebra A* over F a vector space V with F-structure, provided with a cubic form N (the *norm form*), a non-degenerate symmetric bilinear form (the *trace form*), both defined over F and an element $e \in V(F)$, such that the analogues of (i), (ii), (iii), (iv) hold and that this structure is isomorphic over F_s to the one defined above, which we call the *standard Albert algebra* **A**. This definition of an Albert algebra is equivalent to the one given in [Pe] (where further references are given).

From the point of view adopted in [Spr1], an Albert algebra is a vector space with an inverse-like birational map. In the case of **A** this map is given by

$$jx = (N(x))^{-1}n(x),$$

on a dense open subset of V. It follows from (iv) that $N(nx) = N(x)^2$, which implies that j is involutorial. The inverse-like character is shown by the identity (valid on a dense open subset)

$$j(e + x) + j(e + jx) = e,$$

see [loc.cit., p. 58]. In [loc.cit.] a general theory of such structures (called J-structures) is developed.

17.6.2. Consider the standard Albert algebra **A**. Let $H_1 \subset GL(V)$ be the subgroup of $GL(V)$ whose elements fix the cubic form N. It is a linear algebraic group. It follows from the definitions that for $u \in H_1$ we have

$$N(ux, uy) = N(x, y).$$

Let $\tilde{u}$ be the contragredient of u with respect to σ, i.e.

$$\sigma(x, y) = \sigma(ux, \tilde{u}y) \ (x, y, \in V).$$

By property (iii) we then have

$$n(ux) = \tilde{u}(nx).$$

Using (iv) one shows that $u \mapsto \tilde{u}$ defines an automorphism of H_1. Let H be the automorphism group of **A**. Using property (ii) we see that H is the isotropy group of e in H_1, hence is a linear algebraic group.

17.6.3. Theorem. *(i) H_1 is a connected, quasi-simple, simply connected group of type E_6 defined over F and is F-split;*
(ii) H is a connected, simple group of type F_4 defined over F and F-split.

These results are contained in [loc.cit.], as special cases of more general results. The first part of (i) is contained in the results of [loc.cit., 14.20]. That H_1 is defined over F follows from [loc.cit., 14.27 (i), 4.10]. For the first part of (ii) see [loc.cit., 14.24]. That H is defined over F follows from [loc.cit., 14.27 (iii)]. In [loc.cit., 14.21] a maximal torus T of H_1 is described. It is obvious that it is defined over F and F-split. By [loc.cit., 14.24], T contains a maximal torus of H. The latter is defined over F and F-split by 13.2.2 (i), and (ii) follows. □

17.6.4. We have seen in 17.5.1 that $H^1(F, H)$ classifies the forms of F-groups of type F_4. From 17.6.3 (ii) it follows by 11.3.3 that $H^1(F, H)$ also classifies the (isomorphism classes) of Albert algebras over F. There are constructions (first and second Tits constructions) of Albert algebras over F by means of which all Albert algebras over F can be obtained, up to isomorphism (see [Spr1, §5], [Pe] or [Kn, §40]). The ingredients of the first construction are a a central simple algebra K of dimension 9 over F and an element $\alpha \in F^*$. Denote by d and t the reduced norm and reduced trace of K. Define an Albert algebra $A(K, \alpha)$ over F as follows: $V = K^3$ and $e = (1, 0, 0)$. The cubic form N on V is given by

$$N(x) = d(x_0) + \alpha^{-1}d(x_1) + \alpha d(x_2) - t(x_0x_1x_2).$$

Then n and σ are determined by the formulas (ii) and (iii) of 17.6.1 and are, in fact, given by the same formulas as in 17.6.1. Clearly, $\mathbf{A} = A(\mathbf{M}_3, 1)$. Also, $A(K, \alpha)$ is F_s-isomorphic to **A**.

The ingredients of the second Tits construction are a central simple algebra L with an involution of the second kind ι, with center E. Let $L^+ \subset L$ be the fixed point set of ι. Assume that $F = E \cap L^+$, then E is a quadratic extension of F. Let $u \in L^+$ and $\beta \in F$ be such that, d and t now denoting reduced norm and trace of L, we have $d(u) = \beta(\iota\beta)$.

We define an Albert algebra $A(L, u, \beta)$ as follows. We have $V = L^+ \oplus L$, $e = (1, 0)$. The cubic form N is given by

$$N((a, b)) = d(a) + \beta d(b) + \iota(\beta d(b)) - t(abu\iota(b)).$$

Again, n and σ are determined. Then $A(L, u, \beta)$ is isomorphic over E to $A(L, \beta)$.

Let A be an Albert algebra over F, with underlying vector space V and cubic form N. Let $H_1(A)$ and $H(A)$ be, respectively, the groups fixing N and the automorphism group of A. These are F-forms of the groups H_1 and H of 17.6.3. They are semi-simple of types E_6, respectively F_4. We say that A is *reduced* if there exists $x \in V(F) - \{0\}$ with $N(x) = 0$. Otherwise, A is an *Albert division algebra* over F.

17.6.5. Theorem. *(i) A is isomorphic to a first Tits construction $A(K, \alpha)$ or a second Tits construction $A(L, u, \beta)$;*
(ii) A is reduced if and only if $H(A)$ or $H_1(A)$ is isotropic over F.

For (i) see [loc.cit., 15.9] or [Pe, p. 201]. Assume that $H(A)$ or $H_1(A)$ contains a non-trivial F-split torus S. Let $x \in V(F) - \{0\}$ be a weight vector for a non-zero weight of S in V. From the fact that the groups leave invariant N, one deduces that $N(x) = 0$. Hence A is reduced.

Conversely, if this is so one deduces from property (iv) of 17.6.1 that there is $x \in V(F) - \{0\}$ such that $n(x) = 0$. Using part (i) one sees, by working in a first or second Tits construction, that there is also such an element which, moreover, satisfies $\sigma(e, x) \neq 0$. We may assume that $\sigma(e, x) = 1$. One then checks that x is an idempotent element of A, in the sense of [Spr1, p. 45]. By [loc.cit., 10.1] the idempotent element x determines a two dimensional F-split torus in the group generated by $H_1(A)$ and the non-zero scalar multiplications of V, which is a reductive F-group with commutator group $H_1(A)$. Then $H_1(A)$ must contain an F-split subtorus. That the same holds for $H(A)$ can be deduced from the fact that a reduced Albert algebra can be coordinatized by an octonion algebra (see [Pe, p. 199]). We next describe this coordinatization.

17.6.6. Reduced Albert algebras. Let C be an octonion algebra over F, with norm form N_C and let

$$d = \text{diag}(d_1, d_2, d_3)$$

be a diagonal matrix in $GL_3(F)$. Let $A(C, d)$ be the set of 3×3-matrices $x = (x_{ij})$ with entries in the algebra C and diagonal entries in k, which are Hermitian with respect to d, i.e. such that

$$dx = {}^t(d\bar{x}),$$

where $x\bar{x} = N_C(x)$, as in the proof of 17.4.3. Write $x_{12} = c_3$, $x_{23} = c_1$, $x_{31} = c_2$. Define

$$N(x) = x_{11}x_{22}x_{33} - \sum d_h^{-1} d_j x_{ii} N_C(c_i) + N_C(c_1 c_2, \bar{c}_3),$$

where (hij) runs through the three cyclic permutations of (123) and $N_C(\ ,\ )$ is the bilinear form associated to N_C. For $x, y \in A$ we have

$$\sigma(x, y) = \text{trace}(xy).$$

Define n by property (iii) of 17.6.1. Then the ingredients N, n, σ define a structure of Albert algebra $A(C, d)$, which is easily seen to be reduced. See [Pe, p. 198].

17.6.7. Theorem. *A reduced Albert algebra over F is F-isomorphic to an $A(C, d)$; (ii) If $\text{char}(F) \neq 2, 3$, two reduced Albert algebras over F are isomorphic if and only if their trace forms are equivalent over F.*

For (i) see [loc. cit.]. (ii) is a recent result, due to Serre (see [Se3, p.18] or [Pe, p. 202]).   □

17.6.8. Relation with Jordan algebras. Assume that $\text{char}(F) \neq 2$. Let A be an Albert algebra over F. The notations are as before. There is a unique structure of commutative algebra on V with identity e whose squaring operation is given by

$$x^2 = n(x) + \sigma(e, x)x - \sigma(e, nx)e \ (x \in V).$$

It is defined over F. It is a Jordan algebra structure, i.e. we have

$$x(x^2 y) = x^2(xy) \ (x, y, \in V),$$

see [Jac2, p. 21] or [Spr1, §6]. Denote this Jordan algebra over F by $J(A)$.

The last identity holds in any associative algebra B over F with symmetrized product

$$xy = \frac{1}{2}(x \circ y + y \circ x),$$

$\circ$ denoting the product in B. Let $\mathcal{J}(B)$ be the Jordan algebra with underlying space B and this product. The Jordan algebras $J(A)$ are *exceptional* in the sense that they cannot be imbedded as subalgebra in an algebra $\mathcal{J}(B)$, see [Jac2, p. 49-51]. In characteristic 2 one has similar results, involving quadratic Jordan algebras (see [Jac3, p. 1.45, p.1.49]).

17.6.9. Using 17.6.3 (ii) and 17.6.7 (ii) we see that the classification of isotropic groups of type F_4 over F (at least in characteristics $\neq 2, 3$) is reduced to a problem about quadratic forms over F. From 17.6.5 we find that the classification of anisotropic groups of type F_4 is reduced to the classification of Albert division algebras over F. By 17.6.5 (i) these can be obtained by a Tits construction.

17.6.10. Lemma. *$A(K, \alpha)$ $(A(L, u, \beta))$ is an Albert division algebra if and only if $\alpha \notin d(K(F))$ (respectively, $\beta \notin d(F(E))$).*

The notations are as in 17.6.4. The proof follows by a straightforward check, using that reducedness is equivalent to the existence of a non-zero rational solution of $nx = 0$.   □

17.6.11. It follows that an Albert division algebra exists over F if there is a nine dimensional central simple division algebra K with center F such that not every element of F^* is a reduced norm. By 12.3.9 (3) this is equivalent with $H^1(F, \mathbf{SL}_{1,K}) \neq \{1\}$. The classification problem of Albert division algebras over F has not yet been solved.

17.6.12. By 17.4.4 groups of type G_2 are classified by a cohomological invariant $H^3(F, \mu_2)$. In type F_4 there are three cohomogical invariants of an Albert algebra V over F: $f_3(V) \in H^3(F, \mathbf{Z}/2\mathbf{Z})$, $f_5(V) \in H^5(F, \mathbf{Z}/2\mathbf{Z})$, $g_3(V) \in H^3(F, \mathbf{Z}/3\mathbf{Z})$. It is not known whether these determine the isomorphism class of V. But there are partial results: for example, V is reduced if and only if $g_3(V) = 0$, and in that case the invariants $f_3(V)$, $f_5(V)$ characterize the isomorphism class of V. For more details and references see [Pe], [Se3, §9] or [Kn, §41].

17.7. Type E_6

17.7.1. Assume that H is simply connected of type E_6. The root system is described in [Bou2, p. 260-262]. The Dynkin diagram $\mathcal{D}$ is

The cocenter C^* has order 3. The automorphism group of $\mathcal{D}$ has order 2. Its nontrivial element acts on D as the opposition involution ι of 15.5.8. The orbits of ι are

$$\{1, 6\}, \{3, 5\}, \{2\}, \{4\}.$$

In this case the ingredient τ of the index need not be trivial.

If char $F \neq 3$ the center of H has order 3 (as follows from 8.1.12 (8)). It is generated by

$$z = \alpha_1^\vee(\zeta)\alpha_3^\vee(\zeta^{-1})\alpha_5^\vee(\zeta)\alpha_6^\vee(\zeta^{-1}), \tag{77}$$

where ζ is a primitive third root of unity. In fact, it is straightforward to check that $\alpha(z) = 1$ for all $\alpha \in D$. The central isogeny π of 16.5.4 is separable, its kernel is the center of H.

If char $F = 3$ this isogeny is of the form $H \to H/\mathfrak{a}$, as in 12.2, where $\mathfrak{a}$ is the one dimensional central subalgebra of the Lie algebra $\mathfrak{g}$ spanned by

$$Z = X_1 - X_3 + X_5 - X_6, \tag{78}$$

with $X_1 = d\alpha_i^\vee(1)$. We leave it to the reader to check this fact.

17.7.2. Proposition. *(i) In type E_6 we have, apart from $D - D_0 = \emptyset$, D, the following possibilities for $D - D_0$:*

$$\tau \text{ trivial}: \{1, 6\}, \{2, 4\} \text{ (and perhaps } \{2\} \text{ if char } F = 3),$$

$$\tau \text{ non} - \text{trivial}: \{2\}, \{1, 6\}, \{2, 4\}, \{1, 2, 6\};$$

(ii) In the respective cases of (i) the root system $_F R$ is of type

$$A_2, \quad G_2, \quad , \quad BC_1,$$

$$BC_1, \quad BC_1, \quad G_2, \quad BC_2.$$

The highest root is $(1, 2, 2, 3, 2, 1)$. By 15.5.8 we know that D_0 has to be stable under the opposition involution ι. We then use 15.5.6 and 15.5.7, as in 17.5.2 and 17.5.4, and the description of highest root coefficients for types B, C, D given in [Bou2, Planches]. It follows that we can only have the following non-trivial possibilities for $D - D_0$:

$$\tau \text{ trivial}: \{2\}, \{1, 6\}, \{2, 4\}, \{3, 5\}, \{1, 2, 6\}, \{2, 3, 5, \}, $$

$$\tau \text{ non} - \text{trivial}: \{2\}.\{1, 6\}, \{2, 4\}, \{1, 2, 6\}.$$

To prove (i) we may assume that τ is trivial. To prove that the case $\{3, 5\}$ is impossible we apply 16.5.6. With the notations used there we have that πM contains a factor over F of type $\mathbf{SL}_2$. This implies that $H^1(F, \pi M)$ cannot contain anisotropic elements. The cases $\{2, 3, 5\}$ and $\{1, 2, 6\}$ are eliminated in a similar way.

If $D - D_0 = \{2\}$ we have $M \simeq \mathbf{SL}_6$. If char $F \neq 3$, the element z of (77) lies in M and $\pi M \simeq \mathbf{SL}_6/U$, where U is the group of third roots of unity. An anisotropic form of $\mathbf{SL}_6$ must be of the form $\mathbf{SL}_{1,K}$, where K is a central division algebra over F of dimension 36, see 12.3.5 (1) and 12.3.8 (3). Then K defines an element in the Brauer group of F, which has order 6. On the other hand the image in $H^1(F, \mathbf{PGL}_6)$ of an anisotropic element of $H^1(F, \mathbf{SL}_6/U)$ would define an element of order 3 in the Brauer group, which leads to a contradiction. For more details see [Selb, p. 71]. In characteristic 3 this case should also be impossible, but there is no reference. We have proved (i). (ii) follows by an argument as in the proof of 17.5.4 (ii). □

17.7.3. We write again G_{D-D_0} for an inner form with prescribed D_0. If τ is non-trivial let $E = E_\tau$ be the quadratic extension defined by Ker τ. We then write $_E G_{D-D_0}$. We discuss the question of existence.

(a) $_E G_{\{2\}}$. With the notations of 16.5.4 we see from 16.5.5 that $_E G_{\{2\}}$ exists if and only if $H^1(F, (\mathbf{SL}_6/U)_\tau)$ contains anisotropic elements (char $F \neq 3$, U being as in the proof of 17.7.2). By the discussion of 17.1 this amounts to the existence of certain anisotropic hermitian forms over a division algebra with center E with an involution of the second kind fixing F. See [Selb, p. 99, p. 112-113].

(b) An application of 16.5.5 and 16.5.6 shows that for $G_{\{1,6\}}$ to exist over F it is necessary and sufficient that $H^1(F, \mathbf{Spin}_8)$ contains anisotropic elements. Likewise, $_EG_{\{1,6\}}$ exists over F if and only if $H^1(F, (\mathbf{Spin}_8)_\tau)$ contains anisotropic elements. They come from anisotropic quadratic forms Q over F with $\delta(Q) = \delta(\tau)$ (notations as in 17.3). See [loc.cit., p. 98, p.111].

(c) $G_{\{2,4\}}$. Assume first char $F \neq 3$. Then Ker $\pi \subset M$ (use (77)). In fact, we can identify M with $\mathbf{SL}_3 \times \mathbf{SL}_3$ such that Ker π is the subgroup Z of elements (c, c^{-1}), where c lies in the center of $\mathbf{SL}_3$. The image in M of $\mathbf{SL}_3 \times \{1\}$ is a normal subgroup isomorphic to $\mathbf{SL}_3$, and the quotient is isomorphic to $\mathbf{PGL}_3$. Using the exact sequence of 12.3.4 (ii) we conclude that $H^1(F, \pi M)$ contains anisotropic elements if and only if $H^1(F, \mathbf{PGL}_3)$ contains such elements. By a Lie algebra argument, using the element of (78), we see that the same holds in characteristic 3. We conclude from 16.5.5 and 12.3.5 (1) that $G_{\{2,4\}}$ exists over F if and only if there exists a 9-dimensional division algebra with center F. Similarly, $_EG_{\{2,4\}}$ exists if and only if there exists such a division algebra with center E with an involution of the second kind fixing F (see 17.1.4).

(d) $_EG_{\{1,2,6\}}$ exists if and only if $H^1(F, (\mathbf{SL}_4)_\tau)$ contains anisotropic elements. This amounts to the existence of certain anisotropic hermitian forms over a division algebra with center E and an involution of the second kind fixing F, as in the case $_EG_{\{2\}}$. See [loc.cit., p. 98, p.112-113.]

17.7.4. Explicit description. Let A be the standard Albert algebra introduced in 17.6.1. It determines a cubic form N on a 27-dimensional space V, these ingredients being defined over F. By 17.6.3 (i) we may view H as the stabilizer of N in $GL(V)$. It follows from the principle of 11.3.3 that we can identify $H^1(F, H)$ with the set of isomorphism classes of F-forms of N (i.e. cubic forms on V over F which are F_s-isomorphic to N).

As in 17.6.5 one sees that if $H^1(F, H)$ contains an anisotropic element, a corresponding cubic form N' must be anisotropic, in the sense that $N'(x) = 0$ has only the zero solution in $V(F)$. Such an anisotropic form is provided by an Albert division algebra over F. A description of all anisotropic groups of type E_6 does not seem to be known.

17.8. Type E_7

17.8.1. Assume that H is simply connected of type E_7. The root system is described in [Bou2, p. 264-266]. The Dynkin diagram is

The group A is trivial and the cocenter C^* has order 2. If char $F \neq 2$ the center of H has order 2. Its non-trivial element is

$$z = \alpha_2^\vee(-1)\alpha_5^\vee(-1)\alpha_7^\vee(-1).$$

It is easily seen that $\alpha(z) = 1$ for all $\alpha \in D$. In characteristic 2 the isogeny π is of the form $H \to H/\mathfrak{a}$, as in 17.7.1. Now $\mathfrak{a}$ is spanned by

$$Z = X_2 + X_5 + X_7.$$

17.8.2. Proposition. *(i) In type E_7 we have, apart from $D - D_0 = \emptyset$, D the following possibilities for $D - D_0$*

$$\{1\}, \{6\}, \{7\}, \{1, 6\}, \{1, 6, 7\}, \{1, 3, 4, 6\};$$

(ii) In the respective cases of (i) the root system $_F R$ is of type

$$BC_1, BC_1, A_1, BC_2, C_3, F_4.$$

The highest root is $(2, 2, 3, 4, 3, 2, 1)$. Application of 15.5.6 and 15.5.7 leads to a longish list of possibilities for $D - D_0$, viz.

$$\{1\}, \{2\}, \{6\}, \{7\},$$

$$\{1, 2\}, \{1, 3\}, \{1, 5\}, \{1, 6\}, \{1, 7\}, \{2, 3\}, \{2, 5\}, \{2, 6\}, \{2, 7\}, \{3, 6\}, \{5, 6\}, 6, 7\},$$

$$\{1, 2, 6\}, \{1, 2, 7\}, \{1, 6, 7\}, \{2, 6, 7\},$$

$$\{1, 2, 3, 4\}, \{1, 2, 4, 5\}, \{1, 2, 6, 7\}, \{1, 3, 4, 6\}, \{1, 4, 5, 6\}, \{2, 3, 4, 6\}, \{2. 4, 5, 6\}.$$

We have to eliminate the cases not listed in (i). Assume first that $D - D_0$ has at least two elements and contains α_2 (and is not listed in (i)). Application of 16.5.7 with $D_1 = D - \{\alpha_2\}$ leads to an index for type A_6 that violates 15.5.8, except for the case of $\{1, 2, 7\}$. In that case application of 16.5.7 with $D_1 = D - \{\alpha_1\}$ leads to a non-existent index for type D_6.

There remain the cases that

$$D - D_0 = \{2\}, \{1, 3\}, \{1, 5\}, \{1, 7\}, \{3, 6\}, \{5, 6\}, \{6, 7\}, \{1, 4, 5, 6\}.$$

In the first one the group M of 16.5.4 is isomorphic to $\mathbf{SL}_7$, whose cocenter is of order 7, prime to the order 2 of the cocenter of G. By 16.5.6 we have $\pi M \simeq M$, and we can proceed as in previous cases.

In the second case $M \simeq \mathbf{SL}_6$. When char $F \neq 2$ the element z of 17.8.1 does not lie in M, and $\pi M \simeq M$. We can proceed as before. If char $F = 2$ the same is true, via a Lie algebra argument.

In the remaining six cases apply 16.5.7 with $D_1 = D - \{\alpha_i\}$, where $i = 5, 7, 3, 5, 7, 4$ respectively. This leads to indices for groups of respective types A_4, E_6, A_5, A_2, E_6, A_2 which violate 16.5.7. We have established (i).

Except for the case of $\{1, 6, 7\}$, (ii) follows, as before, from the fact that the highest root determines ${}_F R$. In the exceptional case we can only conclude that ${}_F R$ is either of type B_3 or of type C_3. To rule out the first case, one can use the description of the positive roots of a root system of type E_7 given in the last lines of [Bou2, II ,p. 264]. A straightforward check, using the description of root systems of types B_3 and C_3 given in [loc.cit.], shows that we have the second alternative, establishing (ii). $\square$

17.8.3. We discuss the problem of existence. Notations are as in 17.5.5. In the discussion below we assume that char $F \neq 2$. In characteristic 2 the results are the same but the arguments are slightly different (the Lie algebra comes into play). We skip the details.

(a) $G_{\{1\}}$. The group M is now isomorphic to $\mathbf{Spin}_{12}$ and $z \in M$. It follows that $\pi M \simeq M/\{1, z\}$, so is isomorphic to a quotient of $\mathbf{Spin}_{12}$. But πM is not isomorphic to $\mathbf{SO}_{12}$, as z does not lie in the kernel of the isogeny $\mathbf{Spin}_{12} \to \mathbf{SO}_{12}$. The existence of $G_{\{1\}}$ is tantamount to the existence of certain quadratic forms over division algebras. See [Selb, p. 95].

(b) $G_{\{6\}}$. Now $M \simeq \mathbf{Spin}_{10} \times \mathbf{SL}_2$, and $z \in M$. One checks that $\pi M \simeq \mathbf{SO}_{10} \times \mathbf{PGL}_2$, from which it follows that $G_{\{6\}}$ exists if and only if there exists over F a 10-dimensional quadratic form with trivial δ-invariant and a quaternion division algebra.

(c) In the case of $G_{\{7\}}$ the group M is of type E_6. Application of 16.5.6 shows that $\pi M \simeq M$. It follows that $G_{\{7\}}$ exists if and only if $H^1(F, K)$ contains anisotropic elements, where K is split, simply connected of type E_6. This condition was discussed in 17.7.4.

(d) The case of $G_{\{1,6\}}$ is similar to case (b). Now $M \simeq \mathbf{Spin}_8 \times \mathbf{SL}_2$ and $\pi M \simeq \mathbf{SO}_8 \times \mathbf{PGL}_2$. The conclusion is that $G_{\{1,6\}}$ exists if and only if there exists over F an 8-dimensional quadratic form with trivial δ-invariant and a quaternion division algebra.

(e) In the case of $G_{\{1,6,7\}}$ we have $M \simeq \mathbf{Spin}_8$, and $z \notin M$, from which one concludes that $G_{\{1,6,7\}}$ exists if and only if $H^1(F, \mathbf{Spin}_8)$ contains anisotropic elements (like case (d) in 17.5.5).

(f) $G_{\{1,3,4,6\}}$. Now $M \simeq \mathbf{SL}_2 \times \mathbf{SL}_2 \times \mathbf{SL}_2$ and z corresponds to (c, c, c), where c is the central element of $\mathbf{SL}_2$. One shows (compare with 17.7.3, case (c)) that $G_{\{1,3,4,6\}}$ exists if and only there exists a quaternion division algebra with center F.

(g) An exhaustive description of anisotropic forms of G, for arbitrary F, is not known. In [Ti7] one finds a construction of an inner anisotropic form, assuming the existence of certain central division algebras over F with dimension 16, defining an element of order 4 in the Brauer group.

17.8.4. Explicit description. Assume char $F \neq 2$. Let **A** be the standard Albert algebra over F. With the notations of 17.6.1 put

$$M = V \oplus V \oplus k \oplus k.$$

This is a 56-dimensional vector space over k, with an obvious F-structure. For $m = (x, y, \xi, \eta) \in M$ define

$$J(m) = \sigma(nx, ny) - \xi N(x) - \eta N(y) - \frac{1}{4}(\sigma(x, y) - \xi \eta)^2.$$

J is a homogeneous polynomial function on M of degree 4 defined over F.

17.8.5. Proposition. *(char $F = 0$) The identity component of the stabilizer of J in $GL(M)$ is F-isomorphic to H.*

This can be deduced from [Fre, I, p. 228-229]. There is no doubt that the proposition is also true if char $F \neq 2, 3$. $\qquad\square$

It follows by the principle of 11.3.3 that (at least in characteristic 0) $H^1(F, G)$ classifies isomorphism classes of certain F-forms of the quartic form J. It is likely that anisotropic elements of $H^1(F, H)$ come from anisotropic forms of J (defined as in 17.7.4).

17.9. Trialitarian type D_4

17.9.1. We now assume that H is a split simply connected F-group of type D_4. We may assume that it is the spin group **Spin**$_8$. For the root system see 7.4.7 (4), (7). The Dynkin diagram is

Its automorphism group A is isomorphic to the symmetric group S_3. The cocenter C is isomorphic to $(\mathbf{Z}/2\mathbf{Z})^2$, see [Bou2, p. 257].

17.9.2. Proposition. *(i) In the trialitarian D_4-case the possibilities for $D - D_0$ are* $\emptyset$, $\{2\}$, D;

(ii) In the second case the root system $_F R$ is of type BC_1.

The proof is left to the reader.   □

17.9.3. Let E/F be the Galois extension (of degree 3 or 6) defined by Ker τ. The notation $_E G_{D-D_0}$ is as in 17.7.3.
(a) In the case of $_E G_{(2)}$ the group M is isomorphic to $\mathbf{SL}_2^3$. If char $F \neq 2$ the center of H is contained in M. One sees that πM is isomorphic to the quotient of $\mathbf{SL}_2^3$ by the subgroup of order 4 of elements (c_1, c_2, c_3), the c_i being central with $c_1 c_2 c_3 = 1$. There is a similar description in characteristic 2.
By 16.5.5, $H^1(F, (\pi M)_\tau)$ must contain anisotropic elements. An explicit condition for this to be the case does not seem to be in the literature. The same holds for the explicit description of the anisotropic F-forms of H.
(b) The classification of the trialitarian forms of G is dealt with in [Kn, §44].

17.9.4. Explicit description. The trialitarian forms of G are related to certain algebraic structures, resembling the octonion algebra structures of 17.4.1. For details we refer to [Kn, §36] or [SpV, Ch. 4].

Let A be an F-form of the commutative algebra $k \oplus k \oplus k$. Then $A(F)$ is a 3-dimensional separable commutative F-algebra (an example is a separable cubic extension of F). For $x = (x_1, x_2, x_3) \in A$ define

$$n(x) = x_1 x_2 x_3, \quad q(x) = (x_2 x_3, x_3 x_1, x_1 x_2).$$

Then n is the norm map on A; it is defined over F. If x is invertible we have $n(x) \neq 0$ and $x^{-1} = (n(x))^{-1} q(x)$. This implies that the quadratic map $q : A \rightarrow A$ is also defined over F.
A *twisted composition structure over A* is a triple (V, N, Q) consisting of a free A-module V, a cubic map $N : V \rightarrow k$ and a quadratic map $Q : V \rightarrow V$, all ingredients being defined over F, and such that for $x \in A$, $v \in V$:
(1) $Q(xv) = q(x)Q(v)$,
(2) $N \circ Q = q \circ N$,
(3) $N(v, Qv) \in k$.
One shows that the rank of V is 1, 2, 4, 8. From now on we assume that it equals 8, in which case we say that A is of *octonion type*.

17.9.5. Consider the case that $E = A(F)$ is a cyclic extension of F of degree 3, and let σ be a generator of the Galois group of E/F. For $x \in E$ we have $n(x) = x(\sigma x)(\sigma^2 x)$, $q(x) = (\sigma x)(\sigma^2 x)$. In this case, $C = V(F)$ is an 8-dimensional vector space over E, with a non-degenerate quadratic form N. One shows that there is a unique F-bilinear multiplication $(v, w) \mapsto v \star w$ such that for $x, y \in E$, $c, d \in C$ we have $Q(c) = c \star c$ and

(4) $xc \star yd = (\sigma x)(\sigma^2 y)c \star d$,

(5) $N(c \star d) = (\sigma N(c))(\sigma^2 N(d))$,

(6) $N(c, c \star c) \in F$.

The structure $(C, \star, N)$ satisfying (4),(5),(6) is a *twisted composition algebra* (of octonion type).

The following result gives the connection with trialitarian forms of G.

17.9.4. Proposition. *The automorphism group of an octonion twisted composition structure over F is defined over F and is a trialitarian form of H.*

See [Kn, 36.5, 36.13] and [SpV]. If E/F is a cyclic extension we have $A(F) = E$.
□

There is a close relation between twisted composition structures of octonion type and Albert algebras. We refer to [Kn] or [SpV] for more details.

17.10. Special fields

In this final section we give a very brief review of the situation over some special fields.

17.10.1. Fields of dimension ≤ 1. F has dimension ≤ 1 if F is the only central division algebra over F. Examples are finite fields and the field of rational functions $C(t)$. See [Se2, Ch. III, §3].

17.10.2. Theorem. *Assume F to be perfect of dimension ≤ 1. Then $H^1(F, H) = 1$ for any connected F-group H.*

The theorem is due to Steinberg [St2] (reproduced as an appendix in the English translation of [Se2]).
□

If H is reductive the perfectness assumption can be omitted (see [BoS, 8.6]). It follows from the theorem (see [Se2, Ch. III, 2.2]) that over a field of dimension ≤ 1 any semi-simple F-group is quasi-split (which we already knew in the case of finite fields, see 16.2.9 (2)).

Next let $F = \mathbf{R}$. The following theorem goes back to E. Cartan.

17.10.3. Theorem. *A connected semi-simple F-group H has an anisotropic F-form. It is unique up to F-isomorphism.*

For a proof see [Ha1, 3.3.2].
□

Using the theorem one can determine the possible indices of quasi-simple F-groups, via the case by case discussion of this chapter. The list that one obtains is given in [Sat] and [Ti2].

The following result, due to Borel and Serre (see [Se2, Ch. III, 4.5]), describes the Galois cohomology sets.

17.10.4. Theorem. *Let H be an anisotropic semi-simple R-group. There is a bijection of $H^1(\mathbf{R}, H)$ onto the set of conjugacy classes of elements of order ≤ 2 of $H(\mathbf{R})$.*

The group H need not be connected. Applying it to a group $\mathbf{Aut}(G)$ one obtains from 16.4.4 that the **R**-forms of an anisotropic, connected, quasi-simple **R**-group G are classified by conjugacy classes of automorphisms of G of order ≤ 2, a result which also goes back to E. Cartan.

17.10.5. Local fields. Now let F be a field with a non-trivial discrete valuation, for which it is complete. Assume that the residue field $\bar{F}$ is perfect.

A deep study of reductive F-groups was made by Bruhat and Tits [BruT]. The following is proved in [loc. cit., Ch. III, 4.3].

17.10.6. Theorem. *Assume that $\bar{F}$ has dimension ≤ 1. Let G be a connected, semi-simple, simply connected F-group. Then:*
(i) $H^1(F, G) = 1$,
(ii) If, moreover, G is quasi-simple and anisotropic then its root system is of type A_n.

Using part (ii) one can determine the possible indices of quasi-simple F-groups. The list is contained in [Ti2]. The results apply, in particular, in the case that F is a local field (in which case $\bar{F}$ is finite). For further details and references see [Se2, III, §3].

17.10.7. Global fields. Assume that F is a global field. Let V be the set of places (equivalence classes of absolute values) of F and denote by V_∞ the set of archimedean places (it is non-empty if and only if F is an algebraic number field). For $v \in V$ denote by F_v the corresponding completion of F. For any F-group G we have a map

$$r : H^1(F, G) \to \prod_{v \in V} H^1(F_v, G).$$

If G is connected, semi-simple and simply connected then by 17.10.6 all factors in the right-hand side with $v \notin V_\infty$, are 1, so we can take the product over the finite set V_∞.

17.10.8. Theorem. *If G is connected, semi-simple and simply connected then r is a bijection.*

This is the 'Hasse principle.' It suffices to prove this when G is quasi-simple. In that case it was was proved by Harder, except for the case that F is a number field and G is of type E_8. This case was settled by Cernousov. Harder's proof for the number field case involves a case by case analysis (but not in the function field case). See [Ha2], [Ha3] and [Ce]. More details about the number field case can be found in [Pl, Ch. 6].

Notes

The determination of the possible indices of quasi-simple groups is due to Tits, see [Ti2]. This paper does not contain proofs. In many cases these were given in [Selb]. See also [Ti7] and [Sat].

In the account given in the present chapter we have limited ourselves. Our account is fairly complete in the case of classical types. But in the exceptional types we have been more sketchy.

A full treatment of what is known about the classification of quasi-simple groups is beyond the scope of this book, as is a thorough review of what is known about the situation over special fields. We have given a few indications in 17.7.10.

In 17.5.4 and 17.6.11 we briefly mentioned cohomological invariants. These are examples of a general cohomological invariant found by M. Rost, for any simply connected, semi-simple group. They lie in $H^3(F, A)$, where A is a (suitable) finite abelian group. See [Se3, 7.3].

Table of Indices

1. $\quad$ o ---- o —$\overset{d}{\bullet}$— o ---- o —$\overset{n-d}{\bullet}$— o ---- o

2. $\quad$ o ---- o —$\overset{d}{\bullet}$— o ---- o —$\overset{rd}{\bullet}$— o ---- o —$\overset{n-ra}{\bullet}$— o ---- o —$\overset{n-d}{\bullet}$— o ---- o

3. $\quad$ • — • ---- $\overset{r}{•}$ — o ---- o ⇉ o

4. $\quad$ o ---- o —$\overset{d}{\bullet}$— o ---- o —$\overset{rd}{\bullet}$— o ---- o ⇉ o

5. $\quad$ o ---- o —$\overset{d}{\bullet}$— o ---- o —$\overset{rd}{\bullet}$— o ---- o — o
   $\qquad\qquad\qquad\qquad\qquad\qquad\qquad\quad |$
   $\qquad\qquad\qquad\qquad\qquad\qquad\qquad\quad o$

6. $\quad$ • — o — o — o — •
   $\qquad\qquad\quad |$
   $\qquad\qquad\quad o$

7. $\quad$ o — o — • — o — o
   $\qquad\qquad\quad |$
   $\qquad\qquad\quad •$

8. $\quad$ o — o — o — o — o
   $\qquad\qquad\quad |$
   $\qquad\qquad\quad •$

9. $\quad$ • — o — o — o — •
   $\qquad\qquad\quad |$
   $\qquad\qquad\quad •$

10. $\quad$ • — o — o — o — o — o
   $\qquad\qquad\quad |$
   $\qquad\qquad\quad o$

11. $\quad$ o — o — o — o — • — o
   $\qquad\qquad\quad |$
   $\qquad\qquad\quad o$

12. $\quad$ o — o — o — o — o — •
   $\qquad\qquad\quad |$
   $\qquad\qquad\quad o$

13. $\quad$ • — o — o — o — • — o
   $\qquad\qquad\quad |$
   $\qquad\qquad\quad o$

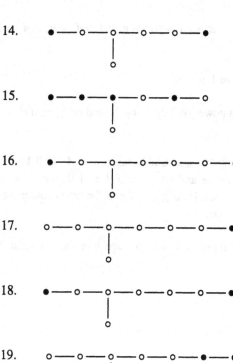

14. ●—○—○—○—○—●
      (with ○ below)

15. ●—●—●—○—●—○
      (with ○ below)

16. ●—○—○—○—○—○—○
      (with ○ below)

17. ○—○—○—○—○—○—●
      (with ○ below)

18. ●—○—○—○—○—○—●
      (with ○ below)

19. ○—○—○—○—○—●—●
      (with ○ below)

20. ●—○—○—○—●—●—●
      (with ○ below)

21.  ○—○⟹○—●

22. ○—●—○
      (with ○ below)

The black vertices of the Dynkin diagrams are those of the possible subsets $D - D_0$ for groups that are not anisoptropic or quasi-split, over some ground field F. They determine the index, except in the case of outer forms, in which case there is still the ingredient τ.

Comments

1. Inner type A_{n-1} ($n \geq 2$) (17.1.3). $d \geq 2$ is a divisor of n.

2. Outer type A_{n-1} $(n \geq 2)$ (17.1.6). d is a divisor of n and $2rd \leq n$; if $d = 1$ then $2r < n$.

3. Type B_n $(n \geq 2)$ (17.2.3). We have $1 \leq r < n$.

4. Type C_n $(n \geq 3)$ (17.2.10). d is a power of 2 dividing $2n$ and $rd \leq n$; if $d = 1$ then $r < n$.

5. Inner and outer type type D_n $(n \geq 4)$, non-trialitarian if $n = 4$ (17.3.13, see also 17.3.14). d is a power of 2 dividing $2n$ and $rd \leq n$; if $d = 1$ then $r < n$. If $rd \leq n - 3$ both inner and outer forms occur. If $rd \geq n - 2$ and the type is outer then $rd = n - 1$, $n - 2$ and $\alpha_{n-1}, \alpha_n \in D - D_0$.

6-9. E_6 (17.7.2). All occur as outer types; 6, 7 (and perhaps 8 in characteristic 3) as inner types.

10-15. Type E_7 (17.8.2).

16-20. Type E_8 (17.5.4).

21. Type F_4 (17.5.2).

22. Trialitarian type D_4 (17.9.2).

Bibliography

[Az]   H. Azad, Structure constants of algebraic groups, *Journal of Algebra* **75** (1982), 209-222.

[Bo1]   A. Borel, Groupes linéaires algébriques, *Annals of Mathematics* **64** (1956), 20-82.

[Bo2]   A. Borel et al., Algebraic groups and related finite groups, *Lecture Notes in Math.*, nr. 131, 2nd ed., Springer-Verlag, 1986.

[Bo3]   A. Borel, *Linear algebraic groups*, 2nd ed., Graduate Texts in Math., vol. 126, Springer-Verlag, 1991.

[BoS]   A. Borel and T. A. Springer, Rationality properties of linear algebraic groups, *Tôhoku Math. Journal* **20** (1968), 443-497.

[BoT1]   A. Borel and J. Tits, Groupes réductifs, *Publications Mathématiques IHES* **27** (1965), 55-150; Compléments, ibid. **41** (1972), 253-276.

[BoT2]   A. Borel and J. Tits, Eléments unipotents et sous-groupes paraboliques de groupes réductifs, *Inventiones Mathematicae* **12** (1971), 95-104.

[BoT3]   A. Borel and J. Tits, Théorèmes de structure et de conjugaison pour les groupes algébriques linéaires, *Comptes Rendus Ac. Sc. Paris* **287** (1978), 55-57.

[Bou1]   N. Bourbaki, *Algèbre commutative*, Hermann, 1961-1965.

[Bou2]   N. Bourbaki, *Groupes et algèbres de Lie*, Hermann, 1971-1975.

[Bru]   F. Bruhat, Représentations induites des groupes de Lie semi-simples connexes, *Comptes Rendus Ac. Sc. Paris* **138** (1954), 437-439.

[BruT]   F. Bruhat and J. Tits, Groupes réductifs sur un corps local, Ch. I, *Publications Mathématiques IHES* **41** (1972), 5-251; Ch. II, ibid. **60** (1984); Ch. III, *Journal Fac. Science University Tokyo* **34** (1987), 671-698.

[Ca]   R. W., Carter, *Finite groups of Lie type, conjugacy classes and complex representations*, Wiley, 1985.

[Car]   P. Cartier, Questions de rationalité des diviseurs en géométrie algébrique, *Bulletin Société Math. de France* **86** (1958), 177-251.

[Ce]   V. I. Cernousov, The Hasse principle for groups of type E_8 (in Russian), *Math. USSR Izv.* **34**(1990), 409-423.

[Ch1]   C. Chevalley, *Théorie des groupes de Lie*, vol. II, Groupes algébriques, Hermann, 1951.

[Ch2]   C. Chevalley, On algebraic group varieties, *Journal Math. Soc. Japan* **6** (1954), 303-324.

[Ch3]   C. Chevalley, *Fondements de la géométrie algébrique*, Paris, 1958.

[Ch4]   C. Chevalley, *Classification des groupes de Lie algébriques*, Séminaire Ecole Normale Supérieure, Paris, 1956-1958.

[Ch5]   C. Chevalley, Certains schémas de groupes semi-simples, Séminaire Bourbaki, exp. 219, Paris, 1960-1961.

[Cu]   C. W. Curtis, Central extensions of groups of Lie type, *Journal f. d. reine u. angewandte Mathematik* **220** (1965), 174-185.

[Del]   P. Deligne, Catégories Tannakiennes, in: P. Cartier et al., *The Grothendieck Festschrift, vol. II*, p. 111-195, Birkhäuser, 1990.

[DelL]   P. Deligne and G. Lusztig, Representations of reductive groups over finite fields, *Annals of Mathematics* **103** (1976), 103-161.

[DelM]   P. Deligne and J. S. Milne, Tannakian categories, in: P. Deligne et al., Hodge cycles, motives and Shimura varieties, *Lecture Notes in Math.*, nr. 900, Springer-Verlag, 1982.

[Dem]   M. Demazure, Désingularisation des variétés de Schubert généralisées, *Annales Ecole Normale Supérieure* **7** (1974), 53-88.

[DG]   M. Demazure and P. Gabriel, *Groupes algébriques I*, Masson/North-Holland, 1970.

[Dieu]   J. Dieudonné, *La géometrie des groupes classiques*, Ergebnissse der Math., Bd. 5, 2nd. ed., Springer-Verlag, 1962.

[EGA]   A. Grothendieck and J. Dieudonné, Eléments de géometrie algébrique, Publications mathématiques IHES, 1960-1967.

[Fre]   H. Freudenthal, Beziehungen der E_7 und E_8 zur Oktavenebene I, II, *Indagationes Mathematicae* **16** (1954), 218-230.

[FK]   I. B. Frenkel and V. Kac, Basic representations of affine Lie algebras and dual resonance models, *Inventiones Mathematicae* **62** (1980), 23-66.

[Gel]   I. M. Gelfand, Automorphic functions and the theory of representations, in: *Proceedings of the International Congress of Mathematicians* (Stockholm, 1962), p. 74-85.

[GN]   I. M. Gelfand and M. A. Neumark, *Unitäre Darstellungen der klassischen Gruppen*, Akademie-Verlag, Berlin, 1957.

[God]   R. Godement, *Théorie des faisceaux*, Hermann, 1958.

[Gor]   D. Gorenstein, *Finite groups*, Harper & Row, 1968.

[Ha1]   G. Harder, Über einen Satz von E. Cartan, Abhandlungen Math. Seminar Univ. Hamburg, **28** (1965), 208-214.

[Ha2]   G. Harder, Über die Galoiskohomologie halbeinfacher Matrizengruppen I, *Math. Zeitschrift.* **90** (1965); II, ibid. **92** (1966), 396-415.

[Ha3]   G. Harder, Chevalley groups over function fields and automorphic forms, *Annals of Mathematics* **100** (1974), 249-306.

[Har]   R. Hartshorne, *Algebraic geometry*, Graduate Texts in Math., vol. 52, Springer-Verlag, 1977.

[Hu1]   J. E. Humphreys, *Linear algebraic groups*, 2nd ed., Graduate Texts in Math., vol. 21, Springer-Verlag, 1981.

[Hu2]   J. E. Humphreys, *Reflection groups and Coxeter groups*, Cambridge University Press, 1990.

[Jac1]   N. Jacobson, *Lie algebras*, Interscience, 1962.

[Jac2]   N. Jacobson, *Structure and Representations of Jordan Algebras*, American Math. Society Colloquium Publications, vol. XXXIX, American Mathematical Society, 1968.

[Jac3]   N. Jacobson, *Lectures on quadratic Jordan algebras*, Tata Institute of Fundamental Research, Bombay, 1969.

[Jac4]   N. Jacobson, *Basic algebra I*, Freeman, 1974.

[Jac5]   N. Jacobson, *Basic Algebra II*, Freeman, 1980.

[Jan1]   J. C. Jantzen, *Representation of Algebraic Groups*, Academic Press, 1987.

[Jan2]   J. C. Jantzen, *Lectures on Quantum Groups*, Graduate Studies in Math., vol. 6, American Mathematical Society, 1996.

[Kam]   T. Kambayashi, M. Miayamishi, M. Takeuchi, Unipotent algebraic groups, *Lecture Notes in Math.*, nr. 414, Springer-Verlag, 1974.

[Kas]   C. Kassel, *Quantum Groups*, Graduate Texts in Math., vol. 155, Springer-Verlag, 1995.

[Kn]   M.-A. Knus, A. Merkurjev, M. Rost, J.-P. Tignol, *The Book of Involutions*, to appear.

[Kol1]   E. R. Kolchin, Algebraic matrix groups and the Picard-Vessiot theory of homogeneous linear ordinary differential equations, *Annals of Mathematics* **49** (1948), 1-42.

[Kol2]   E. R. Kolchin, On certain concepts in the theory of algebraic matric groups, *Annals of Mathematics* **49** (1948), 774-789.

[La1]   S. Lang, Algebraic groups over finite fields, *American Journal of Mathematics* **78** (1956), 555-563.

[La2]   S. Lang, *Algebra*, Addison-Wesley, 1977.

[Laz]   M. Lazard, Sur les groupes de Lie formels à un paramètre, *Bull. Soc. Math. France* **83** (1955), 251-274.

[Ma]   H. Matsumura, *Commutative Algebra*, 2nd ed., Benjamin, 1980.

[Mu1]   D. Mumford, *Abelian Varieties*, Oxford University Press, 1970.

[Mu2]   D. Mumford, The Red Book of varieties and schemes, *Lecture Notes in Math.*, nr. 1358, Springer-Verlag, 1988.

[Od]   T. Oda, *Convex bodies and algebraic geometry-An introduction to the theory of toric varieties*, Ergebnisse der Math. (3), Bd. 15, Springer-Verlag, 1988.

[Oe]   J. Oesterlé, Nombres de Tamagawa et groupes unipotents en caractéristique p, *Inventiones Mathematicae* **78** (1984), 13-88.

[Pe]   H. P. Petersson and M. Racine, Albert algebras, in: *Proceedings Conference on Jordan algebras*, (ed. W. Kaup, K. McCrimmon, H. P. Petersson), p. 197-207, W. de Gruyter, 1994.

[Pl]   V. P. Platonov and A. S. Rapinchuk, *Algebraic groups and number theory* (in Russian), Moscow, 1991 (English translation: Academic Press, 1993).

[Ri]   R. W. Richardson, Conjugacy classes in Lie algebras and algebraic groups, *Annals of Mathematics* **86** (1967), 1-15.

[Ron]   M. Ronan, *Lectures on Buildings*, Academic Press, 1989.

[Ros1]   M. Rosenlicht, Some rationality questions on algebraic groups, *Annali di Mat. Pura Appl.* **43** (1957), 25-50.

[Ros2]   M. Rosenlicht, Questions of rationality for solvable algebraic groups over nonperfect fields, *Annali di Mat. Pura Appl.* **61** (1963), 97-120.

[Rus]   P. Russell, Forms of the affine line and its additive group, *Pacific J. Math.* **32** (1970), 527-539.

[Sat]   I. Satake, *Classification theory of semi-simple algebraic groups* (mimeographed notes), University of Chicago, 1967.

[Scha]   W. Scharlau, *Quadratic and Hermitian Forms,* Grundlehren der math. Wissenschaften, Bd. 270, Springer-Verlag, 1985.

[Selb]   M. Selbach, *Klassifikationstheorie halbeinfacher algebraischer Gruppen*, Bonner mathematische Schriften, Nr. 83, 1976.

[Se1]   J.-P. Serre, Quelques propriétés des variétés abéliennes en caracteristique *p*, *American Journal of Mathematics* **80** (1958), 715-739.

[Se2]   J.-P. Serre, Cohomologie galoisienne, 5me éd., *Lecture Notes in Math.*, nr. 5, Springer-Verlag, 1994 (English translation: Springer-Verlag, 1997).

[Se3]   J.-P. Serre, Cohomologie galoisienne, progrès et problèmes, Séminaire Bourbaki, no. 783, 1993-94.

[SGA3]   M. Demazure, A. Grothendieck, Schémas en groupes I-III, *Lecture Notes in Math.*, nrs. 151-153, Springer-Verlag, 1970.

[Sha]   I. R. Shafarevich, *Basic Algebraic Geometry*, Grundlehren der math. Wissenschaften, Bd. 213, Springer-Verlag, 1974.

[Slo]   P. Slodowy, Simple singularities and simple algebraic groups, *Lecture Notes in Math.*, nr. 815, Springer-Verlag, 1980.

[Spr1]   T. A. Springer, *Jordan algebras and algebraic groups,* Ergebnisse der Math.. Bd. 75, Springer-Verlag, 1970.

[Spr2]   T. A. Springer, Linear algebraic groups, in: *Algebraic Geometry IV* (ed. A. N. Parshin and I. R. Shafarevich), Encyclopedia of Mathematical Sciences, vol. 55, p. 1-121, Springer-Verlag, 1994.

[SpV]   T. A. Springer and F. D. Veldkamp, *Octonions, Jordan algebras and exceptional groups* , to appear.

[St1]   R. Steinberg, Générateurs, relations et revêtements de groupes algébriques, in: *Colloque sur la théorie des groupes algébriques*, p. 113-127, Bruxelles, 1962.

[St2]   R. Steinberg, Regular elements of semisimple algebraic groups, *Publications Mathématiques IHES* **25** (1965), 281-312.

[St3]   R. Steinberg, *Endomorphisms of linear algebraic groups*, Memoirs American Mathematical Society, nr. 80 (1968).

[St4]   R. Steinberg, *Lectures on Chevalley groups*, Yale University, 1968.

[St5]   R. Steinberg, Conjugacy classes in algebraic groups, *Lecture Notes in Math.*, nr. 366, Springer-Verlag, 1974.

[Tak]   M. Takeuchi, A hyperalgebraic proof of the isomorphism and isogeny theorems for reductive groups, *Journal of Algebra* **85** (1983), 179-196.

[Tan]   T. Tannaka, Über den Dualitätssatz der nichtkommutativen topologischen Gruppen, *Tôhoku Mathematical Journal*, **45** (1938), 1-12.

[Ti1]   J. Tits, Normalisateurs de tores I. Groupes de Coxeter étendus, *Journal of Algebra* **4** (1966), 96-116.

[Ti2]   J. Tits, Classification of algebraic semisimple groups, in: Algebraic groups and discontinuous groups, *Proceedings Symp. Pure Math., vol. IX*, p. 33-62, American Math. Society, 1966.

[Ti3]   J. Tits, Algèbres alternatives, algèbres de Jordan et algèbres de Lie exceptionnelles I. Construction, *Indagationes Mathematicae* **28** (1966), 223-237.

[Ti4]   J. Tits, *Lectures on Algebraic Groups*, Yale University, 1968.

[Ti5]   J. Tits, Formes quadratiques, groupes orthogonaux et algèbres de Clifford, *Inventiones Mathematicae* **5** (1968), 19-41.

[Ti6]   J. Tits, Résumé des cours, 1990-1991, 1991-1992, 1992-1993, Collège de France, Paris.

[Ti7]   J. Tits, Strongly inner anisotropic forms of simple algebraic groups, *Journal of Algebra* **131** (1990), 648-677.

[We1]   A. Weil, On algebraic groups and homogeneous spaces, *American Journal of Mathematics* **70** (1955), 493-512.

[We2]   A. Weil, Algebras with involutions and the classical groups, *Journal Indian Mathematical Society* **24** (1960), 589-623.

[We3]   A. Weil, *Foundations of algebraic geometry*, revised ed., American Mathematical Society, 1962.

[We4]   A. Weil, *Adeles and algebraic groups*, 2nd ed., Birkhäuser, 1982.

Index